Revolutionary Guerrilla Warfare

Revolutionary Guerrilla Warfare

Edited by
Sam C. Sarkesian

Precedent Publishing, Inc./Chicago

Copyright © 1975 by Precedent Publishing, Inc.
520 North Michigan Avenue
Chicago, Illinois 60611
All Rights Reserved
Library of Congress Catalog Card Number: 74-12995
ISBN-0-913750-05-0

Manufactured in United States of America

355.02184
R454

96433

Preface

"This is the age of revolution" is a well worn cliche, but it also happens to be near the truth. Not only has this era witnessed a varity of national and social revolutions, but it also has been characterized by population revolution, revolution of ideas, technological revolution, and revolution in education, to mention but a few. The term revolution has become a popular one associated with many kinds of dramatically changing situations and events, while evoking a corresponding variety of emotional and intellectual reactions. Simultaneously, there has been a proliferation of the literature on revolution. Revolution as used in this volume, however, refers primarily to armed struggle and violence aimed at basic change in the structure, policies and leadership of government and society. This narrower definition leaves room for a wide range of viewpoints and interpretations, whether one is studying the older classic revolutions or the more recent "wars of national liberation."

After exploring antecedents, this volume focuses primarily upon latter day revolutionary movements. Historically, these have been most likely to occur in societies attempting to evolve from traditional to modern. Traditional societies, usually peasant-based, ultimately elicited the Maoist model of revolutionary guerrilla warfare. But this does not presume a rigid guideline; rather, it allows and even demands adaptability to the particular characteristics of a particular society.

This volume stems from my experience in the military and in the university. Having taken part in counterrevolutionary guerrilla warfare in Vietnam during 1966-1967, I became painfully aware of its complexity. Moving from military to academic life, I attempted to convey to my students the ambiguous multi-dimensional character of such warfare. In attempting to integrate classroom discussion and reading matter, however, I found that there are few readily available sources which provide a thorough introduction to revolutionary guerrilla warfare as well as the intellectual tools necessary to study in depth. There are a few collections of articles of the "How to do it" approach, or of high ideological content, and many useful and important one author books, but there remains a need for a thorough survey of the many inter-related aspects of this class of phenomena. Not that everything important to the study of revolution will be found in this volume — if indeed they can even be introduced in any one-volume work. And the answers to problems are not here! But it is hoped that some of the tools to find them may be found here.

Little emphasis is placed on military tactics and operations here. Not that they are unimportant; but it is vital to grasp the social, political, and psychological aspects in order to appreciate the role of the military, while military views of revolution abound. Basic concepts and models drawn from the social sciences introduce the problem, and a series of selections suggests the historical development of its analysis. This leads to the study of the Maoist approach, which is based on the understanding of traditional peasant society as the environment of revolutionary guerrilla warfare. It was Mao Tse-tung who developed the first systematic modern analysis of revolutionary guerrilla warfare, while providing a dramatic demonstration of its applicability. In turn, this analysis serves as a reference point for many comparable but sometimes deviant revolutionary situations throughout the world, even, for example, in urban settings.

As the discussions in this volume will attest, revolutionary struggles are more than simply armed internal conflict; they involve the very essence of the political system. The desire to make such phenomena understandable often leads to oversimplification. Attempts to encompass their multi-dimensional nature, on the other hand, can become immersed in complexities, ambiguities, and misinterpretations. The struggle in South Vietnam has added to the difficulties in the study of this type of war. Many persons based their assessment of revolutionary guerrilla warfare and the U.S. involvement in South Vietman on information from the mass media, but these rarely depict the essence of the problem. Refugees, air strikes, body counts, assassinations, and slogans like "winning the hearts and minds of the people" often obscure the social and political difficulties within the society.

Even among those who purport to be serious students, one finds serious gaps. U.S. military men, for example, have long been aware of the nature of revolutionary guerrilla warfare. Few, however, until very recently, have seriously considered it as a part of their traditional responsibilities. Although much emphasis lately has been placed on the study of this type of warfare and of the Maoist approach, there remains a gap between understanding and implementation. Indeed, few military men appreciate the complex nature of revolutionary guerrilla warfare and the limitations it imposes on the application of military power. On the other hand, academics often fail to appreciate its actualities and exigencies. While the military man may be guilty of depreciating the social, the intellectual may misjudge the importance of the military factor. What is needed is an appreciation of the possibilities of revolutionary war, tempered by an understanding that such wars do not exactly follow models. Moreover, the study of revolutionary war must allow for the uniqueness of each society and recognize that there are imponderables and immeasurables that do not fit into intellectual constructs, but may be crucial in the actual course of revolutionary and counterrevolutionary war.

There are those who argue that the effectiveness of the guerrilla is much less than most scholars and practitioners would admit. Indeed, some argue that the day of the guerrilla has ended. This argument is based on the assumption that guerrilla war is primarily a tactic and strategy for anticolonial movements. The decolonization of the world, with the exception of some parts of Black Africa, it is argued, considerably diminishes the utility of revolutionary guerrilla war. As additional evidence, the failures of guerrilla war in Bolivia and other parts of Latin America are cited.

This line of reasoning ignores the very basis of revolutionary guerrilla warfare movements — the political and social dynamics. A variety of forms of struggle can evolve, and these need not culminate in the development of revolutionary conventional forces struggling against conventional forces of the established regime. The regime may topple at any time, or it may accommodate itself to opposition demands. Indeed, some form of revolutionary guerrilla warfare might conceivably be used to attain limited short term gains. In some form it can always be used by a weaker group against a stronger, or even vice versa. As a matter of fact, revolutionary guerrilla warfare does not necessarily have to succeed in order to influence the political system. In the extreme formulation of Frantz Fanon, it may promote the psychological health and self-confidence of the participants regardless of its ultimate success or failure. In any case, political and social ideas and forces can be set in motion.

This type of warfare will probably be common in the foreseeable future, particularly in developing societies. Since

Preface

the end of World War II scores of armed conflicts have taken place, many of them unconventional. The assumption that the end of colonialism or the detente between major powers will bring an end of them is not supported by recent history, nor is it sensitive to the internal instability associated with modernization. The perspective of this volume is relatively simple. Revolution is here to stay indefinitely and revolutionary guerrilla warfare is particularly useful strategy for the weak, the frustrated, the alienated, the seekers of power against existing regimes.

The essays in this volume have been selected only partly with regard to geographical or ideological balance. They have been chosen primarily for their interpretive quality, conceptual clarity, and variety of approach. How clear and informative are the essays? The answer to this question determined their selection. Additionally, no attempt has been made to provide a micro-functional analysis. That is, subjects such as terror, ambush, assassination, etc., are not discussed in themselves, but rather as characteristics of revolutionary guerrilla warfare.

Many persons assisted in the preparation of this volume. I especially thank A. Thomas Ferguson, Jr. of Loyola University, who assisted in its preparation, particularly for his assessment of the usefulness of the readings at the college and university level. I also wish to thank Dr. Henry Cohen for his suggestions and encouragement.

Sam C. Sarkesian
Chicago, Illinois

Contributors

Gil Carl AlRoy is Professor of political science at Hunter College, City University of New York. His publications include *Attitudes Toward Jewish Statehood in the Arab World* and *The Involvement of Peasants in Internal Wars*.

James C. Davies is professor of political science, University of Oregon. He is attempting to develop a theory of the psychological foundations of the origins, growth, maturation, decline, and overthrow of governments. His interests include the psychological analysis of the role of ideology in social and political change.

Dennis J. Ducanson has been a British colonial administrator in Malaya and Hong Kong and an advisor to the South Vietnamese Government. He is now reader in Southeast Asia studies at the University of Kent in England. His chief current research interest is the rise of the Indochina Communist Party.

A. Thomas Ferguson, Jr., is a Ph. D. candidate at Princeton University. He is co-editor of the forthcoming Karl Marx and Friedrich Engels, *The Collected Writings in the New York Daily Tribune*.

W.D. Franklin is associate professor of economics and director of community service and management development at Tennessee Wesleyan College. He is president of Econotec Research Company and a member of various government bodies. He is currently conducting a study of the Mid-East oil situation and its strategic implications for the U.S.

Edward Friedman is associate professor of political science at the University of Wisconsin. He has co-edited *America's Asia* and is preparing two forthcoming books: *Backwards Toward Revolution* and *Asia and America in the Era of Mao Tse-tung*.

Samuel B. Griffith is a retired brigadier general, USMC, and holds a doctorate from Oxford University in Chinese history.

Edward Gude is a fellow of the Adlai Stevenson Institute, Chicago. He is working on a study of the relations between social science and government in Project Camelot.

Contributors

xiv **Ted Robert Gurr** is professor of political science at Northwestern University. He has written extensively on the social problem of violence. His major work, *Why Men Rebel*, received the American Political Science Association's Woodrow Wilson Prize in 1970.

Martha D. Huggins is preparing her doctorate in sociology at the University of New Hampshire. She is studying the relationship between economic dependence and political instability in underdeveloped non-socialist nations.

Chalmers Johnson is professor of political science at the Center for Chinese Studies, University of California, Berkeley. His works include *Peasant Nationalism and Communist Power*, *Revolutionary Change*, and *Conspiracy at Matsukawa*. He has also edited *Change in Communist Systems* and *Ideology and Politics in Contemporary Ctina*.

Amrom H. Katz is a senior staff consultant to the RAND Corporation and consultant to various U.S. government agencies and industrial groups. He was chief physicist at the Air Force Aerial Reconnaissance Laboratory.

Y. Krasin is a contributor to the *World Marxist Review*.

Carlos Marighella was born in 1911. At age twenty-three he became a member of the Brazilian Communist Party. In 1967 he broke with the party and became leader of an urban guerrilla organization. He was killed in an ambush by the Brazilian police in 1969, six months after writing the Minimanual.

Charles Burton Marshall is professor of international politics in the School of Advanced International Studies, Johns Hopkins University. His recent books include *The Exercise of Sovereignty*, *The Cold War: A Concise History*, and *Crisis over Rhodesia*.

Scott G. McNall is associate professor of sociology at Arizona State University. He specializes in the comparative study of internal conflict and of military coups and the changing role of the military.

Contributors

Manus Midlarsky is associate professor of political science at the University of Colorado. He is working on mathematical models of political processes and ethics in research procedures and has completed a forthcoming study of war.

Eduardo Mondlane received his Ph. D. in sociology from Northwestern University and later worked as a research officer for the United Nations. In 1962 FRELIMO (Mozambique Liberation Front) was formed and he was elected president, a position he held until his assassination in 1969.

Jose A. Moreno is associate professor of sociology at the University of Pittsburgh. He is author of *Barrios in Arms: Revolution in Santa Domingo* and is now at work on a study of the Tupamaros.

Robert Moss is a lead writer of *The Economist* (London) and former lecturer at Australian National University. His works include *Urban Guerrilla Warfare* and *Counter-Terrorism*. He is preparing a book on Chile's Marxist experiment and a biography of General Giap.

Lars Rudebeck is research associate, Political Science Institute, University of Uppsala, Sweden. His essay in this volume is a result of a study tour of Guinea-Bissau in November and December 1970. He has written theoretical and empiracal essays on political development.

Sam C. Sarkesian is associate professor of political science at Loyola University of Chicago. He is editor of *The Military Industrial Complex: A Reassessment* and is associate chairman of the Inter-University Seminar on Armed Forces and Society.

James C. Scott is associate professor of political science at the University of Wisconsin. His works include *Political Ideology in Malaysia: Reality and The Beliefs of an Elite* and *Comparative Political Corruption*.

T. Shanin is professor of sociology at Haifa University and has been visiting fellow at St. Anthony's College, Oxford. His publications include *Peasants and Peasant Societies: the Awk-*

ward *Class,* and *The Rules of the Game: Models in Contemporary Scholarly Thought.*

Lawrence Stone is professor of history and director of the Shelby Cullom Davis Center for Historical Studies at Princeton University. He specializes in English history and has written *The Crisis of the Aristocracy, 1558-1641.*

George K. Tanham is vice-president of RAND Corporation. He is working on a book on Thai insurgency and the United States.

Raymond Tanter is professor of political science at the University of Michigan. His interests include the policy relevance of social science research, the Middle East, and threat perception and the sharing of alliance burdens in Europe.

Charles Wolf Jr. is head of RAND Corporation's Economic Department and director of the RAND Graduate Institute for Policy Studies.

Eric R. Wolf is Distinguished Professor of Anthropology at Herbert Lehman College, City University of New York. Among his published works are *Peasants* and *Peasant Wars in the Twentieth Century.*

Acknowledgments

Adelphi Papers, No. 79, International Institute for Strategic Studies, London. Robert Moss, "Urban Guerrilla Warfare," with appendix: "Minimanual of the Urban Guerrilla" by Carlos Marighella.

American Political Science Review, Vol. LXVI, No. 1, March 1972. James Scott, "Patron-Client Politics and Change in Southeast Asia."

Asian Survey. Copyright © 1968 by The Regents of the University of California. Reprinted by permission of The Regents. Chalmers Johnson, "The Third Generation of Guerrilla Warfare," Vol. 8, No. 6, June 1968.

Comparative Studies in Society and History, Vol.12, April 1970. Published by Cambridge University Press. J. Moreno, "Che Guevara on Guerrilla Warfare: Doctrine, Practice, and Evaluation;" Edward Friedman, "Neither Mao, Nor Che: The Practical Evolution of Revolutionary Theory. A Comment on J. Moreno's 'Che Guevara on Guerrilla Warfare.' "

Foreign Affairs. Copyright © Council on Foreign Relations, Inc., New York. Reprinted by permission from *Foreign Affairs*. George K. Tanham and Dennis J. Duncanson, "Some Dilemmas of Counterinsurgency," Vol. 48, October 1969.

International Social Science Journal, Vol. XXI, No. 2, 1969. Reprinted by permission of UNESCO. Eric R. Wolf, "On Peasant Rebellions."

Journal of Conflict Resolution, Vol. II, September 1967. Raymond Tanter and Manus Midlarsky, "A Theory of Revolution."

Journal of Modern African Studies. Published by Cambridge University Press. Lars Rudebeck, "Political Mobilisation in Guinea-Bissau," Vol. 10, No. 1, May 1972.

Military Review, June 1967. Published by the U.S. Command and General Staff College, Fort Leavenworth, Kansas. W. D. Franklin, "Clausewitz on Limited War."

Acknowledgments

Orbis, Summer 1958. Published by the Foreign Policy Research Institute. Copyright © Foreign Policy Research Institute. V. I. Lenin, "Partisan Warfare."

Rebellion and Authority: An Analytic Essay on Insurgent Conflicts, Nathan Leites and Charles Wolf, Jr., (Chicago: Markham Publishing Co., 1970), pp. 28-47. Copyright © Markham Publishing Co.

Selected Works of Mao Tse-Tung, abridged by Bruno Shaw (New York: Harper and Row, 1970), pp. 129-159. Copyright © 1970 by Bruno Shaw. Reprinted by permission of Harper and Row Publishers, Inc.

Sociological Review, Vol. 14, March 1966. T. Shanin, "The Peasantry as a Political Factor."

Southeast Asian Perspectives, No. 4, December 1971. Amrom H. Katz, "An Approach to Future Wars of National Liberation;" Charles Burton Marshall, "Morality and National Liberation Wars."

Sun Tzu, The Art of War, translated and with introduction by Samuel Griffith (New York: Oxford University Press, 1963), pp. 77-84, 96-101, 102-110. Copyright © Oxford University Press, 1963. Reprinted by permission of The Clarendon Press, Oxford.

The Antioch Review, Vol. 26, No. 2, Summer 1966. Copyright © 1966 by *The Antioch Review*, Inc. Reprinted by permission of the editors. Gil Carl Alroy, "Insurgency in the Countryside of Underdeveloped Countries."

Violence in America: Historical and Comparative Perspectives (2 vols © Washington, 1969), Hugh Davis Graham and Ted Robert Gurr, eds.; James C. Davies, *The J-Curve of Rising and Declining Satisfactions as a Cause of Some Great Revolutions and a Contained Rebellion*; Edward E. Gude, *Batista and Betancourt: Alternative Responses to Violence*.

The Struggle for Mozambique, Eduardo Mondlane (London: Penguin Books Ltd., 1969), pp. 130-133, 141-142, 145-151,

Acknowledgments

153-161. Copyright © Eduardo Mondlane, 1969.

World Marxist Review, North American edition, Toronto, Canada, Vol. 15, No. 2, February 1972. Y. Krasin, "Social Revolution as Seen by Bourgeois Ideologists."

World Politics, Copyright © 1966, 1968 by Princeton University Press. Reprinted by permission of Princeton University Press. Lawrence Stone, "Theories of Revolution," Vol. XVIII, No. 2, 1966; Ted R. Gurr, "Psychological Factors in Civil Violence," Vol. XX, No. 2, 1968.

Contents

vii **Preface**

xvii **Contributors**

xxiii **Acknowledgments**

1. Chapter I
Revolutionary Guerrilla Warfare:
An Introduction by Sam C. Sarkesian

23. **Chapter II**
Revolution

27. **Theories in Revolution**
by Lawrence Stone
47. **A Theory of Revolution**
by Raymond Tanter and Manus Midlarsky
75. **Psychological Factors in Civil Violence**
by Ted Gurr
117. **The J-Curve of Rising and Declining Satisfactions as a Cause of Some Great Revolutions and a Contained Rebellion**
by James C. Davies
143. **Social Revolution As Seen by Bourgeois Ideologists**
by Y. Krasin

159. **Chapter III**
Revolutionary Guerrilla Warfare:
The Development of Doctrine

163. **Selections from *Sun Tzu, The Art of War***
translated by Samuel Griffith
179. **Clausewitz on Limited War**
by W. D. Franklin
187. **Partisan Warfare**
by V. I. Lenin
205. **Selections from *Selected Works of Mao Tse-Tung***
abridged by Bruno Shaw

Contents

xxiv

237. **Chapter IV
Modern Revolutionary Guerrilla Warfare
in Macro-View: Causes and Contexts**

241. **Guerrilla Warfare: Predisposing and Precipitating Factors**
by Scott G. McNall and Martha Huggins
257. **Insurgency in the Countryside of Underdeveloped Societies**
by G. C. Alroy
267. **The Peasantry as a Political Factor**
by T. Shanin
291. **On Peasant Rebellion**
by Eric R. Wolf
303. **Patron-Client Politics and Change in Southeast Asia**
by James Scott

353. **Chapter V
Modern Revolutionary Guerrilla Warfare:
Some Micro-views**

357. **The Third Generation of Guerrilla Warfare**
by Chalmers Johnson
375. **The Malayan Emergency: The Roots of Insurgency**
by Sam C. Sarkesian
395. **Che Guevara on Guerrilla Warfare:
Doctrine, Practice, and Evaluation**
by J. Moreno
421. **Neither Mao, Nor Che: The Practical Evolution
of Revolutionary Theory**
by Edward Friedman
431. **Political Mobilisation in Guinea-Bissau**
by Lars Rudebeck
453. **Selections from *The Struggle for Mozambique***
by Eduardo Mondlane

473. **Chapter VI
Urban Guerrilla Warfare**

475. **Urban Guerrilla Warfare**
by Robert Moss
507. **Minimanual of the Urban Guerrilla**
by Carlos Marighella

Contents

533. **Chapter VII**
The Problems of the Defenders

537. **Selections from *Rebellion and Authority:***
An Analytic Essay on Insurgent Conflicts
by Nathan Leites and Charles Wolf, Jr.
557. **Some Dilemmas of Counterinsurgency**
by George K. Tanham and Dennis J. Duncanson
569. **Batista and Betancourt: Alternative Responses to Violence**
by Edward W. Gude
587. **An Approach to Future Wars of National Liberation**
by Amrom H. Katz
603. **Morality and National Liberation Wars**
by Charles Burton Marshall

617. **Chapter VIII**
Sources for the Study of Revolutionary Guerrilla Warfare
by A. Thomas Ferguson, Jr.

Revolutionary Guerrilla Warfare: An Introduction
by Sam C. Sarkesian

"Revolution" is a word that causes fear in some, exhiliration in others, and confusion in most. Originally used to describe a restoration, it eventually came to mean a sweeping, sudden attack upon an existing order. From the time of the rebellion of the American colonies against England in the 18th century, and especially after the revolution in France, the word has suggested not only violence but also great scope and importance. Since the late 19th century, its meaning has expanded to include far reaching change, such as social revolution, Industrial Revolution, revolution of rising expectations, etc., even if not primarily violent. Revolution is now regarded as a complex, multidimensional phenomenon deeply originating in social systems; it cannot be realistically studied apart from the systems and environments from which it evolves.

The purpose of this introduction is to identify the main characteristics of revolution and the guerilla warfare it generates, and suggest an explanatory framework embodying the results of the analysis. Our concern with revolutionary guerrilla warfare stems primarily from its frequency and importance, and from the probability that this type of warfare will be the characteristic form of armed conflict in developing systems.

The recent study of revolution as a special class of phenomena began with historian Crane Brinton's *Anatomy of Revolution*, first published in 1938 and often revised and reprinted. This highly original comparative work stimulated the study of revolutions as a subject in its own right within the various academic fields. Studies in particular disciplines, however, have tended to reflect the biases of those disciplines, as well as those of the individual scholar. The historian is likely to use the great revolutions of the past as criteria for the study of all revolutions. The sociologist is likely to focus on social class and order; the psychologist, on the sense of grievance; while the political scientist tends to direct his attention to power and force in the framework of institutions. Aside from the disciplinary orientation, there is a problem of scope. While some scholars prefer a narrow definition of revolution, restricting its application to civil disorders and violent upheaval, others are inclined to accept a looser definition which encompases basic change in general. Some recent works are eclectic in nature and

Revolutionary Guerrilla Warfare:
An Introduction

borrow from the broad range of the social sciences. The behavioralist approach, emphasizing the revolutionary mentality and a quasi-economic model which suggests that revolutions stem from perceived economic deprivation, has a greater empirical bent than other theories. Currently, much work is going into efforts to develop cross-national analyses of revolution. Finally, there are ideological interpretations representing all shades of political outlook.

Any one-dimensional explanation of revolutions would presuppose unambiguous and clearly differentiated political, social, and economic systems, hardly a realistic expectation. The substantive issues of revolutions are often in themselves difficult enough to identify precisely. As an initial step, however, revolutions can be usefully viewed as struggles at points on a continuum, from the threshold of violence (such as sit-ins and demonstrations) to total civil war. In between are a variety of revolutionary conflicts conducted with varying degrees of intensity, differing in scope and in the character of the participants.

Recognizing the difficulties of generalizing, there are nevertheless common environmental characteristics in post-World War II revolutionary movements. Most of them have developed in peasant societies undergoing the tensions and strains of modernization. Elites have been at the forefront of these political and social movements: that is, the educated from "higher" social classes generally articulate grievances and ideals and become the leaders and organizers. It is this combination of elite alienation and leadership, peasant awakening and mobilization, and organizational and ideological cohesion that is at the base of most successful revolutionary movements. Revolutionary guerrilla warfare, as will be demonstrated later, is particularly suited to peasant societies. This is not to ignore, however, the outbreaks of urban guerrilla warfare in certain areas of the world; although even in some of these the urban populations involved include heavy recent arrivals from the countryside.

Admittedly social upheaval can threaten any society with serious unsolved problems, but most developed countries in our time have shown a relatively low propensity for revolutionary armed struggle. On the other hand, it can be argued that the revolutions of the twentieth century hold more contrasts than

Revolutionary Guerrilla Warfare:
An Introduction

similarities to the classic earlier revolutions. This is not to suggest that historical "maps" do not identify certain common characteristics of revolution. However, the environment of the contemporary emerging nations makes them distinct from the revolutionary societies of terminal Czarist Russia, Bourbon France, or colonial America. The peasant-based revolutionary impulse in a colonial framework, stimulated by modernization and new ideology, developed a high propensity for guerrilla warfare.

Mao Tse-tung provided the first systematic analysis of this type of warfare. Although he is usually identified with peasant war, his earlier experience included urban uprising, first in united front with and then in opposition to the Kuomintang led by Chiang Kai-shek in the China of the 1920s. Mao's approach does not absolutely rule out the strategy of a quick overthrow of the government by the urban masses, which assumes that once the government machinery is captured or dismantled the revolution can work from the cities outward to complete control of the countryside. But this strategy is characteristic of the Russian strategy rather than the Maoist. The Maoist model eschews urban areas, stressing the mobilization of the peasantry; revolution in the countryside is to envelope and engulf the cities. This scheme follows the classic strategy of the Chinese game of "wei-ch'i".[1] It is based on the premise that the strongest area of government is the city and the weakest, the countryside. Maoist logic dictates that the enemy should be attacked at his weakest point first.

Mao's principles developed out of his experience of the failure of urban uprising in the 1920s. In the next decade he perceived the crucial weakness of the Kuomintang, and later the invading Japanese, in the countryside. The antagonists' control of the cities and the major lines of communications and their weak presence elsewhere became the keys to Maoist strategy. The revolutionary forces worked to mobilize the peasants and control the countryside, confining the enemy to the towns and cities and narrowing its access to population and resources. Limited in its movements and denied the foodstuffs and manpower of the rural areas, the government could be starved into surrender or destroyed by a final military offensive.

Guerrilla warfare, colonial as well as purely internal, is armed struggle between pre-existing and challenging political systems. Such wars are primarily socio-political in nature, and although armed conflict is important and may determine the outcome, the "politics" of the conflict are the fundamental causes and determinants. The probability and intensity of revolutionary guerrilla warfare are directly correlated to political instability and social frustration, which in turn are largely determined by the rate of modernization and development. Even in the exceptional cases of its outbreak in developed countries, it is most likely to originate in underdeveloped sectors of the socio-economic order. The propensity for revolutionary war increases as grievances become less susceptible to resolution within the existing political system. The Maoist model, with its emphasis on the role of the peasantry in the successive stages of revolution, has been the most relevant guide to this phenomenon.

2

Recent years have brought a plethora of writings on unconventional wars, ranging from purely military studies at the level of the smallest troop units to the analysis of nation-building. Useful as they have been, they reveal a lack of agreement on basic definitions and terminology.[2] Revolutionary war, civil war, internal war, insurgency, resistance movements, guerrilla war, wars of national liberation, stability operations, internal defense, counter-insurgency—these are but a few examples of the proliferating terminology.[3] To appreciate the confusion it may be useful to cite some examples.

According to a standard reference work, civil war

> . . . is conflict within a society resulting from an attempt to seize or maintain power and sumbols of legitimacy by extra-legal means. It is civil because civilians are engaged in it. It is war because violence is applied by both sides. Civil war is intrasocietal and may take place within a group, some parts of which either desire to maintain or wish ot initiate separate ethnic and/or polticial identity or wish to change the government.[4]

The Law of Land Warfare, an official U.S. Army document, simply terms civil war an "armed conflict not of an international character occurring in the territory of one of the

High Contracting Parties." [5] In another instance, it is defined as a "military conflict between two or more approximately equal governments for sovereignty over people and territory native to both. Civil War is also distinguished from insurrection. The latter is characterized as a struggle from the bottom up—an uprising of a more or less politically unorganized group against an established authority. Civil War is horizontal, insurrection is vertical." [6]

Guerrilla warfare has been defined by Osanka as follows:

> In modern times, the objectives of guerrilla warfare have been more political than military. Since the end of World War II there have been at least ten revolutionary wars using guerrilla warfare as the principle means of violence . . . The most adequate descriptive term would seem to be 'revolutionary guerrilla warfare.' It is revolutionary in that it is used as a means of acquiring national power for the purpose of altering or completely changing the social and political structure of a nation. It is guerrilla warfare in that its participating advocates of change are indigenous civilians waging a small war utilizing principles learned from guerrilla history. [7]

Pike, in contrast, emphasizes external sources of support:

Revolutionary guerrilla warfare should not be confused with older concepts of a similar nature, such as irregular troops in wartime disrupting the enemy's rear, or with civil war, which is between two groups in the same nation (revolutionary warfare is not indigenous), rebellion (militant opposition to authority with the issue quickly settled), revolution (successful rebellion), bandit warfare (plunder as a way of life), or partisan warfare (armed fighting by light troops) . . . . Revolutionary guerrilla warfare as practised in Vietnam was a way of life. Its aim was to establish a totally new social order, thus differing from insurgencies whose objective is either statehood or change of government . . . . It was an imported product, revolution from the outside; its stock in trade, the grievance, was often artificially created; its goal of liberation, a deception. [8]

Revolutionary Guerrilla Warfare:
An Introduction

The term "unconventional warfare," which is commonly used in United States Army doctrine, can also be confusing. In an official Department of the Army publication it is defined as including "the interrelated fields of guerrilla warfare, evasion and escape, and resistance. Such operations are conducted in enemy-held or controlled territory and are planned and executed to take advantage of or stimulate resistance movements or insurgency against hostile governments or forces."[9] The same publication defines guerrilla warfare as "the conduct of combat operations inside a country in an enemy or enemy-held territory on a military or paramilitary basis by units organized from predominantly indigenous personnel," with the primary aim being to "weaken the established government."[10]

The Department of the Army defines insurgency as

> a condition of subversive political activity, civil rebellion, revolt, or insurrection against a duly constituted government or occupying power wherein irregular forces are formed and engage in actions, which may include guerrilla warfare, that are designed to weaken and overthrow the government or occupying power.[11]

Schuman offers definitions by stages, proposing that the term "insurrection" be applied to armed violence in "initial stages of movements of opposition to government." The terms "rebellion" and "revolution," he continues, should be employed only when a "substantial portion of the armed forces of the established government" must be used in defense. "In this sense an insurrection may be thought of as an incipient rebellion or revolution still localized and limited to securing modifications of governmental policy or personnel and not yet a *serious threat* to the state or the government in power."[12]

The problem of definition is suggested in an analysis of the war in Vietnam. A noted British authority writes:

> The point to be stressed is that the war has always remained basically an *insurgency*, boosted by infiltration and aided, to a certain but limited extent, by both invasion and raids.... *People's Revolutionary War* is therefore by nature a *civil* war of a very sophisticated type and using highly refined techniques to seize power and take over a country. The

significant feature of it, which needs to be recognized, is its immunity to the application of power.[13]

Despite the ambiguities, some characteristics are common to all these efforts: the use of force; the objective of changing the composition of government; revolutionary goals; organization; and the fact that the participants are apt to appear to be civilians and avoid conventional battle tactics. Unconventional warfare may include the entire range of activities from sabotage and ambush to operations involving organized units on a large scale, employing tactics of dispersion, rapid assembly, surprise attack, and dispersion. But the decisive element is the departure from the use of always visibly distinguishable combatants in formal battle order. Thus, we come to our working definition:

> Revolutionary guerrilla warfare is the forcible attempt by a politically organized group to gain control or change the structure and/or policies of the government, using unconventional warfare integrated with political and social mobilization, resting on the premise that the people are both the targets and the actors.

We may note that the tactics of guerrilla warfare might be used to achieve limited, short-term goals; but the success of such tactics could lead to the adoption of guerrilla warfare for the purpose of seizing power.

Revolutionary warfare is not simply military action by irregular troops or militia; its main purpose is to supplant the existing government. It combines action in many fields of human endeavor, political, social, economic and psychological, to mobilize the masses in conjunction with military action. The involvement of the people is basic. The individual revolutionary is not merely a conventional soldier, but a political agent. Our definition also implies that the revolutionary has the support or acquiescence of segments of society to which revolutionary ideology and organization represent a legitimate political system, and exercises many if not all the functions of government over population and territory.

3

Mao Tse-tung's contribution to the theory of revolutionary warfare was primarily to synthesize a wide variety of military techniques and concepts into a systematic operational doctrine subordinated to a comprehensive social philosophy. His three stages, based on the ancient dicta of Sun Tzu, offer a relatively simple model of the strategy of revolutionary war.

> It can reasonably be assumed that this protracted war will pass through three stages. The first stage covers the period of the enemy's strategic offensive and our strategic defensive. The second stage will be the period of the enemy's strategic consolidation and our preparation for the counter offensive. The third stage will be the period of our strategic counter offensive and the enemy's strategic retreat. [14]

Implicit throughout Mao's analysis is the stress on the practical relevance of revolutionary ideology, organization, and operations. (For example, terror is useful if it furthers the political purpose of the revolutionaries; indiscriminate terror has no place in Maoist strategy.) Before the first stage of the process is begun, there must be penetration of the target social system by the revolutionary cadre. Only after the revolutionaries have gained a degree of legitimacy and have made sufficient progress in mobilizing the masses is the revolution ready to move into stage one. Social penetration not only prepares the way for stage one; it continues in all stages and ultimately inheres in the control system of the revolutionary government.

The implementation of strategy is conditioned by the particular characteristics of the population and territory. Mao stresses the political factor:

> A national revolutionary war . . . cannot be won without universal and thoroughgoing political mobilization . . . This move is crucial; it is indeed of primary importance, while our inferiority in weapons and other things is only secondary. The mobilization of the common people throughout the country will create conditions that will make up for our inferiority in arms and other things, and create the prerequisites for overcoming every difficulty in the war . . .

Revolutionary Guerrilla Warfare: An Introduction

What is the relationship of guerrilla warfare to the people? Without a political goal, guerrilla warfare must fail, as it must if its political objectives do not coincide with the aspirations of the people and their sympathy; cooperation, and assistance cannot be gained. The essence of guerrilla warfare is thus revolutionary in character. [15]

Mao warns that revolutionary warfare will not produce quick results: it is a protracted test of strength between the enemy and the people. [16] That the enemy (government) will fail is assured because the revolutionaries, according to Mao, are of and for the people and as such have the stronger will, the endurance, and the just cause. Because of the support of the people the enemy can never successfully bring to bear the main source of his strength, his military power.

Trained and dedicated revolutionaries, through a philosophically simple but highly refined and effective technique, implement these theories. They establish organizations in the villages and hamlets (stage one). Using psychological techniques relevant to the political, social and economic environment of the peasantry, the revolutionaries subtly and effectively develop an administrative structure to control and govern them. In short, they create a "shadow" government. In conjunction with political expansion and mobilization, the military arm engages in action short of frontal challenge to the military forces of the existing government.

Sabotage, assassination, propaganda, terror, intimidation are used to subvert the established government's control. The revolutionary military forces, enjoying local and temporary superiority and attacking at times and places of their own choosing, demonstrate to the masses their efficiency and the government's ineffectiveness. With each success additional recruits are gained, equipment captured, and people brought under control. Eventually the growing base areas make possible further expansion of the war. As the revolutionaries expand their control, the government's influence contracts back in from the countryside toward the cities and towns in what becomes a classic situation in revolutionary warfare: isolation of the central government from the peasant masses.

When the revolution moves into stage two the revolutionary military forces are organized into companies and battalions, political cadres extend their influence and activities, and a more sophisticated political and military operational system develops. The revolutionary military arm is now capable of directly challenging government military forces and in many instances faces them on an almost conventional footing. Government units of less than battalion size may become easy prey. Further contraction of government control takes place in an attempt at consolidation and more effective use of thinly-spread military forces. The revolutionaries extend their influence even farther, bringing under control more of the civilian populace and widening the gap between the government and the people. They may establish a formal government, claiming to legitimately represent the people and raising the possibility of foreign support. Unless dramatic successes are achieved by the established regime in organization, morale, and combat skill, the revolution is well on its way to success with the attainment of stage two. To understand how difficult is the situation of the established regime, it should be remembered that the same factors that made it vulnerable to the beginnings of the process tend to continue to prevent a successful response at all stages; a regime capable of generating a successful response would probably not have become vulnerable in the first stage.

The third and final stage is the conversion of the revolutionary military forces to conventional armies mounting large scale frontal attacks. Simultaneously, on a growing scale the political arms mobilize the masses, conduct psychological warfare, and implement revolutionary programs in areas they control. In essence, the revolutionaries create a legitimacy surpassing that of the established government. Once stage three is reached the government has all but lost the war. Revolutionary momentum gathers relentless force and only a "miracle" will forestall government defeat.

Although not specifically so characterized by Mao, consolidation becomes primary after stage three, although it takes place at all stages and at all levels of organization. When the revolutionaries take unchallenged control they can extend the revolution throughout the social system. Consolidation may well be the most protracted and brutal period of all since it is intended to cleanse the political system of any possibility of opposition. The revolution becomes institutionalized, coextensive with the nation and state.

Revolutionary Guerrilla Warfare: An Introduction

This is the Maoist scenario, or model, of revolutionary warfare. In practice the transitions do not necessarily take place smoothly and clearly. Characteristics of all three stages may be present simultaneously. The key factor is the judicious employment of the military arm for political objectives with primary emphasis on the growth and effectiveness of the political forces. In the final analysis, the basis for the success of the revolution is the effectiveness of the political cadres and their influence upon the masses. "Only on a basis of strong political organization could semi-armed organizations be set up firmly and guerrilla groups and guerrilla units organized which have close connection with the revolutionary masses, eventually to further their activities and development." [17]

Revolutionary warfare cannot succeed unless the conditions are correct, as Mao recognizes. This is most likely to occur where an unresponsive, inept or uncompromising government fails to deal with problems at the local level. The necessary conditions include:

> An unstable political situation, marked by sharp social divisions and usually, but not always, by a foundering or stagnant economy. A political objective, based on firm moral and ideological grounds, that can be understood and accepted by the majority as the overriding "cause" of the insurgency, desirable in itself and worthy of any sacrifice. An oppressive government, with which no political compromise is possible. Some form of revolutionary political organization, capable of providing dedicated and consistent leadership toward the accepted goal. There is one final requirement: the clear possibility or even probability of success (for the revolutionaries). [18]

It is appropriate to reiterate the political and social character of revolutionary guerrilla warfare. As one authority in the field notes:

> A revolutionary war is never confined within the bounds of military action. Because its purpose is to destroy an existing society and its institutions and to replace them with a completely new state structure, any revolutionary war is a unity of which the constituent parts, in varying importance,

are military, political, economic, social and psychological. For this reason it is endowed with a dynamic quality and a dimension in depth that orthodox wars, whatever their scale, lack. This is particularly true of revolutionary guerrilla war, which is not susceptible to the type of superficial military treatment frequently advocated by antediluvian doctrinaires. [19]

4

What is the task of the counterrevolution? At the outset, it is to prevent the revolution from carrying out stage one. To undermine the revolutionary appeal, policy must be formulated and programs implemented which will, in effect, create a revolution controlled by the government. More negatively, a "climate of collapse" must be created in the revolutionary program. The former Chief of Staff of the United States Army, Harold K. Johnson, has written:

> One of the key strategies of insurgency, or "wars of national liberation," is initially to create disorder which can later be exploited, penetrating every institution to the maximum degree possible to promote confusion, disagreement, and uncertainty. The counterinsurgent's task is to maintain the established order while in fact waging war against the insurgents who are spread among the population. The counter-insurgent is thus restrained against the use of force which would normally be acceptable against a completely hostile population. The arms of the government must be long enough to reach out to all of the people, firm enough to give them support, and strong enough to protect them from coercion and outside influence. [20]

Johnson emphasizes the political, social, and psychological aspects of policy.

> A close integration of the political, economic, information, security, and military branches of the government is essential to ensure a concentration of effort against an insurgency. One must constantly keep foremost in mind that military action is only a part of counter insurgency and that

a well-integrated "team" can often compound a military success or minimize failure. [21]

Or, in the words of a successful revolutionary and counter-revolutionary:

> Military actions are only an auxiliary part of the main operations . . . It is urgent to lay stress on the fact that to destroy them (guerrillas) with military might . . . will always fail as long as the guerrilla fighters can conceal themselves among the people and call for help; as long as they have their base within the people. It is this that is not primarily a military matter, but largely a matter of political, psychological and socio-economic concern. [22]

The role of the counterrevolution may be much more difficult than that of the revolution, for it must react to revolutionary activity. The revolutionary's role is relatively simple because his prime concern is to demonstrate the ineffectiveness of the government. He selects his own time, place and method of operation, while the counterrevolutionary must provide at all times for a variety of possibilities covering the entire national area. The program of the counterrevolution must respect the aspirations, values and beliefs of the people; it must be legitimate, if there is to be hope of success in earning and keeping popular loyalty. [23] Psychologically, the revolutionaries, on the other hand, can resort to extra-legal means more easily since they are defined as outlaws in the government's eyes, while the very conditions that allow the growth of revolutionary movements tend to "legitimize" extra-legal activities.

In short, insurgency is a manifestation of the political, administrative and social weaknesses of a government. Efficient and responsive government is the best, the most preventive counterrevolutionary measure. In contrast, the society that allows itself to become vulnerable to guerrilla uprising may have to pay heavily.

> This counter-revolutionary strategy requires that in a tough, protracted struggle the governing power first stop the revolutionaries in whatever phase they have reached and then drive them back through the preceding phases . . . The

idea must be to force and anticipate changes in revolutionary operations and strategy. Therefore the most logical solution does lie in developing a counter-revolutionary strategy which applies revolutionary strategy and principles in reverse to defeat the enemy with his own weapons on his own battlefield.

It has been stressed that the counter-revolutionary strategist must apply counter-actions that are appropriate to the various phases of revolutionary warfare. He should bear in mind, however, that these counter-actions are cumulative; that is, in the mobile warfare phase he must be prepared to combat revolutionary guerrilla warfare, terrorism, and organization, etc. This is why counter-revolutionary warfare is progressively so expensive in resources. [24]

5

Since social and political causes are at the roots of revolutionary warfare, a successful revolution must maintain or evolve an alternative political system. Whether this revolutionary political system is symbolic or real is not necessarily crucial in itself; it is the perceptions of those with grievances against the existing system that is the key. An analytical framework, therefore, must comprehend the general character of the political and social system, the relevant variables, the character and appeal of the revolutionary movement, and the response of the existing regime. These elements will be discussed in the following pages. They are interdependent, not separate, and may be broken down as follows:

I. *Roots of Revolution: the intitial given situation*

Social structure; communal structure; value systems and ideologies, institutional structures; constitution; instruments of force; economic structure; distribution of income and wealth; demographic structure; ethnic structure; cleavages and tensions within these categories; their geographic distribution; legitimacy and loyalty.

II. *Revolution*

Leadership; ideology and its dissemination; organization and administration; political and military strategy; tactics; legitimacy and loyalty; effectiveness.

III. *Counter-Revolution*

Leadership; socio-ideological motivation and appeal; organizational structure and administrative efficiency; strategic and tactical perception and response; political and military effectiveness; legitimacy and loyalty.

Adopting Mao's formulation, the revolution moves—ideally—from a societal penetration phase to a first stage of contention, a second stage of balance and a third stage of counter-offensive, followed and concluded by a consolidation phase. Stage two, if reached and passed successfully, is the transition from the defensive to the offensive. It marks the opposite for the forces of the status quo, and the momentum of change, in itself perhaps overwhelming, may be greatly accelerated by rising expectations of revolutionary victory and disastrous loss of morale in the counterrevolution. If stage three is reached, only a near-miracle, or the introduction of some massive new element (such as foreign intervention) can save the regime.

Yet the initial momentum or inertia tend to be maintained and to determine the later course of the struggle.

> The initial decisions of a government faced with an insurrectionary movement are the key decisions which structure the subsequent conflict . . . Thus it is imperative that the initial decisions of the government provide a basis for distinguishing between commitments of an inflexible character and more tentative measures in order to maintain appearances of consistency in subsequent policies while preserving a degree of flexibility in handling special problems. [25]

What occurs after the beginning of the second stage is a fruition of the earlier period. From societal penetration through stage one is usually the critical period for both the revolution and the counterrevolution. [26]

6

Let us see how these factors operated in a comparison, admittedly rudimentary, of the experiences in Malaya and Vietnam, where one can identify comparable periods. [27] In Malaya the critical period extended from the declaration of the Emergency in 1948 to the inception of the Briggs Plan in 1951.

16 The revolutionaries, who were largely ethnically Chinese, engaged in characteristic stage one activities in which a political base was developed, supported by unconventional military operations. As a result of initial problems of recruitment and operations, the revolutionaries shifted tactical emphasis from political organization to intimidation and terror and from larger military concentrations to smaller camps and units, with base areas moved deep into the jungle.

Somewhat haphazard military operations were conducted by the British in Malaya during the first months of the revolution. Experience led to a coordinated plan and organization stressing the integration of civilian-police-military operations under "war committees" at local, state, and federation levels, refined and formalized in the Briggs Plan of 1951. Analysis of directives, battle and casualty reports, and statistics and other data from both sides indicates that the period 1950-1951 was the watershed of the Emergency. It was at this time that the revolutionaries recognized their inability to move into stage two, marking the end of the critical period. After glancing at the Indochina experience for comparison, we will return to the causes of the revolutionary failure in Malaya.

The Second Indochina War began in Vietnam during the years 1957-1958 (scholars still argue over the exact date) with President Diem's harshly implemented campaign to rid the countryside of persons associated directly or indirectly with the Viet Minh, and of many who were anti-Communist but also anti-Diem. This campaign provided favorable conditions for a second revolutionary war. The Viet Minh organization, dormant since the Geneva Agreements of 1954 had halted hostilities and brought about the French military withdrawal, resumed revolutionary activity under the name Viet Cong. North Vietnamese support followed, although during this period most of the recruits and material came from the southern part of Vietnam. The success of the Viet Cong led to expansion of the revolutionary political base and military strength. This was stage one.

Diem's authoritarian methods and narrow leadership, based on relatives and personal followers, increased the alienation of the mass of peasants and urban nationalistic groups and led to the disaffection of some civil servants and military men. The roots of this alienation may be found in archaic and unjust land tenure systems, disregard of traditional political organization, favoritism to Diem's relatives and the Catholic minority, and an administration largely inherited from the French. These tensions and conflicts, and the increasing isolation of the ruler and his circle from the people, culminated in the assassination of Diem in 1963. By then the Viet Cong had organized much of the countryside and moved into stage two, marking the end of the critical period in the Second Indochina War. The Viet Cong would almost certainly have triumphed had not the United States intervened.

Comparison of experience in Malaya and Vietnam clarifies the reasons for British success and South Vietnamese failure in the critical period. Historically, the British had been tactfully paternalistic in their relationship with the Malays, protective of the traditional society with which they worked. Consequently there was not a deep antagonism against the British. The Chinese experience in Malaya had been quite different. Arriving like the British in the nineteenth century, but in much larger numbers, they eventually furnished a highly competitive middle class as well as an adaptable labor force. Dominant in local, internal trade, they considered the Malays culturally inferior. The antagonisms between Malays and Chinese were not only cultural and economic, but political and social: they were rival national and racial groups, in effect. The revolutionary organization originated in these communal factors. Most of the revolutionaries were Chinese; relatively few were Malays, Indians or Pakistanis.

The experienced British colonial structure, augmented by plantation interests in many remote areas and staffed down to the village level, provided the necessary communications and administration. Furthermore, the major sources of recruits and supplies for the insurgents were the rural Chinese squatters who had been forced into remote districts by the severity of Japanese rule during World War II. This made it possible for the British to isolate the revolutionary movement. Finally, the British agreed to Malayan independence, negating much of the nationalistic appeal of the revolutionaries. Thus, the British recognized the basic objectives of the revolution and fashioned programs to counter them, taking advantage of their administrative structure and knowledge of the country, the geographic isolation of the insurgents, and the support of most of the Malays. The result was an enlightened political policy supported by flexible emergency powers and an integrated civilian-military war structure which was able to counter the revolution without seriously disrupting the social and political fabric of the country.

In Vietnam an almost opposite reaction characterized the Diem government. A small "Mandarin" elite, engaged in a highly complex political game, employed dubious and unpopular measures to maintain control. Policy largely disregarded the wishes of the peasants: land distribution programs were announced but not carried through, or indifferently administered, perpetuating absentee landlordism. The Buddhist majority was harassed by the Catholic-dominated government. Thus the Viet Cong was able to gain the adherence of many nationalists, whom the French had disastrously alienated, as the regime lost much of its legitimacy. The administrative structure, much of which was destroyed during the First Indochina War, was unable to respond to normal demands, much less the demands of revolutionary war: it was weak at the province and district level and a minimal effort by the Viet Cong eroded the government's control of the countryside. The South Vietnamese army was unprepared for revolutionary war: France had left a heritage of conventional military thought and organization which was perpetuated by the United States. Road-bound regiments and divisions were no match for the fast-moving Viet Cong and their unconventional tactics. Only massive intervention by the United States saved the government of South Vietnam as the revolution moved through stage two.

The revolutionary environment in Malaya was at best tenuous. The character and structure of the society permitted the formation of a revolutionary movement, while at the same time limiting its scope. On the other hand, Vietnamese society not only provided a suitable environment for the growth of a revolutionary movement, it nurtured it; and the ruling elite was inept.

7

In summary, from among the various models of revolution we have suggested the Maoist as the most relevant. Focusing primarily on peasant societies caught up in the complex transition of modernization, we find the Maoist model provides a framework for identifying the characteristics of societies having a high propensity for revolution, and for the systematic study of revolution. The critical period in the revolutionary struggle is a useful concept, not only to understand the characteristics of revolution, but also to project the likely course of the revolutionary struggle, for success or failure depends on the ability of the revolution to move into stage two (in Mao's terminology).

We stressed the importance of the "politics" of revolution, i.e., the factors that create and express dissatisfaction with the existing regime. Military considerations are important, but only within the context of the political, social, economic, and psychological variables. The outcome rests on how effectively each side can operate as a total, multi-dimensional political system. Thus there is no simple solution to the problem of revolution.

The selections in the remainder of this volume examine the important factors in revolutionary guerrilla warfare. They enable the reader to develop an understanding of revolution, trace the growth of guerrilla doctrine, and study the specifics of revolutionary and counterrevolutionary guerrilla warfare.

Notes

1. Scott A. Boorman, *The Protracted Game: A Wei'Ch'i Interpretation of Maoist Revolutionary Strategy* (New York: Oxford University Press, 1969).
2. There are distinctions between definitions, concepts, model-building, and theories. The search here is for a working definition of revolution. Theories of revolution go beyond definitions by presenting comprehensive intellectual constructs of causes, actors, and course, of revolution; for this a simple definition is inadequate. As a beginning it is useful, however, to review definitions, since it reveals much about the literature and nature of revolutionary guerrilla war.
3. For a discussion of problems of definition see Carl Leiden and Karl M. Schmitt, *The Politics of Violence: Revolution in the Modern World* (Englewood Cliffs, N.J.: Prentice-Hall, 1968).
4. J.K. Zawodny, "Civil War," in *International Encyclopedia of the Social Sciences* (New York: Macmillan, 1968), Vol. 7, p. 499.
5. Department of the Army, *The Law of Land Warfare, FM 27-10* (Washington, D.C.: U.S. Government Printing Office, 1956), p. 9.
6. Lyford P. Edwards, "Civil War," in *Encyclopedia of the Social Sciences* (New York: Macmillan, 1935), Vol. 3, p. 523.
7. Franklin Mark Osanka, "Guerrilla Warfare," in *International*

Encyclopedia, Vol. 7, p. 503.
8. Douglas Pike, *Viet Cong* (Cambridge: M.I.T. Press, 1966), pp. 32-33.
9. Department of the Army, *Special Warfare* (Washington, D.C.: U.S. G.P.O., 1962), p. 8.
10. *Ibid.* See also Department of the Army, *Dictionary of U.S. Army Terms, AR 310-25* (Washington, D.C.: U.S. G.P.O., March 1969), p. 433.
11. *Special Warfare*, p. 8.
12. Frederick L. Schuman, "Insurrection," in *Encyclopedia of the Social Sciences*, Vol. 8, p. 116.
13. Sir Robert Thompson, *No Exit From Vietnam* (New York: McKay, 1969), p. 45.
14. Mao Tse-tung, *Selected Military Writings* (San Francisco: China Books and Periodicals, 1963), pp. 210-211. Truong Chinh, a North Vietnamese leader and ardent student of Mao, has amplified the three stage concept in his *Primer for Revolt* (New York: Praeger, 1963), p. 146:

Our long resistance will pass through three stages:
(1) The stage of contention
(2) The stage of equilibrium
(3) The stage of general counter-offensive

(a) First stage—strategy: defensive; tactics: attack.
(b) Second stage—strategy: stiff resistance (preparing for offensive); tactics; attack.
(c) Third stage—strategy and tactics: counter-offensive.

15. *Ibid.*, p. 288.
16. Mao Tse-tung, *Selected Works, Vol. II* (New York: International Publishers, 1954), p. 188.
17. General Vo Nguyen Giap, *People's War People's Army* (New York: Praeger, 1962), pp. 77-78.
18. Robert Taber, *The War of the Flea* (New York: Lyle Stuart, 1965), p. 156. Taber also notes (p. 29) that:

In the end it will be a question whether the government falls before the military is destroyed in the field, or whether the destruction of the military brings about the deposition of the political regime. The two processes are complementary. Social and political dissolution bleeds the military, and the protracted and futile campaign in the field contributes to the process of social and political dissolution, creating what I have elsewhere called the "climate of collapse." This is the grand strategic objective of the guerrilla: to create the "climate of collapse." It may be taken as the key to everything he does.

19. *Mao Tse-tung on Guerrilla Warfare*, translated and with introduction by Samuel B. Griffith (New York: Praeger, 1961), p. 7. This passage is from the introduction.
20. Richard L. Clutterbuck, *The Long, Long War* (New York: Praeger, 1966), pp. viii-ix.
21. *Ibid.*, p. ix.
22. Abdul Haris Nasution, *Fundamentals of Guerrilla Warfare* (New York: Praeger, 1965), p. 59. The author fought as a member of revolutionary guerrilla units against the Dutch, and later as a counterrevolutionary in suppression of an uprising against the Indonesian government.
23. Robert Thompson, *Defeating Communist Insurgency* (New York: Praeger, 1967), pp. 63-60.
24. John J. McCuen, *The Art of Counter-Revolutionary War* (Harrisburg: Stackpole, 1966), pp. 78-80.
25. Lucian W. Pye, "Lessons from the Malayan Struggle Against Communism," unpublished MS., p. 57.
26. Recently there has been some interesting and useful work on political violence and revolution, particularly on concepts of relative deprivation and on the application of behavioralist techniques to the study of violence. One of the best treatments is Ted Gurr, *Why Men Rebel* (Princeton: Princeton U. Press, 1970). Gurr proposes a number of premises and correlations, some of which are related to the concept of the critical period in this introduction. Although I agree with Gurr's general analysis, I disagree with his timing. By the time many of his determinants are operative it is already late in the revolutionary struggle.
27. The summaries of the Vietnam and Malayan insurgencies are drawn from Sam C. Sarkesian, "The Emergency in Malaya"(unpublished Ph. D. dissertation, Columbia University, New York, 1969).

Chapter II
Revolution

We begin by exploring some models and theories used to explain revolution. Models are frameworks used to organize thought and direct research, providing foundations for the development of theory. Resting on categories and assumptions, they serve to identify, explain, and predict relationships between variables. Theories are models, or combinations of models, which "purport to explain, predict, or prescribe in any field of inquiry" (from Jack C. Plano and Robert E. Riggs, *Dictionary of Political Analysis* (Hinsdale, Ill.: Dryden Press, 1973), pp. 48 and 101, which contains detailed explanations of these concepts).

It is difficult to focus on a theory, for there is much disagreement and divergence of approach about the meaning of revolution and its causes. The selections in this chapter are representative of recent authorities in the field. They range from a historian's perspective to the empiricism of the behavioralist, and employ techniques derived from sociology, psychology, and economics. These essays include discussions of traditional as well as recent approaches, and focus on a variety of theories.

The essay by Lawrence Stone is a review and critique of theories of revolution. It is a useful starting point, since the author ranges across the spectrum of social science. Stone points out problems encountered in attempting to establish basic definitions. He identifies what various scholars consider common elements of revolutionary situations. As a historian, Stone acknowledges the contributions of other disciplines to the study of revolutions, but cautions against unqualified acceptance of some recent writings, calling them "ingenious feats of verbal juggling in an esoteric language, performed around the totem pole of an abstract model, surrounded as far as the eye can see by the arid wastes of terminological definitions and mathematical formulae."

The selection by Tanter and Midlarsky is more empirically focused and is an exposition of a particular theory. The measurement of the duration and intensity of domestic violence is at the base of their argument that revolutions are closely correlated to economic and educational factors. They conclude that "revolutions occur more frequently in societies with lowest levels of economic attainment." Their case is reinforced by empirical studies of a number of revolutions of the 'fifties.

Ted Robert Gurr argues that "many of the variables and

relationships identified in social psychological research on frustration-aggression appear to underlie the phenomenology of civil violence." The author suggests that civil violence is a pressing area of inquiry for social scientists; his essay grows out of the investigations of violence in America that were set in motion by the assassinations and racial uprisings of the 'sixties. Gurr proposes a wide-ranging approach to civil violence, of which he regards revolution as but one form. In his study of "illicit" violence he focuses on three theories of aggression, concluding that frustration theory is the most useful. Frustration is caused by relative deprivation and is based on "perception of discrepancy between value expectation and their environment's apparent value capabilities." This leads to anger, which can develop into civil violence.

The next selection, by James Davies, offers a useful model, the J-curve. Davies suggests that "revolution is most likely to take place when a prolonged period of rising expectations and rising gratifications is followed by a short period of sharp reversals during which the gap between expectations and gratification quickly widens and becomes intolerable." From these frustrations violent action develops, which in turn can become revolution if it is focused on the institutions of government and is coherent, purposeful, widespread and intense. Davies applies this approach to a variety of historical uprisings, distinguishing between revolution and rebellion. While he points out that if basic needs are not fulfilled no social convention will limit the violent action of individuals, he also suggests that the J-curve is basically a psychological phenomenon.

The final selection in this chapter is a Soviet attack upon Western studies of revolution. Labelling them pseudo-scientific, vague in terminology and definition, Y. Krasin charges they are capitalistic-imperialistic assessments of a profound and complex social phenomenon that cannot be understood in Western social science terms. He suggests that most of these studies are ethnocentric and biased, and evade social issues by focusing on the abstract mechanics of models. The author also attacks "left radical concepts of revolution" which "are not a step forward but a big step back in social thought."

Theories of Revolution
by Lawrence Stone

In attacking the problem of revolution, as most others of major significance in history, we historians should think twice before we spurn the help offered by our colleagues in the social sciences, who have, as it happens, been particularly active in the last few years in theorizing about the typology, causes, and evolutionary patterns of this particular phenomenon. The purpose of this article is not to advance any new hypothesis, but to provide a summary view and critical examination of the work that has been going on.

The first necessity in any inquiry is a careful definition of terms: What is, and what is not, a revolution? According to one view, it is change, effected by the use of violence, in government, and/or regime, and/or society.[1] By *society* is meant the consciousness and the mechanics of communal solidarity, which may be tribal, peasant, kinship, national, and so on; by *regime* is meant the constitutional structure—democracy, oligarchy, monarchy; and by *government* is meant specific political and administrative institutions. Violence. it should be noted, is not the same as force; it is force used with unnecessary intensity, unpredictably, and usually destructively.[2] This definition of revolution is a very broad one, and two historians of the French Revolution, Crane Brinton and Louis Gottschalk, would prefer to restrict the use of the word to the major political and social upheavals with which they are familiar, the "Great Revolutions" as George S. Pettee calls them.[3]

Even the wider definition allows the historian to distinguish between the seizure of power that leads to a major restructuring of government or society and the replacement of the former elite by a new one, and the coup d'état involving no more than a change of ruling personnel by violence or threat of violence. This latter is the norm in Latin America, where it occurred thirty-one times in the ten years 1945–1955. Merle Kling has arrived at a suggestive explanation of this Latin American phenomenon of chronic political instability, limited but frequent use of violence, and almost complete lack of social or institutional change. He argues that ownership of the principal economic resources, both agricultural and mineral, is concentrated in the hands of a tiny, very stable, elite of enormously wealthy monoculture landlords and mining capitalists. This elite is all-powerful and cannot be attacked by opposition groups within the country; externally, however, it is dependent on foreign interests for its markets and its capital. In this colonial situation of a foreign-supported closed plutocracy, the main avenue of rapid upward social mobility for nonmembers of the elite leads, via the army, to the capture of the government machine, which is the only accessible source of wealth and power. This political instability is permitted by the elite on the condition that its own interests are undisturbed. Instability, limited violence, and the absence of social or institutional change are therefore all the product of the contradiction between the realities of a colonial economy run by a plutocracy and the facade of political sovereignty—between the real, stable power of the economic elite and the nominal, unstable control of politicians and generals.[4]

The looser definition of revolution thus suits both historians of major social change and historians of the palace coup. It does, however, raise certain difficulties. Firstly, there is a wide range of changes of government by violence which are neither a mere substitution of personalities in positions of power nor a prelude to the restructuring of society; secondly, conservative counterrevolutions become almost impossible to fit into the model; and lastly, it remains hard to distinguish between colonial wars, civil wars, and social revolution.

To avoid these difficulties, an alternative formulation has recently been put forward by a group of social scientists working mainly at Princeton. They have dropped the word "revolution" altogether and put "internal war" in its place.[5] This is defined as any attempt to alter state policy, rulers, or institutions by the use of violence, in societies where violent competition is not the norm and where well-defined institutional patterns exist.[6] This concept seems to be a logical consequence of the preoccupation of sociologists in recent years with a model of society in a stable, self-regulating state of perpetual equipoise. In this utopian world of universal harmony, all forms of violent conflict are anomalies, to be treated alike as pathological disorders of a similar species. This is a model which, although it has its uses for analytical purposes, bears little relation to the reality familiar to the historian. It looks to a society without change, with universal consensus on values, with complete social harmony, and isolated from external threats; no approximation to such a society has ever been seen. An alternative model, which postulates that all societies are in a condition of multiple and perpetual tension held in check by social norms, ideological beliefs, and state sanctions, accords better with historical fact, as some sociologists are now beginning to realize.[7]

The first objection to the all-embracing formula of internal war is that, by covering all forms of physical conflict from strikes and terrorism to civil war, it isolates the use of violence from the normal processes of societal adjustment. Though some of the users of the term express their awareness that the use of violence for political ends is a fairly common occurrence, the definition they have established in fact excludes all times and places where it *is* common. It thus cuts out most societies the world has ever known, including Western Europe in the Middle Ages and Latin America today. Secondly, it isolates one particular means, physical violence, from the political ends that it is designed to serve. Clausewitz's famous definition of external war is equally applicable to internal war, civil war, or revolution: "War is not only a political act, but a real political instrument; a continuation of political transactions, an accomplishment of them by different means. That which remains peculiar to war relates only to the peculiar nature of its means."[8]

It is perfectly true that any means by which society exercises pressure or control, whether it is administrative organization, constitutional law, economic interest, or physical force, can be a fruitful field of study in its own right, so long as its students remain aware that they are looking at only one part of a larger whole. It is also true that there is something peculiar about violence, if only because of man's highly ambivalent attitude towards the killing of his own species. Somehow, he regards physical force as different in kind from, say, economic exploitation or psychological manipulation as a means of exercising power over others. But this distinction is not one of much concern to the historian of revolution, in which violence is a normal and natural occurrence. The concept of internal war is too broad in its comprehension of all types of violence from civil wars to strikes, too narrow in its restriction to normally nonviolent societies, too limited in its concern with one of many means, too arbitrary in its separation of this means from the ends in view, and too little concerned with the complex roots of social unrest to be of much practical value to him.

The most fruitful typology of revolution is that of Chalmers Johnson, set out in a pamphlet that deserves to be widely read.[9] He sees six types, identified by the targets selected for attack, whether the government personnel, the political regime, or the community as a social unit; by the nature of the carriers of revolution, whether as a mass or an elite; and particularly by the goals and the ideologies, whether reformist, eschatological, nostalgic, nation-forming, elitist, or nationalist. The first type, the *Jacquerie*, is a spontaneous mass peasant rising, usually carried out in the name of the traditional authorities, Church and King, and with the limited aims of purging the local or national elites. Examples are the Peasant Revolt of 1381, Ket's Rebellion of 1549, and the Pugachev rebellion in Russia in 1773–1775. The second type, the *Millenarian Rebellion*, is similar to the first but with the added feature of a utopian dream, inspired by a living messiah. This type can be found at all times, in all parts of the world, from the Florentine revolution led by Savonarola in 1494, to the Anabaptist Rebellion in Münster led by John Mathijs and John Beukels in 1533–1535, to the Sioux Ghost-Dance Rebellion inspired by the Paiute prophet Wovoka in 1890. It has attracted a good deal of attention from historians in recent years, partly because the career of Hitler offered overwhelming proof of the enormous historical significance of the charismatic leader, and partly because of a growing interest in the ideas of Max Weber.[10] The third type is the *Anarchist Rebellion*, the nostalgic reaction to progressive change, involving a romantic idealization of the old order: the Pilgrimage of Grace and the Vendée are examples.

The fourth is that very rare phenomenon, the *Jacobin Communist Revolution*. This has been defined as "a sweeping fundamental change in political organization, social structure, economic property control and the predominant myth of a social order, thus indicating a major break in the continuity of development."[11]

This type of revolution can occur only in a highly centralized state with good communications and a large capital city, and its target is government, regime, and society—the lot. The result is likely to be the creation of a new national consciousness under centralized, military authority, and the erection of a more rational, and hence more efficient, social and bureaucratic order on the ruins of the old ramshackle structure of privilege, nepotism, and corruption.

The fifth type is the *Conspiratorial Coup d'Etat*, the planned work of a tiny elite fired by an oligarchic, sectarian ideology. This qualifies as a revolutionary type only if it in fact anticipates mass movement and inaugurates social change—for example the Nasser revolution in Egypt or the Castro revolution in Cuba; it is thus clearly distinguished from the palace revolt, assassination, dynastic succession-conflict, strike, banditry, and other forms of violence, which are all subsumed under the "internal war" rubric.

Finally, there is the *Militarized Mass Insurrection*, a new phenomenon of the twentieth century in that it is a deliberately planned mass revolutionary war, guided by a dedicated elite. The outcome of guerrilla warfare is determined by political attitudes, not military strategy or matériel, for the rebels are wholly dependent on broad popular support. In all cases on record, the ideology that attracts the mass following has been a combination of xenophobic nationalism and Marxism, with by far the greater stress on the former. This type of struggle has occurred in Yugoslavia, China, Algeria, and Vietnam.

Although, like any schematization of the historical process, this sixfold typology is concerned with ideal types, although in practice individual revolutions may sometimes display characteristics of several different types, the fact remains that this is much the most satisfactory classification we have so far; it is one that working historians can recognize and use with profit. The one obvious criticism is semantic, an objection to the use of the phrase "Jacobin Communist Revolution." Some of Johnson's examples are Communist, such as the Russian or Chinese Revolutions; others are Jacobin but not Communist, such as the French Revolution or the Turkish Revolution of 1908–1922. It would be better to revert to Pette's category of "Great Revolutions," and treat Communist revolutions as a subcategory, one type, but not the only type, of modernizing revolutionary process.

Given this classification and definition of revolution, what are its root causes? Here everyone is agreed in making a sharp distinction between long-run, underlying causes—the preconditions, which create a potentially explosive situation and can be analyzed on a comparative basis—and immediate, incidental factors—the precipitants, which trigger the outbreak and which may be nonrecurrent, personal, and fortuitous. This effectively disposes of the objections of those historians whose antipathy to conceptual schematization takes the naïve form of asserting the uniqueness of each historical event.

One of the first in the field of model-building was Crane Brinton who, as long ago as 1938, put forward a series of uniformities common to the four great Western revolutions: English, French, American, and Russian. These included an economically advancing society, growing class and status antagonisms, an alienated intelligentsia, a psychologically insecure and politically inept ruling class, and a governmental financial crisis.[12]

The subjectivity, ambiguity, and partial self-contradiction of this and other analyses of the causes of specific revolutions— for example the French Revolution—have been cruelly shown up by Harry Eckstein.[13] He has pointed out that commonly adduced hypotheses run the spectrum of particular conditions, moving from the intellectual (inadequate political socialization, conflicting social myths, a corrosive social philosophy, alienation of the intellectuals) to the economic (increasing poverty, rapid growth, imbalance between production and distribution, long-term growth plus short-term recession) to the social (resentment due to restricted elite circulation, confusion due to excessive elite recruitment, anomie due to excessive social mobility, conflict due to the rise of new social classes) to the political (bad government, divided government, weak government, oppressive government). Finally there are explanations on the level of general process, such as rapid social change, erratic social change, or a lack of harmony between the state structure and society, the rulers and the ruled. None of these explanations are invalid in themselves, but they are often difficult or impossible to reconcile one with the other, and are so diverse in their range and variety as to be virtually impossible to fit into an ordered analytical framework. What, then, is to be done?

Fundamental to all analyses, whether by historians like Brinton and Gottschalk or by political scientists like Johnson and Eckstein, is the recognition of a lack of harmony between the social system on the one hand and the political system on the other. This situation Johnson calls *dysfunction*, in a word derived from the structural-functional equilibrium model of the sociologists. This dysfunction may have many causes, some of which are merely cyclical, such as may develop because of personal weaknesses in hereditary kingships or single-party regimes. In these cases, the revolution will not take on serious proportions, and will limit itself to attacks on the governing institutions, leaving regime and society intact. In most cases, however, including all those of real importance, the dysfunction is the result of some new and developing process, as a result of which certain social subsystems find themselves in a condition of relative deprivation. Rapid economic growth, imperial conquest, new metaphysical beliefs, and important technological changes are the four commonest factors involved, in that order. If the process of change is sufficiently slow and sufficiently moderate, the dysfunction may not rise to dangerous levels. Alternately, the elite may adjust to the new situation with sufficient rapidity and skill to ride out the storm and retain popular confidence. But if the change is both rapid and profound, it may cause the sense of deprivation, alienation, anomie to spread into many sectors of the society at once, causing what Johnson calls multiple dysfunction, which may be all but incurable within the existing political system.

In either case the second vital element in creating a revolutionary situation is the condition and attitude of the entrenched elite, a factor on which Eckstein rightly lays great stress. The elite may lose its manipulative skill, or its military superiority, or its self-confidence, or its cohesion; it may become estranged from the nonelite, or overwhelmed by a financial crisis; it may be incompetent, or weak, or brutal. Any combination of two or more of these features will be dangerous. What is ultimately fatal, however, is the compounding of its errors by intransigence. If it fails to anticipate the need for reform, if it blocks all peaceful, constitutional means of social adjustment, then it unites the various deprived elements in single-minded opposition to it, and drives them down the narrow road to violence. It is this process of polarization into two coherent groups or alliances of what are naturally and normally a series of fractional and shifting tensions and conflicts within a society that both Peter Amman and Wilbert Moore see as the essential preliminary to the outbreak of a Jacobin Revolution.[14] To conclude, therefore, revolution becomes *possible* when a condition of multiple dysfunction meets an intransigent elite: just such a conjunction occurred in the decades immediately before the English, the French, and the Russian Revolutions.

Revolution only becomes *probable* (Johnson might say "certain"), however, if certain special factors intervene: the "precipitants" or "accelerators." Of these, the three most common are the emergence of an inspired leader or prophet; the formation of a secret, military, revolutionary organization; and the crushing defeat of the armed forces in foreign war. This last is of critical importance since it not only shatters the prestige of the ruling elite, but also undermines the morale and discipline of the soldiers and thus opens the way to the violent overthrow of the existing government.

The first defect of Johnson's model is that it concentrates too much on objective structural conditions, and attempts to relate conditions directly to action. In fact, however, as Eckstein points out, there is no such direct relationship; historians can point to similar activity arising from different conditions, and different activity arising from similar conditions. Standing between objective reality and action are subjective human attitudes. A behaviorist approach such as Brinton's, which lays equal stress on such things as anomie, alienation of the intellectuals, frustrated popular aspirations, elite estrangement, and loss of elite self-confidence, is more likely to produce a satisfactory historical explanation than is one that sticks to the objective social reality. Secondly, Johnson leaves too little play for the operation of the unique and the personal. He seems to regard his accelerators as automatic triggers, ignoring the area of unpredictable personal choice that is always left to the ruling elite and to the revolutionary leaders, even in a situation of multiple dysfunction exacerbated by an accelerator. Revolution is never inevitable—or rather the only evidence of its inevitability is that it actually happens. Consequently the only way to prove this point is to indulge in just the kind of hypothetical argument that historians prudently try to avoid. But it is still just possible that modernization may take place in Morocco and India without revolution. The modernization and industrialization of Germany and Britain took place without revolution in the nineteenth century (though it can be argued that in the latter case the process was slow by twentieth-century standards, and that, as is now becoming all too apparent, the modernization was far from complete). Some think that a potentially revolutionary situation in the United States in the 1930s was avoided by political action.

Lastly it is difficult to fit into the Johnson model the fact that political actions taken to remedy dysfunction often themselves precipitate change. This produces the paradoxical hypothesis that measures designed to restore equilibrium in fact upset equilibrium. Because he begins with his structural-functional equilibrium model, Johnson is a victim of the fallacy of intended consequences. As often as not in history it is the *unintended* consequences that really matter: to mention but one example, it was Louis XVI's belated and half-hearted attempts at reform that provoked the aristocratic reaction, which in turn opened the way to the bourgeois, the peasant, and the sans-culotte revolutions. Finally the dysfunction concept is not altogether easy to handle in a concrete historical case. If societies are regarded as being in a constant state of multiple tension, then some degree of dysfunction is always present. Some group is always in a state of relative deprivation due to the inevitable process of social change.

Recognition of this fact leads Eckstein to point out the importance of forces working *against* revolution. Historians, particularly those formed in the Western liberal tradition, are reluctant to admit that ruthless, efficient repression—as opposed to bumbling, half-hearted repression—involving the physical destruction of leading revolutionaries and effective control of the media of communication, can crush incipient revolutionary movements. Repression is particularly effective when governments know what to look for, when they have before their eyes the unfortunate example of other governments overthrown by revolutionaries elsewhere. Reaction, in fact, is just as infectious as revolution. Moreover diversion of energy and attention to successful—as opposed to unsuccessful—foreign war can ward off serious internal trouble. Quietist—as opposed to activist—religious movements may serve as the opiate of the people, as Halévy suggested about Methodism in England. Bread and circuses may distract popular attention. Timely—as opposed to untimely—political concessions may win over moderate opinion and isolate the extremists.

Basing himself on this suggestive analysis, Eckstein produces a paradigm for universal application. He sees four positive variables—elite inefficiency, disorienting social process, subversion, and available rebel facilities—and four negative variables—diversionary mechanisms, available incumbent facilities, adjustive mechanisms, and effective repression. Each type of internal war, and each step of each type, can, he suggests, be explained in terms of these eight variables. While this may be true, it is fair to point out that some of the variables are themselves the product of more deep-seated factors, others mere questions of executive action that may be determined by the accidents of personality. Disruptive social process is a profound cause; elite inefficiency a behavior pattern; effective repression a function of will; facilities the by product of geography. One objection to the Eckstein paradigm is therefore that it embraces different levels of explanation and fails to maintain the fundamental distinction between preconditions and precipitants. Secondly, it concentrates on the factors working for or against the successful manipulation of violence rather than on the underlying factors working to produce a revolutionary potential. This is because the paradigm is intended to apply to all forms of internal war rather than to revolution proper, and because all that the various forms of internal war have in common is the use of violence. It is impossible to tell how serious these criticisms are until the paradigm has been applied to a particular historical revolution. Only then will its value become apparent.

If we take the behaviorist approach, then a primary cause of revolutions is the emergence of an obsessive revolutionary mentality. But how closely does this relate to the objective material circumstances themselves? In every revolutionary situation one finds a group of men—fanatics, extremists, zealots—so convinced of their own righteousness and of the urgent need to create a new Jerusalem on earth (whether formally religious or secular in inspiration is irrelevant) that they are prepared to smash through the normal restraints of habit, custom, and convention. Such men were the seventeenth-century English Puritans, the eighteenth-century French Jacobins, the twentieth-century Russian Bolsheviks. But what makes such men is far from certain. What generates such ruthlessness in curbing evil, such passion for discipline and order? Rapid social mobility, both horizontal and vertical, and particularly urbanization, certainly produces a sense of rootlessness and anxiety. In highly stratified societies, even some of the newly-risen elements may find themselves under stress.[15] While some of the *arrivistes* are happily absorbed in their new strata, others remain uneasy and resentful. If they are snubbed and rebuffed by the older members of the status group to which they aspire by reason of their new wealth and position, they are likely to become acutely conscious of their social inferiority, and may be driven either to adopt a pose *plus royaliste que le Roi* or to dream of destroying the whole social order. In the latter case they may try to allay their sense of insecurity by imposing their norms and values by force upon society at large. This is especially the case if there is available a moralistic ideology like Puritanism or Marxism to which they can attach themselves, and which provides them with unshakable confidence in their own rectitude.

But why does the individual react in this particular way rather than another? Some would argue that the character of the revolutionary is formed by sudden ideological conversion in adolescence or early adult life (to Puritanism, Jacobism, or Bolshevism) as a refuge from this anxiety state.[16] What is not acceptable is the fashionable conservative cliché that the revolutionary and the reformer are merely the chance product of unfortunate psychological difficulties in childhood. It is possible that this is the mechanism by which such feelings are generated, though there is increasing evidence of the continued plasticity of human character until at any rate post-adolescence. The main objection to this theory is that it fails to explain why these particular attitudes become common only in certain classes and age groups at certain times and in certain places. This failure strongly suggests that the cause of this state of mind lies not in the personal maladjustment of the individuals or their parents, but in the social conditions that created that maladjustment. Talcott Parsons treats disaffection or "alienation" as a generalized phenomenon that may manifest itself in crime, alcoholism, drug addiction, daytime fantasies, religious enthusiasm, or serious political agitation. To use Robert Merton's formulation, Ritualism and Retreatism are two possible psychological escape-routes; Innovation and Rebellion two others.[17]

Even if we accept this behaviorist approach (which I do), the fact remains that many of the underlying causes both of the alienation of the revolutionaries and of the weakness of the incumbent elite are economic in origin; and it is in this area that some interesting work has centered. In particular a fresh look has been taken at the contradictory models of Marx and de Tocqueville, the one claiming that popular revolution is a product of increasing misery, the other that it is a product of increasing prosperity.

Two economists, Sir Arthur Lewis and Mancur Olson, have pointed out that because of their basic social stability, both preindustrial and highly industrialized societies are relatively free from revolutionary disturbance.[18] In the former societies, people accept with little question the accepted rights and obligations of family, class, and caste. Misery, oppression, and social injustice are passively endured as inevitable features of life on earth. It is in societies experiencing rapid economic growth that the trouble usually occurs. Lewis, who is thinking mostly about the newly emerging countries, primarily of Africa, regards the sense of frustration that leads to revolution as a consequence of the dislocation of the old status patterns by the emergence of four new classes—the proletariat, the capitalist employers, the urban commercial and professional middle class, and the professional politicians—and of the disturbance of the old income patterns by the sporadic and patchy impact of economic growth, which creates new wealth and new poverty in close and conspicuous juxtaposition. Both phenomena he regards as merely transitional, since in a country fully developed economically there are strong tendencies toward the elimination of inequalities of opportunity, income, and status.

This model matches fairly well the only detailed analysis of a historical revolution in which a conscious effort has been made to apply modern sociological methods. In his recent study of the Vendée, Charles Tilly argues that a counterrevolutionary situation was the consequence of special tensions created by the immediate juxtaposition of, on the one hand, parish clergy closely identified with the local communities, great absentee landlords, and old-fashioned subsistence farming, and, on the other, a large-scale textile industry on the putting-out system and increasing bourgeois competition.[19] Though the book is flawed by a tendency to take a ponderous sociological hammer to crack a simple little historical nut, it is nonetheless a suggestive example of the application of new hypotheses and techniques to historical material.

Olson has independently developed a more elaborate version of the Lewis theory. He argues that revolutionaries are déclassé and freed from the social bonds of family, profession, village or manor; and that these individuals are the product of rapid economic growth, which creates both *nouveaux riches* and *nouveaux pauvres*. The former, usually middle-class and urban artisans, are better off economically, but are disoriented, rootless, and restless; the latter may be workers whose wages have failed to keep pace with inflation, workers in technologically outdated and therefore declining industries, or the unemployed in a society in which the old cushions of the extended family and the village have gone, and in which the new cushion of social security has not yet been created. The initial growth phase may well cause a decline in the standard of living of the majority because of the need for relatively enormous forced savings for reinvestment. The result is a revolution caused by the widening gap between expectations—social and political for the new rich, economic for the new poor—and the realities of everyday life.

A sociologist, James C. Davies, agrees with Olson that the fundamental impetus toward a revolutionary situation is generated by rapid economic growth but he associates such growth with a generally rising rather than a generally falling standard of living, and argues that the moment of potential revolution is reached only when the long-term phase of growth is followed by a short-term phase of economic stagnation or decline.[20] The result of this "J-curve," as he calls it, is that steadily soaring expectations, newly created by the period of growth, shoot further and further ahead of actual satisfaction of needs. Successful revolution is the work neither of the destitute nor of the well-satisfied, but of those whose actual situation is improving less rapidly than they expect.

These economic models have much in common, and their differences can be explained by the fact that Lewis and Olson are primarily concerned with the long-term economic forces creating instability, and Davis with the short-term economic factors that may precipitate a crisis. Moreover their analyses apply to different kinds of economic growth, of which three have recently been identified by W. W. Rostow and Barry Supple: there is the expansion of production in a pre-industrial society, which may not cause any important technological, ideological, social, or political change; there is the phase of rapid growth, involving major changes of every kind; and there is the sustained trend toward technological maturity.[21] Historians have been quick to see that these models, particularly that of Rostow, can be applied only to a limited number of historical cases. The trouble is not so much that in any specific case the phases—particularly the last two—tend to merge into one another, but that changes in the various sectors occur at irregular and unexpected places on the time-scale in different societies. Insofar as there is any validity in the division of the stages of growth into these three basic types, the revolutionary model of Olson and Lewis is confined to the second; that of Davis is applicable to all three.

The Davis model fits the history of Western Europe quite well, for it looks as if in conditions of extreme institutional and ideological rigidity the first type of economic growth may produce frustrations of a very serious kind. Revolutions broke out all over Europe in the 1640's, twenty years after a secular growth phase had come to an end. [22] C. E. Labrousse has demonstrated the existence of a similar economic recession in France from 1778,[23] and from 1914 the Russian economy was dislocated by the war effort after many years of rapid growth. Whatever its limitations in any particular situation, the J-curve of actual satisfaction of needs is an analytical tool that historians can usefully bear in mind as they probe the violent social upheavals of the past.

As de Tocqueville pointed out, this formula of advance followed by retreat is equally applicable to other sectors. Trouble arises if a phase of liberal governmental concessions is followed by a phase of political repression; a phase of fairly open recruitment channels into the elite followed by a phase of aristocratic reaction and a closing of ranks; a phase of weakening status barriers by a phase of reassertion of privilege. The J-curve is applicable to other than purely economic satisfactions, and the apex of the curve is the point at which underlying causes, the preconditions, merge with immediate factors, the precipitants. The recipe for revolution is thus the creation of new expectations by economic improvement and some social and political reforms, followed by economic recession, governmental reaction, and aristocratic resurgence, which widen the gap between expectations and reality.

All these attempts to relate dysfunction to relative changes in economic prosperity and aspirations are hampered by two things, of which the first is the extreme difficulty in ascertaining the facts. It is never easy to discover precisely what is happening to the distribution of wealth in a given society. Even now, even in highly developed Western societies with massive bureaucratic controls and quantities of statistical data, there is no agreement about the facts. Some years ago it was confidently believed that in both Britain and the United States incomes were being levelled, and that extremes of both wealth and poverty being steadily eliminated. Today, no one quite knows what is happening in either country.[24] And if this is true now, still more is it true of societies in the past about which the information is fragmentary and unreliable.

Secondly, even if they can be clearly demonstrated, economic trends are only one part of the problem. Historians are increasingly realizing that the psychological responses to changes in wealth and power are not only not precisely related to, but are politically more significant than, the material changes themselves. As Marx himself realized at one stage, dissatisfaction with the status quo is not determined by absolute realities but by relative expectations. "Our desires and pleasures spring from society; we measure them, therefore, by society, and not by the objects which serve for their satisfaction. Because they are of a social nature, they are of a relative nature."[25] Frustration may possibly result from a rise and subsequent relapse in real income. But it is perhaps more likely to be caused by a rise in aspirations that outstrips the rise in real income; or by a rise in the *relative* economic position in society of the group in question, followed by a period in which its real income continues to grow, but less fast than that of other groups around it. Alternatively it may represent a rise and then decline of status, largely unrelated to real income; or if status and real income are related, it may be inversely. For example, social scientists seeking to explain the rise of the radical right in the United States in the early 1950s and again in the early 1960s attribute it to a combination of great economic prosperity and an aggravated sense of insecurity of status.[26] Whether or not this is a general formula for right-wing rather than left-wing revolutionary movements is not yet clear.

Moreover the problem is further complicated by an extension of the reference-group theory.[27] Human satisfaction is related not to existing conditions but to the condition of a social group against which the individual measures his situation. In an age of mass communications and the wide distribution of cheap radio receivers even among the impoverished illiterate of the world, knowledge of high consumption standards elsewhere spreads rapidly, and as a result the reference group may be in another, more highly developed, country or even continent. Under these circumstances, revolutionary conditions may be created before industrialization has got properly under way.

The last area in which some new theoretical work has been

42 done is in the formulation of hypotheses about the social stages of a "Great Revolution." One of the best attacks on this problem was made by Crane Brinton, who was thinking primarily about the French Revolution, but who extended his comparisons to the three other major Western revolutionary movements. He saw the first phase as dominated by moderate bourgeois elements; their supersession by the radicals; a reign of terror; a Thermidorian reaction; and the establishment of strong central authority under military rule to consolidate the limited gains of the revolution. In terms of mass psychology he compared revolution with a fever that rises in intensity, affecting nearly all parts of the body politic, and then dies away.

A much cruder and more elementary model has been advanced by an historian of the revolutions of 1848, Peter Amman.[28] He sees the modern state as an institution holding a monopoly of physical force, administration, and justice over a wide area, a monopoly dependent more on habits of obedience than on powers of coercion. Revolution may therefore be defined as a breakdown of the monopoly due to a failure of these habits of obedience. It begins with the emergence of two or more foci of power, and ends with the elimination of all but one. Amman includes the possibility of "suspended revolution," with the existence of two or more foci not yet in violent conflict.

This model admittedly avoids some of the difficulties raised by more elaborate classifications of revolution: how to distinguish a coup d'état from a revolution; how to define the degrees of social change; how to accommodate the conservative counterrevolution, and so on. It certainly offers some explanation of the progress of revolution from stage to stage as the various power blocs that emerge on the overthrow of the incumbent regime are progressively eliminated; and it explains why the greater the public participation in the revolution, the wider the break with the habits of obedience, and therefore the slower the restoration of order and centralized authority. But it throws the baby out with the bathwater. It is impossible to fit any decentralized traditional society, or any modern federal society, into the model. Moreover, even where it might be applicable, it offers no framework for analyzing the roots of revolution, no pointers for identifying the foci of power, no means of distinguishing between the various revolutionary types, and its notion of "suspended revolution" is little more than verbal evasion.

Though it is set out in a somewhat confused, overelaborate, and unnecessarily abstract form, the most convincing description of the social stages of revolution is that outlined by Rex D. Hopper.[29] He sees four stages. The first is characterized by indiscriminate, uncoordinated mass unrest and dissatisfaction, the result of dim recognition that traditional values no longer satisfy current aspirations. The next stage sees this vague unease beginning to coalesce into organized opposition with defined goals, an important characteristic being a shift of allegiance by the intellectuals from the incumbents to the dissidents, the advancement of an "evil men" theory, and its abandonment in favor of an "evil institutions" theory. At this stage there emerge two types of leaders: the prophet, who sketches the shape of the new utopia upon which men's hopes can focus, and the reformer, working methodically toward specific goals. The third, the formal stage, sees the beginning of the revolution proper. Motives and objectives are clarified, organization is built up, a statesman leader emerges. Then conflicts between the left and the right of the revolutionary movement become acute, and the radicals take over from the moderates. The fourth and last stage sees the legalization of the revolution. It is a product of psychological exhaustion as the reforming drive burns itself out, moral enthusiasm wanes, and economic distress increases. The administrators take over, strong central government is established, and society is reconstructed on lines that embody substantial elements of the old system. The result falls far short of the utopian aspirations of the early leaders, but it succeeds in meshing aspirations with values by partly modifying both, and so allows the reconstruction of a firm social order.

Some of the writings of contemporary social scientists are ingenious feats of verbal juggling in an esoteric language, performed around the totem pole of an abstract model, surrounded as far as the eye can see by the arid wastes of terminological definitions and mathematical formulae. Small wonder the historian finds it hard to digest the gritty diet of this neo-scholasticism, as it has been aptly called. The more historically-minded of the social scientists, however, have a great deal to offer. The history of history, as well as of science, shows that advances depend partly on the accumulation of factual information, but rather more on the formulation of hypotheses that reveal the hidden relationships and common properties of apparently

44 distinct phenomena. Social scientists can supply a corrective to the antiquarian fact-grubbing to which historians are so prone; they can direct attention to problems of general relevance, and away from the sterile triviality of so much historical research. They can ask new questions and suggest new ways of looking at old ones. They can supply new categories, and as a result may suggest new ideas.[30]

Notes

* I am grateful to Professors Cyril E. Black, Arno J. Mayer, and John W. Shy for some very helpful criticisms of this article.
1. Chalmers Johnson, *Revolution and the Social System*, Hoover Institution Studies 3 (Stanford 1964).
2. Sheldon S. Wolin, "Violence and the Western Political Tradition," *American Journal of Orthopsychiatry*, xxxii (January 1963), 15-28.
3. Brinton, *The Anatomy of Revolution* (New York 1938); Gottschalk, "Causes of Revolution," *American Journal of Sociology*, L (July 1944), 1-8; Pettee, *The Process of Revolution* (New York 1938).
4. "Toward a Theory of Power and Political Instability in Latin America," *Western Political Quarterly*, IX (1956).
5. Harry Eckstein, ed., *Internal War* (New York 1964), and "On the Etiology of Internal War," *History and Theory*, IV, No. 2 (1965), 133-63. I am grateful to Mr. Eckstein for allowing me to read this article before publication.
6. The formula has been used by a historian, Peter Paret, in *Internal War and Pacification: The Vendée, 1793-96* (Princeton 1961).
7. Barrington Moore, "The Strategy of the Social Sciences," in his *Political Power and Social Theory* (Cambridge, Mass., 1958); Ralph Dahrendorf, "Out of Utopia: Toward a Reorientation of Sociological Analysis," *American Journal of Sociology*, LXIV (September 1958), 115-27; C. Wright Mills, *The Sociological Imagination* (New York 1959); Wilbert E. Moore, *Social Change* (Englewood Cliffs 1963). It should be noted that both the equilibrium and the conflict views of society have very respectable ancestries. The equilibrium model goes back to Rousseau—or perhaps Aquinas; the conflict model to Hobbes, Hegel, and Marx.
8. Quoted in Edward Mead Earle, ed., *Makers of Modern Strategy* (Princeton 1943), 104-5.
9. *Revolution and the Social System*.
10. N. R. C. Cohn, *Pursuit of the Millennium* (New York 1961); Eric J. Hobsbawm, *Primitive Rebels* (Manchester 1959); S. L. Thrupp. *Millennial Dreams in Action*, Supplement II,

Comparative Studies in Society and History (The Hague 1962); A. J. F. Köbben, "Prophetic Movements as an Expression of Social Protest," *Internationales Archiv Fur Ethnographie*, XLIX, No. 1 (1960), 117-64.
11. Sigmund Neumann, quoted in Chalmers, 2.
12. *Anatomy of Revolution*.
13. "On the Etiology of Internal War."
14. Amman, "Revolution: A Redefinition," *Political Science Quarterly*, LXXVII (1962).
15. Emile Durkheim, *Suicide* (Glencoe 1951), 246-54; A. B. Hollingshead, R. Ellis, and E. Kirby, "Social Mobility and Mental Illness," *American Sociological Review*, XIX (1954).
16. Michael L. Walzer, "Puritanism as a Revolutionary Ideology," *History and Theory*, III, No. 1 (1963), 59-90.
17. Parsons, *The Social System* (Glencoe 1951); Merton, *Social Theory and Social Structure* (Glencoe 1957), chap. 4.
18. W. Arthur Lewis, "Commonwealth Address," in *Conference Across a Continent* (Toronto 1963), 46-60; Olson, "Rapid Growth as a Destabilizing Force," *Journal of Economic History*, XXIII (December 1963), 529-52. I am grateful to Mr. Olson for drawing my attention to Sir Arthur Lewis's article, and for some helpful suggestions.
19. *The Vendée* (Cambridge, Mass., 1964).
20. "Toward a Theory of Revolution," *American Sociological Review*, XXVII (February 1962), 1-19, esp. the graph on p. 6.
21. Rostow, *The Stages of Economic Growth* (Cambridge, Mass., 1960); Supple, *The Experience of Economic Growth* (New York 1963), 11-12.
22. Hobsbawm, "The Crisis of the Seventeenth Century," in T. H. Aston, ed., *Crisis in Europe, 1560-1660* (London 1965), 5-58.
23. *La Crise de l'Économie française à la fin de l'Ancien Régime et au début de la Révolution* (Paris 1944).
24. Gabriel Kolko, *Wealth and Power in America* (New York 1962); Richard M. Titmuss, *Income Distribution and Social Change* (London 1962).
25. Davis, 5, quoting Marx, *Selected Works in Two Volumes* (Moscow 1955), 1, 947.
26. Daniel Bell, ed., *The Radical Right* (Garden City 1963).
27. Merton, chap. 9.
28. "Revolution: A Redefinition."
29. "The Revolutionary Process," *Social Forces*, XXVII (March 1950), 270-79.
30. See Werner J. Cahnman and Alvin Boskoff, eds., *Sociology and History: Theory and Research* (New York 1964); H. Stuart Hughes, "The Historian and the Social Scientist," *American Historical Review*, LXVI, No. 1 (1960), 20-46; A. Cobban, "History and Sociology," *Historical Studies*, III (1961), 1-8; M. G. Smith, "History and Social Anthropology," *Journal of the*

Royal Anthropological Institute, XCII (1962); K. V. Thomas, "History and Anthropology," *Past and Present,* No. 24 (April 1963), 3-18.

A Theory of Revolution
by Raymond Tanter and Manus Midlarsky

Overview

The focus of this study is an empirical examination of some causes of revolution.[1] While prior studies have delineated dimensions of conflict behavior within nations (Rummel, 1963, 1966, 1967; Tanter, 1965, 1966) in order to discover the variables most representative of internal conflict, a purpose of the present inquiry is to discover their theoretical significance. A typology of revolution is presented and two possible causes of revolution—changes in economic development and level of education—are examined regarding the extent of their association with certain characteristics of revolution. Regional differences are found which imply a fundamental distinction between at least two of the categories in the typology. Finally, hypotheses are offered to account for some of the findings of this study.

A Categorization of Revolutions

Prior to a discussion of theoretical significance, it may be useful to inquire into the meaning of revolution itself. This exercise in definition appears necessary primarily because prior usage of the term has been somewhat ambiguous. To the Hegelian, the revolutionary idea is equated with irresistible change—a manifestation of the world spirit in an unceasing quest for its own fulfillment. Similarly, the Marxist, although opposed to Hegelian idealism, sees revolution as a product of irresistible historical forces, which culminate in a struggle between the bourgeoisie and the proletariat. Hannah Arendt (1965, pp. 34-40), on the other hand, interprets the revolutionary experience as a kind of restoration, whereby the insurgents attempt to restore liberties and privileges which were lost as the result of the government's temporary lapse into despotism. Indeed, aspects of the American Revolution as well as some recent anti-colonial revolutions may be amenable to Arendt's interpretation. The insurgents might view the colonial elite as strangers who have usurped the freedoms which, mythically or otherwise, once belonged to the people now in revolt.

Tocqueville (1955, p. 8), on the other hand, has employed a more empirical approach to the problem of revolution, and has defined it as an overthrow of the legally constituted elite, which initiated a period of intense social, political, and economic change. Crane Brinton (1952, pp. 3-4) has continued this em-

Characteristics of Four Types of Revolution

Type of revolution	Mass Participation	Duration	Domestic violence	Intentions of the insurgents
Mass revolution	High	Long	High	Fundamental changes in the structure of political authority and the social system.
revolutionary coup	Low	Short to Moderate	Low to moderate	Fundamental changes in the structure of political authority and possibly some change in the social system.
Reform coup	Very low	Short, sometimes moderate	Low	Moderate changes in the structure of political authority.
Palace revolution	None	Very short	Virtually none	Virtually no change.

pirical thrust by differentiating between the *coup d'état*, as a simple replacement of one elite by another, and major revolutions such as the French or Russian, which were accompanied by social, political, and economic changes. Similarly, with respect to Latin America, George Blanksten (1962, p. 72) suggests that we should distinguish between the *coup d'état* and revolutions such as the Mexican experience, which eventually had profound consequences for the structure of that society.

The distinction between two forms of revolution may provide a basis for the development of further classifications. For example, Harold Lasswell and Abraham Kaplan (1950, p. 252) present a further refinement in the classification of revolution by the introduction of a three-category typology in which they differentiate between palace revolutions, political revolutions, and social revolutions. Edwin Lieuwen (1960, pp. 22-24) constructs a similar classification, but instead of the palace revolution he discusses *"caudillismo"* (predatory militarism), which is a common form of the *coup d'état* in Latin America. These three forms of revolution appear to reflect an increasing degree of change initiated by the successful insurgents, and may be placed on a rank-order of increasing political or social change. James Rosenau, in fact, constructs such a classification of revolution.

Personnel wars are defined by Rosenau as those which are fought over the occupancy of existing roles in the structure of political authority (Rosenau, 1964, pp. 63-64). An example of this type is the palace revolution or Latin American *caudillismo*. A second category is what Rosenau calls the authority wars, or those in which the insurgents compete not only for the occupancy of roles in the political structure, but for their arrangement as well. Struggles to replace dictatorships with democracies would be classified as authority wars. The final classification is that of the structural wars, in which the goal of the insurgents is the introduction of social and economic changes in the society. Wars involving Communist factions would fall under this heading. In addition, Rosenau notes that structural wars contain elements of both personnel and authority wars. Changes in the occupancy of government roles, as well as the arrangement of these roles, would automatically take place if the insurgents were successful in a structural war. Similarly, the authority war is personnel-oriented, because the arrangement of political roles would seldom, if ever, be altered without a change in the occupancy of these roles. In Rosenau's ranking of the three types of internal war, personnel wars would occupy the lowest rank with regard to the degree of societal change; the authority wars would occupy an intermediate rank, and the structural wars would receive the highest rank.

Samuel Huntington (1962, pp. 23-24) has suggested a classification of revolution in which four categories are enumerated: the internal war, the revolutionary coup, the reform coup, and the palace revolution. Huntington's use of the concept "internal war" differs from the meaning attributed to that concept in earlier systematic studies. For that reason the term mass revolution[2] will be substituted for internal war as used by Huntington. The terms mass revolution and palace revolution correspond respectively to Rosenau's structural and personnel wars, while the revolutionary and reform coups both may be placed under the heading of the authority wars. Kemal Ataturk's revolution in Turkey, for example, illustrates what Huntington might call a revolutionary coup, whereas the 1955 coup in Argentina might be classified as a reform coup. The major difference between the two forms is in the degree of change initiated in the structure of the political authority. The "Young Turks" implemented a complete revision of the political authority which led to a truncation of the Ottoman Empire and the establishment of a republic. The revolt against Peron, on the other hand, was an attempt at reform, in that Peron's mismanagement of the economy and the dissatisfaction of major political forces, such as the Roman Catholic Church, led to a revolt against what had become an oppressive political executive.

The existence of several *types* of revolution suggests that we might be able to isolate different *characteristics* of revolution. Karl Deutsch (1964, pp. 102-104) proposes that the degree of mass participation in a revolution, as well as its duration, may be essential to an adequate description of the revolutionary experience. A third characteristic may be the number killed as a result of the revolution. Given a high degree of commitment by the insurgents and the incumbents, the number of persons killed both during and after the revolution may be a measure of intensity. This measure will be discussed more fully at a later point. Finally, the intentions of the insurgents may be critical to the form of the revolution as well as to its eventual outcome. If the successful insurgents are ideologically committed to certain goals, then they may initiate changes in the societal structure to effect the realization of these goals. If, on the other hand, the insurgents have no particular ideological orientation, then they might intend to replace the incumbents in the structure of political authority without recourse to changes in the societal structure.

Table 1 contains a rank-order of the categories of revolution based on their position with respect to mass participation, duration, domestic violence, and the intentions of the insurgents. We may now suggest a definition of revolution which would be designed to include the categories in Table 1, as well as to allow for the possibility of a continuum. A revolution may be said to exist when a group of insurgents illegally and/or forcefully challenges the governmental elite for the occupancy of roles in the structure of political authority. A successful revolution occurs when, as a result of a challenge to the governmental elite, insurgents are eventually able to occupy principal roles within the structure of the political authority. This is not to say that once the insurgents have occupied these roles, the structure of political authority will remain unchanged. As suggested above, changes in the personnel of the governmental elite often are the precondition for meaningful changes in the political and social

A Theory of Revolution

51

Four Types of Revolution[a]

Mass revolution	Revolutionary coup	Reform coup	Palace revolution
French (1789)[b]	Turkish (1919)	Argentinian (1955)	Venezuelan (1948)
American (1776)	Nazi (1933)	Syrian (1956)	Brazilian (1955)
Russian (1917)	Egyptian (1952)	Jordanian (1957)	Columbian (1953, 1957)
Chinese (1949)	Iraqi (1958)	Thai (1957, 1958)	Honduran (1956)
Viet Minh (1954)		Burmese (1958)	Guatemalan (1957)
Cuban (1959)		French (1958)	Haitian (1957)
Algerian (1962)		Pakistani (1958)	El Salvador (1960)
		Sudanese (1958)	
French/German/(1848)		Venezuelan (1958)	
Austrian/		Turkish (1960)	
French (1871)		Dominican Republic (1963)	
Huk (1948)			
Malayan (1956)			
Hungarian (1956)			

(a) Unsuccessful revolutions are indicated below.
(b) The date in parentheses refers to the year in which the insurgents were either defeated or seized the roles in the structure of political authority.
(c) The Algerian revolution of 1962 refers to the anti-colonial revolt of the Algerian independence movement, whereas the French revolution of 1958 applies to the fall of the French Fourth Republic.

structure. If the insurgents intend major political and social changes, they must first occupy these roles within the political structure. This definition, then, sets a lower bound or minimum criterion for the existence of revolution.

Table 2 provides examples of the four types of revolution. This table is not meant to be exhaustive, but serves to provide a listing of those revolutions judged by the authors to be illustrative of the particular categories.[3] The dotted line in Column 1 separates successful mass revolutions from those which were unsuccessful. In addition, instances of both successful and unsuccessful anti-colonial revolutions are included to indicate the possible similarity of this type to the general mass revolution. Having provided such illustrative material, a next step is to provide the operational referents for the examination of causes of revolution.

Empirical Referents

Two of the four measures—duration and domestic violence—may be interpreted as defining characteristics of revolution, and taken together may be seen as a measure of revolutionary intensity.[4] Domestic violence, for example is present in the majority of revolutions. If the population is indifferent to changes in the personnel of the governmental elite, then a palace revolution may be initiated by insurgents. They may encounter opposition only from the incumbent elite. Because of the lack of large-scale, organized conflict behavior, the number of deaths would be at a minimum, and the duration of such a revolution would be short. If, however, the insurgents proclaim their intention to initiate changes in the social structure, then large segments of the population may be alienated from the insurgent cause. The number of deaths, and possibly the duration of the revolution, may increase proportionally to the degree of societal change envisaged by the insurgents. Similarly, the duration of a revolution has been cited as a possible measure of revolutionary intensity (Rosenau, 1964, pp. 76-77). Many Latin American palace revolutions are of short duration, and the number killed is relatively small. Indeed, bloodless coups often occur in which no one is killed, but a new set of elites assumes the major roles in the political authority structure. Revolutions such as the French and the Russian, however, are of longer duration and a greater number of persons are killed.

Deaths and Duration for Successful Revolutions, 1955-60

Revolution[a]	Deaths as a result of domestic group violence[b]	Duration (in days)[c]
Cuba (1959)	2900	2190
Iraq (1958)	344	3
Colombia (1957)	316	7
Argentina (1955)	217	86
Burma (1958)	152	1
Honduras (1956)	111	90
Venezuela (1958)	111	21
Guatemala (1957)	57	10
Syria (1956)	44	2
Haiti (1957)	16	36
Pakistan (1958)	9	19
Jordan (1957)	7	15
Thailand (1957)[d]	3	1
Thailand (1958)[d]	3	1
El Salvador (1960)	2	50
Brazil (1955)	1	2
Turkey (1960)	0.9	33
France (1958)	0.3	19

(a) The data sources for Column 1 are the raw data for Rummel (1963), Tanter (1965), and the *New York Times*. Two countries which experienced successful revolutions, Sudan (1958) and Laos (1960), were excluded because of missing data on deaths.

(b) Per million population, 1950-62 (Russett et al., 1964, pp. 97-100).

(c) Data relevant to the duration of revolutions were drawn from the *New York Times*.

(d) Both the 1957 and 1958 Thai revolts were initiated by the same individual, the commander-in-chief of the army, Sarit Thanarat. In addition, the 1958 coup took place with the express agreement of the "ousted" Kittikachorn government. For these reasons, this latter instance is omitted from the subsequent analysis although, for the sake of completeness, it is included in this listing.

The two measures—duration and deaths from domestic violence—are presented in Table 3. The first column lists all successful revolutions which occurred between 1955 and 1960. The criterion for the inclusion of a revolution was that the revolt ended sometime within that period.[5] Column 2 lists the number killed as a result of domestic group violence, per million population. These figures are for the period 1950-62. An alternative would be to construct an index on the basis of the number killed during the period of the revolution itself.

54 However, the number killed during a revolutionary coup lasting a few days may not be indicative of the characteristics of that revolution. For example, thirty persons were killed during the Iraqi revolution of 1958. Two of these were the king, Faisal, and his prime minister, Nuri. The intensity of feeling against the government was manifested when the populace dragged the king's body through the streets of Baghdad. The death of thirty people would seem to be an inadequate reflection of this degree of intensity. However, if we examine the number of deaths *prior* to the revolution and *after* its occurrence, we find that Iraq experienced the second highest number of deaths per million population of any revolution during the period 1955-60. Similarly, Venezuela, which experienced the overthrow of the Jiminez dictatorship, also had one of the highest numbers of deaths as a result of domestic violence. The inclusion of the number of deaths prior to and after a revolution might be indicative of the frustration built up after a long period of repression. A revolution generally occurs after a period of instability, and it is suggested in this paper that the form of a revolution is dependent on the degree of political instability which exists prior to its occurrence.

Two Correlates of Revolution

Having provided operational referents for the concept of revolution, we may now turn to an examination of possible correlates. The problem of identifying the preconditions of revolution has been of long concern to political theorists. For example, Plato (1951, p. 54) proposed that differences in economic interests led to factionalism in politics and contributed to the instability of the city-state. Poverty, according to Plato, produces revolution, meanness, and villainy, while riches produce luxury, idleness, and villainy. In substantial agreement with this position, Aristotle (1962, p. 59) also proposed that poverty may be a cause of political revolution. Tocqueville (1955, pp. 22-23), however, dissented from this emphasis on poverty. He suggested that the French peasant prior to 1789 enjoyed a considerably higher degree of economic independence than did the remainder of the European peasantry. Because of this independence and security, those aspects of feudalism still remaining in French society, such as the *corvée* (a form of periodic forced labor), appeared all the more odious and contemptible.

Indeed, *long-term* economic depressions were not present prior to the outbreaks of either the American or Russian revolutions. After an exhaustive study of four major revolutions, Crane Brinton (1952, p. 264) concluded that one uniformity in the occurrence of these revolutions was that the societies under investigation were all on the upgrade economically before the outbreak of revolution. Some statistics relevant to the French economy prior to 1789 may be instructive. In the *élection* of Melun from 1783-85, the amount of uncultivated land was reduced from 14,500 to 10,000 *arpents*. Rouen in 1787 produced cotton cloth worth 50,000 *livres*, which was double the production of a generation before. In the dozen years since the death of Louis XIV, French trade had increased nearly 100,000,000 *livres* (Brinton, 1952, p. 31).

The English Revolution of the 17th century and the American Revolution were preceded by similar economic experiences. According to Brinton, early Stuart England was notably prosperous, as were the American colonies prior to 1775 (1952, p. 32). Russia also was making significant economic progress prior to the outbreak of World War I. Table 4 illustrates this growth in the Russian economy.

Russian Exports and Imports per Capita: 1895-1913[a]

Period	Exports (in rubles)	Imports (in rubles)
1895-99	5.5	4.7
1900-04	6.3	4.6
1905-09	7.4	5.1
1910-13	9.1	7.1

(a) Mazour, 1962, p. 323.

The theories of Plato and Aristotle appear to contradict those of Tocqueville and Brinton. The first set asserts that poverty leads to revolution while the second set claims that revolutions are preceded by a significant increase in economic development. James C. Davies (1962, pp. 6-7) suggests that a partial synthesis of these approaches may provide a more comprehensive explanation than either of the two taken alone; major revolutions may be preceded by steady long-term increases in economic development, followed by a sharp reversal just before the outbreak of the revolution. In more general terms, economic development might be viewed as a single aspect of a society's achievement, That is, political, economic, and cultural development may together comprise a single concept which, for our purposes, may be called *achievement*. A second concept closely associated with achievement is termed *aspirations*. The results of political, economic, or cultural attainment ordinarily are visible to the general populace. Newly granted freedoms, as well as roads, factories, museums, and a higher standard of living, which generally accompany a high rate of achievement, also may tend to increase aspirations. It is posited that the *rates of change* of the achievements and aspirations are correlated. If achievement is increasing at a given rate, then the populace would most likely aspire to the acquisition of social commodities at the same rate as they had been previously acquired.

If we were to plot the increase in aspirations over time, the *slope* (rate of change) of this variable would approximate the rate of change of achievement. The similarities in the slope of the two plots is illustrated in Figure 1. Now if, instead of increasing, the rate of achievement were to decrease, we may posit a third concept, *expectations*. The rate of change of expectations would approximate that of the decrease in the rate of achievement. Rates of expectations may be affected more by immediate reality than are rates of aspirations. Expectations represent a change in outlook caused primarily by an *immediate* decrease in the production of social commodities. Aspirations are more in the nature of a hope and an optimism generated by *long-term* past performance.[6] *The distance between the two concepts (revolutionary gap) may be seen as a measure of the potentiality for the occurrence of a violent revolution. The larger the revolutionary gap, the longer and the more violent the revolution may be.* This hypothesis is represented in Figure 1.

As evidence for the low rate of expectations prior to major revolutions, Davies cites the poor agricultural harvests of 1788--89 in France which followed the economic advances of the prior

Relationship Between Rate of Achievement, Aspirations, and Expectations

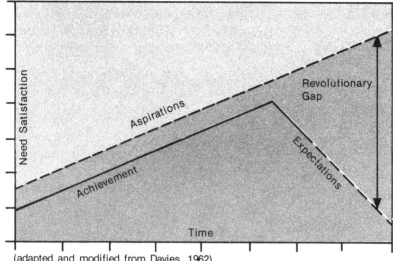

(adapted and modified from Davies, 1962)

decade. The English economy had suffered a reversal prior to 1688, while there was a tightening up of business conditions in America in 1774–75. Conditions in Tsarist Russia in 1917 provide a striking example of a breakdown in the economic machinery immediately prior to the outbreak of a revolution. Industrial production had declined rapidly. Coal mining had fallen to a level that threatened the nation's transportation. The peasants had grown sick of the war, and instead of transporting much-needed grain to Russia's urban centers, they were hoarding food in the expectation of future disasters. The economy was in a state of virtual collapse when the revolution broke out (Mazour, 1962, pp. 553-55). Apparently, these four major revolutions provide evidence for Davies' hypothesis. But whether this hypothesis holds true in all such instances is a question to be investigated by means of a systematic analysis across a larger universe of revolutions.

A second correlate of revolution might be the level of educational attainment prior to the outbreak of the disturbance. Seymour M. Lipset (1959, pp. 73-80) has found that the occurrence of revolutions is associated negatively with the level of educational attainment. Lipset categorized nations according to their stability and degree of democratic or totalitarian control and found that Latin American dictatorships, which are notoriously amenable to the occurrence of palace revolutions, are also societies which rank lowest on his scale of educational attainment. Evidence supporting Lipset's finding is provided in the *Dimensions of Nations* (Rummel, Sawyer, Tanter, and Guetzkow, forthcoming, 1967). These writers find that three educational variables are correlated negatively with the occurrence of revolution. One of the three—the ratio of number enrolled in primary schools to the total population aged 5--14— gave the highest negative correlation ($r = -.84$, $N = 66$). It appears from the data of the mid-1950s that revolutions will occur more frequently in societies with the lowest levels of educational attainment.

Having stated the principal concepts and their measures, it may be useful to summarize the definitions prior to a presentation of the results:

(1) A revolution is operationally defined by domestic violence and duration.

(2) Achievement and aspirations are defined by the rate of change of GNP/CAP over time (gross national product divided by population).

(3) Expectations are defined by the drop or reversal in the rate of change of GNP/CAP.

(4) Educational level is defined by the primary school ratio—number enrolled in primary school divided by total population aged 5--14.

Given these definitions, the following hypotheses are suggested:

(1) The higher the rate of increase of GNP/CAP preceding the revolution and the sharper the reversal *immediately* prior to the revolution, the greater the duration and violence of the revolution.

(2) The lower the level of educational attainment prior to the revolution, the greater the duration and violence of the revolution.

These hypotheses were tested across seventeen cases of successful revolutions for 1955-60.

Data and Results

Data for the years 1955–60 are presented in Table 5. Unsuccessful and colonial revolutions are excluded. The variation over time in GNP/CAP is presented in Column 3. The figures in this column are regression coefficients which were calculated for the seven-year period preceding each revolution.[7] Here, the

Domestic Violence, Time Rate of Change of GNP/CAP, and Primary Enrollment Ratio

Revolution	Domestic group violence [a]	Time rate of change [b] of GNP/capita	Primary enrollment ratio [c]
Cuba (1959)	2900	—[d]	43
Iraq (1958)	344	6.86[e]	19
Colombia (1957)	316	2.79	30
Argentina (1955)	217	−1.64	68
Burma (1958)	152	4.36	16
Honduras (1956)	111	9.25	43
Venezuela (1958)	111	1.64	27
Guatemala (1957)	57	.64	23
Syria (1956)	44	1.60[f]	37
Haiti (1957)	16	—[g]	20
Pakistan (1958)	9	.96	—
Jordan (1957)	7	—	42
Thailand (1957)	3	2.69	54
El Salvador (1960)	2	2.18	33
Brazil (1955)	1	2.89	34
Turkey (1960)	0.9	2.43	33
France (1958)	0.3	4.11	78

(a) Same variable as in Table 3, Column 2.
(b) The sources for this data column are the UN *Statistical Yearbooks*, 1955-1962. The data are presented in time-series form with base year 1953 = 100.
(c) Primary enrollment divided by total population aged 5-14. These ratios were drawn from the *World Survey of Education II*, UNESCO, 1958, Table 15, pp. 58-60. The majority of these values are for the years 1950-1954, with Brazil, Cuba and Iraq for 1950-1953, Haiti and Jordan for 1952-1954, and Turkey for 1950-1952.
(d) The rate of change of GNP/CAP for Cuba can be seen in Figure 2.
(e) The values for Iraq and El Salvador were not included in the standardized tabular form in the data sources, and were therefore calculated separately by means of net national product and population data.
(f) The values for Syria and Thailand are based on four and six data points respectively, due to limited data availability.
(g) The dashes in the columns indicate that the data were not available.

regression coefficient is a quantitative representation of a given variable's rate of change over time (slope).[8] Gross national product per capita was plotted against time for each nation, and the slopes were approximately linear. The values of these coefficients are comparable, since the same base year (1953 = 100) was chosen for all cases.[9] Column 4 lists the primary school ratios.

The product-moment correlation coefficients are presented in Table 6. These coefficients were calculated first for the total number of cases in Colunm 1, and then separately for the Asian and Latin American cases. The number of cases on which each coefficient is based is written in parentheses below the value of that coefficient. Before calculating the coefficients, curves were plotted and, where necessary, domestic violence was logarithmically transformed to compensate for the effect of an outlier (Cuba).

Since these calculations are based on an entire population of revolutions in the given period, it must be noted that the sample of cases is not random. Although the conventional significance levels are thus not applicable, they are utilized to provide some criterion for the tenability of hypotheses. All correlation coefficients are considered significant at the .05 level of confidence; i.e., five times in 100 the coefficient could be due to chance. A second interpretation of the correlation coefficient is the percentage of variance explained—the square of the coefficient. This mode of presentation is also employed in the following discussion.

Correlation Coefficients Between Changes in Wealth, Level of Education, and Domestic Violence[a]

	All cases	Latin America	Asia	Middle East
Time rate of GNP/CAP	.22 (14)	−.12 (7)	.94 (6)	.96 (4)
Primary enrollment ratio	−.31 (16)	.33 (9)	−.76 (6)	−.92 (4)

[a] Domestic violence was logarithmically transformed for the first two values in the second row. No transformations were necessary for the calculation of the remaining values.

The dependent variable in the calculation of these coefficients was domestic violence.

The values in the first two data columns of Table 6 are statistically nonsignificant. The figures in Column 3, however, are somewhat more indicative of the presence of relationships. As Hayward R. Alker (Russett et al., 1964, pp. 322-23) has suggested, cultural differences between regions may account for an apparent lack of significant relationships. When the total number of cases is considered, domestic violence in Asian countries in general, and Middle Eastern countries in particular, may have an entirely different political meaning from the same degree of violence in Latin American countries. This division of the countries in Table 5 into Latin American, Asian, and Middle Eastern is a first approximation to a control for cultural differences. Only France, of the countries included in Table 5, did not fall into any one of these categories.

When Asian revolutions are considered, we find that the degree of association between rate of change of GNP/CAP and domestic violence is significant at $p < .01$ and accounts for approximately 88 percent of the variance. The proportion of students enrolled in primary schools is correlated negatively with domestic violence, and the coefficient is nonsignificant at $p < .10$. Approximately 58 percent of the variance is accounted for by this relationship. For the Middle Eastern countries, the correlation between the rate of change of GNP/CAP and domestic violence is significant at $p < .05$ and accounts for approximately 92 percent of the variance. The rate of change of GNP/CAP prior to the revolution is almost perfectly associated with the level of domestic violence. The correlation between the primary enrollment ratio and domestic violence is nonsignificant at $p < .10$ and accounts for 85 percent of the variance.

The coefficients for the Asian and Middle Eastern revolutions appear to provide some of the explanation of variance for all the cases. However, because of the small number of cases, these results should be accepted with caution. Before we can assume that 92 percent of the variation of domestic violence in the Middle East is accounted for by the rate of change of economic development, we clearly must have further verification efforts based on a longer time-period.

Deaths from domestic violence is one measure of revolution; another is the number of days of the revolution. A set of correlation coefficients was calculated for duration, the rate of change of GNP/CAP, and the level of the primary enrollment ratio; the results were not significant at $p < .05$. Perhaps one reason for this lack of a relationship is the extreme difficulty in measuring duration. As Brinton (1952, pp. 238-50) notes in his study of the Russian Revolution, the notion of duration may be better suited to the making of qualitative judgments rather than quantitative distinctions. For example, there is a large degree of uncertainty as to the duration of the Russian Revolution. On the one hand, the period in the early 1920s may be viewed as a "Thermidor," hence a denouement of the revolution. Later, on the other hand, the Soviet Union under Stalin experienced a reign of terror which rivalled the intensity of the French Revolution at its height. The Soviets may have undergone a "permanent revolution" throughout the Stalin period, and only after Khrushchev's condemnation of Stalin did the revolution subside into a period of normalcy. Thus there are at least two possible measures of the duration of the Russian Revolution. In his attempt to define the duration of the English Revolution, Brinton encountered similar difficulties. Because of the close-knit series of events prior to the outbreak of the Civil War in 1642, the duration of this revolution is essentially a matter of subjective determination (Brinton, 1952, p. 75).

Brinton's analysis and the results of this study suggest that the duration variable is difficult to measure. Moreover, when comparing durations we often commit the fallacy of employing different units of analysis in the same investigation. By recording durations we are, in effect, ignoring the events which occurred during these periods. A three-day revolutionary coup which is fought with bitter intensity should not be equated with a three-day palace revolution which is bloodless and leads only to a change in the personnel of the governmental elite. A more adequate measure might be the number of events of a similar type compressed into a given time-interval. Rather than treating duration as a continuous variable, however, we may more profitably view it as discrete at a given threshold. Below a certain value of this variable, the duration of a revolution may be irrelevant to the explanation of either the antecedents or the consequences of its occurrence. Above that value, the duration of a revolution may be a meaningful characteristic.

The longer the revolution persists, the greater the polarization of the society around the warring factions. Tradi-

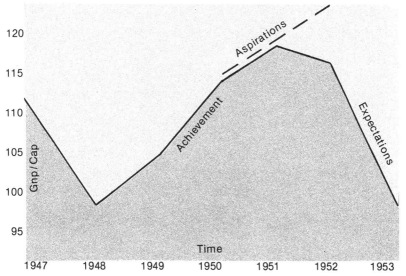

Plot of Gross National Product per Capita (GNP/CAP) Versus Time for Cuba, 1947-53

(In the construction of this figure, the inception of the Cuban revolution was taken to be Castro's *first* armed landing in 1953.)

tional associations are broken. Factions which at the start of the conflict might have been able to settle their differences by negotiation now find that, with the passage of time, enmities have increased to the extent that only unconditional surrender by one of the factions can terminate the revolution. Reference to Table 3 reveals that the Cuban revolution, with a duration of six years, is at the same time the only instance of mass revolution in the chosen period. The duration of the Cuban revolution is more than ten times greater than any of the other values in Column 3, and the "threshold" may be somewhere between one and five years. Future attempts at the determination of this value may be fruitful for a better explanation of the revolutionary process.

A Mass Revolution: The Case of Cuba

We may recall that Davies' hypothesis suggests that the

higher the long-term rate of achievement and aspirations preceding the revolution, and the sharper the reversal (expectations) immediately prior to the revolution, the greater the intensity as measured by duration and domestic violence. Rate of achievement is defined as rate of change of GNP/CAP, whereas the reversal is equated with expectations. The rate of change of GNP/CAP for Cuba was not included in Table 5. Instead, Figure 2 plots GNP/CAP against time for the seven-year period preceding Castro's first armed incursion against Batista in 1953.

If we compare Figure 1 (p. 271) with Figure 2 we see a marked similarity. A high rate of achievement is followed by a reversal which, according to the theory, corresponds to the expectation of the society. In addition, the Cuban revolution, with 2,900 deaths per million and a duration of six years, falls into the mass revolution category—the only instance of such a revolution in the period 1955--60. Although Cuba is a single case, it compares favorably with Brinton's analysis, in that the four countries in his study experienced economic conditions similar to that of Cuba prior to the onset of the revolution.

The Cuban revolution is the only instance in the period under consideration where a severe reversal actually occurred immediately prior to the revolution. We thus find a confirmation of the existence of a "revolutionary gap" in the Cuban example—our one example of a mass revolution—but are left without an adequate test for the other instances of revolution. How do we measure expectations in cases of revolutionary coups, reform coups, and palace revolutions?

Differences in Expectations: The Gini Index

An operational definition of expectations might be the degree of inequality in income or land distribution.[10] If the rate of achievement is high and unequally distributed, large sectors of the population might be aroused by this appearance of new wealth in a limited segment of the population. Aspirations, which are based on long-term achievements, may tend to be higher, whereas expectations—more a function of immediate reality—might be lower because of the prevailing economic inequalities. Where the "revolutionary gap" between aspirations and expectations is largely due to a high level of inequality in land distribution, we would expect an increase in the probability of revolution.

The Gini index is a measure of inequality. It measures the area between the Lorenz curve of any distribution and the line of perfect equality. For high inequality, the area between the Lorenz curve and the line of perfect equality would be correspondingly large; hence, the Gini index would also have a large value. For lower degrees of inequality, the reverse is true.

Table 7 presents the Gini indices for all nations included in the data source (Russett et al., 1964 p., 239ff). Table 7(a) lists the values for nations which experienced successful revolutions in the period 1955--60, whereas Table 7(b) presents the values for those nations which did *not* have successful revolutions. The indices are for land, whose distribution is probably correlated with income distribution. A one-tailed t-test of the significance of difference was performed and a t equal to 2.641 was calculated. With 48 degrees of freedom, this value is significant at $p < .05$. Thus degree of land inequality might be interpreted as an operational form of expectations; at any rate, revolutions occurred in those societies with a higher degree of land inequality. Russett (1964, p. 272) also reports a relationship between the Gini index of land inequality and domestic violence of $r = .44$ across 44 cases.

Conclusions

The analysis of inequalities of land distribution and the Cuban case study provide empirical support for the hypothesized relationship between the revolutionary gap and the domestic violence characteristic of revolutions. However, while the correlational analysis provides support for the hypothesis in the case of Asian revolutions, it was not substantiated for Latin American revolutions in this period.

The results suggest two alternatives: (1) If we assume that the level of domestic violence is indeed an indicator of the type of revolution in Latin America, then we are forced to conclude that in Latin America there is virtually no relationship between the rate of change of GNP/CAP and the form of revolution. This alternative leads to a reliance on an explanation based on regional differences. (2) If we assume that domestic violence bears little relationship to the palace revolution in Latin America (most of the Latin American revolutions in Table 5 are of the palace variety), then the relationship between the rate of change of GNP/CAP and the other forms of revolution is not invalidated. Those revolutions occurring in Asia are either reform coups or

Gini Indices of Land Inequality[a]

(a) Successful revolutions

Venezuela	.909	Brazil	.837
Iraq	.881	El Salvador	.828
Argentina	.863	Cuba	.792
Guatemala	.860	Honduras	.757
Colombia	.849	France	.583

(b) No successful revolutions

Bolivia	.938	United States	.705
Chile	.938	Libya	.700
Australia	.929	West Germany	.674
Costa Rica	.892	South Vietnam	.671
Peru	.875	Norway	.669
Ecuador	.864	Taiwan	.652
Jamaica	.820	Luxembourg	.638
Uruguay	.817	Netherlands	.605
Italy	.803	Finland	.599
Dominican Republic	.795	Ireland	.598
Nicaragua	.757	Belgium	.587
Spain	.780	Sweden	.577
New Zealand	.773	Philippines	.564
Greece	.747	India	.522
Austria	.740	Switzerland	.498
Egypt	.740	Canada	.497
Puerto Rico	.738	Japan	.470
Panama	.737	Denmark	.458
United Kingdom	.710	Poland	.450
Surinam	.709	Yugoslavia	.437

[a]From Russett et al., 1964, pp. 239-40.

revolutionary coups; thus we may suggest that the palace revolution is little affected by domestic group violence, whereas the other forms of revolution may indeed be so affected. The rate of change of GNP/CAP and level of education may then be theoretically significant for the study of revolutions other than the palace variety. Now we may ask why the palace revolution should be relatively unaffected by domestic violence.

George Blanksten (1958, p. 143) has suggested that one reason for the repeated "palace revolutions" in Latin America is that the class-oriented social structure provides the upper class with a virtual monopoly of participation in the political authority. Indians are almost entirely excluded from participation in politics, and the *mestizos* (descendants of Spanish-Indian marriages) are also inactive in politics but emerge at times as followers of a particular political faction. Class distinctions in Latin America then act as a relatively impermeable barrier between the elite (the socioeconomic class from which holders of high political office are drawn) and the mass (all those who are excluded from this highest socioeconomic category). Because of this barrier, disturbances in the society may seldom affect the "normal" operations of government in most Latin American nations. Thus palace revolutions may frequently take place unconnected with violent changes within the society. We find that the rate of change of GNP/CAP bears little or no relationship to domestic violence in Latin America, but—at the same time—domestic violence bears little relationship to the palace revolution, which until now has been the most frequent form of revolution in this region.

The Asian countries included in this study may not experience the same degree of class rigidity as do the Latin American nations. Barriers between the mass and the elite might not be so rigid or impermeable. Therefore, economic changes which affect domestic violence also have an effect on the form of the revolution. A similar argument can be developed for the effect of education on the form of revolution. As a general proposition we suggest that the more impermeable the barriers between the mass and the elite, the less the effect of domestic violence on the form of revolution. The converse may be equally valid. The more permeable the barriers between the mass and the elite, the greater the effect of mass activity on the form of the revolution.

The social structure of France prior to 1789 provides an example of these relationships. According to Tocqueville (1955, pp. 97-108) the seventeenth and eighteenth centuries witnessed a hardening of class lines between the nobility, the small but growing bourgeoisie, and the peasantry. Before 1780, class lines had become so rigid that Tocqueville referred to the nobility as a caste within the general class structure of French society. We may also note that France during this period was free of revolutionary disturbances. At the same time, the English were thought of as politically unstable, whereas the French were considered to be fortunate in having a government which exemplified political stability.

In the decade prior to 1789, however, there was a significant decrease in the rigidity of barriers between classes. Local *parlements* were once again assigned a political function which they had virtually relinquished centuries before. The meetings of the *parlements* provided a social setting for this increasing fluidity across class lines. The case of the French Revolution may provide a comparison within a single country: during one period, barriers between the elite and the mass prevented social disturbances from disrupting the "normal" operations of government; later, according to Tocqueville, this barrier became increasingly permeable and mass revolution occurred.

Recent theories of mass society have also emphasized the importance of the relationship between the elite and the mass. LeBon (1947, pp. 214*ff*) and Ortega y Gassett (1940, pp. 97*ff*) stress that for the governmental elite to perform its proper function of judicious decision-making, it should be protected from the effects of mass persuasion, which frequently manifests itself in violent forms. On the other hand, Hannah Arendt's view (1958, pp. 315-24) is that the mass, if unprotected from the elite, may undergo a process of "atomization" in which individuals are dissociated from each other and become easy prey for the totalitarian designs of the elite.

William Kornhauser (1959, pp. 39*ff*) suggests that if organized group activity in a pluralistic society can act as an intermediary between the elite and the mass, then neither will be accessible to undue penetration by the other. The intermediary group associations can then be viewed as a second form of the social barrier between the elite and the mass.

Both notions, that of class distinctions and that of the pluralistic group process, may be amenable to future research as to their effect on the incidence and form of revolution. With the possible exception of France, none of the countries in Table 5 have strongly developed pluralistic societies. However, the *state* of their pluralistic development may be related to the form of revolution.

Finally, what do these results and our interpretation say for the scheme of classification of revolutions presented initially? That is, we suggested that revolutions may fall on a rank-order: mass revolution, revolutionary coup, reform coup, and palace revolution. In Asian societies, such an ordering was suggested by the relationship between changes in wealth, in educational level, and in domestic violence. Latin American societies, with a preponderance of palace revolutions, did not exhibit these relationships. The palace revolution appears, then, to be a breed apart from the others.

Our definition was predicated on the palace revolution as a lower bound above which lies a rank-order or continuum of more intense changes in both government and society. Since the data appear to exclude the palace revolution from the concept of a continuum, the definition might have to be revised so that the "lower bound" is raised and the palace revolution is excluded from the definition.

This would appear reasonable if we examine the insurgents' orientations. A continuum of revolution is essentially one of commitment to change. Revolutions are not pieces of historical machinery, as the Hegelians or Marxists may contend, but are deeply associated with the ideological orientations of the participants. Indeed, this study suggests that domestic violence may vary with the degree of change intended by the insurgents. Therefore, the continuum we speak of is essentially a continuum of human desires for change, which—translated into a specific form of political action—make a revolution. The insurgents in a palace revolution, however, intend only to occupy roles in the existing authority structure. Changes in the social system or in the structure of political authority are not included in their program. Thus, if we are measuring a continuum of political and social change, the palace revolution may not contribute to these measurements.

References

ARENDT, HANNAH. *The Origins of Totalitarianism.* Cleveland: Meridian, 1958.
ARENDT. *On Revolution.* New York: Viking, 1963.
ARISTOTLE. *Politics.* (Ernest Barker, ed. and translator.) New York: Oxford University Press, 1962.
BLANSKTEN, GEORGE. "Revolutions." In H. E. DAVIS (ed.), *Government and Politics in Latin America.* New York: Ronald Press, 1958, 119–46.
BLANKSTEN. "Latin American Revolutions." In *The 1962 Carolina Symposium: Today's Revolutions.* Chapel Hill: University of North Carolina, 1962, 71–79.
BRINTON, CRANE. *The Anatomy of Revolution.* New York: Vintage, 1952.

DAVIES, JAMES C. "Toward a Theory of Revolution," *The American Sociological Review*, 27 (Feb. 1962), 5–13.
DEUTSCH, KARL W. "External Involvement in Internal Wars." In H. ECKSTEIN (ed.), *Internal War*. Glencoe: Free Press, 1964, 100–10.
FEIERABAND, IVO K., and ROSALIND L. FEIERABAND. "Agressive Behavior Within Polities, 1948--62 ; A. Cross-National Study," *Journal of Conflict Resolution*, 10, 3 (Sept. 1966), 149–69.
GROSS, FELIX. *The Seizure of Political Power*. New York: Philosopical Library, 1958.
HUNTINGTON, Samuel. "Patterns of Violence in World Politics." In S. HUNTINGTON (ed.), *Changing Patterns of Military Politics*. New York: Free Press, 1962, 17--50.
KORNHAUSER, WILLIAM. *The Politics of Mass Society*. Glencoe: Free Press, 1959.
LASSWELL, HAROLD, and ABRAHAM KAPLAN. *Power and Society*. New Haven: Yale University Press, 1950.
LEBON, GUSTAVE. *The Crowd: A Study of the Popular Mind*. London: T. Fisher Unwin, 1896.
LIEUWEN, EDWIN. *Arms and Politics in Latin America*. New York: Praeger, 1960.
LIPSET, SEYMOUR M. "Some Social Requisites of Democracy," *American Political Science Review*, 53 (March 1959), 69--105.
MAZOUR, ANTOLE G. *Russia: Tsarist and Communist*. New York: Van Nostrand, 1962.
ORTEGA Y GASSET, JOSE. *The Revolt of the Masses*. New York: W. W. Norton, 1940.
PLATO. *The Republic*. (A. D. Lindsay, translator.) New York: Dutton, 1951.
ROSENAU, JAMES N. "Internal War as an International Event." In J. N. ROSENAU (ed.), *International Aspects of Civil Strife*. Princeton: Princeton University Press, 1964, 45--91.
RUMMEL, RUDOLPH J. "Dimensions of Conflict Behavior Within and Between Nations," *General Systems Yearbook*, 8 (1963), 1-50.
RUMMEL, "Dimensions of Conflict Behavior Within Nations, 1946--59," *Journal of Conflict Resolution*, 10, 1 (March 1966), 65--73.
RUMMEL, JACK SAWYER, RAYMOND TANTER, and HAROLD GUETZKOW. *Dimensions of Nations*, forthcoming, 1967.
RUSSETT, BRUCE M., HAYWARD R. ALKER, JR., KARL DEUTSCH, and HAROLD LASSWELL. *World Handbook of Political and Social Indicators*. New Haven: Yale University Press, 1964.
SILVERT, KALMAN H. *Reaction and Revolution in Latin America*. New Orleans: Hauser Press, 1961.
STONE, LAWRENCE. "Theories of Revolution," *World Politics*,

18 (Jan. 1966), 159–76.
TANTER, RAYMOND. *Dimensions of Conflict Behavior Within and Between Nations, 1958--60,"* Journal of Conflict Resolution, 10, 1 (March 1966), 41--64.
TANTER. "Dimensions of Conflict Behavior Within Nations, 1955--60: Turmoil and Internal War," *Papers, Peace Research Society,* 3 (1965), 159–83.
TOCQUEVILLE, ALEXIS DE. *The Old Regime and the French Revolution.* New York: Doubleday, 1955.

Notes

1. This study is a part of a series on revolutions and their implications for international relations and especially some consequences for international business communities. Acknowledgements are due to Harold Guetzkow of the International Relations Program at Northwestern University for his suggestions during formative stages of this inquiry. Thanks are due to the Ford Foundation for a fellowship given to the junior author and to the National Science Foundation for its assistance to the senior author. Helpful comments were made by Richard A. Brody, James C. Davis, and Robert C. North.
2. One may also distinguish between various types of mass revolution. Lawrence Stone (1966, pp. 162–63), for example, presents a typology which differentiates among the Jacquerie, the Anarchistic Rebellion, the Jacobin Communist Revolution and the Militaristic Mass Insurrection. However, a purpose of this study is to compare the larger categories of revolution (e.g., mass revolution vs. palace revolution) rather than to engage in a detailed examination of subcategories.
3. Other writers such as Gross (1958), Huntington (1962), and Silvert (1961) are in substantial agreement with this classification.
4. The degree of mass participation as well as the intentions of the insurgents are less satisfactory as empirical referents. Mass participation might range from popular dissatisfaction without direct involvement, on the one hand, to overt complicity and active support of the insurgents, on the other. Thus, measurement of mass participation presents a great deal of difficulty unless the researcher is present at the site of the revolution and is able to measure public opinion at the time the revolution takes place. Similarly, the intentions of the insurgents, although important as a descriptive variable, nevertheless present important problems of measurement. Intentions and ideological directions are subject to changes during the process of revolution. Fidel Castro, for example, proclaimed his support for the goals of the Cuban middle class during the recent Cuban revolution, yet once in power he allied

himself with the Communists.
5. The duration of a revolution is bounded by two points in time: (1) when active hostilities first broke out against the regime in power, and (2) when the insurgents occupied the roles in the structure of political authority. A period of unrest prior to the victory of the insurgents was included in the determination of the values in Column 3. Riots and clashes between the police and populace are examples of these disturbances. In addition, if two or more changes in the personnel of the governmental elite took place within a three-month period of time, the period between changes was included in the duration.
6. The distinction between aspirations and expectations may be analogous to the Feierabends' (1966, pp. 256–57) use of the terms "want formation" and "want satisfaction."
7. The period of seven years prior to the outbreak of the revolution is an arbitrary choice, determined chiefly by the availability of comparable data. It is possible that the investigation of longer periods might yield different results.
8. The data in Table 5 are presented with certain reservations. The regression coefficients in the second data column were calculated from UN statistical yearbooks, and in that source it is noted that gross national product data are indicative only of general trends. Indeed, different yearbooks reported varying values of GNP/CAP for the *same* country and the *same* year. In order to partially resolve these inconsistencies, the regression coefficients were calculated using a single time series (from one yearbook) for each coefficient. In addition, the criterion for choosing a time series was that it was reported in the yearbook published one year subsequent to the occurrence of the rovolution (e.g., the 1959 *Yearbook* for Pakistan [1958]). For those cases in which the time series was incomplete, such as Syria or Thailand, a subsequent yearbook was chosen which contained the most complete time series.
9. The term "base year" refers to the basis of the UN *Statistical Yearbook's* calculation of the time-series figures. In all of the series, 1953 is set equal to 100 and the values for the other years are expressed relative to that year. This term does not refer to the starting point for the calculation of the regression coefficients. The choice of a starting year for the calculation of these coefficients is based solely on the fact that the chosen year is seven years prior to the outbreak of the revolution.
10. It should be noted that this treatment of "expectations" is cross-sectional in design, as opposed to the longitudinal nature of Davies' theory.

Psychological Factors in Civil Violence
by Ted Gurr

Until recently many political scientists tended to regard violent civil conflict as a disfigurement of the body politic, neither a significant nor a proper topic for their empirical inquiries. The attitude was in part our legacy from Thomas Hobbes's contention that violence is the negation of political order, a subject fit less for study than for admonition. Moreover, neither the legalistic nor the institutional aproaches that dominated traditional political science could provide much insight into group action that was regarded by definition as illegal and the antithesis of institutionalized political life. The strong empirical bent in American political science led to ethnocentric inquiry into such recurring and salient features of American political life as voting and legislative behavior. The American Revolution and Civil War appeared as unique events, grist for exhaustive historical inquiry but unlikely subjects for systematic comparative study or empirical theory. Representative of the consequences of these attitudes is a recent judgment that political violence "by its very nature [is] beyond any simple or reasonable laws of causation."[1]

This article proposes, first, that civil violence *is* a significant topic of political inquiry and, second, not only that it is capable of explanation, but that we know enough about the sources of human violence to specify in general, theoretical terms some of the social patterns that dispose men to collective violence.

The proposition that civil violence is important as a genus is widely but not yet universally accepted, even by scholars concerned with some of its forms, revolution in particular.[2] This is the case, one suspects, because revolutions have traditionally been regarded as the most significant form of civil strife, because the universe of such events has been defined by reference to their consequences rather than their common characteristics or preconditions, and because the older theoretical generalizations have emphasized primarily the processes of such events and categorization of their concomitants at a low level of generality.[3] But the evidence both of recent history and of systematic attempts at specifying the incidence of civil strife suggests that revolutions are but one of an extraordinarily numerous variety of interrelated forms of strife;[4] that some of these forms, among them coups d'état, guerrilla war, and massive rioting, can alter political processes and social institutions as drastically as any of the classic revolutions; and that the forms themselves are mutable, or rather, that by reifying our arbitrary distinctions among forms of strife we have overlooked some fundamental similarities.[5] Examination of those special conditions and processes that lead from turmoil to revolution provides a partial understanding of revolution *per se*, but for a sufficient explanation we require a more general theory, one capable of accounting for the common elements of that much larger class of events called civil strife.

The resort to illicit violence is the defining property that distinguishes these collective events from others. We can regard this as just a definitional point,[6] but it has a crucial theoretical consequence: to direct attention to psychological theories about the sources of human aggression.

Some types of psychological theories about the sources of aggressive behavior can be eliminated at the outset. There is little value in pseudopsychological speculation about revolutionaries as deviants, fools, or the maladjusted. Psychodynamic explanations of the "revolutionary personality" may be useful for microanalysis of particular events but scarcely for general theory. Aggression-prone victims of maladaptive socialization processes are found in every society, and among the actors in most outbreaks of civil violence, but they are much more likely to be mobilized by strife than to be wholly responsible for its occurrence. Nor can a general theory of civil strife rest on culturally specific theories of modal personality traits, though it might well take account of the effects of these traits. Some cultures and subcultures produce significantly more aggression-prone than cooperative personalities, but an explanation of this order says little of the societal conditions that elicit aggression from the aggression-prone, and nothing at all of the capacity for civil violence of even the most apparently quiescent populations.

The only generally relevant psychological theories are those that deal with the sources and characteristics of aggression in all men, regardless of culture. Such psychological theories do not directly constitute a theory of civil strife. They do offer alternative motivational bases for such a theory and provide means for identifying and specifying the operation of some crucial explanatory variables. As is demonstrated in the following section, one or another of these theories is implicit in most theoretical approaches to civil strife that have no explicit motivational base, although only one of them appears highly plausible in the light of empirical evidence.

Psychological Theories of Aggression

There are three distinct psychological assumptions about the generic sources of human aggression: that aggression is solely instinctual, that it is solely learned, or that it is an innate response activated by frustration.[7] The instinct theories of aggression, represented, among others, by Freud's attribution of the impulse to destructiveness to a death instinct and by Lorenz's view of aggression as a survival-enhancing instinct, assume that most or all men have within them an autonomous source of aggressive impulses, a drive to aggress that, in Lorenz's words, exhibits "irresistible outbreaks which recur with rhythmical regularity."[8] Although there is no definitive support for this assumption, and much evidence to the contrary, its advocates, including Freud and Lorenz, have often applied it to the explanation of collective as well as individual aggression.[9] The assumption is evident in Hobbes's characterization of man in the state of nature and is perhaps implicit in Nieburg's recent concern for "the people's capability for outraged, uncontrolled, bitter, and bloody violence,"[10] but plays no significant role in contemporary theories of civil strife.

Just the opposite assumption, that aggressive behavior is solely or primarily learned, characterizes the work of some child and social psychologists, whose evidence indicates that some aggressive behaviors are learned and used strategically in the service of particular goals—aggression by children and adolescents to secure attention, by adults to express dominance strivings, by groups in competition for scarce values, by military personnel in the service of national policy.[11]

The assumption that violence is a learned response, rationalistically chosen and dispassionately employed, is common to a number of recent theoretical approaches to civil strife. Johnson repeatedly, though not consistently, speaks of civil violence as "purposive," as "forms of behavior *intended* to disorient the behavior of others, thereby bringing about the demise of a hated social system."[12] Parsons attempts to fit civil violence into the framework of social interaction theory, treating the resort to force as a way of acting chosen by the actor(s) for purposes of deterrence, punishment, or symbolic demonstration of their capicity to act.[13] Schelling is representative of the conflict theorists: he explicitly assumes rational behavior and interdependence of the adversaries' decisions in all types of conflict.[14] Stone criticizes any emphasis on violence as a distinguishing or definitional property of civil strife on grounds that it is only a particular means, designed to serve political ends.[15]

The third psychological assumption about aggression is that it occurs primarily as a response to frustration. A "frustration" is an interference with goal-directed behavior; "aggression" is behavior designed to injure, physically or otherwise, those toward whom it is directed. The disposition to respond aggressively when frustrated is considered part of man's biological makeup; there is an innate tendency to attack the frustrating agent. Learning can and does modify the tendency: what is perceived to be frustrating, modes of aggressive response, inhibition through fear of retaliation, and appropriate targets are all modified or defined in the learning process, typically but not solely during socialization.

Frustration-aggression theory is more systematically developed, and has substantially more empirical support, than theories that assume either that all men have a free-flowing source of destructive energy or that all aggression is imitative and instrumental. Moreover, the kinds of evidence cited in support of theories of the latter type appear to be subsumable by frustration-aggression theory, whereas the converse is not the case.

One crucial element that frustration-aggression theory contributes to the study of civil violence concerns the drive properties of anger. In the recent reformulation of the theory by Berkowitz, the perception of frustration is said to arouse anger, which functions as a drive. Aggressive responses tend not to occur unless evoked by some external cue, but their occurrence is an inherently satisfying response to that anger.[16] Similarly, Maier has amassed extensive evidence that the innate frustration-induced behaviors (including regression, fixation, and resignation, as well as aggression) are for the actor ends in themselves, unrelated to further goals and qualitatively different from goal-directed behavior.[17]

To argue that aggression is innately satisfying is not incompatible with the presence of learned or purposive components in acts of individual or collective aggression. Cues that determine the timing, forms, and objects of aggression are learned, just as habits of responding aggressively to moderate as well as severe frustration can be learned. The sense of frustration may result from quite rational analysis of the social universe. Leaders can put their followers' anger to rational or rationalized uses. If anger is sufficiently powerful and persistent it may function as an autonomous drive, leading to highly rational and effective efforts by both leaders and the led to satisfy anger aggressively. The crucial point is that rationalization and organization of illicit violence are typically subsequent to, and contingent upon, the existence of frustration-induced anger. Collective violence may be a calculated strategy of dispassionate elite aspirants, and expectations of gains to be achieved through violence may be present among many of its participants. Nonetheless the implication of frustration-aggression theory is that civil violence almost always has a strong "appetitive," emotional base and that the magnitude of its effects on the social system is substantially dependent on how widespread and intense anger is among those it mobilizes.

If anger implies the presence of frustration, there is compelling evidence that frustration is all but universally characteristic of participants in civil strife: discontent, anger, rage, hate, and their synonyms are repeatedly mentioned in studies of strife. Moreover, the frustration assumption is implicit or explicit in many theoretical analyses of the subject. Smelser's concept of "strain" as one of the major determinants of collective behavior, particularly hostile outbursts and value-oriented movements (revolutions), can be readily reformulated in terms of perceived frustration.[18] So can Willer and Zollschan's notion of "exigency" as a precursor of revolution.[19] Ridker characterizes the consequence of failure to attain economic expectations as "discontent," analogous in source and consequence to anger.[20] In Davies' theory of revolution, the reversal of a trend of socioeconomic development is said to create frustration, which instigates revolution.[21] Galtung's theory of both intranational and international aggression recognizes that "the external conditions leading to aggression . . . probably have to pass through the minds of men and precipitate as perceptions with a high emotive contei t before they are acted out as aggression."[22]

In none of these approaches to theory, however, has frustration-aggression theory been systematically exploited nor have its variables been taken into account.[23] The primary object of this article is to demonstrate that many of the variables and relationships identified in social psychological research on the frustration-aggression relationship appear to underlie the phenomenology of civil violence. Juxtaposition of these two diverse types of material provides a basis for an interrelated set of propositions that is intended to constitute the framework of a general theory of the conditions that determine the likelihood and magnitude of civil violence. These propositions are of two types, whose proposed relationships are diagrammed in Figure 1: (1) propositions about the operation of *instigating variables*, which determine the magnitude of anger, and (2) propositions about *mediating variables*, which determine the likelihood and magnitude of overt violence as a response to anger.[24]

This approach does not deny the relevance of aspects of the social structure, which many conflict theorists have held to be crucial. The supposition is that theory about civil violence is most fruitfully based on systematic knowledge about those properties of men that determine how they react to certain characteristics of their societies.

Relative Deprivation: Variables Determining the Magnitude of Anger

My basic premise is that the necessary precondition for violent civil conflict is relative deprivation, defined as actors' perception of discrepancy between their *value expectations* and their environment's apparent *value capabilities*.[25] Value expectations are the goods and conditions of life to which people believe they are justifiably entitled. The referents of value capabilities are to be found largely in the social and physical environment: they are the conditions that determine people's perceived chances of getting or keeping the values they legitimately expect to attain. In a comparable treatment, Aberle defines relative deprivation as "a negative discrepancy between legitimate expectation and actuality," viewing expectations as standards, not mere prophecies or hopes.[26] For purposes of general theoretical specification I assume that perceived discrepancies between expectations and capabilities with respect to any collectively sought value—economic, psychosocial, political—constitute relative deprivation. The extent to which some values may be more salient than others is a subject of

FIGURE I. VARIABLES DETERMINING THE LIKELIHOOD AND MAGNITUDE OF CIVIL VIOLENCE

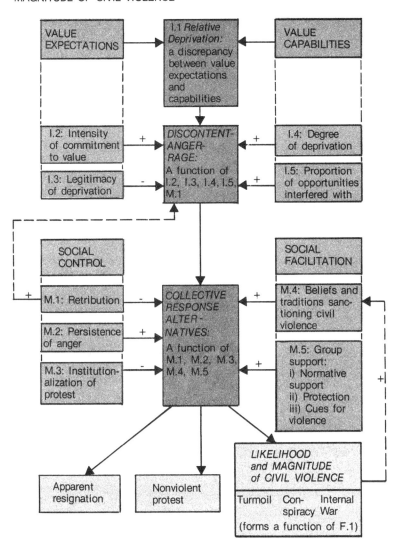

a. The direction(s) of proposed effects on magnitude of civil violence are indicated by + and -.

theoretical and empirical inquiry not evaluated here.

Relative deprivation can be related to the concept of frustration by extending Yates's distinction between the frustrating situation and the frustrated organism.[27] A frustrating situation is one in which an actor is, by objective standards, thwarted by some social or physical barrier in attempts to attain or continue enjoyment of a value. The actor can be said to be frustrated, however, only when he is aware of interference or thwarting. The awareness of interference is equivalent to the concept of relative deprivation as defined above.

A further distinction is necessary between two general classes of deprivation. those that are personal and those that are group or category experiences.[28] For given groups, and for some classes of societies, it is possible to identify events and patterns of conditions that are likely to be widely seen as unjust deprivation. Such events may occur abruptly—for example, the suppression of a political party or a drastic inflation—or slowly, like the decline of a group's status relative to other social classes. Such conditions can be called collective frustrations[29] to distinguish them from such unexpected personal frustrations as failure to obtain an expected promotion or the infidelity of a spouse, which may be relatively common but randomly incident in most populations.

Whether empirical research ought to focus on conditions defined as collectively frustrating or directly on perceived deprivation is an operational question whose answer depends on the researcher's interest and resources, not upon the following theoretical formulation. Survey techniques permit more or less direct assessment of the extent and severity of relative deprivation.[30] To the extent that the researcher is prepared to make assumptions about measurable conditions that are collectively frustrating in diverse nations, cross-national aggregate data can be used in correlational studies.[31]

The basic relationship can be summarized in a proposition analogous to and assuming the same basic mechanism as the fundamental theorem of frustration-aggression theory.[32]

> *Proposition I.1*: The occurrence of civil violence presupposes the likelihood of relative deprivation among substantial numbers of individuals in a society; concomitantly, the more severe is relative deprivation, the greater are the likelihood and intensity of civil violence.

This proposition may be truistic, although theories were noted above which attempt to account for civil strife without reference to discontent. Moreover, relative deprivation in some degree can be found in any society. The usefulness of the basic proposition is best determined by reference to the set of propositions that qualify it. These propositions specify the conditions that determine the severity and in some cases the occurrence of deprivation, whether or not it is likely to lead to civil violence, and the magnitude of violence when it does occur. The fundamental question, which is susceptible to a variety of empirical tests, is whether the proposed precise relationship between severity of deprivation, as determined by variables I.2 through I.5, and magnitude of violence does hold when the effects of the mediating variables M.1 through M.5 are taken into account.

Definitions and Qualifications

Civil violence and relative deprivation are defined above. If relative depriviation is the perception of frustrating circumstances, the emotional response to it tends to be anger. Obviously there are degrees of anger, which can usefully be regarded as a continuum varying from mild dissatisfaction to blind rage. The severity of relative deprivation is assumed to vary directly with the modal strength of anger in the affected population; the determinants of strength of anger are specified in propositions I.2 to I.5, below.

The concept of magnitude requires elaboration. Various measures of quantity or magnitude of aggression are used in psychological research on the frustration-aggression relationship—for example, the intensity of electric shocks administered by frustrated subjects to a supposed frustrater, numbers of aggressive responses in test situations, or the length of time frustrated children play destructively with inanimate objects. A consideration of theory, however, suggests that no single measure of magnitude of aggression is *prima facie* sufficient. Assuming the validity of the basic frustration-aggression postulate that the greater the strength of anger, the greater the quantity of aggression, it seems likely that strong anger can be satisfied either by inflicting severe immediate damage on the source of frustration or by prolonged but less severe aggression, and that

84 either of these tactics is probably more or less substitutable for the other. Which alternative is taken may very well be a function of opportunity, and while opportunities can be controlled in experimental situations, in civil violence they are situationally determined. Hence neither severity nor duration alone is likely to reflect the modal strength of collective anger or, consequently, to constitute an adequate measure of magnitude of civil violence.

Moreover, there are evidently individual differences—presumably normally distributed—in the strength of anger needed to precipitate overt aggression. Hence the proportion of a population that participates in collective violence ought to vary with the modal strength of anger: discontent will motivate few to violence, anger will push more across the threshold, rage is likely to galvanize large segments of a collectivity into action. This line of argument suggests that magnitude of civil violence has three component variables: the degree of participation within the affected population, the destructiveness of the aggressive actions, and the length of time violence persists.

Frustration-aggression theory stipulates a set of variables that determine the strength of anger or discontent in response to a given frustration. Dollard and others initially proposed that the strength of instigation to aggression (anger) varies with "(1) the strength of instigation to the frustrated response, (2) the degree of interference with the frustrated response, and (3) the number of frustrated response-sequences."[33] The first of these variables, modified in the light of empirical evidence, provides the basis for propositions about characteristics of value expectations that affect the intensity of anger. The second and third variables, similarly modified, suggest several propositions about value capabilities.

Before the propositions are presented, two qualifications of the classic behaviorist conceptualization of frustration as interference with specific, goal-directed responses must be noted. First, it appears from examination of specific outbreaks of civil violence that abrupt awareness of the likelihood of frustration can be as potent a source of anger as actual interference. The Vendée counterrevolution in eighteenth-century France was triggered by the announcement of military conscription, for example.[34] A survey of twentieth-century South African history shows that waves of East Indian and Bantu rioting historically have coincided with the parliamentary discussion of restrictive legislation more than with its actual imposition. The Indian food riots in the spring of 1966 were certainly not instigated by the onset of starvation but by its anticipation.

Second, it seems evident that the sense of deprivation can arise either from interference with goal-seeking behavior or from interference with continued enjoyment of an attained condition. As an example from psychological experimentation, frustration is often operationalized by insults; it seems more likely that the insults are a threat to the subject's perceived level of status attainment or personal esteem than they are an interference with behavior directed toward some as-yet-unattained goal. Several examples from the history of civil violence are relevant. A student of the coup d'état that overthrew the Peron regime in Argentina states that the crucial events that precipitated the anti-Peronists into action were Peron's public insults to the Catholic hierarchy and isolated physical depredations by his supporters against Church properties—events symbolizing an attack on the moral foundations of upper-middle-class Argentine society.[35] In Soviet Central Asia, according to Massell, the most massive and violent resistance to Sovietization followed systematic attempts to break Muslim women loose from their slavish subordination to Muslim men.[36] The two kinds of interference may have differential effects on the intensity and consequences of anger; the point to be made here is that both can instigate violence.

Consequently, analysis of the sources of relative deprivation should take account of both actual and anticipated interference with human goals, as well as of interference with value positions both sought and achieved. Formulations of frustration in terms of the "want: get ratio," which refers only to a discrepancy between sought values and actual attainment, are too simplistic. Man lives mentally in the near future as much as in the present.[37] Actual or anticipated interference with what he has, and with the act of striving itself, are all violatile sources of discontent.

Value Expectations

The propositions developed here concern the effects on preceived deprivation of the salience of an expectation for a group, rather than the absolute level of the expectation.[38] The first suggestion derived from psychological theory is that the more intensely people are motivated toward a goal, or committed to an attained level of values, the more sharply is interference resented and the greater is the consequent instigation to aggression. One can, for example, account for some of the efficacy of ideologies in generating civil violence by reference to this variable. The articulation of nationalistic ideologies in colonial territories evidently strengthened preexisting desires for political independence among the colonial bourgeoisie at the same time that it inspired a wholly new set of political demands among other groups. Similarly, it has been argued that the desire of the nineteenth-century European factory worker for a better economic lot was intensified as well as rationalized by Marxist teachings.

Experimental evidence has suggested qualifications of the basic proposition which are equally relevant. One is that the closer men approach a goal, the more intensely motivated toward it they appear to be.[39] This finding has counterparts in observations about civil violence. Hoffer is representative of many theorists in noting that "discontent is likely to be highest when misery is bearable [and] when conditions have so improved that an ideal state seems almost within reach. . . . The intensity of discontent seems to be in inverse proportion to the distance from the object fervently desired."[40] The intensity of motivation varies with the perceived rather than the actual closeness of the goal, of course. The event that inflicts the sense of deprivation may be the realization that a goal thought to be at hand is still remote. The mechanism is clearly relevant to the genesis of post-independence violence in tropical Africa. Failure to realize the promises of independence in the Congo had extraordinarily virulent results, as is evident in a comparison of the intensive and extensive violence of the uprisings of the "Second Independence" of 1964--1965 with the more sporadic settling of accounts that followed the "First Independence" of 1960.[41]

The proposition relates as well to the severity of discontent in societies in the full swing of socioeconomic change. The rising bourgeoisie of eighteenth-century France, for example, individually and collectively had a major commitment to their improving conditions of life, and great effort invested in them. Many felt their aspirations for political influence and high social status to be close to realization but threatened by the declining responsiveness of the state and by economic deprivations inherent in stumbling state efforts to control trade and raise taxes.[42]

Although much additional evidence could be advanced, the relationships cited above are sufficient to suggest the following proposition and its corollaries:

Proposition I.2: The strength of anger tends to vary directly with the intensity of commitment to the goal or condition with regard to which deprivation is suffered or anticipated.
I.2a: The strength of anger tends to vary directly with the degree of effort previously invested in the attainment or maintenance of the goal or condition.
I.2b: The intensity of commitment to a goal or condition tends to vary inversely with its perceived closeness.

It also has been found that, under some circumstances, anticipation or experience of frustration tends to reduce motivation toward a goal. This is particularly the case if frustration is thought to be justified and likely.[43] Pastore, for example, reports that when subjects saw frustration as reasonable or justifiable, they gave fewer aggressive responses than when they perceived it to be arbitrary. Kregarman and Worchel, however, found that the reasonableness of a frustration did not significantly reduce the aggression and that anticipation of frustration tended not to reduce anger but rather to inhibit external aggressive responses.[44]

The low levels of motivation and the moderate nature of interference that characterize these studies make generalization to "real," collective situations doubtful. If applied to a hypothetical example relevant to civil strife—say, the effects of increased taxation on a population under conditions of varying legitimacy attributed to the action—the experimental findings suggest three alternatives: (1) that anger varies inversely with the legitimacy attributed to interference; (2) that anger is constant, but inhibition of its expression varies directly with legitimacy; or (3) that no systematic relationship holds between the two. If the sources of legitimacy are treated in Merelman's learning-theory terms, the first of these alternatives appears most likely: if legitimacy is high, acceptance of deprivation (compliance) provides symbolic substitute rewards.[45] It may also be that the first alternative holds in circumstances in which legitimacy is high, the second in circumstances in which it is moderate. The first relationship can be formulated in propositional form, with the qualification that evidence for it is less than definitive:

Proposition I.3: The strength of anger tends to vary inversely with the extent to which deprivation is held to be legitimate.

Value Capabilities

The environment in which people strive toward goals has two general characteristics that, frustration-aggression theory suggests, affect the intensity of anger: the degree of interference with goal attainment and the number of opportunities provided for attainment.

Almost all the literature on civil strife assumes a causal connection between the existence of interference (or "frustration," "cramp," or "disequilibrium") and strife. "Discontent" and its synonyms are sometimes used to symbolize the condition of interference without reference to interference *per se*. A direct relationship between degree of interference and intensity of strife is usually implicit but not always demonstrated. Rostow has shown graphically that poor economic conditions—high wheat prices, high unemployment— corresponded with the severity of overt mass protest in England from 1790 to 1850.[46] Variations in bread prices and in mob violence went hand in hand in revolutionary France.[47] There is correlational evidence that the frequency of lynchings in the American South, 1882–1930, tended to vary inversely with indices of economic well-being.[48] From cross-national studies there is suggestive evidence also—for example, Kornhauser's correlation of $-.93$ between per capita income and the Communist share of the vote in sixteen Western democracies in 1949.[49] The Feierabends devised "frustration" measures, based on value capability characteristics of sixty-two nations, and correlated them with a general measure of degree of political stability, obtaining a correlation coefficient of $.50$.[50]

As far as the precise form of the relationship between extent of interference and intensity of aggression is concerned, the experimental results of Hamblin and others are persuasive. Three hypotheses were tested: the classic formulation that instigation to aggression varies directly with the degree of interference, and the psychophysical hypotheses that aggression ought to be a log or a power function of interference. The data strongly support the last hypotheses, that aggression is a power function of degree of interference—i.e., if magnitude of aggression is plotted against degree of interference, the result is a sharply rising "J-curve." Moreover, the power exponent—the sharpness with which the J-curve rises—appears to increase with the strength of motivation toward the goal with which interference was experienced.[51] It is at least plausible that the J-curve relationship should hold for civil strife. Compatible with this interference, though not bearing directly on it, is the logarithmic distribution curve that characterizes such cross-polity measures of intensity of civil violence as deaths per 100,000 population.[52] It also may account for the impressionistic observation that moderate levels of discontent typically lead to easily quelled turmoil but that higher levels of discontent seem associated with incommensurately intense and persistent civil violence. In propositional form:

Proposition I.4: The strength of anger tends to vary as a power function of the perceived distance between the value position sought or enjoyed and the attainable or residual value position.[53]

Experimental evidence regarding the hypothesis of Dollard and others that the greater the number of frustrations, the greater the instigation to aggression is somewhat ambiguous. Most people appear to have hierarchies of response to repeated frustration, a typical sequence being intensified effort, including search for alternative methods or substitute goals, followed by increasingly overt aggression as other responses are extinguished, and ultimately by resignation or apparent acceptance of frustration. Berkowitz suggests that most such evidence, however, is congruent with the interpretation that "the probability of emotional reactions is a function of the degree to which all possible nonaggression responses are blocked, more than to the interference with any one response sequence."[54]

The societal equivalents of "all possible nonaggressive responses" can be regarded as all normative courses of action available to members of a collectivity for value attainment, plus all attainable substitute value positions. Relevant conditions are evident in the portraits of "transitional man" painted by Lerner and others. Those who are committed to improving their socioeconomic status are more likely to become bitterly discontented if they have few rather than many prospective employers, if they can get no work rather than some kind of work that provides a sense of progress, if they have few opportunities to acquire requisite literacy and technical skills, if associational means for influencing patterns of political and economic value distributions are not available, or if community life is so disrupted that hearth and kin offer no surcease from frustration for the unsuccessful worker.[55] All such conditions can be subsumed by the rubric of "opportunities for value attainment," with the qualification that perception of opportunities tends to be more crucial than actual opportunities.

Much evidence from studies of civil strife suggests that the greater are value opportunities, the less intense is civil violence. The argument appears in varying guises. Brogan attributes the comparative quiescence of mid-nineteenth-century English workers vis-à-vis their French counterparts in part to the proliferation in England of new cooperatives, friendly and building societies, and trade unions, which provided positive alternatives to violent protest.[56] The first of the American Negro urban rebellions in the 1960's occurred in a community, Watts, in which by contemporary accounts associational activity and job-training programs had been less effective than those of almost any other large Negro community. Cohn explains the high participation of unskilled workers and landless peasants in the violent millenarian frenzies of medieval Europe by reference to the lack of "the material and emotional support afforded by traditional social groups; their kinship groups had disintegrated and they were not effectively organised in village communities or in guilds; for them there existed no regular, institutionalised methods of voicing their grievances or pressing their claims."[57] Kling attributes the chronic Latin American pattern of coup d'état to the lack of adequate alternatives facing elite aspirants with economic ambitions; political office, seized illicitly if necessary, provides opportunity for satisfying those ambitions.[58]

More general observations also are relevant. Economists suggest that government can relieve the discontents that accompany the strains of rapid economic growth by providing compensatory welfare measures—i.e., alternative means of value satisfaction.[59] Numerous scholars have shown that migration is a common response to deprivation and that high emigration rates often precede outbreaks of civil violence. In a cross-national study of correlates of civil violence for 1961–1963, I have found a rather consistent association between extensive educational opportunities, proportionally large trade union movements, and stable political party systems on the one hand and low levels of strife on the other, relationships that tend to hold among nations whatever their absolute level of economic development. Education presumably increases the apparent rage of opportunity for socioeconomic advance, unionization can provide a secondary means for economic goal attainment, and parties serve as primary mechanisms for attainment of participatory political values.[60] Hence:

> *Proposition I.5*: The strength of anger tends to vary directly with the proportion of all available opportunities for value attainment with which interference is experienced or anticipated.

The Mediation of Anger: The Effects of Social Control and Social Facilitation

For the purpose of the theoretical model I assume that the average strength of anger in a population is a precise multiple function of the instigating variables. Whether or not civil violence actually occurs as a response to anger, and its magnitude when it does occur, are influenced by a number of mediating variables. Evidence for these variables and their effects is found both in the psychological literature and in studies of civil violence *per se*. It is useful to distinguish them according to whether they inhibit or facilitate the violent manifestation of anger.

Social Control: The Effects of Retribution

The classic formulation is that aggression may be inhibited through fear of "such responses on the part of the social environment as physical injury, insults, ostracism, and deprivation of goods or freedom."[61] Good experimental evidence indicates that anticipation of retribution is under some circumstances an effective regulator of aggression.[62] Comparably, a linear relationship between, on the one hand, the capacity and willingness of government to enforce its monopoly of control of the organized instrumentalities of force and, on the other, the likelihood of civil violence is widely assumed in the literature on civil strife. Strong apparent force capability on the part of the regime ought to be sufficient to deter violence, and if violence should occur, the effectiveness with which it is suppressed is closely related to the likelihood and intensity of subsequent violence. Smelser states that a major determinant of the occurrence of civil strife is declining capacity or loyalty of the police and military control apparatus.[63] Johnson says that "the success or failure of armed insurrection and . . . commonly even the decision to attempt revolution rest . . . upon the attitude (or the revolutionaries' estimate of that attitude) that the armed forces will adopt toward the revolution."[64] In Janos' view, the weakening of law enforcement agencies "creates general disorder, inordinate concrete demands by various groups, and the rise of utopian aspirations."[65] Military defeat is often empirically associated with the occurrence of revolution. Race riots in the United States and elsewhere have often been associated with tacit approval of violence by the authorities.[66] Paret and Shy remark that "terror was effective in Cyprus against a British government without sufficient political strength or will; it failed in Malaya against a British government determined and able to resist and to wait." [67]

It also has been proposed, and demonstrated in a number of experimental settings, that if aggression is prevented by fear of retribution or by retribution itself, this interference is frustrating and increases anger. Maier, for example, found in animal studies that under conditions of severe frustration, punishment increased the intensity of aggression.[68] Walton inferred from such evidence that a curvilinear relationship ought to obtain between the degree of coerciveness of a nation and its degree of political instability, on the argument that low coerciveness is not frustrating and moderate coerciveness is more likely to frustrate than deter, while only the highest levels of coerciveness are sufficient to inhabit men from civil violence. A permissiveness-coerciveness scale for eighty-four nations, based on scope of political liberties, has been compared against the Feierabends' political stability scale, and the results strongly support the curvilinearity hypothesis.[69] Bwy, using a markedly different measure of coerciveness—one based on defense expenditures—found the same curvilinear relationship between coerciveness and "anomic violence" in Latin America.[70] Some theoretical speculation about civil strife implies the same relationship—for example, Lasswell and Kaplan's stipulation that the stability of an elite's position varies not with the actual use of violence but only with ability to use it,[71] and Parsons' more detailed "power deflation" argument that the repression of demands by force may inspire groups to resort to increasingly intransigent and aggressive modes of making those demands.[72]

One uncertainty about the curvilinear relationship between retribution and aggression is whether or not it holds whatever the extent of initial deprivation-induced anger. It is nonetheless evident that the threat or employment of force to suppress civil violence is by no means uniform in its effects, and that it tends to have a feedback effect that increases the instigation to violence. Such a relationship is diagrammed in Figure 1 and is explicit in the following proposition and its corollary:

> *Proposition M.1*: The likelihood and magnitude of civil violence tend to vary curvilinearly with the amount of physical or social retribution anticipated as a consequence of participation in it, with likelihood and magnitude greatest at medium levels of retribution.
>
> *M.1a*: Any decrease in the perceived likelihood of retribution tends to increase the likelihood and magnitude of civil violence.

These propositions and corollaries, and all subsequent propositions, hold only, of course, if deprivation-induced anger exists. If the modal level of collective discontent is negligible, a condition that holds for at least some small, although few large, collectivities, the mediating variables have no inhibiting or facilitating effects by definition.

The propositions above do not exhaust frustration-aggression evidence about effects of retribution. Experimental evidence further indicates that a delay in the expression of the aggressive response increases its intensity when it does occur.[73] Observations about civil violence also suggest that the effects of feared retribution, especially external retribution, must take account of the time variable. The abrupt relaxation of authoritarian controls is repeatedly associated with intense outbursts of civil violence, despite the likelihood that such realization reduces relative deprivation. Examples from recent years include the East German and Hungarian uprisings after the post-Stalin thaw, the Congo after independence, and the Dominican Republic after

94 Trujillo's assassination.

A parsimonious way to incorporate the time dimension into frustration-aggression theory is to argue that in the short run the delay of an aggressive response increases the intensity of anger and consequently the likelihood and magnitude of aggression, but that in the long run the level and intensity of expectations decline to coincide with the impositions of reality, and anger decreases concomitantly. Cognitive dissonance theory would suggest such an outcome: men tend to reduce persistent imbalances between cognitions and actuality by changing reality, or, if it proves intransigent, by changing their cognitive structures.[74] The proposed relationship is sketched in Figure 2.

One example of experimental evidence to this point is the finding of Cohen and others that once subjects became accustomed to certain kinds of frustration—withdrawal of social reinforcement in the experimental situation used—they were less likely to continue to seek the desired value or condition.[75] One can, moreover, speculate that the timescale is largely a function of the intensity of commitment to the frustrated response or condition. The effects of South Africa's apartheid policies and the means of their enforcement offer an example. These policies, which impose substantial and diverse value-deprivations on nonwhites, especially those in urban areas, were put into effect principally in the 1950's. Violent protests over their implementation were repressed with increasing severity, culminating in the Sharpeville massacre of 1960 and a series of strikes and riots. By the mid-1960's, when deprivation was objectively more severe than at any time previously in the twentieth

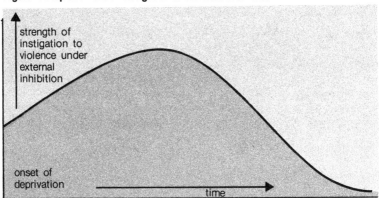

Figure 2. Displacement of Instigation to Violence over Time

century, levels of civil strife were very low, inferentially the result of very high levels of deterrence (feared retribution). Since deprivation remains severe and has affected a wide range of values, avoidance of violence in this case probably would require the maintenance of very high and consistent deterrence levels beyond the active life span of most of those who have personally experienced the initial value-deprivation. Any short-run decline in the perceived likelihood or severity of retribution, however, is highly likely to be followed by intense violence. In propositional form:

> *Proposition M.2*: Inhibition of civil violence by fear of external retribution tends in the short run to increase the strength of anger but in the long run to reduce it.
> *M.2a*: The duration of increased anger under conditions of inhibition tends to vary with the intensity of commitment to the value with respect to which deprivation is suffered.

Social Control: The Effects of Institutionalized Displacement

On the evidence, the effects of repression in managing discontent are complex and potentially self-defeating in the short run. Displacement theory suggests means that are considerably more effective. Several aspects of displacement theory are relevant for civil violence. Among Miller's basic propositions about object and response generalization is the formulation that the stronger the fear of retribution relative to the strength of anger, the more dissimilar will the target of aggression be from the source of interference and the more indirect will be the form of aggression.[76] With reference to object generalization, Berkowitz has proposed and demonstrated that hostility tends to generalize from a frustrater to previously disliked individuals or groups.[77] A counterpart of this thesis is that displaced aggressive responses tend to be expressed in previously used forms.

Examples of object generalization in civil violence are legion. Several studies have shown positive relationships between poor economic conditions and lynchings of Negroes in the American South.[78] An initial reaction of urban white colonialists to African rural uprisings in Madagascar in 1947 and Angola in 1961 was vigilante-style execution of urban Africans who had no connections with the rebellions. English handweavers, when their livelihood was threatened by the introduction of new weaving machines, destroyed thousands of machines in the Luddite riots of 1811--1816, but almost never directly attacked the employers who installed the machines and discharged superfluous workers.[79]

Object generalization is a crucial variable in determining who will be attacked by the initiators of particular acts of civil violence, but is only peripheral to the primary concern of the theory, the determination of likelihood and magnitude of violence as such. Most important in this regard is the psychological evidence regarding response generalization. Experimental evidence suggests that only a narrow range of objects provides satisfying targets for men's aggressive responses, but that almost any form of aggression can be satisfying so long as the angry person believes that he has in some way injured his supposed frustrater.[80]

By extension to the collectivity, insofar as adequate response displacement options are available, much anger may be diverted into activity short of civil violence. The evidence is diverse and extensive that participation in political activity, labor unions, and millenarian religious movements can be a response to relative deprivation which permits more or less nonviolent expression of aggression. Studies of voting in the United States show that politicians in farm states are rather consistently voted out of office after periods of low rainfall and that the occurrence of natural disasters may lead to hostility against officials.[81] Extremist voting—which may be regarded as nonviolent aggression—in nine European countries during the Depression has been shown to correlate $+.85$ with the percentage of the labor force unemployed.[82] Studies of labor movements repeatedly document the transformation of labor protest from violent to nonviolent forms as unionization increases. Comparative and case studies similarly document the development of aggressive millenarian religious movements as a response to natural disaster or political repression, in places and eras as diverse as medieval Europe, colonial Africa, and among the indigenous peoples of the Americas and the South Pacific.[83]

This is not to imply that displacement is a sole or exclusive function of such institutions. Their instrumental functions for participants (Proposition I.5) can be crucial: peaceful political and union activism are alternative means to goals whose attainment by other means is often impaired; religious chiliasm provides hope and belief for those whose social universe has been destroyed. But insofar as men become accustomed to express discontents through such institutional mechanisms, the likelihood that anger will lead to civil violence is diminished. In propositional form:

> *Proposition M.3*: The likelihood and magnitude of civil violence tend to vary inversely with the availability of institutional mechanisms that permit the expression of nonviolent hostility.

Social Facilitation: Cognitive Factors

Experimental, developmental, and field studies of the effects of rewarding individual aggression demonstrate that habitual aggression may be developed and maintained through intermittent rewards and may also be generalized to situations other than those in which the habits were acquired.[84] A number of experiments indicate that the presence of cues or stimuli associated with anger instigators is necessary for most aggressive responses to occur. A summary proposition is that "a target with appropriate stimulus qualities 'pulls' (evokes) aggressive responses from a person who is ready to engage in such actions either because he is angry or because particular stimuli have acquired cue values for aggressive responses from him."[85]

For members of a collectivity a variety of common experiences can contribute to the acquisition of aggressive habits and the recognition of aggression-evoking cues. Among them are socialization patterns that give normative sanction to some kinds of aggressive behavior; traditions of violent conflict; and exposure to new generalized beliefs that justify violence. The literature on civil violence suggests at least four specific modes by which such experiences facilitate violent responses to deprivation. They can (1) stimulate mutual awareness among the deprived, (2) provide explanations for deprivation of ambiguous origin, (3) specify accessible targets and appropriate forms of violence, and (4) state long-range objectives to be attained through violence.

Subcultural traditions of violent protest are well documented in European history. The frequency with which

Parisian workers and shopkeepers took to the streets in the years and decades following the great *journées* of 1789 is one example. At least 275 food riots, most of them similar in form and sequence, took place in rural England between 1725 and 1800 in close correlation with harvest failures and high food prices.[86] Hobsbawm points out that in Southern Italy "every political change in the nineteenth century, irrespective from what quarter it came, automatically produced its ceremonial marches of peasants with drums and banners to occupy the land," while in Andalusia "millenarian revolutionary waves occurred at roughly ten-year intervals for some sixty of seventy years."[87] Lynching as a Southern white response to Negro transgressions and the mobbing of white policemen by Negroes are comparable expressions of subcultural traditions that facilitate civil violence.

The theoretical point is that the initial occurrences of civil violence among some homogenous groups of the deprived—those events that set the pattern—tend to be nonrational responses to extreme deprivation. If violence provides a satisfactory outlet for tensions or if it motivates authorities to remedy the sources of deprivation, civil violence tends to become a sanctioned group activity. The fact that normative support for violence thus develops does not mean that violence subsequently occurs without instigation. Deprivation remains a necessary precondition for violence; the strength of anger at which it is likely to occur is lowered.

A related source of attitudinal support for collective violence is the articulation of ideology or, more generally, what Smelser calls generalized belief among the deprived. Such beliefs, ranging from rumors to fully articulated ideologies, are said to develop in situations characterized by social strain that is unmanageable within the existing framework for social action.[88] It is evident that in many social settings relative deprivation is manifest but its sources obscure. In psychological terms, no cues associated with the anger instigator are present. The agency responsible for an unwanted tax increase is apparent to the most ignorant villager; the causes of economic depression or of the disintegration of traditional mores are often unclear even to economists and sociologists. A new ideology, folk-belief, or rumor can serve to define and explain the nature of the situation, to identify those responsible for it, and to specify appropriate courses of action.

Moreover, there usually are a number of competing

generalized beliefs circulating among the deprived. Those most likely to gain acceptance tend to be those with substantial aggressive components, i.e., those that rationalize and focus the innate drive to aggression. Cohn's comparative study of the waves of chiliastic excitement that swept medieval Europe in times of plague and famine, for example, documents the fact that the heresies that most effectively mobilized the deprived were those that suited or could be molded to their states of mind: "when . . . eschatological doctrines penetrated to the uprooted and desperate masses in town and country they were re-edited and reinterpreted until in the end they were capable of inspiring revolutionary movements of a peculiarly anarchic kind."[89]

Some of these observations can be summarized in this proposition and its corollary:

Proposition M.4: The likelihood and magnitude of civil violence tend to vary directly with the availability of common experiences and beliefs that sanction violent responses to anger.

M.4a: Given the availability of alternative experiences and beliefs, the likelihood that the more aggressive of them will prevail tends to vary with the strength of anger.

Social Facilitation: Sources of Group Support for Violence

A classic subject of social psychological theory is the extent to which some social settings facilitate overt aggression. It is incontrovertible that individuals tend to behave in crowds differently from the way they act alone. The crowd psychologies of scholars such as Le Bon and Sorokin have emphasized the "unconscious" nature of crowd behavior and its "deindividuating" effects.[90] It appears more fruitful to examine experimentally identified variables that contribute to the "crowd behavior" phenomenon. From this point of departure one can distinguish at least three modes by which groups affect individuals' disposition to violence: (1) by providing normative support, (2) by providing apparent protection from retribution, and (3) by providing cues for violent behavior.

1. *Normative support.* There is good experimental evidence

that individuals alone or in poorly cohesive groups are less likely to express hostility than those in highly cohesive groups. Members of highly cohesive friendship groups respond to external frustrations with greater hostility than randomly formed groups. Similarly, if individuals believe that their peers generally agree with them about a frustrater, their public display of antagonism more closely resembles their privately expressed antagonism than if they do not perceive peer agreement.[91]

Theoretical and empirical studies of civil violence repeatedly refer to the causal efficacy of comparable conditions. Social theorists describe the perception of anonymity and of universality of deprivation characteristic of riotous crowds. Hopper's classic picture of group interaction under conditions of relative deprivation in the early stages of the revolutionary process is relevant: by participating in mass or shared behavior, discontented people become aware of one another; "their negative reactions to the basic factors in their situations are shared and begin to spread. . . . Discontent . . . tends to become focalized and collective."[92] Comparative studies of labor unrest show that the most strike-prone industries are those whose workers are relatively homogeneous and isolated from the general community.[93] Some of the efficacy of revolutionary brotherhoods and tightly knit bands of rebels in prosecuting civil violence can be interpreted in terms of the reinforcement of mutual perception of deprivation and the justification of violence as a response to it.

2. *Protection from retribution.* Groups appear capable of reducing fears of external retribution for violence in at least three ways. Crowd situations in particular provide members with a shield of anonymity. In an experimental study by Meier and others, two-thirds of subjects who were prepared to join a lynching mob said, *inter alia*, that they would do so because in the crowd they could not be punished. The same relationship is apparent in the handful of studies made of riot participants: crowd members usually feel insulated from retribution.[94]

Organized groups can provide apparent protection from retribution by acquiring sufficient force capability to prevent the agents for retribution—i.e., military and internal security forces—from effectively reaching them. Increases in the relative force capability of a deprived group may also reinforce rationalization for violence by raising hopes of success or may merely facilitate the expression of rage by providing desperate men with the means to strike at tormentors who had previously been unassailable.

A third aspect of group protectiveness is the perceived effect of hierarchical organization and the presence of highly visible leaders. Leaders of revolutionary organizations, in addition to their other manifest functions, not only foment but assume responsibility for illicit violence. Their followers tend to see such leaders as the likely objects of retaliatory efforts and hence feel less personal risk.

3. *Cues for violence.* The transition from anger to aggression is not automatic or even abrupt. Laboratory studies of imitative behavior repeatedly document the significance of aggression-releasing cues provided by social models. The act of punishing aggression itself can serve as a model for imitation by the person punished. Aggression-releasing cues need not necessarily originate with high-status persons. Polansky and others found that when frustrations were imposed on groups of children, "impulsive" but low-status children were both initiators and ready followers of aggressive behavioral contagion. On the other hand, not any aggressive model evokes aggression from angered subjects; the models that evoke greatest aggression are those associated with the subjects' present situation or with settings in which they were previously aggressive.[95]

Angry crowds of men also appear to require some congruent image or model of violent action before they will seize cobblestones or rope or rifles to do violence to fellow citizens. Such models may be symbolic: invocation of a subcultural tradition of violence by a leader, or articulation of a new generalized belief that is explicit in its prescription of violence. In general, however, a "call to arms" or an appeal to a tradition of violence appears less effective by itself than when accompanied by the sight or news of violence. The calculated use of terrorism by rebels can have such an effect, and so can a soldier's random shot into a crowd of demonstrators. Many specific cases of civil violence have been set off by comparable acts of violence elsewhere. "Revolutionary contagion" is evident in the 1830 and 1848 waves of European revolutionary upheavals and in the post-Stalin uprisings in Eastern Europe and Siberia. The same phenomenon is apparent in the initiation of innumerable cases of small-scale, unstructured violence. Series of riots in rural France and England have graphically been shown to spread outward from one or a few centers, riots occurring in the furthest villages days or weeks after the initial incident. Such patterning is evident, to mention a few cases, in the French Corn Riots of 1775, the "Plug-Plot" riots around Manchester in 1842, and the incidence of farmers' protest meetings and riots in Brittany in the summer of 1961.[96] The demonstration effect apparent in such series of events appears to have affected their form and timing more than the likelihood of the occurrence of strife. The people who responded to the events were already angered; they probably would have erupted into violence in some form sometime in the proximate future.

These three modes of group facilitation of civil violence can be summarized in propositional form:

> *Proposition M.5*: The likelihood and magnitude of civil violence tend to vary directly with the extent to which the deprived occupy organizational and/or ecological settings that provide (1) normative support through high levels of interaction, (2) apparent protection from retribution, and (3) congruent models for violent behavior.

The Forms of Civil Violence

The theoretical framework comprising the ten propositions is formally restricted to physically violent collective behavior. It is likely that it is as applicable to a still larger class of events, including those characterized by the threat of violence or by high levels of verbal aggression—for example, bloodless coups, demonstrations, and political strikes. Violent events tend to be more salient for the political system, however, and for most operational purposes constitute a more workable and clearly defined universe.

I have not discussed the propositions with reference to specific forms of civil violence on grounds that all of the variables specified are relevant to each form specified in current typologies.[97] It is nonetheless likely that the propositions are of differential weight for different forms, and it is useful to demonstrate how variations in form may be generally accounted for in the context of the theoretical model. The first question to be asked is how detailed a listing of forms one should attempt to account for. A series of factor analytic studies provide a systematic, empirical answer to that question. In each of eleven studies, data on the incidence and characteristics of various types of strife were collected and tabulated, by country, and the "country scores" (number of riots, assassinations, deaths from civil violence, coups, mutinies, guerrilla wars, and so on, in a given time period) were factor analyzed. Whatever the typology employed, the period of reference, or the set of countries, essentially the same results were obtained. A strong *turmoil* dimension emerges, characterized by largely spontaneous strife such as riots, demonstrations, and nonpolitical clashes, quite distinct from what we may call a *revolutionary* dimension, characterized by more organized and intense strife. This revolutionary dimension has two components, appearing in some analyses as separate dimensions: *internal war*, typically including civil war, guerrilla war, and some coups; and *conspiracy*, typically including plots, purges, mutinies, and most coups.[98] Events within each of the three types tend to occur together; events within any two or all three categories are less likely to do so. The implication is that they are substantively distinct forms of strife for each of which separate explanation is required.

Two complementary approaches to accounting for these three basic types of civil violence can be proposed within the context of the theoretical model. The first is that the two major dimensions, turmoil and revolution, reflect the varying class incidence of deprivation among societies. The defining characteristic of "turmoil" events is mass participation, usually rather spontaneous, disorganized, and with low intensity of violence; the forms of "revolution" reflect organized, often instrumental and intense, application of violence. The ability to rationalize, plan, and put to instrumental use their own and others' discontent is likely to be most common among the more skilled, highly educated members of a society—its elite aspirants. Thus if the incidence of mass deprivation is high but elite deprivation low, the most likely form of civil violence is turmoil. But if severe discontent is common to a substantial, alienated group of elite aspirants, then organized, intensive strife is likely.

The forms of revolution differ principally in their scale and tactics: internal wars are large-scale, and their tactics are typically to neutralize the regime's military forces; conspirators, usually few in number, attempt to subvert the regime by striking at its key members.

104   The differences between internal war and conspiracy can be accounted for by several characteristics. If severe deprivation is restricted largely to elite aspirants, the consequence is likely to be "conspiracy" phenomena such as plots, coups d'état, and barracks revolts. If discontent is widespread among substantial numbers of both mass and elite aspirants, the more likely consequence is large-scale, organized violence—civil and guerrilla war. The strategic position of the discontented elite aspirants may be relevant as well. If they are subordinate members of the existing elite hierarchy, they are likely to attack the regime from within, hence coups, mutinies, and plots. If they are instead excluded from formal membership in the elite though they possess elite qualities—acquired, for example, through foreign education—they must organize violent resistance from without. These are essentially Seton-Watson's explanations for the relative frequency of conspiracy in underdeveloped societies compared with the frequency of massive revolutionary movements in more developed states. In summary, "it is the combination of backward masses, extremist intellectuals and despotic bureaucrats which creates the most conspiratorial movements."[99]

These observations are of course only the beginning of an accounting of the forms of civil strife. They are intended to demonstrate, however, that such a theoretical explanation not only is compatible with but can be formulated within the framework of the theoretical model by showing the loci of deprivation in a society. They can be stated thus in propositional form:

> *Proposition F.1*: The characteristic form of civil violence tends to vary with the differential incidence of relative deprivation among elite aspirants and masses: (1) mass deprivation alone tends to be manifested in large-scale civil violence with minimal organization and low intensity; (2) elite-aspirant deprivation tends to be manifested in highly organized civil violence of high intensity.
>
> *F.1a*: Whether organized and intense civil violence is large-scale or small-scale is a joint function of the extent of mass deprivation and the strategic access of deprived elite aspirants to the incumbent political elite.

Conclusion

I have advanced eleven general propositions about the variables operative in generating and structuring violent political unrest. They are based on the assumption that the frustration-aggression mechanism, however culturally modified, is the source of most men's disposition to illicit collective violence. The propositions do not constitute a theory of the revolutionary process or of the outcomes of strife, but of the conditions that determine the *likelihood* and *magnitude* of strife. On the other hand, the variables stipulated by the propositions are not irrelevant to revolutionary processes. Process models can be formulated wholly or partly in terms of changing patterns of weights on the component variables.

It is likely that most "causes" and "correlates" of the occurrence and intensity of civil strife can be subsumed by these variables, with one exception: foreign intervention. This exception is no oversight but simply recognition that decisions to intervene are external to domestic participants in civil strife. The effects of foreign intervention can be readily interpreted by reference to the model, however: intervention on behalf of the deprived is likely to strengthen group support (M.5) and may, as well, heighten and intensify value expectations (I.2). Foreign assistance to a threatened regime is most likely to raise retribution levels (M.1), but may also alter aspects of value capabilities (I.4, I.5) and strengthen justification for violence among the deprived, insofar as they identify foreigners with invaders (M.4).

The framework has not been elaborated merely to provide a satisfying theoretical reconstruction of the general causes of civil violence. It is intended primarily as a guide for empirical research using the techniques of both case and comparative studies. The framework stipulates the variables for which information should be sought in any thorough case study of the origins of an act of civil strife.[100] For purposes of comparative analysis it stipulates relationships that should hold among cultures and across time. Its most important objectives are to encourage empirical validation of its component propositions in a variety of contexts by a variety of operational means, and specification of their separate weights and interacting effects in those contexts.[101]

Notes

This article is a revision of a paper read to the panel on "The Psychology of Political Unrest," at the Annual Meeting of the American Psychological Association, New York, September 2-6, 1966. Harry Eckstein's careful and helpful evaluation of draft versions of this paper is gratefully acknowledged. Others who have provided useful, though not always satisfiable, criticism of the theoretical model include Leonard Berkowitz, Alfred de Grazia, Mohammed Guessous, Marion J. Levy, Jr., John T. McAlister, Jr., Mancur L. Olson, Jr., Joel Prager, Bryant Wedge, and Oran R. Young. Theoretical work was supported by an award from a National Science Foundation institutional grant to New York University and by the Center for Research on Social Systems (formerly SORO) of American University.

1. Arnold Forster, "Violence on the Fanatical Left and Right," *Annals of the American Academy of Polictical and Social Science*, CCCLXIV (March 1966), 142.
2. For example, Lawrence Stone, "Theories of Revolution," *World Politics*, XVIII (January 1966), 159-76, advances the curious argument that collective violence generally cannot be the object of useful theorizing because it is at the same time both pervasive and somehow peripheral.
3. The emphasis on processes is evident in the major theoretical analyses of the "classic" revolutions, including Lyford P. Edwards, *The Natural History of Revolutions* (Chicago 1927); Crane Brinton, *The Anatomy of Revolution* (New York 1938); George S. Pette, *The Process of Revolution* (New York 1938); Louis R. Gottschalk, "Causes of Revolution," *American Journal of Sociology*, L (July 1944), 1-9; and Rex D. Hopper, "The Revolutionary Process: A Frame of Reference for the Study of Revolutionary Movements," *Social Forces*, XXVIII (March 1950), 270–279.
4. A great many counts of the incidence of civil strife events have recently been reported. Harry Eckstein reports 1,632 "internal wars" in the period 1946-1959 in "On the Etiology of Internal Wars," *History and Theory*, IV, No. 2 (1965), 133-63. Rummel and Tanter counted more than 300 "domestic conflict events" per year during the years 1955-1960, including an annual average of 13 guerrilla wars and 21 attempted overthrows of government; see Raymond Tanter, "Dimensions of Conflict Behavior Within Nations, 1955-60: Turmoil and Internal War," *Peace Research Society Papers*, III (1965), 159-84. Most important to the argument that civil strife is a single universe of events are results of Rudolph Rummel's factor analysis of 236 socioeconomic and political variables, including nine domestic conflict measures, for a large number of nations. Eight of the conflict measures—e.g., number of riots, of revolutions, of

purges, of deaths from group violence—are strongly related to a single factor but not significantly related to any others, strong empirical evidence that they comprise a distinct and interrelated set of events. See *Dimensionality of Nations Project: Orthogonally Rotated Factor Tables for 236 Variables*, Department of Political Science, Yale University (New Haven, July 1964), mimeographed.
5. The "French Revolution" was a series of events that would now be characterized as urban demonstrations and riots, peasant uprisings, and a coup d'état. It is called a revolution in retrospect and by virtue of the Duc de Liancourt's classic remark to Louis XVI. The American Revolution began with a series of increasingly violent urban riots and small-scale terrorism that grew into a protracted guerrilla war.
6. The universe of concern, civil violence, is formally defined as *all collective, non-governmental attacks on persons or property, resulting in intentional damage to them, that occur within the boundaries of an autonomous or colonial political unit*. The terms "civil strife," "violent civil conflict," and "civil violence" are used synonymously in this article. The universe subsumes more narrowly defined sets of events such as "internal war," which Harry Eckstein defines as "any resort to violence within a political order to change its constitution, rulers, or policies" (in "On the Etiology of Internal Wars," 133), and "revolution," typically defined in terms of violently accomplished fundamental change in social institutions.
7. Bryant Wedge argues (in a personal communication) that much human aggression, including some civil strife, may arise from a threat-fear-aggression sequence. Leonard Berkowitz, however, proposes that this mechanism can be subsumed by frustration-aggression theory, the inferred sequence being threat (anticipated frustration)-fear-anger-aggression, in *Aggression: A Social Psychological Analysis* (New York 1962), chap. 2. It may be conceptually useful to distinguish the two mechanisms; it nonetheless appears likely that most variables affecting the outcome of the frustration-aggression sequence also are operative in the postulated threat-aggression sequence.
8. Konrad Lorenz, *On Aggression* (New York 1966), xii.
9. Sigmund Freud, *Civilization and Its Discontents*, trans. Joan Riviere (London 1930); Lorenz, chaps. 13, 14. Freud's instinctual interpretation of aggression is advanced in his later works; his early view was that aggression is a response to frustration of pleasure-seeking behavior. For a review and critique of other instinct theories of aggression, see Berkowitz, chap.I.
10. H. L. Nieburg, "The Threat of Violence and Social Change," *American Political Science Review*, LVI (December 1962), 870.

11. A characteristic study is Albert Bandura and Richard H. Walters, *Social Learning and Personality Development* (New York 1963). For a commentary on instrumental aggression, see Berkowitz, esp. 30-32, 183-83, 201-2.
12. Chalmers Johnson, *Revolutionary Change* (Boston) 1966), 12, 13, italics added.
13. Talcott Parsons, "Some Reflections on the Place of Force in Social Process," in Harry Eckstein, ed., *Internal War: Problems and Approaches* (New York 1964), 34-35.
14. Thomas C. Schelling, *The Strategy of Conflict* (Cambridge, Mass., 1960), 4.
15. P. 161.
16. The most influential and systematic statement of the theory is John Dollard and others, *Frustration and Aggression* (New Haven 1939). Two important recent syntheses of the evidence are Berkowitz, *Aggression,* and Aubrey J. Yates, *Frustration and Conflict* (New York 1962). Also see Leonard Berkowitz, "The Concept of Aggressive Drive: Some Additional Considerations," in Berkowitz, ed., *Advances in Experimental Psychology,* Vol. II (New York 1965), 307-22.
17. Norman R. F. Maier, *Frustration: The Study of Behavior Without a Goal* (New York 1949), 92-115, 159-61. Maier postulates a frustration threshold that may open the way to any of four classes of "goal-less" behavior of which aggression is only one. His findings have not been related adequately to the body of research on the frustration-aggression relationship. One can suggest, however, that the nonaggressive responses—fixation, regression, and apparent resignation—can be treated as more or less innate responses in a response hierarchy which are resorted to in the absence of aggression-evoking cues.
18. Neil J. Smelser, *Theory of Collective Behavior* (New York 1963).
19. David Willer and George K. Zollschan, "Prolegomenon to a Theory of Revolutions," in George K. Zollschan and Walter Hirsch, eds., *Explorations in Social Change* (Boston 1964), 125-51.
20. Ronald G. Ridker, "Discontent and Economic Growth," *Economic Development and Cultural Change,* XI (October 1962), 1-15.
21. James C. Davies, "Toward a Theory of Revolution," *American Sociological Review,* XXVII (February 1962), 5-19.
22. Johan Galtung, "A Structural Theory of Aggression," *Journal of Peace Research,* II, No. 2 (1964), 95.
23. Ivo K. and Rosalind L. Feierabend, in "Aggressive Behaviors Within Polities, 1948-1962: A Cross-National Study," *Journal of Conflict Resolution,* X (September 1966), 249-71, have formally equated political instability with aggressive behavior and have derived and tested several hypotheses about stability from frustration-aggression theory. They have attempted no general

theoretical synthesis, however.

24. The term "instigating" is adapted from the behavioristic terminology of Dollard and others. Instigating variables determine the strength of instigation, i.e., stimulus or motivation, to a particular kind of behavior. Mediating variables refer to intervening conditions, internal or external to the actors, which modify the expression of that behavior.
25. The phrase "relative deprivation" was first used systematically in Samuel A. Stouffer and others, *The American Soldier: Adjustment During Army Life,* Vol. I (Princeton 1949), to denote the violation of expectations. J. Stacy Adams reviews the concept's history and some relevant evidence and suggests that feelings of injustice intervene between the condition of relative deprivation and responses to it, in "Inequity in Social Exchange," in Berkowitz, ed., *Advances in Experimental Psychology,* 267-300. The "injustice" aspect is implicit in my definition and use of relative deprivation as *perceived* discrepancy between what people think they will get and what they believe they are entitled to. The Stouffer concept has been related to levels of social satisfaction and to anomie, but has not, so far as I know, been associated with the discontent-anger-rage continuum in the frustration-aggression relationship.
26. David F. Aberle, "A Note on Relative Deprivation Theory," in Sylvia L. Thrupp, ed., *Millennial Dreams in Action: Essays in Comparative Study* (The Hague 1962), 209-14. Bert Hoselitz and Ann Willner similarly distinguish between expectations, regarded by the individual as "what is rightfully owed to him," and aspirations, which represent "that which he would like to have but has not necessarily had or considered his due," in "Economic Development, Political Strategies, and American Aid," in Morton A. Kaplan, ed., *The Revolution in World Politics* (New York 1962), 363.
27. Pp. 175-78.
28. Aberle, 210.
29. The Feierabends use the comparable term "systemic frustration" to describe the balance between "social want satisfaction" and "social want formation."
30. Hadley Cantril's work offers examples, especially *The Pattern of Human Concerns* (New Brunswick 1965).
31. This approach is exemplified by the Feierabends' work and by Bruce M. Russett, "Inequality and Instability: The Relation of Land Tenure to Politics," *World Politics,* XVI (April 1964), 442-54.
32. The basic postulate of Dollard and others is that "the occurrence of aggressive behavior always presupposes the existence of frustration and, contrariwise, that the existence of frustration always leads to some form of aggression" (p. I). It is evident from context and from subsequent articles that this statement

was intended in more qualified fashion.
33. *Ibid.*, 28.
34. Charles Tilly, *The Vendée* (Cambridge, Mass., 1964).
35. Reuben de Hoyos, personal communication.
36. Gregory Massell, "The Strategy of Social Change and the Role of Women in Soviet Central Asia: A Case Study in Modernization and Control," Ph.D. diss., Harvard University, 1966.
37. For this kind of approach, see Daniel Lerner, "Toward a Communication Theory of Modernization: A Set of Considerations," in Lucian W. Pye, ed., *Communications and Political Development* (Princeton 1963), 330-35.
38. This general statement of theory is concerned with specification of variables and their effects, not with their content in specific cases; hence the conditions that determine the *levels* of expectation and changes in those levels are not treated here, nor are the conditions that affect perceptions about value capabilities. For some attempts to generalize about such conditions see Ted Gurr, "The Genesis of Violence: A Multivariate Theory of Civil Strife," Ph.D. diss., New York University, 1965, esp. chaps. 6-8. For empirical evaluation or application of the theory, it is of course necessary to evaluate in some way levels of expectation in the population(s) studied. Some approaches to evaluation are illustrated in Ted Gurr with Charles Ruttenberg, *The Conditions of Civil Violence: First Tests of a Causal Model,* Center of International Studies, Princeton University, Research Monograph No. 28 (Princeton 1967), and Ted Gurr, "Explanatory Models for Civil Strife Using Aggregate Data," a paper read at the Annual Meeting of the American Political Science Association, 1967.
39. See Berkowitz, *Aggression*, 53-54.
40. Eric Hoffer, *The True Believer* (New York 1951), 27-28.
41. Compare Crawford Young, *Politics in the Congo* (Princeton 1965), chap. 13, with commentaries on the Kwilu and Stanleyville rebellions, such as Renée C. Fox and others, " 'The Second Independence': A Case Study of the Kwilu Rebellion in the Congo," *Comparative Studies in Society and History,* VIII (October 1965), 78-109; and Herbert Weiss, *Political Protest in the Congo* (Princeton 1967).
42. See, among many other works, Georges Lefebvre, *The Coming of the French Revolution* (Princeton 1947), Part II.
43. Value expectations are defined above in terms of the value positions to which men believe they are justifiably entitled; the discussion here assumes that men may also regard as justifiable some types of interference with those value positions.
44. Nicholas Pastore, "The Role of Arbitrariness in the Frustration-Aggression Hypothesis," *Journal of Abnormal and Social Psychology,* XLVII (July 1952), 728-31; John J. Kregarman and Philip Worchel, "Arbitrariness of Frustration and Aggression,"

Journal of Abnormal and Social Psychology, LXIII (July 1961), 183-87.

45. The argument is that people comply "to gain both the symbolic rewards of governmental action and the actual rewards with which government originally associated itself" and rationalize compliance with "the feeling that the regime is a morally appropriate agent of control . . ." (Richard M. Merelman, "Learning and Legitimacy," *American Political Science Review,* LX [September 1966], 551). The argument applies equally well to compliance, including acceptance of deprivation, with the demands of other social institutions.
46. Walt W. Rostow, *British Economy of the Nineteenth Century* (Oxford 1948), chap. 5.
47. George Rudé, "Prices, Wages, and Popular Movements in Paris During the French Revolution," *Economic History Review,* VI (1954), 246-67, and *The Crowd in History, 1730-1848* (New York 1964), chap. 7.
48. Carl Hovland and Robert Sears, "Minor Studies in Aggression, VI: Correlation of Lynchings with Economic Indices," *Journal of Psychology,* IX (1940), 301-10.
49. William Kornhauser, *The Politics of Mass Society* (New York 1959), 160.
50. "Aggressive Behaviors Within Polities."
51. Robert L. Hamblin and others, "The Interference-Aggression Law?" *Sociometry,* XXVI (1963), 190-216.
52. Bruce M. Russett and others, *World Handbook of Political and Social Indicators* (New Haven 1963), 97–100.
53. There is a threshold effect with reference to physical well-being. If life itself is the value threatened and the threat is imminent, the emotional response tends to be fear or panic; once the immediate threat is past, anger against the source of threat tends to manifest itself again. See n. 7 above, and Berkowitz, *Aggression,* 42-46.
54. Leonard Berkowitz, "Repeated Frustrations and Expectations in Hostility Arousal," *Journal of Abnormal and Social Psychology,* LX (May 1960), 422-29.
55. See, for example, Daniel Lerner, *The Passing of Traditional Society* (Glencoe 1958).
56. Denis W. Brogan, *The Price of Revolution* (London 1951), 34.
57. Norman R. C. Cohn, *The Pursuit of the Millennium,* 2d ed. rev. (New York 1961), 315.
58. Merle Kling, "Toward a Theory of Power and Political Instability in Latin America," *Western Political Quarterly,* IX (March 1956), 21-35.
59. Ridker, 15; Mancur Olson, Jr., "Growth as a Destabilizing Force," *Journal of Economic History,* XXIII (December 1963), 550-51.
60. Gurr with Ruttenberg.

61. Dollard and others, 34.
62. For summaries of findings, see Richard H. Walters, "Implications of Laboratory Studies of Aggression for the Control and Regulation of Violence," *Annals of the American Academy of Political and Social Science,* CCCLXIV (March 1966), 60-72; and Elton D. McNeil, "Psychology and Aggression," *Journal of Conflict Resolution,* III (September 1959), 225-31.
63. Pp. 231-36, 261-68, 332, 365-79.
64. Chalmers Johnson, *Revolution and the Social System* (Stanford 1964), 16-17.
65. Andrew Janos, *The Seizure of Power: A Study of Force and Popular Consent,* Center of International Studies, Princeton University, Research Monograph No. 16 (Princeton 1964), 5.
66. See, for example, H. O. Dahlke, "Race and Minority Riots: A Study in the Typology of Violence," *Social Forces,* XXX (May 1952), 419-25.
67. Peter Paret and John W. Shy, *Guerrillas in the 1960's,* rev. ed. (New York 1964), 34-35.
68. *Frustration,* passim.
69. Jennifer G. Walton, "Correlates of Coerciveness and Permissiveness of National Political Systems: A Cross-National Study," M.A. thesis, San Diego State College, 1965. Douglas Bwy, "Governmental Instability in Latin America: The Preliminary Test of a Causal Model of the Impulse to 'Extra-Legal' Change," paper read at the Annual Meeting of the American Psychological Association, 1966.
70. Douglas Bwy, "Governmental Instability in Latin America: The Preliminary Test of a Causal Model of the Impulse to 'Extra-Legal' Change," paper read at the Annual Meeting of the American Phychological Association, 1966.
71. Harold Lasswell and Abraham Kaplan, *Power and Society: A Framework for Political Inquiry* (New Haven 1950), 265-66.
72. "Some Reflections on the Place of Force."
73. J. W. Thibaut and J. Coules, "The Role of Communication in the Reduction of Interpersonal Hostility," *Journal of Abnormal and Social Psychology,* XLVII (October 1952), 770-77.
74. See Leon Festinger, *A Theory of Cognitive Dissonance* (Evanston 1957).
75. Arthur R. Cohen and others, "Commitment to Social Deprivation and Verbal Conditioning," *Journal of Abnormal and Social Psychology,* LXVII (November 1963), 410-21.
76. Neal E. Miller, "Theory and Experiment Relating Psychoanalytic Displacement to Stimulus-Response Generalization," *Journal of Abnormal and Social Psychology,* XLIII (April 1948), 155-78.
77. *Aggression,* chap. 6.
78. See n. 48 above.
79. Rudé, *The Crowd in History,* chap. 5. The high levels of verbal aggression directed against the employers suggest that

displacement was involved, not a perception of the machines rather than employers as sources of deprivation. In the Luddite riots, fear of retribution for direct attacks on the owners, contrasted with the frequent lack of sanctions against attacks on the machines, was the probable cause of object generalization. In the Madagascar and Angola cases structural and conceptual factors were responsible: the African rebels were not accessible to attack but local Africans were seen as like them and hence as potential or clandestine rebels.

80. Some such evidence is summarized in Berkowitz, "The Concept of Aggressive Drive," 325-27.
81. A Critical and qualifying review of evidence to this effect is F. Glenn Abney and Larry B. Hill, "Natural Disasters as a Political Variable: The Effect of a Hurricane on an Urban Election," *American Political Science Review*, LX (December 1966), 974-81.
82. Kornhauser, 161. For interview evidence on the motives of protest voting, see Hadley Cantril, *The Politics of Despair* (New York 1958).
83. Representative studies are Cohn; James W. Fernandez, "African Religious Movements: Types and Dynamics," *Journal of Modern African Studies*, II, No. 4 (1964), 531-49; and Vittorio Lanternari, *The Religions of the Oppressed* (New York 1963).
84. Summerized in Walters.
85. Leonard Berkowitz, "Aggressive Cues in Aggressive Behavior and Hostility Catharsis," *Psychological Review*, LXXI (March 1964), 104-22, quotation from 106.
86. Rudé, *The Crowd in History*, 19-45.
87. E. J. Hobsbawm, *Social Bandits and Primitive Rebels*, 2nd ed. (Glencoe 1959), 63-64.
88. Chap. 5.
89. P. 31.
90. Gustave Le Bon, *The Psychology of Revolution* (London 1913); Pitirim Sorokin, *The Sociology of Revolutions* (Philadelphia 1925).
91. Representative studies include J. R. P. French, Jr., "The Disruption and Cohesion of Groups," *Journal of Abnormal and Social Psychology*, XXXVI (July 1941), 361-77; A. Pepitone and G. Reichling, "Group Cohesiveness and the Expression of Hostility," *Human Relations*, VIII, No. 3 (1955), 327-37; and Ezra Stotland, "Peer Groups and Reactions to Power Figures," in Dorwin Cartwright, ed., *Studies in Social Power* (Ann Arbor 1959), 53-68.
92. Pp. 272-75, quotation from 273.
93. Clark Kerr and Abraham Siegel, "The Isolated Mass and the Integrated Individual: An International Analysis of the Inter-Industry Propensity to Strike," in Arthur Kornhauser and others, eds., *Industrial Conflict* (New York 1954), 189-212.

94. Norman C. Meier and others, "An Experimental Approach to the Study of Mob Behavior," *Journal of Abnormal and Social Psychology*, XXXVI (October 1941), 506-24. Also see George Wada and James C. Davies, "Riots and Rioters," *Western Political Quarterly*, X (December 1957), 864-74.
95. See Walters; Norman Polansky and others, "An Investigation of Behavioral Contagion in Groups," *Human Relations*, III, No. 3 (1950), 319-48 and Leonard Berkowitz and Russell G. Geen, "Film Violence and the Cue Properties of Available Targets," *Journal of Personality and Social Psychology*, III (June 1966), 525-30.
96. Rudé, *The Crowd in History*; Henri Mendras and Yves Tavernier, "Les Manifestations de juin 1961," *Revue française des sciences politiques*, XII (September 1962), 647-71.
97. Representative typologies are proposed by Johnson, *Revolution and the Social System*, 26-68; Rudolph J. Rummel, "Dimensions of Conflict Behavior Within and Between Nations," *Yearbook of the Society for General Systems Research*, VIII (1963), 25-26; and Harry Eckstein, "Internal Wars: A Taxonomy," unpubl. (1960).
98. Two summary articles on these factor analyses are Rudolph J. Rummel, "A Field Theory of Social Action With Application to Conflict Within Nations," *Yearbook of the Society for General Systems Research*, X (1965), 183-204; and Tanter. What I call internal war is referred to in these sources as subversion; I label conspiracy what these sources call revolution. My terminology is, I believe, less ambiguous and more in keeping with general scholarly usage.
99. Hugh Seton-Watson, "Twentieth Century Revolutions," *Political Quarterly*, XXII (July 1951), 258.
100. For example, it has been used by Bryant Wedge to analyze and compare interview materials gathered in the study of two Latin American revolutions, in "Student Participation in Revolutionary Violence: Brazil, 1964, and Dominican Republic, 1965," a paper read at the Annual Meeting of the American Political Science Association, 1967.
101. Studies based on this theoretical model and using cross-national aggregate data include Ted Gurr, *New Error-Compensated Measures for Comparing Nations: Some Correlates of Civil Strife*, Center of International Studies, Princeton University, Research Monograph No. 25 (Princeton 1966); Gurr with Ruttenberg; Gurr, "Explanatory Models for Civil Strife"; and Gurr, "Why Urban Disorders? Perspectives From the Comparative Study of Civil Strife," *American Behavioral Scientist* (forthcoming).

The J-Curve of Rising and Declining Satisfactions as a Cause of Some Great Revolutions and a Contained Rebellion
by James C. Davies

The J-curve is this: revolution is most likely to take place when a prolonged period of rising expectations and rising gratifications is followed by a short period of sharp reversal, during which the gap between expectations and gratifications quickly widens and becomes intolerable. The frustration that develops, when it is intense and widespread in the society, seeks outlets in violent action. When the frustration becomes focused on the government, the violence becomes coherent and directional. If the frustration is sufficiently widespread, intense, and focused on government, the violence will become a revolution that displaces irrevocably the ruling government and changes markedly the power structure of the society. Or the violence will be contained within the system, which it modifies but does not displace. This latter case is rebellion. The following chart (figure 19-1) shows what happens as a society heads toward revolution.

This is an assertion about the state of mind of individual people in a society who are likely to revolt. It says their state of mind, their mood, is one of high tension and rather generalized hostility, derived from the widening of the gap between what they want and what they get. They fear not just that things will no longer continue to get better but—even more crucially—that ground will be lost that they have already gained. The mood of rather generalized hostility, directed generally outward, begins to turn toward government. People so frustrated not only fight

Need Satisfaction and Revolution

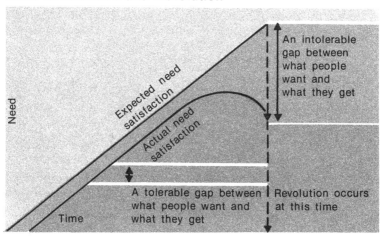

with other members of their families and their neighbors. They also jostle one another in crowds and increase their aggressiveness as pedestrians and bus passengers and drivers of cars. When events and news media and writers and speakers encourage the direction of hostilities toward the government, the dispersed and mutual hostility becomes focused on a common target. The hostility among individuals diminishes. The dissonant energy becomes a resonant, very powerful force that heads like a great tidal wave or forest fire toward the established government, which it may then engulf.

This phenomenon of synergic unification of a public when frustration becomes widespread and deep is awesome in its tendency to erase hostility between people. It is akin to the feeling that develops in a quarrelsome household at times. A fighting family may just barely manage to hold together. The father may be unemployed and frequently drunk, the mother worn to a frazzle, the children quarrelsome as they displace the tensions generated by poverty and the frustrations of their fighting parents. The father, no longer able to provide for his family, may lose his authority within the family and strike out at those nearest to him. But when the landlord knocks on the door and announces that the rent must be paid by 10 o'clock tomorrow morning on pain of eviction, the family suddenly stops its own fighting, beats up the landlord, and throws him out on the street.

Such tension within the family is a microcosm of the tension within the national community; that is, among the individual members of the political society and among its conflicting regional, religious, racial, and socioeconomic groups. When the various segments of a deeply divided society suddenly sense that they all have the same enemy, the government, they can spontaneously unite for long enough to overthrow it.

Causes of Revolution and Rebellion Psychologically Considered

Revolutions and rebellions differ in result but have like origins. And the differences in origin are less ones of kind than of degree. Revolutions involve more segments of the population than do rebellions. The intensity of feeling in revolutions is probably greater and has taken a longer time to develop than in rebellions. The violent phase of a revolution is longer and more savage. The bitterness that lingers after the violence is likely to endure for decades or centuries after a revolution.

The difference is not in causes and the violent action or even in the long-range consequences so much as it is in the immediate result. Rebellions do not remove the established government but instead are contained, partly as the consequence of the use of violent force in large enough amounts to override the rebels' anger at the government. The rebels may choose to live with their frustrations rather than endure the blows of the police and the army and the dull, sick anguish of imprisonment. But rebellions also are contained within the established system to the extent that the government pays heed to the grievances that led to the rebellion. If the only response to extralegal violence is legal violence, then hatred of oppression becomes deeply imprinted in the minds of the momentarily silenced rebels. The hatred lingers and deepens like embers in dry tinder after firefighters have tried to beat to death a small fire.

As the American Declaration of Independence said in 1776, people do not for "light and transient causes" make a rebellion or revolution. What then are the grave and enduring conditions that produce frustrations in a broad and varied citizenry, that in turn produce the revolutionary state of mind?

The common condition appears to be the denial of satisfaction of those needs that are basic to all human beings. Not all needs (as for a new automobile) are basic, and not all basic needs are of equal revolutionary potential. Abraham Maslow has argued that man's basic needs are arranged in sequence, from the most to the least powerful. The most potent are the physical needs, which must continuously be satisfied for all people during their entire life. But when a person gains their satisfaction—as an infant, a child, and an adult—he does not then, animal-like, remain content with satisfying just these bodily needs. Soon after birth he demands affection and, if he gets it, he reciprocates affection toward others. But his physical needs persist, and if he is forced to choose, he will first satisfy his need for food and then his need for affection.

In early childhood the individual who has been regularly gratified in his physical and affectional needs does not then rest content with this mental state of affairs. He begins, usually no later than when he is 8 or 10 years old, to demand recognition as an individual who is worthy of his own regard for himself and of others' regard for him. In early childhood people begin to demand that others accord them respect. The respect of others is necessary if people are to acquire self-respect.

It is this kind of demand that lies so close to the surface of the Declaration of Independence, in the statement that all men are created equal and in the specific indictments of British rule—for example, in the great indignation expressed at the quartering of troops in private homes without the consent of the homeowners, and at the removal of trials at law from the Colonies to England. This demand is evident in the Declaration's "decent respect to the opinions of mankind," whose approval the American rebels sought.

And once these successive needs—the physical, the social-affectional, and the equal esteem or dignity needs—are sufficiently gratified, humans are not even then content: they then begin to look for that kind of activity that is particularly suited to them as unique individuals. Whether their competence is to be a ditchdigger, a powershovel operator, a construction foreman, a civil engineer or a building contractor, an architect, a mother, a writer, or a politician—they must do these things when they have become rather sure in the gratification of their even more basic physical, social, and esteem needs.

The crucial point is this: no human being so long as he lives is ever completely gratified in the satisfaction of his needs. Up to the moment of his death, he must eat and sleep, he must be with people; he has to be acknowledged as a distinct person; and he must realize his individual potential. When he ceases to do these things, he ceases to live. All of these needs of his have got to be gratified; they ultimately can be denied only by natural or by violent death. Armies and police forces can quash these natural and irrepressible human needs only by reducing human beings to animals and then killing them. The logic of this was stated in fictional form by George Orwell, in describing what was necessary for the perpetuation of dictatorship: "a boot, stamping in a human face, forever."

The Maslow need hierarchy is a necessary part of a psychological explanation of the causes of revolution. Marx to the contrary, revolutions are made not only by economically depressed classes and their leaders but by the joint effort of large numbers of those people in all social groups who are experiencing frustration of different basic needs. People deprived of career opportunities may join in revolt with people who have suffered indignities at the hands of employers, landlords, police, or military troops. They also may join with people who have suffered no indignities but are for the moment simply hungry.

The common characteristic of potential revolutionaries is that each of them individually senses the frustration of one or more basic needs and each is able to focus his frustration on the government. After this need frustration is generated, people begin to share their discontents and to work together. But preceding this joint action, there is no more conspiracy than there is among trees when they burst into flame during a forest fire.

The J-Curve and Particular Revolutions

On the level of general theory, one can say precisely the same thing—in abstract terms—about each revolution and rebellion. But in some ways each revolution is unlike every other revolution. And from the practical research standpoint, directly comparable data are not available for all revolutions, particularly when they took place decades or centuries ago. In many nations now, the seeds of revolution are sprouting. But established governments in these nations are not likely to welcome social scientists in search of data by conducting public opinion surveys inquiring about attitudes toward the government.

In the interest of arriving at some conclusions and of arriving at the understanding that they are tentative, we can profitably consider particular revolutions.

The French Revolution of 1789

The French Revolution is the first of the great modern postindustrial revolutions. It is the first grand revolution after that grandest of all modern revolutions, the 16th-century Protestant Reformation.

The position of the various major social classes in France gives a major clue as to how the revolution came about. The relationship between these classes helps explain also why liberty, equality, and fraternity did not arrive on the day they were declared to be human rights. The major segments of French society in the late 18th century were the well-known three estates: the clergy, the nobility, and then everybody else, who collectively were called the third estate for lack of a more precise term.

What is less well known are the proportions that each of these estates comprised of the total French population of about 23 million. There were, according to George Lefebvre, perhaps the greatest historian in 20-century France of the French Revolution, about 100,000 Frenchmen in the clergy (less than half of 1 percent); about 400,000 in the nobility (about 2 percent); and over 23 million in the third estate. The third estate included the high bourgeoisie, an economically, socially, and politically active group of merchants, bankers, and manufacturers. Also in the third estate were the petty bourgeoisie—small merchants, bankers, artisans in wood and metal, and the growing body of skilled government bureaucrats. Finally, the third estate included workingmen, many of them the sons of peasants, and also the vast body of peasants. France, beginning its industrialization somewhat later than England, was still overwhelmingly an agricultural nation.

Even less well known than the proportions of each of the three estates in the total population is the proportion of land which each estate owned. Again, according to Lefebvre, the clergy owned about 10 percent of the land, the nobility somewhat less than 20 percent, the bourgeoisie about 20 percent, and the peasants all the rest. The heavy imbalance of landownership reflects only the most evident part of the land-tenure picture. Anywhere from 20 to 75 percent of the rural households in France before the revolution did not own any land. These peasants were either working as tenants to save money to buy land, or they had given up and were working as paid farm laborers. And in massive numbers they were drifting into cities to find work.

Those peasants who aspired to landownership or who had achieved it saw themselves as facing an unending struggle to survive and to get a little ahead. As the industrial economy began to develop rapidly, the demand for farm products increased because so many people who once tilled and lived off the soil now worked for money in cities, which—as everywhere in developing nations undergoing industrialization—made ever-increasing, insatiable demands on the countryside to feed their people. Peasants, seeing the chance thus to move up the ladder from farm

The J-Curve of Rising and Declining Satisfactions

labor or land rental, were beset by a variety of inhibitors. There were the feudal dues (payment to landlords for the use of his flour mill, the exclusive right of landlords to hunt and fish, the reversion of land to a landlord if the peasant died without proper heirs, etc.); the duty to perform physical labor for public purposes (building public roads and other structures); the tithe (a 10 percent tax due to the church); and a variety of taxes payable directly to the national government through its local representatives and more specifically to the local collector, who took his lawful share of what he was able to extract from the peasant.

Peasants did not, in short, believe that they were beloved objects of solicitude of other segments of French society. Neither did their sons who went into the cities to work in factories and small shops or into mines to dig coal. Wages went up slowly in the 18th century, as we shall see later, and prices went up rather rapidly. Better off than when they left the countryside, they were nevertheless gradually getting worse off than they had been in years past in the city.

That portion of the bourgeoisie containing skilled artisans suffered some of the same taxing pains as did the landowning peasants. Their guilds were heavily taxed and so were their incomes. The high bourgeoisie, growing in wealth and power, suffered the disadvantage that the more systematically they ran their enterprises and kept record of profits and losses, the more they had to pay in taxes. And they believed the government was becoming increasingly subservient to the nobility.

The nobility saw the government as increasingly subservient to the bourgeoisie. With no respect for the dignity of inherited title, the government for a price was adding pseudonoble titles to wealthy men of no family, arrogantly designating these arrivistes as "nobles de robe" to distinguish them from the natural-born "nobles d'épée." The old nobles observed the new nobles buying country estates from increasingly vestigial but still very sworded noblemen. The nobles of the robe were enfolding, smothering, the nobles of the sword.

Old nobles, looking through dusty old documents, discovered a way to be with but not be of the modern mercantile-industrial world. They found that services and payments in kind were due them from peasants, many of whom had for centuries been free peasants. (Serfdom was first abolished, according to Tocqueville, in Normandy in the 13th century and was virtually nonexistent on the eve of the great 18th century revolution.) In short, landlords, seeing their economic advancement, their political power, and their prestige all threatened and actually diminished by the energetic and of course unhonorable bourgeoisie, began in the mid-18th century to reassert long-dead "rights" against peasants, who thereby saw not only their freedom, power, and prestige but also their economic welfare threatened and diminished.

The sworded nobles furthermore had little to do—few, if any functions in society to give their lives meaning. The government gradually was taking away—efficiently, effectively, and thoroughly—such governmental powers as nobles had possessed before kings could successfully establish national power. They were no longer needed to keep the peace, to adjudicate disputes among vassals and serfs, and most particularly to protect from violence their people, their peasants, their onetime serfs. Now the government acted, or tried or professed to act directly in the behalf of the population at large. So the old nobles, sensing their loss of position in society as the new nobles of the robe began to emulate the sword-bearing style of life, began to emulate the new nobles in their wealth. And this meant evading such taxes as the capitation, by law payable by all people with incomes. It also meant using documents to enforce feudal dues, in many cases centuries after the reason for the dues had been reduced to legal paper, which now was hard to read and harder to justify.

The clergy, that one-half of 1 percent of the population, had a few functions to perform. They kept records of births and deaths. They baptized. They warned souls of the need for grace and invoked God's grace. And they prayed, managed estates, bottled wine, and extracted the tithe from the peasants. And, for their recordkeeping and their divine intervention, they were freed of any tax payments.

These then were the major segments of society, each of which eyed every other segment and its members with suspicion and envy. Was that a tax collector coming? Was that a secret hoard of grain which that peasant or that landlord so hurriedly covered up? Where was that set of books of the merchant that were a true report of how rich he was getting? How much did that fat father pay the government to get his bright but unprincipled son the job as secretary to the resident government commissioner?

If individuals in each estate tended to suspect and envy individuals in all other estates, they all mistrusted and condemned the government. The monarchy asserted in the mid-15th century (during the reign of Charles VII) the power to tax anyone without the consent of any estate. In the 18th century, the crown was intermittently, and more frequently, beginning to use the power. It had to. France engaged in a nearly unbroken series of expensive wars in the 18th century. Good for members of all estates, as businessmen, landowning lords and peasants and the small class of workers, the wars were bad for businessmen and peasants as taxpayers. Starting in 1781, the government increased—but with an infirm hand—its efforts to collect taxes, demanding even that the nobility actually pay the taxes nominally due from it.

The government was thus disappointing the popular expectation of continued prosperity without cost. And until the French intervention in the American War of Independence, the wars were lost. The intervention in America gave France pride in somewhat vicariously defeating England, which by 1763 had virtually knocked France out of North America except for Louisiana. The financial crisis—which threatened and actually deprived high bourgeoisie, nobles, and now even clergy of wealth they had come to expect as their due—got worse. Inflation intensified. Lefebvre has calculated the rise in the cost of living thus: in about 50 years before the revolution, prices went up some 65 percent and wages went up some 22 percent. Whether rich or poor, most people had enough excuse to displace at least some of their inter-estate hostility onto the government. And in addition they had reason enough also to dislike and condemn the government, which either lost its wars or was unable to pay for the one war that it assisted in winning.

These growing tensions, increasingly directed toward government, were aggravated by events that amounted to bad luck at best and gullibility at worst. In 1786 France made a trade treaty with the England it had helped to defeat 3 years earlier in America. France agreed to reduce the tariff on textiles, which helped the then more efficient and mass-market-oriented English mills. In return, England agreed to reduce tariffs on wines and brandies, which England did not produce anyhow but imported from Portugal and France. The trade treaty went into effect in 1788.

In 1787, the French harvest of grain suffered from bad weather. That is, the weather was bad for grain and good for the vineyards. There was thus a nearly catastrophic shortage of grains for bread and a large surplus of wine to flood the English market after the lowering of the trade barriers. And the opening of war of Turkey against Russia and Austria diminished these countries as markets for French textiles, which now faced competition in French stores from cheap English cloth.

Unemployment rose along with the reduced demand for textiles. The abundant grape harvest dropped wine prices somewhat. Peasants who produced wine had to buy their food at higher prices. These economic dislocations chain reacted to reduce the demand for everything but jobs and bread—the two goods that were in scarcest supply. Bread had never been so expensive since the end of the reign of Louis XIV in 1715, and so bread riots broke out in the major cities, and people in the cities began restlessly roaming out into the countryside to get food. Long-term rising expectations of a prosperous and peaceful economy and effective government were quickly disappointed. In the spring of 1789 and into the summer, the growing interclass hostility and growing hostility to government quickly burst into revolution, when the fear of physical deprivation quickly became real and immediate. The 18th-century developments are shown in figure 19-2.

The J-curve helps explain the French Revolution. The growing frustration of the land expectations of peasants, of the dignity expectations of landlords who wanted the status-wealth of the high bourgeoisie, and the dignity and power expectations of the high bourgeoisie are all closely comparable to developments in other nations that have had revolutions. And so is the effect of sudden economic dislocation following long-term economic growth.

But the J-curve is not a total explanation of the French Revolution. At least in its intensity, the interclass hostility in France, not as such related to the J-curve, was unique. More or less independently of frustrated rising expectations in the 18th century, French society was already deeply fragmented. The internal war of all against all had already begun. The absorption of these forces, in conflict between classes and individuals, did temporarily deflect them from the government. But in the end the sheer hostility, as it became more intense, turned toward the government.

And in addition to the disappointed expectations and deflected interclass hostility, there was in France in 1789 the visible and exciting example of the successful American revolution. In 1968 French students followed the example of students in Japan, America, and elsewhere by rebelling against French universities and government. But this was only the second time the American revolutionary example had been followed: it had already been followed almost 180 years before. . . .

The Black Rebellion of the 1960's

The black uprisings in America in the 1960's clearly amount to a rebellion, but they are not in any precise sense a revolution like those of France in 1789 or Russia in 1905 and 1917. However, the differences between these revolutions and the black rebellion are largely quantitative. The latter involves a widespread joint commitment to rather fundamental change among all segments of Negro society in the country: change in the political power structure of the country in all political units—cities, counties, states, and nation. And these changes, involving all blacks and all parts of the political community, have been accompanied by the violence that is a universal element of revolution.

The differences between revolution and the black rebellion derive from several facts: Negroes constitute only about 11 percent of the national population and therefore are numerically incapable of enforcing changes to the same degree as in a nation where a substantial majority is frustrated by the established government. The constitutions and laws of the national and most state governments have not contained many restrictions that discriminate against people on racial grounds—quite the contrary.

So there has been no basic change in principles. The Constitution and law of the land have been used or developed in ways that make them instruments to achieve changes in the nonlegal social practices and customs of discrimination. And there has been a sufficiently developed sense of commitment to equality as a major social value to make the equalizing of opportunities for black people a process against which most whites could not readily fight. They could not readily deny the applicability of their principles to those who newly have demanded equality. And that portion of the nation which frankly accepted the principle of racial inequality lost the savage Civil War. The very slow struggle for racial equality and human dignity for blacks commenced with Lincoln's 1863 Emancipation Proclamation and the military defeat of the South 2 years later.

What is striking, in a comparison of this (and other) contained rebellions with the great revolutions, is that the Negro rebellion appears to have been preceded by the same J-curve of expectations that are at first gratified and then frustrated. The same reaction patterns of this level of analysis appear to have developed in the minds of American blacks as have developed in the minds of those who have become revolutionaries in previous eras and other countries.

The difficulty in seeing this likeness relates to the vast gap between what whites and blacks have gotten in America. This gap has made it hard to see just what advances blacks have made and when. Those who as blacks and whites believe in equality have emphasized the vast and continuing inequalities. Black or white, they do not see what advances have been made. In the 1960's, when conditions were better than in the 1860's or than in the 1930's, the expressions of discontent have been at their maximum. The new words and deeds of discontent can be understood only if one appreciates that profoundly deprived people are often incapable of expressing their discontent. In short, to understand why the black rebellion has occurred, it is necessary to see how black people had already developed.

At the end of the Civil War, Negroes were perhaps as near to minimal survival in the psychological sense, as human beings, as they had been since their initial transportation from Africa. They had lost the security of provision for food, clothing, shelter, and physical safety that had been fairly well assured them as long as they docilely accepted their position as slaves. They could no longer be sure that the master would provide for them. They had, often, to forage for themselves, like war refugees everywhere when crops have been destroyed and normal patterns of collaboration in productive work have been shattered. Overjoyed at their emancipation, they could use their freedom no more effectively than could concentration camp inmates in Germany when the doors at last swung open in early 1945. They could concern themselves really with only the satisfaction of their physical needs, which freedom is not and equality and dignity are not.

The J-Curve of Rising and Declining Satisfactions

Those who must concentrate only on survival usually do not revolt: they are too hungry. This preoccupation simply with staying alive if anything strengthened in the late decades of the 19th century as the practice of lynching—killing by mobs rather than by lawfully or other systematically employed force—continued. Between 1882, when records of lynchings were first kept, down to 1941, lynchings averaged 78 per year. The constant fear that one might be arbitrarily killed, maimed, or injured was one of the day-to-day facts of life for most blacks, particularly until the early 1920's. Lynching and physical injury could be said to have declined to a relatively minor worry—comparable perhaps to the level of worry about automobile accidents in the 1960's—in the late 1930's and the 1940's: the average for 1937–42 was five per year and for 1943–48 less than three per year. But the level of general health remained low and so did life expectancy.

The process of moving up off the even, flat plane of survival itself was of course continuous. But it was so slow that it seems best to date the first major upturn, from concern for mere survival for most blacks, at the beginning of the Second World War. Responding to the threat of a large demonstration, a repetition of something akin to the 1932 veterans' march on Washington, Franklin Roosevelt in 1941 issued an Executive order prohibiting discriminatory hiring practices in all defense industries, and establishing the Fair Employment Practices Commission (FEPC) to administer the order. Though it worked unevenly and in many cases not at all, it nevertheless was a major basis for advance above subsistence for Negroes. By war's end, some 2 million blacks were employed in war industry, and the FEPC reported that 1,300,000 of these had gotten jobs in consequence of its efforts.

What could have been a cataclysmic frustration of rising expectations for blacks at war's end turned out not to be. The successful efforts to avoid a postwar recession, which would have witnessed the old (and still common) practice of discharging Negroes first, benefited blacks as well as whites. There was no widespread and sudden drop in Negro employment. Instead, the pace of rising economic opportunity continued. In 1946 the CIO and AFL trade union organizations undertook a drive to organize Negro workers in the South and to integrate them into existing unions. By 1948, FEPC legislation had been passed in six states, taking up some of the slack when the ending of war contracts removed the protection of the wartime FEPC. Symbolically, and a bit more, the first Negro was admitted in 1947 to major league baseball, Jackie Robinson; there were 14 major league Negro players in 1951; by 1954 all but 3 of the 16 major league teams were integrated. Racially integrated low-cost public housing after the war began the breakdown of discrimination in this basic concern of life. In 1956 all public housing in Washington, D.C., was desegregated. In 1962 President Kennedy issued an order prohibiting discrimination in any housing that was either financed or had mortgage insurance under a Government program. It was estimated that this affected a fourth of all future housing construction in the nation.

These advances relate to jobs and housing and therefore to the physical needs, but they also—notably in the case of sport participations—have overtones of equal dignity. Advances that more directly related to this profound, non-physical need for equality included the following:

> The admission of Negroes into the category of commissioned officer: 500 Negro officers in the Army in 1943, 7,500 by war's end; and 28 officers in the Navy in 1944, 50 by war's end;
>
> The integration of 90 percent of all Negro army personnel into unsegregated units by 1953 and complete integration a year later;
>
> The first desegregation of interstate buses in 1946, of railway dining cars in 1950, and of railway passenger cars in 1952;
>
> The long series of steps designed to desegregate education, commencing with the court order to the University of Oklahoma in 1948 to admit on a segregated basis a graduate student who was black, to the University of Texas in 1950 to admit on a non-segregated basis a Negro to the law school, down to and beyond the landmark 1954 case which ordered the integration of public secondary and primary schools "with all deliberate speed";
>
> The similarly long series of steps to end discrimination in the voting process, starting with the court invalidation in 1944 of the white primary closed to blacks and continuing with the 1954, 1964, and later civil-rights acts, which in-

creasingly protected and enforced the right of blacks to register and vote in all elections.

The range and number of national and state legislative and judicial and administrative efforts to see that black people were accorded equal dignity is very large indeed. Repeatedly in the 1940's, 1950's and early 1960's it gave evidence to Negroes that progress was being made. Their expectations inevitably rose from the near-ground level before the second World War to what proved increasingly to be excessively optimistic. Acts of legislatures, court, and administrative agencies—and of private groups and citizens—to equalize life opportunities for black people have never quite fulfilled their initial purpose. This brings us to the matter of promise and performance, to assessment of the gap between the expectations aroused by legislation, Executive order, and court decision, on the one hand, and realization of equality, on the other.

The killings by lynch mobs dwindled to one case in 1947 and two in 1948. A new kind of killing of blacks began and at times something like the old lynch mob operated again. In 1952 a top state NAACP official in Florida who organized a campaign to secure the indictment of a sheriff charged with killing a Negro prisoner was killed by a bomb. After the 1954 commencement of public school integration, there were some 530 cases of violence (burning, bombing, and intimidation of children and their parents) in the first 4 years of integration. Schools, churches, and the homes of black leaders were bombed and many people were killed in these bombings. Federal troops were brought into Little Rock in September 1957 to integrate the high school; during the following school year (1958–59), public schools were closed in Little Rock.

In short—starting in the mid-1950's and increasing more or less steadily into the early 1960's—white violence grew against the now lawful and protected efforts of Negroes to gain integration. And so did direct action and later violence undertaken by blacks, in a reciprocal process that moved into the substantial violence of 1965–67. That 3-year period may be considered a peak, possibly the peak of the violence that constituted the black rebellion. It was violence mostly against white property and black people. It merits reemphasis that during this era of increased hostility, progress continued to be made. Indeed, the occurrence of some progress intensified both the white reaction to it and the black counteraction to the reaction, because every time a reaction impeded the progress, the apparent gap widened between expectations and gratifications.

Direct (but not violent) action by Negroes began in late 1956 with the bus boycott in Montgomery, Alabama, which endured for over a year and succeeded. It was precipitated when a Negro woman got on a city bus, sat down in a front seat, was ordered to give up her seat to a white man, and refused. The bus boycott soon came under the leadership of Dr. Martin Luther King, Jr., whose belief in nonviolent resistance—and the mild temper of blacks in Montgomery at the time—succeeded in keeping the action relatively peaceful.

Direct violent action began in April 1963 in Birmingham, Alabama, in what may be called the first full-scale concerted violent encounter of blacks and whites in recent years. Seeking integration of such facilities as lunch counters, parks, and swimming pools, the blacks in Birmingham, most of them young, were met with water hoses, police dogs, and violent acts of police and white people. The numbers of demonstrators increased to some 3,000 and there were 1,000 arrests. The repressiveness of the police united a hitherto divided black community in Birmingham. And it produced perhaps the first major case since the second World War in which Southern blacks threw rocks and bottles at police. From this time on, violence deepened and spread among blacks. The Birmingham riots immediately touched off a response in other cities—according to one estimate, 758 demonstrations in the 10 weeks following the Birmingham violence. And in 6 weeks of that 1963 summer, blacks (in Birmingham and elsewhere) succeeded in getting some 200 lunch counters and other public facilities desegregated.

The combined effect of substantial, though slow, progress in employment, housing, education, and voting did not have the effect of quieting blacks or stopping the Negro rebellion of the 1960's. The full-fledged riots of Los Angeles in 1965 and Newark and Detroit in 1967 have been amply studied, at least from the descriptive viewpoint. But there is a tendency to see these events in isolation. It is recognized that riots in one place will touch off riots in another or—more likely—in several others, but the social-contagion theory (including the contagion of seeing African nations liberated after the Second World War) by no means gets to the roots of the rebellion. And neither does the notion that blacks are frustrated and are striking out rather blindly at the centuries of repression. If 300 years of repression have been too much, why were 200 or 280 not enough to produce rebellion?

What is striking is the time sequence of events. As in major historic revolutions, the events relating to the 1960's rebellion consist of a rather long period of rising expectations followed by a relatively brief period of frustration that struck deep into the psyches of black people. And I suggest that from the 17th to the early 20th century there has been very little development beyond mere physical survival for virtually all black people in America (and in Africa). It is significant to note that in the prosperous 1960's, there was no sharp or sudden rise in unemployment of blacks. There was no market deprivation of material goods to which blacks had become accustomed. But there was, starting notably in 1963, not the first instance of violence against blacks but a sudden increase in it. This resurgence of violence came after, and interrupted, the slow but steady progress since 1940. It quickly frustrated rising expectations.

This increase in violence, commencing so to speak with the firehoses and police dogs in May 1963 in Birmingham, affronted not only the physical safety of the demonstrators, thereby reactivating anxiety and fear of bodily harm itself—the most basic of human concerns. This increase of violence also affronted the dignity of black people as human beings. Black people sensed that their various and continuously rising expectations, now confronted with violence, were to rise no more.

In addition to this violence between whites against blacks and of blacks against whites, there has been an explosive growth of private acts of violence of blacks against blacks. This has newly activated the fear for physical safety itself. And the ever-growing congestion in the slums has worsened housing conditions.

White people who fail to understand their own past and their own ever-rising expectations (if we have one car, we must have two; if we finished 4 years of college, our son must become a doctor or a lawyer) are puzzled at the dissatisfaction of blacks who have made such considerable progress since the Second World War. But what would be odd about blacks, and indicate that they indeed had some special nature, would be for them to be satisfied in present circumstances. The very rapidity of their advance makes them expect to continue its pace. The very low point from which they started makes them expect to reach equality within a few years or to the very most a few decades.

134 Their mental processes are operating in an altogether normal manner. They would be less than human if they acted otherwise.

Figure 19-5 and Table 19-1, the latter devised by Harmon Zeigler with the assistance of Jerry B. Jenkins, represent one striking index of the origin and time sequence of black frustrations. He chooses, as the items to form his index, average family income and average years of schooling. He divides the former by the latter, for the total U.S. population and for the nonwhite population (which is about 95 percent black), from

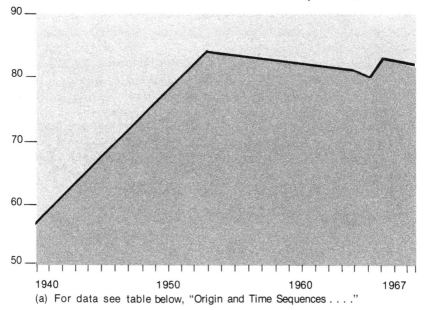

Index of Nonwhite Economic Satisfaction, 1940-67[a]

(a) For data see table below, "Origin and Time Sequences...."

1940 to 1967, using U.S. Census Bureau data. The increase and decrease in the gap between what an average family and a nonwhite family of given educational level get in income becomes Zeigler's measure of frustration.

If black and white workers with the same amount of education were earning the same income, there would be no difference in the indexes between the two categories of people. As the chart indicates, nonwhites were closest to earning the same amount as the total population in 1952. They rose from 58 percent in 1940 to 86 percent in 1952, but declined after 1952 to a low of 74 percent in 1962. They did not return to their relative status of 1940, but they lost substantial ground compared with where they were in 1952.

If the education-income relationship were the only one involved in producing frustration in people as their expectations and gratifications diverge, we could have expected a peak of unrest sometime in the mid-1950's. It came later—by my reckoning in 1963. This suggests that the gap formed from the increased incidence of violence on the part of police and white citizens provided the quantum of energy necessary to raise black frustrations to the point of rebellion. Figure 19.6 shows the developments.

Two ways are possible of resolving the problem that arises when the expectations and gratifications gap develops—and perhaps there are others. One way to close the growing gap is to attempt to deprive blacks in America of all the gains that they have made since at least the beginning of the Second World War. These gains have been mostly in the satisfaction of their physical needs (in jobs and housing); their social and their dignity needs are beginning to gain prominence. In George Orwell's phrase, we may call this the technique of the boot stamping in a human face forever. If white people were to attempt and even succeed in so reducing black people to a life that consisted of trying to stay alive—the life they lived under slavery and, most of them, for two to four generations after emancipation—black rebelliousness could be contained. In the process white people would be reduced to the same animal-like behavior that they themselves were imposing on blacks, just as concentration camp guards and concentration camp inmates came to resemble each other in appearance and behavior.

A second way to resolve the problem is to recognize and help them to satisfy their expectations, which fundamentally are the expectations which degraded white people in decades and centuries long past have themselves achieved—notably the recognition of their equal dignity and worth. It is not to be supposed or hoped that black people then will at last become satisfied, any more than white people who achieve dignity become satisfied. But at least those blacks who have achieved dignity will then be that much closer to becoming fulfilled human beings, able at last to realize themselves in the climate of self-respect that is necessary for people to grow.

Student Rebellions of the 1950's and 1960's

These rebellions seemingly commenced first among university students in Japan and then began in the United States and in Western and Central Europe. Partly because of space limitations, it is impossible to assess the J-curve as a device for explaining these particular rebellions. Part of the reason for deferring an attempt to do so lies in the continuing complexity of related events.

It is true, for example, that living quarters and classrooms for French students were getting progressively worse as the postwar French wave of newborn children reached university age. Expectations rose and were frustrated, contributing to the May 1968 riots. It is true that the 1964 Berkeley riots followed soon after enforcement of a previously unenforced rule against on-campus fund solicitation for off-campus organizations. Expectations had risen and were frustrated. And it is true that American university students who were raised with expectations of a bright future have seen their hopes disappointed when faced with military service in Vietnam. These explanations may be the central ingredients of the rebellious mood.

But these phenomena are not quite new. Students have previously been crowded and otherwise disappointed. What is new is the occasional use of violence. The reasons for resort to this technique are not clear. The amount of violence—as distinguished from nonviolent direct action, which people tend incorrectly to read as violence—is not great but it is real. When black people, who have been the victims of violence for centuries, use violence, it is comprehensible. When white students use it, the reasons remain obscure enough to cause at least this writer to postpone an attempt to explain.

Some Tentative Conclusions

We have seen that the J-curve is a necessary though not sufficient part of the explanation of at least several revolutions and some contained rebellions. This J-curve is a psychological, not a sociological explanation. The units of analysis are individual human beings. They may fall into visible categories (like blacks or students or working men or peasants), but their mental processes that relate to frustration and aggression are fun-

Origin and Time Sequences of Black Frustrations

Col. 1: $\dfrac{\text{Social want satisfaction}}{\text{Social want formation}}$ = systematic frustration

Col. 2: $\dfrac{\text{Social want satisfaction (nonwhite)}}{\text{Social want formation (nonwhite)}}$ = frustration (nonwhite)

Col. 3: Index of satisfaction (nonwhite)

Col. 4: Index of frustration (nonwhite)

	1	2	3	4
	Average family income divided by average years of schooling for—		Nonwhite satisfaction	Nonwhite frustration
Year	Total population	Nonwhite population	Nonwhite percentage of total population frustration level (100 percent would represent equality of want satisfaction relative to want formation between nonwhites and total)	Percentage difference between nonwhite population and that of total population (derived by subtracting col. 3 from 100 percent)
1940[a]	$\dfrac{\$1{,}231}{8.4} = 146.3$	$\dfrac{\$489}{5.8} = 84.3$	$\dfrac{84.3}{146.3} = 57.5$	42.5
1947	$\dfrac{\$3{,}031}{9.0} = 336.8$	$\dfrac{\$1{,}614}{6.9} = 233.9$	$\dfrac{233.9}{336.8} = 69.4$	30.6
1950	$\dfrac{\$3{,}319}{9.3} = 356.9$	$\dfrac{\$1{,}869}{6.8} = 274.9$	$\dfrac{274.9}{385.1} = 77.0$	23.0
1952	$\dfrac{\$3{,}890}{10.1} = 385.1$	$\dfrac{\$2{,}338}{7.1} = 329.3$	$\dfrac{329.0}{385.1} = 85.5$	14.5
1957	$\dfrac{\$4{,}971}{10.6} = 469.0$	$\dfrac{\$2{,}764}{7.7} = 359.0$	$\dfrac{359.0}{469.0} = 76.5$	23.5
1960	$\dfrac{\$5{,}620}{10.6} = 530.2$	$\dfrac{\$3{,}233}{8.2} = 394.3$	$\dfrac{394.3}{530.2} = 74.4$	25.6
1962	$\dfrac{\$5{,}956}{11.4} = 522.5$	$\dfrac{\$3{,}330}{8.6} = 387.2$	$\dfrac{387.2}{522.5} = 74.1$	25.9
1964	$\dfrac{\$6{,}559}{11.7} = 560.6$	$\dfrac{\$3{,}839}{8.9} = 431.3$	$\dfrac{431.3}{560.6} = 77.0$	23.0
1967			78.9	20.2

Frustration index and tables by H. Zeigler with the help of J. B. Jenkins. Reprinted unedited.

1940 income figures are actually for 1939, and are for families and unrelated individuals.
SA: 1952. (73d ed.), p. 111: median school years (1947, 1957, and 1964).
SA: 1965 . (86th ed.), p. 112: median school years 1960.
SA: 1966 . (87th ed.), p. 340: median family income: (1947, 1950, 1952, 1957, 1959, 1960, 1962, and 1964).

Revolution

138

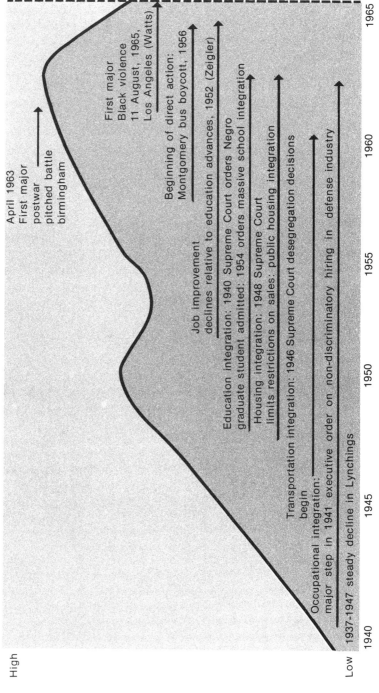

damentally the same. That is, we are positing that anyone deprived of food—whether his normal circumstances include the simple diet of poor people or the elaborate one of rich people—will suddenly become inclined to break any social convention to get food. We are also supposing that anyone who is physically secure in the provision of food, clothing, health, and physical safety will seek to establish and strengthen social ties and then to seek equal dignity. The demand for these things is so profound that constitutions and laws have to be made to adapt to the demands—not the demands to constitutions and laws.

If the ever-emerging expectations of people are gratified without too much resistance by those whose similar expectations have already been gratified, then revolution and rebellion are unlikely. If they are not, orderly political processes are displaced by violence. It was that way with our ancestors; it is that way now. And it is that universal a phenomenon. Lawmakers as well as clerks, businessmen as well as laborers, professors as well as students would react the same if suddenly deprived of the goods and dignity they had come to expect in the normal course of life. They would be less than human if they too did not become angry.

References

REFERENCES ON THE BLACK REBELLION

There are excellent sources of bibliography in such primary sources as the *Report of the National Advisory Commission on Civil Disorders*. Data used here came from the *Encyclopaedia Britannica Book of the Year* (1938-68) and from the U.S. Census Bureau's *Current Population Reports*.

REFERENCES ON THE FRENCH REVOLUTION

Ralph W. Greenlaw, ed., *The Economic Origins of the French Revolution,* Boston: D.C. Health, 1958. A very useful collection of essays.

Georges Lefebvre, *The Coming of the French Revolution* (1939), New York; Vintage Books, 1957. Probably the best single volume to date on the causes of the French Revolution.

Alexis de Tocqueville, *The Old Regime and the French Revolution* (1856), New York: Doubleday Anchor Books, 1955. This hundred-year-old classic is still the best starting point for studying origins of the French Revolution.

GENERAL REFERENCES ON REVOLUTION AND REBELLION

Crane Brinton, *The Anatomy of Revolution* (1938), New York: Vintage Books, 1955. This oft-revised classic is the best introduction to the study of revolution.

James C. Davies, "Toward a Theory of Revolution," *American Sociological Review*, vol. XXVII (Feb. 1962), pp. 5-19. This is the original statement of the J-curve.

James C. Davies, *Human Nature in Politics*, New York: Wiley, 1963. The first two chapters develop a politically oriented theory of human needs; the concluding chapter relates to growth processes that relate to revolution.

James C. Davies, "The Circumstances and Causes of Revolution," *Journal of Conflict Resolution*, vol. XI (June 1967), pp. 247-257. A critical view of the state of knowledge about revolution in the mid-1960's.

John Dollard et. al., *Frustration and Aggression*, New Haven: Yale University Press, 1939. This is the initial and still basic statement of the frustration-aggression nexus, necessary for understanding revolution psychologically.

Harry Eckstein, "On the Etiology of Internal Wars," in George H. Nadel, ed., *Studies in the Philosophy of History*, New York: Harper Torchbooks, 1965, pp. 117-147. Like the Laqueur essay below, a good brief starting point.

Frantz Fanon, *The Wretched of the Earth*, New York: Grove Press, 1968. The most intimate analysis of the roots of political violence that I have seen. Based on the author's psychiatric work with patients in Algeria who had experienced or used violence during the final years before Algeria's independence.

Walter Laqueur, "Revolution," *International Encyclopedia of the Social Sciences*, vol. 13, New York: Macmillan, 1968, pp. 501-507. A good, brief starting place.

O. Mannoni, *Prospero and Caliban: The Psychology of Colonization*, New York: Praeger, 1964. A brilliant study of the tensions resulting from the confrontation of individuals from an advanced cultural state with those from an undeveloped condition. The effects on both the dominant and the subordinate individuals are elegantly analyzed.

Abraham H. Maslow, "A Theory of Human Motivation," *Psychological Review*, vol. L (July 1943), pp. 370-396. The first statement of the hierarchy of basic needs, a theory that is very helpful in understanding why people varying from poor to rich, educated and uneducated, can join together to overthrow a government.

Mancur Olson, Jr., "Rapid Growth as a Destabilizing Force," *Journal of Economic History*, vol. XXIII (Dec. 1963), pp. 529-552. An economist's statement of what happens when an economy causes people to get dislocated and less poor because the economy is developing rapidly. Applies to some of the effects not only on American blacks but also on developing nations.

Thomas F. Pettigrew, "Social Evaluation Theory: Consequences and Applications." in David Levine, ed., *Nebraska Symposium on Motivation*, Lincoln: University of Nebraska Press, 1967, pp. 241-315. A synthesis of research and theory relative to how people see each other in the eyes of others. If self-esteem is indeed not separable from others' esteem of self, then this is a major theoretical contribution. It applies to the black rebellion particularly and to the mental outlook of potential revolutionaries, anywhere and any time.

George Wada and James C. Davies, "Riots and Rioters," *Western Political Quarterly,* vol. X (Dec. 1957) pp. 864-874. Based on riots that took place in the relocation centers for Japanese in this United States during the Second World War, this study distinguishes the riot participants from the non-participants on grounds of the social marginality of the former.

Social Revolution as Seen by Bourgeois Ideologists
by Y. Krasin

'The revolution became a *fact*,' Lenin wrote when, in the summer of 1905, the revolutionary movement in Russia was on the rise. 'It was no longer necessary to be a revolutionary to acknowledge the revolution.' The revolution became a much-discussed topic and even spokesmen of the liberal bourgeoisie had to recognize it. 'They do so,' Lenin noted, 'not because they are revolutionaries, but despite the fact that they are *not* revolutionaries' (*Two Tactics of Social-Democracy in the Democratic Revolution, Coll. Works*, Vol. 9, pp. 125-126).

The above suggests the reasons for the bourgeois ideologists' increased interest in the problems of social revolution today. The world revolutionary process is of such a depth and breadth that no one can ignore it.[1]

Today it is not only capitalist ideologists but also politicians who speak of 'revolution.' In his State of the Union message, delivered on January 22, 1971, President Nixon called for 'opening the way to a new American revolution, a peaceful revolution.' What Nixon means by a 'new American revolution' is a certain extension of state and city powers, chiefly greater state and municipal budgets. Similar statements are made by the leaders of other capitalist countries. They use the term 'revolution' to give weight to insignificant reforms of a system of government upholding monopoly capitalist interests.

We will deal with some current trends of bourgeois social thought as reflected in writings published on the threshold of the seventies.

Pointless Abstractions

'The bourgeoisie's recognition of the revolution,' Lenin wrote, 'cannot be sincere, irrespective of the personal integrity of one bourgeois ideologist or another' (op. cit.). One recalls this statement as one reads the numerous works of bourgeois sociologists and politologists. Bourgeois thinking distorts the nature and laws of social revolution beyond recognition. It often reduces the very concept to a pointless abstraction.

The 'sociology of revolution' is one of the main trends of bourgeois social thought today. It arose at the turn of the century and broke with the democratic traditions of bourgeois political theories of the revolutionary period of the 17th and 18th centuries. Works by B. Adams, G. Le Bon, C. Ellwood, A. Bauer and others, who dreaded the approaching proletarian revolution, offered recipes for perpetuating the capitalist system. This trend became active as a result of the victory of the October Revolution and the emergence of socialism as a reality. L. Edwards, P. Sorokin, G. Pettee, C. Brinton and other authors, professing to make an unbiased investigation of social revolution, misrepresented its laws and virtually denied its historic role. G. Pettee wrote that 'Western' (that is, bourgeois) thinking had taken a sharp turn to anti-revolution.'[2] Bourgeois sociologists and politologists today carry on this line of their predecessors.

A favorite device used by 'classics' of the 'sociology of revolution' was to substitute artificial patterns for the laws of revolutionary development. Present-day 'experts' go much further. C. Brinton's *The Anatomy of Revolution*, brought out in several editions, contained—along with absurd deductions, such as the one about 'Thermidor' being inevitable—some correct ideas, reflecting the actual course of bourgeois revolutions of the past. Today's bourgeois sociologists and politologists often have no such link with reality and indulge in a pseudo-scientific playing with terms. Here are a few illustrations borrowed from recent books.

Why Men Rebel by Ted Robert Gurr, an American professor of politics, is concerned with the cause of political violence, including revolutions. The key point of his concept is 'relative deprivation,' or RD, which he defines as a perceived discrepancy between men's value expectations and their value capabilities. 'Value expectations,' he writes, 'are the goods and conditions of life to which people believe they are rightfully entitled. Value capabilities are the goods and conditions they think they are capable of attaining or maintaining, given the social means available to them. Societal conditions that increase the average level or intensity of expectations without increasing capabilities increase the intensity of discontent.' Furthermore, 'the greater the intensity of discontent, the more likely is violence.' These trivial contentions, which divert one from investigation of the causes of revolution, make up the substance of Gurr's concept.

The text, almost 400 pages long and illustrated with graphs and formulas that are evidently expected to lend the book academic respectability, bristles with banalities such as these: 'The potential for collective violence varies strongly with the intensity and scope of relative deprivation among members of a collectivity.' 'Any decrease in the average level of value capabilities in a collectivity without an accompanying decrease in value expectations increases the intensity of relative deprivation.' 'Regime coercive control varies strongly with the loyalty of coercive forces to the regime.'[3]

Gurr's would-be scientific speculations are no exception in bourgeois sociology. 'The higher (lower) the social want formation in any given society and the lower (higher) the social want satisfaction,' write J. and R. Feierabend, two American sociologists, 'the greater (the less) the systematic frustration and the greater (the less) the impulse to political instability.'[4] This vague thesis is supplied with numerous formulas and tables. It all looks like bona fide research. In reality neither the thesis nor the far-fetched set of arguments trotted out in support of it throws any light on the causes of revolutions or other forms of socio-political contest between social forces.

Just as pseudo-scientific is *A Study of Revolution* by Peter Calvert, a British bourgeois sociologist. Take, for example, the formula 'quantification of revolution'—'$Rn^2 - Bm^2$' where n and m respectively stand for the numbers of revolutionary and government forces, and R and B for their respective 'fighting values.' This is unscientific and useless for the student of real processes.

The salient characteristic of the book, as indeed of other books on 'revolution,' is probably the effort to class a vast number of insignificant political coups under the head of 'social revolution.' Calvert puts the October Revolution on the same footing with the military coups in Latin America. Surely it is not difficult to substantiate any deduction by using arbitrarily selected data as does Calvert, saying that 363 revolutions' took place between 1901 and 1960. All these 'sociological' exercises could be dismissed as a curious fact had they not pursued a very definite class aim—distorting the meaning of social revolution. Revealingly, the author's preface says that the more this concept has been used, 'the vaguer its meaning has become.'[5]

A tendency to submerge the concept of social revolution in loose definitions is evident in a series of investigations carried out by the Center of International Studies at Princeton University and recorded in the book *Internal War*. Its editor, Prof. H. Eckstein, defines 'internal wars' as 'attempts to change by violence, or threat of violence, a government's policies, rulers, or 'organization.'[6] This definition covers almost any government crisis, even one followed by a negligible change of policy. The author lumps together real revolutionary changes, coups d'état and cabinet reshuffles.

Deliberately vague definitions of revolution are occasionally used with an eye to robbing it of all positive meaning. Things are carried to a point where the distinction between revolution and counter-revolution disappears. 'The dividing line,' J. Meisel writes, 'has become blurred, almost invisible. ' A revolution begins with the demand for radical change and ends in the 'conservative' stage of 'consolidation.' 'Counter-revolution turns into another revolution.'[7] The objective purpose of mixing up concepts, as bourgeois sociologists do, is to lump together progressive and regressive changes so as to reject all change.

In defining the nature of revolution, they do not generally spell out its economic or class aspects. They say nothing about change in property forms or the transfer of political power to the new, rising class. This is inevitable since bourgeois sociology generally approaches society from an abstract standpoint.

The meaning and function of social revolution can be understood, provided the history of society is seen as the development and succession of socio-economic formations. Social revolution is a form of transition from one formation to another. Its economic purpose is to end antiquated property relations hampering social progress. In an antagonistic society, this is done through class struggle. The issue of power is central to any social revolution and transfer of power to the progressive class is the content of the decisive, political stage of a social revolution.

The class struggle may go through periods of retreat of the revolutionary classes, restoration and counter-revolution, and political compromise and fresh offensive. This is why a social revolution often extends over a long period and is not accompanied by one but several political revolutions each of which defeats the tendency of the reactionary classes towards restoration and completes what was begun by the previous revolution. Disregard of this dialectic of social revolution is the fundamental methodological drawback of bourgeois concepts, which do not help one in studying revolutionary processes, any more than in understanding their role and function in social development.

'Modernization' Instead of Revolution

Another typical tendency of bourgeois sociology and politology is to substitute technological for social revolution. It is on this methodological basis that various 'modernization' theories are conceived.

Prof. R. Tucker of the USA reduces the Marxist-Leninist theory of revolution to an aspect of the more general theory of 'modernization.' According to this theory, Tucker writes, 'what is most important in Marx is not his theory of the decline and fall of capitalism. Rather, it is his account of the rise and early development of bourgeois society.'[8] It will be seen that bourgeois professors are willing to acknowledge the importance of Marxism, but only insofar as it defines the role of capitalism in 'modernizing' society. They flatly reject the theory of the socialist revolution. Tucker makes Leninism out to be a version of the 'modernization' theory for economically backward countries. A. Said and D. Collier reason along similar lines.'[9] These views are spearheaded against the theory of socialist revolution. They virtually deny that it has a social content of its own and present it as a means of accomplishing the historical task fulfilled by capitalism in the West. From what bourgeois sociologists say, the purpose of the social revolution is not to effect the transition to socialism but to end economic backwardness.

The effort to present the vanguard of modern social revolution, the Communist movement, as a political current typical of economically underdeveloped countries is prompted by the 'doctrine of backwardness.' 'The countries with large Communist parties, however,' writes Prof. S. Lipset of Harvard, 'remain among the less modernized of the big nations.'[10] The idea is clear—contrary to the facts, the author would like the Communist movement to be seen as an exponent of working-class interests at only a relatively early stage of capitalist development. Bourgeois ideologists claim that this movement loses its social roots as industrial progress goes on. They seem to forget that mass Communist parties are gaining in membership and strength today and that this is also the case in many economically developed countries.

'Modernization' in this sense rules out social revolution. The latter is robbed of its subject and hence becomes impossible. Social evolution is regarded as the only method of change in a developed 'industrial society.'

The author of another version of the 'modernization' theory, which misinterprets the content of the present epoch, is Prof. M. Halpern of Princeton University. In his opinion. the great revolutions of the past (meaning the French bourgeois revolution and the October Socialist Revolution) were mere 'symptoms of a more profound, underlying revolution of modernization.'

Now what is 'modernization' supposed to mean? Due to the high rate of social development, Halpern writes, 'man and all the psychological systems by which he has so far organized his life, are persistently being rendered incoherent. Elements are being destroyed and the linkages between them disconnected.' The need is to combine stable social links with rapid and constant changes. 'The revolution of modernization,' Halpern maintains, 'for the first time raises the opportunity and need . . . to create systems which derive their stability from their intrinsic capacity to generate and absorb continuing transformation.'[11] It follows that the important thing is not the socio-economic content of systems but a mechanism capable of reacting promptly to scientific and technological changes.

Zbigniew Brzezinski performs the same operation by substituting a technological content for the social content of the present epoch. He describes the meaning of the changes taking place in the world, with all the difficulties and problems attending them, as emerging from the industrial age to enter the age of 'technetronics.' 'Today,' he writes, 'the most industrially advanced countries, in the first instance the United States, are beginning to emerge from the industrial stage of their development. They are entering an age in which technology and especially electronics—hence my neologism "Technetronic"—are increasingly becoming the principal determinants of social change, altering the mores, the social structure, the values, and the global outlook of society.'

Brzezinski says that technology and electronics directly determine changes in the social structure and superstructure of society. In this way he dissociates the system of production relations, the economic basis of society, from the mechanism of social progress. Through this arbitrary operation he removes the objective criterion of class distinctions and socio-economic systems. Technology and electronics are supposed to progress outside existing social class relations and the struggle of social forces, including the ideological sphere, which expresses class interests.

Modern technology and electronics, Brzezinski says, confront people everywhere with common problems. This is why 'the concern with ideology is yielding to a preoccupation with ecology' and a 'new global consciousness' comes into being.[12] In Brzezinski's view, technology overshadows every sphere of social being and consciousness. Technological progress is becoming the main factor and substance of current history. Processes lose their social class content.

To be sure, the technological revolution raises a number of new problems, including those of mastering new control systems, regulating the growing flow of information and protecting the natural environment. What man needs at this stage is, among others, a flexible technological economic mechanism that can effectively assimilate scientific and technological changes. However, success in accomplishing this task hinges primarily on socioeconomic changes and the establishment of socialist forms of social organization.

Yet it is only from the standpoint of technological progress that bourgeois ideologists tend to view capitalist contradictions today and the prospects of solving them. Everything is reduced to the distinction between 'industrial' and 'post-industrial' society. The former is said to be affected with all the evils of capitalism while the latter is alleged to be free from them. One finds these ideas, for example in *The Year 2000* by Herman Kahn, director of the Hudson Institute, and Anthony J. Wiener, one of his associates.

The two authors attribute the inherent contradictions of capitalism not to capitalism itself, but to industrialism, to the fact that the economy absorbs the whole energy of society. 'Post-industrial society' relieves all its members of material cares and de-emphasizes the economic criterion, so that genuinely human values are regenerated. As a result, there would be no more 'obvious "contradictions" from which revolution presumably arises.'[13] Presenting capitalism, even if it has made economic progress as against the previous stage, as a society free from contradictions is at best wishful thinking or utopia. However, it would be more accurate to describe that as a misrepresentation of capitalism.

The temptation to blame the technological revolution for the evils of capitalism is great. R. Wielder, an American sociologist, alleges that rapid industrial progress contributes to the spread of a spiritual malady—more and more people suffer from a lack of purpose. Technological progress, while leading to important improvements, creates psychological difficulties, causes dissatisfaction and anxiety and arouses the desire to find a scapegoat.[14]

In other words, Wielder traces the roots of general dissatisfaction, political instability and revolutionary sentiment to technological advance.

A. Toffler's *Future Shock* is one of the latest sociological bestsellers. Its author advocates what he calls the 'super-industrial revolution.' In his view, mankind today is faced with the unprecedented problem of social flux. The scientific and technological revolution is challenging man's very ability to adapt to a fast-changing situation. This is the main reason for the 'future shock' at the root of crisis phenomena in American society.

Toffler strips the crisis of contemporary capitalism of its social meaning. He contends that as the 'super-industrial revolution' goes on the very concept of property is becoming meaningless and the conflict between capitalism and communism is reduced to 'comparative insignificance.' In other words, the social revolution, as Toffler sees it, is losing its significance, being entirely superseded by the 'super-industrial revolution.'[15]

Arthur Schlesinger, the American historian, follows the same line, affirming that our age breeds problems irrespective of property system or ideology. 'If the crisis seems today more acute in the United States than anywhere else,' he writes, 'it is not because of the character of our economic system; it is because the revolutions wrought by science and technology have gone farther here than anywhere else. As the nation at the extreme frontier of technological development, America has been the first to experience the unremitting shock and disruptive intensity of accelerated change. The crises we are living through are the crises of modernity. Every nation, as it begins to reach a comparable state of technical development, will have to undergo comparable crises.' Schlesinger sees in unrest, confusion and even violence the price that has to be paid for progress, the 'birth-pangs of a new epoch in the history of man.'

This argument mixes up different aspects of the problem.

Social Revolution as Seen by Bourgeois Ideologists

Indeed, the United States is ahead of other capitalist countries for many technological revolution indices and U.S. experience will be valuable to countries taking the same road. It is the country where the contradictions that plague capitalism in the technological revolution are more obvious and more pronounced than anywhere else. Hence, its socio-political crises require closer study if the processes that sooner or later will involve other developed capitalist countries are to be understood. In this sense, the historical experience of present-day America is instructive for a whole zone of the globe and undoubtedly sheds light on the contradictions of our epoch.

On the other hand, contrary to Schlesinger's view, the United States is not leading in social progress. This is indirectly admitted by Schlesinger himself, who writes that the factors brought into being by the technological revolution 'are placing increasing strain on inherited ideas, institutions and values.' He notes that to solve the crisis 'means social justice as well as racial justice; it means a far broader measure of participation in our great organizations and institutions; it means the determination to enable all Americans to achieve a sense of function, purpose and potency in our national life.'[16] But surely this is impossible without radical socio-economic changes, without a social revolution. The scientific and technological revolution, far from eliminating this necessity, makes it imperative by aggravating capitalism's contradictions.

Flight From Reality

Theories expressing the views and sentiments of radical petty bourgeois, intellectuals and students have gained some currency in bourgeois sociology and politology. The concept of Herbert Marcuse, the American sociologist, is fairly well known. It sharply critizes modern capitalism—the methods it uses to manipulate people's consciousness and to impose on society consumer values, a conformist atmosphere and spiritual emptiness. However, the ideology of petty-bourgeois radicalism can grasp neither the dialectics of the world revolutionary process nor the historic role of social forces, first of all the international working class and socialism as a reality, which blaze new trails.

Because he denies the revolutionary character of the proletariat and the gains of socialism, Marcuse cannot grasp the objective content of the social revolution of our age. Profound pessimism is typical of petty-bourgeois radicalism. The road to liberation is the 'negation of the entire Establishment,' which has enslaved man. But how is this to come about if the mass forces opposed to capitalism have, according to Marcuse, become victims of their own 'repressive' instinctive requirements and values and have been neutralized? Marcuse withdraws from the political sphere to the domain of aesthetic values accessible to a narrow intellectual elite. He urges apolitical protest in 'surrealist forms.' He makes 'revolution' conditional on forming 'the human sensibility which rebels against the dictates of repressive reason, and, in doing so, invokes the sensuous power of the imagination.'[17] Surely this flight from the real class struggle to the realm of illusion means complete surrender to modern capitalism. Any objective analyst is bound to come to this conclusion.

Marcuse's ideas are carried forward in Charles Reich's theory, which may be called the 'revolution of consciousness.' 'There is a revolution coming,' he writes. 'It will not be like revolutions of the past. . . . It will not require violence to succeed, and it cannot be successfully resisted by violence.' Its task is to destroy the power of the 'Corporate State.' In describing the power system of this state, Reich in effect castigates state-monopoly capitalism. He vividly portrays modern bourgeois society's lack of ideas, its spiritual emptiness, the deep crisis of the American way of life. 'In this world, to one beginning a life, there are no open roads for the body, the mind, or the spirit, only long, hard, paved freeways to nowhere.' Reich reveals the decay of the vital values under capitalism, which arouses in people one passion only—the thirst for gain—and cultivates philistine, indifferent, apolitical attitudes.

The inference is evident: there is a need for far-reaching socioeconomic changes that will alter the course of social development and put the gigantic resources of scientific and technological progress at the service of society and not the state-monopoly elite. But Reich draws a different conclusion. He calls for a revolution in consciousness, which means that every individual must change his perception of reality and his world outlook. The new consciousness would influence society, not 'by direct political means, but by changing culture and the quality of individual lives, which in turn change politics and, ultimately, structure.' Reich believes consciousness to be the 'moving part of the engine of change.' And this is how a 'new revolution' that 'will not be like revolutions of the past' will come about, or so Reich thinks.

Of course, consciousness plays a tremendous role in social revolution. In stressing this, Lenin said that without a revolutionary theory there can be no revolutionary movement. Thus the question is of consciousness as a reflection of the conditions of the life and struggle of the progressive class and a generalization of its historical activity on the international scene. Yet to Reich the new consciousness is the unformed expression of a spontaneous rebellion of young people against the 'Corporate State.' In essence this is not even consciousness but a behavior rejecting the habitual way of life and making an anarchist challenge to the system: extravagant clothes, pop music, the ability simply to live 'one's own life according to one's own needs.' 'The new generation,' Reich writes, 'must make their revolution by the yeast theory; they must spread their life.' He goes so far as to describe even marijuana as a 'maker of revolution,' saying that 'it takes people outside the enclosed system,' releases them from domination of their thought and 'makes *unreal* what society takes most seriously.'

It is noteworthy that theorists of Left radicalism and even some liberal bourgeois ideologists regard this point as the most vulnerable in Reich's concept. Commenting on it, Marcuse wrote that the 'goal cannot be attained by an ostrich policy.' T. Hayden of the 'New Left' remarked that 'to believe a system will change simply through a new "Consciousness" is utopian. There is no change possible without program, organization, struggle and conflict.' John Galbraith, for his part, thinks Reich's ideas may become 'a cover for self-indulgent idleness.'[18]

Advanced social consciousness only becomes revolutionary in intimate connection with the class struggle, with politics. It cannot of itself alter existing social relations. To do this, it must become the consciousness of the social force which its objective situation in society enables to sweep away an outworn order and clear the ground for the new system through revolutionary action. Consciousness is a factor of struggle. Reich, however, suggests keeping out of struggle and politics. He says that political contest 'is a fight on the enemy's ground, a fight which the Corporate State is sure to win.'

Just as Herbert Marcuse retreats before the alleged insurmountability of the conformist trends of 'consumer' capitalism, so Charles Reich bows before the alleged omnipotence of the 'Corporate State' machine. Reich pins all his hopes on the prospect of this machine destroying itself, failing automatically. 'It is our theory,' he writes, 'that the State itself is now bringing about its own destruction. The machine itself has begun to do the work of revolution.'

The fundamental error of Reich as a Left radical theorist is that he sees no class forces or interests in today's bourgeois state. To his mind, this state operates by the force of its own inertia, as it were, and is hostile to all social strata. This makes the class struggle pointless, there being no enemy to fight against. 'We,' he writes, 'are all employees of the Corporate State and, what is more, exploited employees who sacrifice ourselves, our environment and our community, for the sake of irrational production. There is no class struggle; today there is only one class. In Marx's terms, we are all the proletariat, and there is no longer any ruling class except the machine itself.' This approach clearly rules out social revolution in its scientific meaning. Nor is it accidental that, in speaking of the new consciousness, Reich implies society as a whole, ignoring class distinctions. Rid of its secondary trappings, his concept may be put as follows: people must be told that the existing society is poorly organized, and when they have realized this everything will change of itself.

Petty-bourgeois radicalism, rejecting the Marxist-Leninist theory of revolution, takes us back to naive and utopian, pre-Marxist theories that have long been refuted by the historical experience of the revolutionary movement.

As for the practical program put forward by Reich, it boils down to the idea of refusing to submit to the 'Corporate State.' The whole system, he writes, 'collapses the moment people refuse to obey. . . . When Americans refuse to buy what the State wants to sell—economically or politically—and refuse to produce by striving for the goals set by the State's organizations, the wheel will have no power to turn, and revolution by consciousness will have become a reality.' Such is the logic of petty-bourgeois radicalism, which is divorced from political reality and rejects struggle based on knowledge of the objective logic of development of the class society.

To this we may reply that the monstrous and heartless

machine so vividly described by Reich serves and submits to the interests of monopoly capital. What is needed to resist it is not the isolated efforts of individuals who have come to realize its injustice but a mass organized political army of revolutionary forces capable of effecting far-reaching economic, social and political changes. Such an army can only be brought into being, consolidated and trained for decisive action in the course of practical political struggles, which is what Communists call for and what experience has shown.

There is virtually nothing revolutionary in Charles Reich's appeal 'to work for the betterment or transformation of society through almost any job, even though the job is located right within the Establishment. [19] What can be achieved by operating only within the system of state-monopoly capitalism is, at best, partial reforms. It can only amount to a slight refurbishing and modernization of the system of capitalist exploitation and will bring about no revolutionary changes, which require an all-out offensive by the masses against the Establishment from without. It is only in close connection with preparations for this offensive that the activity of progressive forces within the Establishment has any positive meaning.

Petty-bourgeois radical concepts, while outwardly revolutionary, are a far cry from revolution. They fully transplant the impulse for revolutionary change from the sphere of material production and social class relations to that of pure consciousness. Revolution is made the mission of a thinking elite. The masses, who are in the final analysis the only motor of social revolution, are debarred from it.

Left radical concepts of revolution are not a step forward but a big step back in social thought. They may have some progressive significance only as a theoretical reflection of the search for a path to the scientific revolutionary ideology by sizable sections of the radical intelligentsia.

Lenin stressed that 'revolution is a profound, difficult and complex science' (Vol. 27, p. 198). Mastering it in theory and practice is no easy task. Marxists have to accomplish this task through grim ideological and political struggles. Bourgeois sociology tries to refute the Marxist-Leninist science of revolutionary transformation of the world by resorting to vague abstractions or travesty of revolution, and often launching an open attack. Both theoretically and in terms of practical policy, sound criticism of bourgeois concepts from Marxist-Leninist positions is essential if the revolutionary forces of today are to accomplish the tasks facing them.

Notes

1. Dozens of books on social revolution have been published in the West in recent years. Here is a sampling: C. Johnson, *Revolutionary change,* 1966; J. H. Meisel, *Counter-Revolution, How Revolutions Die,* 1966; *Revolution,* Ed. by C. J. Friedrich, 1967; R. Wielder, *Progress and Revolution,* 1967; R. C. Tucker, *The Marxian Revolutionary Idea,* 1969; T. R. Gurr, *Why Men Rebel,* 1970; P. Calvert, *A Study of Revolution,* 1970; C. A. Reich, *The Greening of America,* 1970; R. Nungesser, *Pour une nouvelle société, La révolution qu'il faut faire,* 1970; *When Men Revolt and Why,* Ed. by C. Davies, 1971; *Towards Revolution, The Revolutionary Reader,* Vol. I-II, 1971; *Revolution, Readings in Politics and Society,* Ed. by K. Kumar, 1971; A. Said and D. Collier, *Revolutionism,* 1971. In addition there appear new editions of such 'classics' of the 'sociology of revolution' as L. P. Edwards, *The Natural History of Revolution,* and C. Brinton, *The Anatomy of Revolution.*
2. *Revolution,* Ed. by C. J. Friedrich. New York, 1967, p. 30.
3. T. R. Gurr, *Why Men Rebel,* New Jersey, 1970, pp. 13, 24, 60, 251.
4. *When Men Revolt and Why,* Ed. by C. Davies, London, 1971, p. 236.
5. P. Calvert, *A Study of Revolution.* Oxford, 1970, pp. 43, VII.
6. H. Eckstein, *Internal War,* New York, 1964, p. 1.
7. J. H. Meisel, *Counter-Revolution, How Revolutions Die,* New York, 1966, pp. 31, 32.
8. R. C. Tucker, *The Marxian Revolutionary Idea,* New York, 1969, p. 108.
9. A. A. Said and D. M. Collier, *Revolutionism,* Boston, 1971, pp. 76-80, 84.
10. S. Lipset, *Revolution and Counterrevolution,* London, 1969, p. 235.
11. M. Halpern, *Journal of International Affairs,* 1969, No. 1, p. 57; *Revolution,* Ed. by C. J. Friedrich, New York, 1967, p. 183.
12. Z. Brzezinski, *Between Two Ages, America's Role in the Technetronic Era,* New York, 1970, pp. XIV, 61.
13. H. Kahn and A. J. Wiener, *The Year 2000,* New York, 1968, p. 379.
14. R. Wielder, *Progress and Revolution,* New York, 1967, p. 287.
15. A. Toffler, *Future Shock,* New York, 1971, pp. 186, 220-221.
16. *Newsweek,* July 6, 1970, pp. 29-30.
17. H. Marcuse, *An Essay on Liberation,* Boston, 1969, pp. 25, 30.

18. The Con III: The Controversy: The Critics Look at the Greening of America, New York, 1971, pp. 17, 22, 19.
19. C. A. Reich, *The Greening of America*, London, 1971, pp. 2, 138, 13, 245, 217, 219, 191, 223, 139, 228, 231, 234, 272.

# Chapter III
Revolutionary Guerrilla Warfare: The Development of Doctrine

Major General Edward Lansdale, one of the first modern American counter-insurgency experts, recently wrote:

> In the next people's wars in countries we are aiding, I hope that we wouldn't forget the political basis of such conflicts and mistakenly place our main reliance on military, police, and economic actions without recognizing that they are merely implements of political will.*

As the selections in this chapter will demonstrate, "politics" is the essence of the struggle. From Sun Tzu, writing 2½ millennia ago, political and psychological themes recur as the originating and controlling elements in armed struggle. This chapter provides selections from works by and assessments of prominent theorists of unconventional warfare, designed to sketch in rough outline the development of doctrine. (There is no real substitute, of course, for detailed study of the original works.)

The Art of War was written in China about 500 B.C. and was introduced to the West shortly before the French Revolution. It has had much influence on Chinese and Japanese military thought: Mao's theories, for example, draw heavily on Sun Tzu's work, which contains incisive insights into war in general and sets forth some basic premises of guerrilla warfare. These excerpts include illustrative remarks by later Chinese commentators.

Sun Tzu felt that the commitment of the army to battle should be a final act against an enemy virtually defeated by political and psychological means: "to subdue the enemy without fighting is the acme of skill." Flexibility of operations, reliable intelligence, psychological maneuvering, attack at vulnerable points of one's own choosing, stealth, ruse, surprise — all contribute to demoralizing and defeating the enemy. In one of Sun Tzu's most famous passages lay the seed, perhaps, of the strategy which was to characterize most of Mao's guerrilla campaigns:

> Now an army may be likened to water, for just as flowing water avoids the heights and hastens to the lowlands, so an army avoids strengths and weaknesses. And as water shapes its flow in accordance with the ground, so an army

manages its victory in accordance with the situation of the enemy. And as water has no constant form, there are in war no constant conditions.

Writing in a professional military journal, an American author attempts to distill the philosophy of Karl von Clausewitz as it bears upon the American role in Vietnam. W. D. Franklin suggests that Clausewitz's greatest contribution to military thought was to show that there can be no single set of rules by which victory can be assured, in part because "war can never be separated from political intercourse." Franklin claims that Clausewitz provided "the first serious attempt to blueprint the characteristics of irregular operations." Writing in the early 1800's, Clausewitz, recounting his experience in the Napoleonic wars, felt that a populace resisting an enemy did not present a "military" target and was thus virtually immune from conventional military attacks, although its position required that it "must only nibble at the surface and the edges."

Lenin's "On Partisan Warfare" first appeared in 1906. It is reprinted here with notes by Stefan T. Possony. It sets forth a basic operational doctrine for opposing the Czarist regime, one which Communist movements have often adapted for use elsewhere. Lenin justifies or rationalizes terrorist and criminal methods as necessary to combat the terror of the Czarist establishment; but he also condones it as an offensive tactic. Replying to critics, he argues that armed struggle will proceed through various phases and can take many forms. There was some degree of ambivalence on Lenin's part regarding the use of terror and "expropriation," which official Soviet policy since has often disclaimed. Yet Lenin's conclusions at the time of writing are clear.

The selections from the works of Mao Tse-tung include some of his most famous formulations. After arguing for guerrilla war against Japan he sets forth its main principles, including "the use of initiative, flexibility and planning in conducting offensives...battles of quick decision within protracted war...-coordination with regular warfare; the establishment of base areas; strategic defensive and strategic offensive; the development of guerrilla warfare into mobile warfare, and correct relationships of command." Contrasting the power and capacities of Japan and China, Mao concludes that only the use of China's peasant masses can defeat her militarily advanced invader, and this can be done only through a three-stage protracted war. Mao stresses the "politics" of the struggle: "there has never been a war that did not have a political character."

*   Edward Geary Lansdale, *In the Midst of Wars* (New York: Harper and Row, 1972), p. 374.

Selections from *Sun Tzu, The Art of War*
translated by Samuel Griffith

III. Offensive Strategy

Sun Tzu said:

1. Generally in war the best policy is to take a state intact; to ruin it is inferior to this.

Li Ch'üan: Do not put a premium on killing.

2. To capture the enemy's army is better than to destroy it; to take intact a battalion, a company or a five-man squad is better than to destroy them.

3. For to win one hundred victories in one hundred battles is not the acme of skill. To subdue the enemy without fighting is the acme of skill.

4. Thus, what is of supreme importance in war is to attack the enemy's strategy;

Tu Mu: . . . The Grand Duke said: 'He who excels at resolving difficulties does so before they arise. He who excels in conquering his enemies triumphs before threats materialize.'
Li Ch'üan: Attack plans at their inception. In the Later Han, K'ou Hsün surrounded Kao Chun. Chun sent his Planning Officer, Huang-fu Wen, to parley. Huang-fu Wen was stubborn and rude and K'ou Hsün beheaded him, and informed Kao Chun: 'Your staff officer was without propriety. I have beheaded him. If you wish to submit, do so immediately. Otherwise defend yourself.' On the same day, Chun threw open his fortifications and surrendered.

All K'ou Hsun's generals said: 'May we ask, you killed his envoy, but yet forced him to surrender his city. How is this?'

K'ao Hsun said: 'Huang-fu Wen was Kao Chun's heart and guts, his intimate counsellor. If I had spared Huang-fu Wen's life, he would have accomplished his schemes, but when I killed him, Kao Chun lost his guts. It is said: "The supreme excellence in war is to attack the enemy's plans." '

All the generals said: 'This is beyond our comprehension.'

5. Next best is to disrupt his alliances:

Tu Yu: Do not allow your enemies to get together.
Wang Hsi: . . . Look into the matter of his alliances and

cause them to be severed and dissolved. If an enemy has alliances, the problem is grave and the enemy's position strong; if he has no alliances the problem is minor and the enemy's position weak.

6. The next best is to attack his army.

Chia Lin: . . . The Grand Duke said: 'He who struggles for victory with naked blades is not a good general.'
Wang Hsi: Battles are dangerous affairs.
Chang Yü: If you cannot nip his plans in the bud, or disrupt his alliances when they are about to be consummated, sharpen your weapons to gain the victory.

7. The worst policy is to attack cities. Attack cities only when there is no alternative.

8. To prepare the shielded wagons and make ready the necessary arms and equipment requires at least three months; to pile up earthen ramps against the walls an additional three months will be needed.

9. If the general is unable to control his impatience and orders his troops to swarm up the wall like ants, one-third of them will be killed without taking the city. Such is the calamity of these attacks.

Tu Mu: . . . In the later Wei, the Emperor T'ai Wu led one hundred thousand troops to attack the Sung general Tsang Chih at Yu T'ai. The Emperor first asked Tsang Chih for some wine. Tsang Chih sealed up a pot full of urine and sent it to him. T'ai Wu was transported with rage and immediately attacked the city, ordering his troops to scale the walls and engage in close combat. Corpses piled up to the top of the walls and after thirty days of this the dead exceeded half his force.

10. Thus, those skilled in war subdue the enemy's army without battle. They capture his cities without assaulting them and overthrow his state without protracted operations.

Li Ch'üan: They conquer by strategy. In the Later Han the Marquis of Tsan, Tsang Kung, surrounded the 'Yao' rebels at Yüan Wu, but during a succession of months was unable to take the city. His officers and men were ill and covered with ulcers. The King of Tung Hai spoke to Tsang Kung, saying: 'Now you have massed troops and encircled the enemy, who is determined to fight to the death. This is no strategy! You should lift the siege. Let them know that an escape route is open and they will flee and disperse. Then any village constable will be able to capture them!' Tsang Kung followed this advice and took Yüan Wu.

11. Your aim must be to take All-under-Heaven intact. Thus your troops are not worn out and your gains will be complete. This is the art of offensive strategy.

12. Consequently, the art of using troops is this: When ten to the enemy's one, surround him;

13. When five times his strength, attack him;

Chang Yü: If my force is five times that of the enemy I alarm him to the front, surprise him to the rear, create an uproar in the east and strike in the west.

14. If double his strength, divide him.

Tu Yu: . . . If a two-to-one superiority is insufficient to manipulate the situation, we use a distracting force to divide his army. Therefore the Grand Duke said: 'If one is unable to influence the enemy to divide his forces, he cannot discuss unusual tactics.'

15. If equally matched you may engage him.

Ho Yen-hsi: . . . In these circumstances only the able general can win.

16. If weaker numerically, be capable of withdrawing;

Tu Mu: If your troops do not equal his, temporarily avoid his initial onrush. Probably later you can take advantage of a soft spot. Then rouse yourself and seek victory with determined spirit.

Chang Yü: If the enemy is strong and I am weak, I temporarily withdraw and do not engage. This the case when the abilities and courage of the generals and the efficiency of troops are equal.

If I am in good order and the enemy in disarray, if I am energetic and he careless, then, even if he be numerically stronger, I can give battle.

17. And if in all respects unequal, be capable of eluding him, for a small force is but booty for one more powerful.

Chang Yü: . . . Mencius said: 'The small certainly cannot equal the large, nor can the weak match the strong, nor the few the many.'

18. Now the general is the protector of the state. If this protection is all-embracing, the state will surely be strong; if defective, the state will certainly be weak.

Chang Yü: . . . The Grand Duke said: 'A sovereign who obtains the right person prospers. One who fails to do so will be ruined.'

19. Now there are three ways in which a ruler can bring misfortune upon his army:

20. When ignorant that the army should not advance, to order an advance or, ignorant that it should not retire, to order a retirement. This is described as 'hobbling the army'.

Chia Lin: The advance and retirement of the army can be controlled by the general in accordance with prevailing circumstances. No evil is greater than commands of the sovereign from the court.

21. When ignorant of military affairs, to participate in their administration. This causes the officers to be perplexed.

Ts'ao Ts'ao: . . . An army cannot be run according to rules of etiquette.
Tu Mu: As far as propriety, laws, and decrees are concerned, the army has its own code, which it ordinarily follows. If these are made identical with those used in governing a state the officers will be bewildered.
Chang Yü: Benevolence and righteousness may be used to govern a state but cannot be used to administer an army. Expediency and flexibility are used in administering an army, but cannot be used in governing a state.

22. When ignorant of command problems to share in the exercise of responsibilities. This engenders doubts in the minds of the officers.

Wang Hsi: . . . If one ignorant of military matters is sent to participate in the administration of the army, then in every movement there will be disagreement and mutual frustration and the entire army will be hamstrung. That is why Pei Tu memorialized the throne to withdraw the Army Supervisor; only then was he able to pacify Ts'ao Chou.

Chang Yü: In recent times court officials have been used as Supervisors of the Army and this is precisely what is wrong.

23. If the army is confused and suspicious, neighboring rulers will cause trouble. This is what is meant by the saying: 'A confused army leads to another's victory.'

Meng: . . . The Grand Duke said: 'One who is confused in purpose cannot respond to his enemy.'
Li Ch'üan: . . . The wrong person cannot be appointed to command. . . . Lin Hsiang-ju, the Prime Minister of Chao, said: 'Chao Kua is merely able to read his father's books, and is as yet ignorant of correlating changing circumstances. Now Your Majesty, on account of his name, makes him the commander-in-chief. This is like glueing the pegs of a lute and then trying to tune it.'

24. Now there are five circumstances in which victory may be predicted:

25. He who knows when he can fight and when he cannot will be victorious.

He who understands how to use both large and small forces will be victorious.

Tu Yu: There are circumstances in war when many cannot attack few, and others when the weak can master the strong. One able to manipulate such circumstances will be victorious.

27. He whose ranks are united in purpose will be victorious.

Tu Yu: Therefore Mencius said: 'The appropriate season is not as important as the advantages of the ground; these are not as important as harmonious human relations.'

28. He who is prudent and lies in wait for an emeny who is not, will be victorious.

Ch'en Hao: Create an invincible army and await the enemy's moment of vulnerability.
Ho Yen-hsi: . . . A gentleman said: 'To rely on rustics and not prepare is the greatest of crimes; to be prepared beforehand for any contingency is the greatest of virtues.'

29. He whose generals are able and not interfered with by the sovereign will be victorious. . . .

Wang Hsi: . . . A sovereign of high character and intelligence must be able to know the right man, should place the responsibility on him, and expect results.

Ho Yen-hsi: . . . Now in war there may be one hundred changes in each step. When one sees he can, he advances; when he sees that things are difficult, he retires. To say that a general must await commands of the sovereign in such circumstances is like informing a superior that you wish to put out a fire. Before the order to do so arrives the ashes are cold. . . .

To put a rein on an able general while at the same time asking him to suppress a cunning enemy is like tying up the Black Hound of Han and then ordering him to catch elusive hares. What is the difference?

30. It is in these five matters that the way to victory is known.

31. Therefore I say: 'Know the enemy and know yourself; in a hundred battles you will never be in peril.

32. When you are ignorant of the enemy but know yourself, your chances of winning or losing are equal.

33. If ignorant both of your enemy and of yourself, you are certain in every battle to be in peril.' . . .

VI. Weaknesses and Strength

Sun Tzu said:

1. Generally, he who occupies the field of battle first and awaits his enemy is at ease; he who comes later to the scene and rushes into the fight is weary.

2. And therefore those skilled in war bring the enemy to the field of battle and are not brought there by him.

3. One able to make the enemy come of his own accord does so by offering him some advantage. And one able to prevent him from coming does so by hurting him.

Tu Yu: . . . If you are able to hold critical points on his strategic roads the enemy cannot come. Therefore Master

Wang said: 'When a cat is at the rat hole, ten thousand rats dare not come out; when a tiger guards the ford, ten thousand deer cannot cross.'

4. When the enemy is at ease, be able to weary him; when well fed, to starve him; when at rest, to make him move.

5. Appear at places to which he must hasten; move swiftly where he does not expect you.

6. That you may march a thousand *li* without wearying yourself is because you travel where there is no enemy.

Ts'ao Ts'ao: Go into emptiness, strike voids, bypass what he defends, hit him where he does not expect you.

7. To be certain to take what you attack is to attack a place the enemy does not protect. To be certain to hold what you defend is to defend a place the enemy does not attack.

8. Therefore, against those skilled in attack, an enemy does not know where to defend; against the experts in defense, the enemy does not know where to attack.

9. Subtle and insubstantial, the expert leaves no trace; divinely mysterious, he is inaudible. Thus he is master of his enemy's fate.

Ho Yen-hsi: . . . I make the enemy see my strengths as weaknesses and my weaknesses as strengths while I cause his strengths to become weaknesses and discover where he is not strong. . . . I conceal my tracks so that none can discern them; I keep silence so that none can hear me.

10. He whose advance is irresistible plunges into his enemy's weak positions; he who in withdrawal cannot be pursued moves so swiftly that he cannot be overtaken.

Chang Yü: . . . Come like the wind, go like the lightning.

11. When I wish to give battle, my enemy, even though protected by high walls and deep moats, cannot help but engage me, for I attack a position he must succour.

12. When I wish to avoid battle I may defend myself simply by drawing a line on the ground; the enemy will be unable to attack me because I divert him from going where he wishes.

Tu Mu: Chu-ko Liang camped at Yang P'ing and ordered

Wei Yen and various generals to combine forces and go down to the east. Chu-ko Liang left only ten thousand men to defend the city while he waited for reports. Ssŭ-ma I said: 'Chu-ko Liang is in the city; his troops are few; he is not strong. His generals and officers have lost heart.' At this time Chu-ko Liang's spirits were high as usual. He ordered his troops to lay down their banners and silence their drums, and did not allow his men to go out. He opened the four gates and swept and sprinkled the streets.

Ssŭ-ma I suspected an ambush, and led his army in haste to the Northern Mountains.

Chu-ko Liang remarked to his Chief of Staff: 'Ssŭ-ma I thought I had prepared an ambush and fled along the mountain ranges.' Ssŭ-ma I later learned of this and was overcome with regrets.

13. If I am able to determine the enemy's dispositions while at the same time I conceal my own then I can concentrate and he must divide. And if I concentrate while he divides, I can use my entire strength to attack a fraction of his. There, I will be numerically superior. Then, if I am able to use many to strike few at the selected point, those I deal with will be in dire straits.

Tu Mu: ... Sometimes I use light troops and vigorous horsemen to attack where he is unprepared, sometimes strong crossbowmen and bow-stretching archers to snatch key positions, to stir up his left, overrun his right, alarm him to the front, and strike suddenly into his rear. ...

14. The enemy must not know where I intend to give battle. For if he does not know where I intend to give battle he must prepare in a great many places. And when he prepares in a great many places, those I have to fight in any one place will be few.

15. For if he prepares to the front his rear will be weak, and if to the rear, his front will be fragile. If he prepares to the left, his right will be vulnerable and if to the right, there will be few on his left. And when he prepares everywhere he will be weak everywhere.

Chang Yü: ... His force will be scattered and weakened and his strength divided and dissipated, and at the place I engage him I can use a large host against his isolated units.

Selections from *Sun Tzu, The Art of War*

16. One who has few must prepare against the enemy; one who has many makes the enemy prepare against him.

17. If one knows where and when a battle will be fought his troops can march a thousand *li* and meet on the field. But if one knows neither the battleground nor the day of battle, the left will be unable to aid the right, or the right, the left; the van to support the rear, or the rear, the van. How much more is this so when separated by several tens of *li*, or, indeed, by even a few!

> *Tu Yu:* Now those skilled in war must know where and when a battle will be fought. They measure the roads and fix the date. They divide the army and march in separate columns. Those who are distant start first, those who are near by, later. Thus the meeting of troops from distances of a thousand *li* takes place at the same time. It is like people coming to a city market.

18. Although I estimate the troops of Yüeh as many, of what benefit is this superiority in respect to the outcome?

19. Thus I say that victory can be created. For even if the enemy is numerous, I can prevent him from engaging. . . .

20. Therefore, determine the enemy's plans and you will know which strategy will be successful and which will not;

21. Agitate him and ascertain the pattern of his movement.

22. Determine his dispositions and so ascertain the field of battle.

23. Probe him and learn where his strength is abundant and where deficient.

24. The ultimate in disposing one's troops is to be without ascertainable shape. Then the most penetrating spies cannot pry in nor can the wise lay plans against you.

25. It is according to the shapes that I lay the plans for victory, but the multitude does not comprehend this. Although everyone can see the outward aspects, none understands the way in which I have created victory.

26. Therefore, when I have won a victory I do not repeat my tactics but respond to circumstances in an infinite variety of ways.

27. Now an army may be likened to water, for just as

flowing water avoids the heights and hastens to the lowlands, so an army avoids strength and strikes weakness.

28. And as water shapes its flow in accordance with the ground, so an army manages its victory in accordance with the situation of the enemy.

29. And as water has no constant form, there are in war no constant conditions.

30. Thus, one able to gain the victory by modifying his tactics in accordance with the enemy situation may be said to be divine.

31. Of the five elements, none is always predominant; of the four seasons, none lasts forever; of the days, some are long and some are short, and the moon waxes and wanes.

VII. Maneuver

Sun Tzu said:

1. Normally, when the army is employed, the general first receives his commands from the sovereign. He assembles the troops and mobilizes the people. He blends the army into a harmonious entity and encamps it.

Li Ch'üan: He receives the sovereign's mandate and in compliance with the victorious deliberations of the temple councils reverently executes the punishments ordained by Heaven.

2. Nothing is more difficult than the art of maneuver. What is difficult about maneuver is to make the devious route the most direct and to turn misfortune to advantage.

3. Thus, march by an indirect route and divert the enemy by enticing him with a bait. So doing, you may set out after he does and arrive before him. One able to do this understands the strategy of the direct and the indirect.

Ts'ao Ts'ao: . . . Make it appear that you are far off. You may start after the enemy and arrive before him because you know how to estimate and calculate distances.

Tu Mu: He who wishes to snatch an advantage takes a devious and distant route and makes of it the short way. He turns misfortune to his advantage. He deceives and fools the enemy to make him dilatory and lax, and then marches on speedily.

4. Now both advantage and danger are inherent in maneuver.

Ts'ao Ts'ao: One skilled will profit by it; if he is not, it is dangerous.

5. One who sets the entire army in motion to chase an advantage will not attain it.

6. If he abandons the camp to contend for advantage the stores will be lost.

Tu Mu: If one moves with everything the stores will travel slowly and he will not gain the advantage. If he leaves the heavy baggage behind and presses on with the light troops, it is to be feared the baggage would be lost.

7. It follows that when one rolls up the armour and sets out speedily, stopping neither day nor night and marching at double time for a hundred *li*, the three commanders will be captured. For the vigorous troops will arrive first and the feeble straggle along behind, so that if this method is used only one-tenth of the army will arrive. . . .

8. In a forced march of fifty *li* the commander of the van will fall, and using this method but half the army will arrive. In a forced march of thirty *li*, but two-thirds will arrive.

9. It follows that an army which lacks heavy equipment, fodder, food and stores will be lost.

Li Ch'üan: . . . The protection of metal walls is not as important as grain and food.

10. Those who do not know the conditions of mountains and forests, hazardous defiles, marshes and swamps, cannot conduct the march of an army;

11. Those who do not use local guides are unable to obtain the advantages of the ground.

Tu Mu: The *Kuan Tzu* says: 'Generally, the commander must thoroughly acquaint himself beforehand with the maps so that he knows dangerous places for chariots and carts, where the water is too deep for wagons; passes in famous mountains, the principal rivers, the locations of highlands and hills; where rushes, forests, and reeds are luxuriant; the road distances; the size of cities and towns; well-known cities and abandoned ones, and where there are flourishing orchards. All this must be known, as well as the way boundaries run in and out. All these facts the general must store in his mind; only then will he not lose the advantage of the ground.'

Li Ching said: '. . . We should select the bravest officers and those who are most intelligent and keen, and using local guides, secretly traverse mountain and forest noiselessly and concealing our traces. Sometimes we make artificial animals' feet to put on our feet; at others we put artificial birds on our hats and quietly conceal ourselves in luxuriant undergrowth. After this, we listen carefully for distant sounds and screw up our eyes to see clearly. We concentrate our wits so that we may snatch an opportunity. We observe the indications of the atmosphere; look for traces in the water to know if the enemy has waded a stream, and watch for movement of the trees which indicates his approach.'

Ho Yen-hsi: . . . Now, if having received instructions to launch a campaign, we hasten to unfamiliar land where cultural influence has not penetrated and communications are cut, and rush into its defiles, is it not difficult? If I go with a solitary army the enemy awaits me vigilantly. For the situations of an attacker and a defender are vastly different. How much more so when the enemy concentrates on deception and uses many misleading devices! If we have made no plans we plunge in headlong. By braving the dangers and entering perilous places we face the calamity of being trapped or inundated. Marching as if drunk, we may run into an unexpected fight. When we stop at night we are worried by false alarms; if we hasten along unprepared we fall into ambushes. This is to plunge an army of bears and tigers into the land of death. How can we cope with the rebels' fortifications, or sweep him out of his deceptive dens? . . .

12. Now war is based on deception. Move when it is advantageous and create changes in the situation by dispersal and concentration of forces.[2]

13. When campaigning, be swift as the wind; in leisurely march, majestic as the forest; in raiding and plundering, like fire; in standing, firm as the mountains. As unfathomable as the clouds, move like a thunderbolt.

14. When you plunder the countryside, divide your forces. When you conquer territory, divide the profits.

15. Weigh the situation, then move.

16. He who knows the art of the direct and the indirect approach will be victorious. Such is the art of maneuvering.

17. The Book of Military Administration says: 'As the voice cannot be heard in battle, drums and bells are used. As troops cannot see each other clearly in battle, flags and banners are used.'

18. Now gongs and drums, banners and flags are used to focus the attention of the troops. When the troops can be thus united, the brave cannot advance alone, nor can the cowardly withdraw. This is the art of employing a host.

> *Tu Mu:* . . . The Military Law states: 'Those who when they should advance do not do so and those who when they should retire do not do so are beheaded.'
>
> When Wu Ch'i fought against Ch'in, there was an officer who before battle was joined was unable to control his ardour. He advanced and took a pair of heads and returned. Wu Ch'i ordered him beheaded.
>
> The army Commissioner admonished him, saying: 'This is a talented officer; you should not behead him.' Wu Ch'i replied: 'I am confident he is an officer of talent, but he is disobedient.'
>
> Thereupon he beheaded him.

19. In night fighting use many torches and drums, in day fighting many banners and flags in order to influence the sight and hearing of our troops. . . .

20. Now an army may be robbed of its spirit and its commander deprived of his courage.[3]

> *Ho Yen'hsi:* . . . Wu Ch'i said: 'The responsibility for a martial host of a million lies in one man. He is the trigger of its spirit.'
>
> *Mei Yao-ch'en:* . . . If an army has been deprived of its

morale, its general will also lose his heart.

Chang Yü: Heart is that by which the general masters. Now order and confusion, bravery and cowardice, are qualities dominated by the heart. Therefore the expert at controlling his enemy frustrates him and then moves against him. He aggravates him to confuse him and harasses him to make him fearful. Thus he robs his enemy of his heart and of his ability to plan.

21. During the early morning spirits are keen, during the day they flag, and in the evening thoughts turn toward home.

22. And therefore those skilled in war avoid the enemy when his spirit is keen and attack him when it is sluggish and his soldiers homesick. This is control of the moral factor.

23. In good order they await a disorderly enemy; in serenity, a clamorous one. This is control of the mental factor.

Tu Mu: In serenity and firmness they are not destroyed by events. . . .

24. Close to the field of battle, they await an enemy coming from afar; at rest, an exhausted enemy; with well-fed troops, hungry ones. This is control of the physical factor.

25. They do not engage an enemy advancing with well-ordered banners nor one whose formations are in impressive array. This is control of the factor of changing circumstances.

26. Therefore, the art of employing troops is that when the enemy occupies high ground, do not confront him; with his back resting on hills, do not oppose him.

27. When he pretends to flee, do not pursue.

28. Do not attack his *élite* troops.

29. Do not gobble proffered baits.

Mei Yao-ch'en: The fish which covets bait is caught; troops who covet bait are defeated.
Chang Yü: The 'Three Strategies' says: 'Under fragrant bait there is certain to be a hooked fish.'

30. Do not thwart an enemy returning homewards.

31. To a surrounded enemy you must leave a way of escape.

Tu Mu: Show him there is a road to safety, and so create in his mind the idea that there is an alternative to death. Then strike.

Ho Yen-hsi: When Ts'ao Ts'ao surrounded Hu Kuan he issued an order: 'When the city is taken, the defenders will be buried.' For month after month it did not fall. Ts'ao Jen said: 'When a city is surrounded it is essential to show the besieged that there is a way to survival. Now, Sir, as you have told them they must fight to the death everyone will fight to save his own skin. The city is strong and has a plentiful supply of food. If we attack them many officers and men will be wounded. If we persevere in this it will take many days. To encamp under the walls of a strong city and attack rebels determined to fight to the death is not a good plan!' Ts'ao Ts'ao followed this advice, and the city submitted.

32. Do not press an enemy at bay.

Tu Yu: Prince Fu Ch'ai said: 'Wild beasts, when at bay, fight desperately. How much more is this true of men! If, they know there is no alternative they will fight to the death. . . .

Notes

1. Exchange of gifts and compliments was a normal preliminary to battle.
2. Mao Tse-tung paraphrases this verse several times.
3. Or 'of his wits', I am not sure which.

Clausewitz on Limited War
by W.D. Franklin

The strategic genius of Karl von Clausewitz is one that is often quoted, seldom read, and little understood. Many viewed "classical" strategy, as interpreted by Clausewitz, as a Kafkaesque landscape of violence and brutality, when, in actuality, it was a conscientious effort to translate the surrealistic tableau of war to the concrete reality of political intercourse.

Clausewitz was essentially a student of war, and after his death his collected works were published in 10 volumes, the first three of which contain his masterpiece *Vom Kriege* or *On War*. By the beginning of the 20th century, his influence had become so pervasive that his ideas, and even his phrases, had found their way into the thinking and writing of the general staffs of all the great armies of the world.

The greatest contribution which he made to military thought was to show there can be no single tactical pattern or strategic system by which victory can be insured. Much of the blame for the misunderstanding of Clausewitz must rest with those individuals who read his startling sentences out of their context and without the qualifications that invariably accompanied them.

Karl von Clausewitz was born in 1780 and entered the Prussian Army as an ensign in 1792. He served in the Rhine Campaign of 1793--94 and then entered the Berlin Military Academy in 1801. He served in the Prussian Army until the outbreak of the Russian campaign of 1812. He then transferred to the Russian Army, and, during Napoleon Bonaparte's retreat from Moscow, he negotiated the Convention of Tauroggen which led to the War of Liberation. During his long military career, he was present at numerous historic battles whose raw material he distilled into his strategic theories. In 1831 he died of cholera.

His penetrating analysis of the relationship of war and policy has never been excelled and is, perhaps, more important today than when first expounded. Many are familiar with his statement that ". . . War is an act of force, and to the application of the force there is no limit." The depressing result is that words like these have been construed as not only justifying ruthlessness in certain cases, but actually advocating it as the most natural form of warfare.

This form of completely unrestrained violence naturally fits into the theoretical framework of total thermonuclear conflict wherein destruction is the strategic object, but it is incompatible with a strategic concept of limited war. But was unrestrained violence the only alternative offered by Clausewitz? It is interesting to examine carefully exactly what Clausewitz meant when he indicated that war is a "continuation of political intercourse, a carrying out of the same by other means." He actually drew an almost perfect blueprint of modern day irregular operations, and his theories have exerted tremendous influence on almost all major strategic thought—both that of the West and of the Communists.

Clausewitz believed that war is a serious means to a serious end. It always arises from a political condition and is called forth by a political motive. It is, therefore, a political act. We have to think of war not as an independent thing, but as a political instrument.

No war is begun, Clausewitz thought, or at least no war should be begun if people acted wisely—without first finding an answer to the question of what was to be attained by and in war. War never breaks out suddenly, and its spreading is not the work of a moment. But we must sometimes choose war, and thus also make preparations beforehand, because peace is not always an acceptable answer.

War can never be separated from political intercourse. It has, to be sure, its own grammar, but not its own logic. Wars are, in reality, only the manifestations of policy itself.

Philosopher Immanuel Kant said that innocence is a splendid thing, only it has the misfortune not to keep very well and to be easily misled. In an age like ours, states, classes, passions, and interests clash in such confusion that war and not peace seems to be the natural order of things.

War has, nevertheless, always caught Americans unprepared intellectually, emotionally, and materially. This innocence, unfortunately, does not correlate with the harsh face of reality. Clausewitz goes out of his way to explain that war is not made with an abstraction, but with a reality. The advantage of a neo-Clausewitzian type of analysis is a preservation of that elusive quality of reality. There is no doubt that Clausewitz was a realist.

All civilizations—the Greek city-states no less than the Italian cities of the renaissance or the nation-states of Europe—have had the same task: to limit violence. The method of the second half of the 20th century is the differentiation between types of war. The fragmentation of the diplomatic field has a military equivalent in the diversity of wars possible in our time. "Theory has, therefore," says Clausewitz, "to consider the nature of means and ends."

With the possession of thermonuclear weapons and means for their delivery, victory, in one sense of the word, is no longer attainable. It is sometimes argued that limited war, which involves nuclear powers even indirectly, is impossible because each side, rather than lose, would expand the scope and character of the conflict until it would end in mutual nuclear destruction.

Victory in limited war is not gained by putting the existence of the opposing state—and our own too—in issue. It does not seek unconditional surrender. The aim of limited war is to stop the infringement upon our interests. The aim of war, according to its inception, is always supposed to be the overthrow of the enemy. Clausewitz claimed that this need not always imply the complete conquest of the enemy's country. He felt that, if our opponent is to do our will, we must put him in a position more disadvantageous to him than the sacrifice would be that we demanded.

If the aim of the military action is an equivalent for the political object, that action will, in general, diminish as the political object diminishes. The more the object comes to the front, the more this will be so. This explains how, according to Clausewitz, there can be wars of all degrees from one of extermination down to a mere state of armed observation.

Not every war admits of a complete decision and settlement. Discussion must contemplate a vast spectrum of violence—at one end, the destruction which one thermonuclear power may hurl at another, to the hard and bitter fighting now involved in southeast Asia. War does not consist in killing as many men as possible at the smallest cost, nor is it merely reciprocal slaughter. War's effect is more a killing of the enemy's courage than of the enemy's soldiers, but still blood is always its price.

The aim of the West is not simply to avoid war, but to do so without losing vital positions, without allowing the Communist giants to expand continuously into the rimlands. What combination of political and military means will prevent total war and allow an effective conduct of diplomacy under thermonuclear conditions?

The conduct of the conflict in Vietnam has followed a neo-

Clausewitzian strategy particularly suited to conditions of limited war. Clausewitz said that there were two things which, in practice, could take the place of the impossibility of further resistance as motives for making peace. The first was the improbability of success; the second an excessive price to pay for it. A war need not, therefore, always be fought out until one of the parties is overthrown.

Clausewitz applied himself to the question of how to influence the enemy's expenditure of strength—that is to say, how to raise for him the price of success. He concluded there were three special ways of directly increasing the enemy's expenditure of force. The first was invasion; the second was to direct enterprises preferably at those points which do the enemy the most harm; and third, and by far the most important to Clausewitz, was the wearing out of the enemy. The idea of wearing out in a struggle implies a gradual exhaustion of the physical powers and the will by the long continuance of action.

Dr. Henry A. Kissinger has pointed out that no conditions should be sought for which one is not willing to fight indefinitely, and that the side which is willing to outwait its opponent—which is less eager for a settlement—can tip the psychological balance whatever the outcome of the physical battle. In any concept of limited war, according to Dr. Kissinger, it is imperative to find a mode of operation and to create a psychological framework in which our impetuosity does not transform time into an enemy ally. Henceforth, patience and subtlety must be as important components of our strategy as power.

General William C. Westmoreland, US commander of military forces in Vietnam, says that our strategy in that conflict consists of three parts: sustained operations against Viet Cong main-force and North Vietnamese Army units in South Vietnam; support of the government of Vietnam's nationbuilding process, and the bombing campaign against military targets in North Vietnam.

According to General Westmoreland, "The enemy thinks in terms of protracted conflict." He gives a further indication of neo-Clausewitzian concepts of a wearing out of the enemy by adding, ". . . I am confident that we must gear ourselves for the long pull."

The first serious attempt to blueprint the characteristics of irregular operations was undertaken by Clausewitz. An inner front was added to the outer front. He points out that, although the influence on war of a single inhabitant is barely perceptible, the total influence of the inhabitants of a country in war is anything but imperceptible.

Clausewitz believed that a resistance so widely distributed is not suited to great blows requiring concentrated action in time and space. Its action, like the process of evaporation, depends on the extent of the surface exposed; the greater this is, the greater the contact. The conditions under which Clausewitz thought insurgency could become effective were that:

- The war is carried out in the interior of the country.
- It is not decided by a single catastrophe.
- The theater of war embraces a considerable extent of country.
- The national character supports the measures.
- The country is of a broken and inaccessible nature either from being mountainous, or by reason of woods and marshes, or from the peculiar mode of cultivation in use.

In sketching the operational limitations of the insurgent, Clausewitz observed that "a poor population accustomed to hard work and privation usually shows itself more vigorous and better suited to war." He draws a picture that could almost be viewed whole cloth as that existing in southeast Asia today. He illustrates the advantages and limitations of what he calls the "people's war" conducted within the inner boundaries of a nation. He felt that:

Masses of armed peasants cannot, and should not, be employed against the main body of the enemy's army, or even against any considerable forces; they must not attempt to crunch the core; they must only nibble at the surface and the edges.

It was Clausewitz' belief that armed peasants must seize the enemy's lines of communication and prey upon the vital thread by which his existence is supported.

Armed insurgents make the march of every small body of troops in a mountainous, thinly wooded, or otherwise difficult country become very dangerous, for at any moment the march may become an engagement. They should, like a kind of nebulous vapory essence, nowhere condense into a solid body; otherwise an adequate force can be sent to crush it. But it is neccessary that this mist should according to Clausewitz:

. . . gather at some points into denser masses and form

threatening clouds from which now and again a formidable flash of lightning may burst forth, and serve to create a feeling of uneasiness and dread.

The enemy has no means to prevent this action except the detachment of numerous troops to furnish escorts for convoys and to occupy military stations, defiles, and bridges. For example, the magnitude of such operations can be illustrated by the situation in South Vietnam. By July, 120 South Vietnamese infantry battalions are slated to fan out into the countryside in small units to protect some 3,000 hamlets from guerrillas in areas that US troops have cleared of major enemy forces. The easiest way to support insurgency activities is to send small detachments from the army.

"Without such support of a few regular troops as an encouragement," said Clausewitz, "the inhabitants generally lack the impulse and the confidence to take up arms." This is the thinking behind the military operations in Vietnam; if the support of the regular army units from North Vietnam can be curtailed, the inner conflict will gradually abate, and eventual cessation of overt military operations in South Vietnam will occur.

For insurgency to succeed, situations must never develop into decisive engagements. Clausewitz believed that the insurgents should, therefore, defend the approaches to mountains, the dikes of a swamp, and the passages over a river, as long as possible. But when an engagement was broken, he felt that they should disperse, and continue their defense by unexpected attacks rather than concentrate and allow themselves to be shut up in some narrow, last refuge in a regular defensive position.

The psychological and organizational characteristics of armed civilian groups suggest to Clausewitz that, although they form a weapon of strategic defense, they generally or even always must be tactically on the offensive. As Mao Tse-tung was to write a century later, "The ability to run away is the very characteristic of the guerrilla."

The influence of Clausewitz on Communist military thought is profound. War is not a last resort to be invoked if all else fails; rather, it is one form of a continuing struggle. Karl Marx wrote that Communists everywhere support revolutionary movement against the existing social and political order. According to him, the communists openly declare that their ends can be attained only by the forcible overthrow of existing social conditions.

Soviet military doctrine rejects the notion that there is such a thing as purely military considerations. "War," wrote Nikolai Lenin, "is part of the whole. The whole is politics. . . . Appearances are not reality. Wars are most political when they seem most military."

Lenin, like Friedrich Engels and Marx, was fascinated by Clausewitz' war theories; he not only studied them with insight, but annotated his books extensively. That Engels, Marx, Mao, and Lenin, the most noted exponents of the Communist philosophy, acknowledged their debt to Clausewitz, who was a non-Communist thinker, is undoubtedly the highest compliment ever paid to his insight on the nature of war.

The dialectic quality of Clausewitz' argumentation attracted Lenin to him. The passage which most appealed to him concerned the relationship of war to politics. This was emphasized by Joseph Stalin in 1946 as a cardinal tenet of Marxist thought. It has also been subjected to typical verbal inversion by a leading Soviet military authority who said that, if war is a continuation of politics by other means, so also is peace a continuation of struggle by other means.

Communist statecraft turns in peacetime to what are, in effect, lesser points on the conflict spectrum—namely, subversion, sabotage, colonial rebellion, and satellite aggression. They have become masters in combining and operating various nonmilitary forms of war—political, economic, and psychological. Mao Tse-tung has said that, without a political goal, guerrilla warfare must fail. . . .

Karl von Clausewitz was a profound military intellectual who applied himself to the pure theory of war. He visualized a spectrum of conflict that moved all the way from simple, unarmed belligerence through the brutality of total war. Further study indicates that he offered another alternative to unrestrained violence—that of limited war. His work was an early blueprint for insurgency and counter-insurgency of the type being conducted today in Vietnam.

Partisan Warfare
by V.I. Lenin

The question of partisan actions has aroused great interest within the Party and among the workers.[1] We have mentioned this topic repeatedly before. Our present intention is to redeem our promise and summarize our position on this subject.

Let us start from the beginning. What are the basic questions every Marxist must ask when he analyzes the problem of the types of struggle?[2] First of all, unlike primitive forms of socialism, Marxism does not tie the movement to any particular combat method. It recognizes the possibility that struggle may assume the most variegated forms. For that matter, Marxism does not "invent" those forms of struggle. It merely organizes the tactics of strife and renders them suitable for general use. It also renders the revolutionary classes conscious of the forms of the clashes which emerge spontaneously from the activities of the movement. Marxism rejects all abstract thinking and doctrinaire prescriptions about types of struggle. It calls for a careful study of the *mass struggle* which actually is taking place. As the movement develops, as the consciousness of the masses grows, and as the economic and political crises are becoming more intense, ever new and different methods of defense and attack will be used in the conflict. Hence, Marxism never will reject any particular combat method, let alone reject it forever. Marxism does not limit itself to those types of struggle which, at a given moment, are both practical and traditional. It holds that, due to changes in social conditions, new forms of battle will arise *inevitably*, although no one can foresee what the character of these future encounters will be. In this field, if we may say so, Marxism is *learning* from the practice of the masses. It is far from claiming that it should *teach* the masses tactics elaborated in the abstract by strategists of the pen. We know, as Kautsky stated when he was analyzing the different forms of social revolution, that the coming crisis will present us with new and unpredictable forms of action.

Second, Marxism asks that the various types of struggle be analyzed within their *historical* framework. To discuss conflict outside of its historical and concrete setting is to misunderstand elementary dialectic materialism. At various junctures of the economic evolution, and depending upon changing political, national, cultural, social and other conditions, differing types of struggle may become important and even predominant. As a result of those [sociological] transformations, secondary and subordinate forms of action may change their significance. To try and answer positively or negatively the question of whether a certain tactic is usable, without at the same time studying the concrete conditions confronting a given movement at a precise point of its development, would mean a complete negation of Marxism.

Those are the two basic concepts which must serve as our guide. The soundness of this approach has been confirmed by numerous examples from the history of Western European Marxism. At present, European socialists regard parliamentarism and trade unionism as their main method of struggle. Previously, they favored the armed uprising.[3] Contrary to the opinion of liberal-bourgeois politicians like the Russian Cadets and the Bessaglavtsi,[4] the European socialists are perfectly willing to favor the uprising again should the situation change in the future.

During the 1870's, social democrats rejected the idea that the general strike could be used as a panacea tactic and as a non-political method suitable for the immediate overthrow of the bourgeoisie. But after the experience of 1905,[5] the social democrats fully recognized the political mass strike as *a* means which, under *certain* conditions, could become necessary. Similarly, during the 1840's the social democrats recognized the utility of barricades. By the end of the nineteenth century, conditions had changed and the socialists rejected the barricades as unsuitable. However, after the experience of the Moscow rising, which, in Kautsky's words, demonstrated new tactics of barricade fighting, they were willing to revise their position and again acknowledge the usefulness of barricades.[6]

II

After this exposition of general Marxist doctrine, we want to discuss the Russian revolution. Let us consider the historical development of the various action types to which the revolution gave rise. First, there occurred economic strikes by the workers (1896--1900), then political demonstrations by workers and students (1901--1902), peasant unrest (1902), subsequently the beginnings of political mass strikes variously connected with demonstrations (Rostov 1902, strikes during summer of 1903, the affair of January 22, 1905[7]), political general strike with local barricade fighting (October 1905), mass barricade battles waged by large numbers '[of revolutionaries], as well as armed uprising (December 1905), peaceful parliamentary struggle (April--July 1906), local military uprisings (June 1905--June 1906), and local peasant uprisings (fall 1905--fall 1906).

Such was the development of the struggle before the autumn of 1906. Absolutism opposed these types of struggle with Black Hundreds pogroms.[8] These pogroms were initiated in spring 1903 at Kishinev and ended with the Siedliec pogrom in 1906. During this period, the organizing of Black Hundreds pogroms and the tormenting of Jews, students, revolutionaries and class-conscious workers continued unabated and steadily increased in ferocity. Mob violence was paired with military violence perpetrated by reactionary troops. Artillery was used on villages and cities. Punitive expeditions were dispatched, and all over the railroads there were moving trains crowded with political prisoners.

This, then, has been the general background of the situation. From this background there has emerged the phenomenon of *armed struggle*.[9] Our paper is devoted to the study and evaluation of this new occurrence. Although merely a secondary and incidental part of the whole, armed struggle has been pushed into the foreground. What is armed struggle? What are its forms and its causes? When did it originate? What has been the frequency of its occurrence? What is its significance for the general course of the revolution? What is its connection with the proletarian class struggle organized and waged by social democracy? After having described the general background of the problem, we shall now address ourselves to these questions.

Armed struggle is waged by small groups and individuals, some of whom are members of revolutionary parties. In certain regions of Russia, however, the *majority* [of the partisans] are not affiliated with any revolutionary organization. Armed struggle aims at *two different* objectives which must be distinguished *sharply* from one another. The first objective is to kill individuals such as high officials and lower-ranking members of the police and army.[10] The second objective is to confiscate money from the government as well as from private persons. Portions of the captured money are used for party purposes, other portions for arms and the preparation of the rising, and the rest for the sustenance of persons engaging in the struggle described by us.[11]

The money seized in the great expropriations (more than 200,000 rubles in the Caucasus and 875,000 rubles in Moscow) was allocated to the revolutionary parties primarily.[12] Smaller expropriations were used mainly, and sometimes exclusively, for the livelihood of the "expropriators." This type of struggle came into widespread use during 1906, after the December uprising [at Moscow]. The aggravation of the political crisis to the point of armed insurrection, and especially the ever growing pauperization, famine and unemployment in villages and cities were among the most potent causes leading to the emergence of armed combat. The *declassé* elements of the population, the *Lumpenproletariat* and anarchist groups, chose this struggle as the main and even *only* form of the social war. Autocracy answered with the tactics of martial law, conscription of younger military classes, Black Hundreds pogroms (Siedliec) and court martials.

III

Armed struggle often is considered to be anarchism, Blanquism, old-style terrorism and, at any rate, an activity perpetrated by isolated individuals out of touch with the masses. The acts of armed struggle are judged to demoralize the workers. Allegedly they divorce broad strata of the population from the toilers, disorganize the revolutionary movement and hurt the revolutionary cause. Examples supporting this type of evaluation are drawn easily from the daily press.

But how good are these examples? Let us look at one case. Partisan struggle reached its *greatest* popularity in the Lettish districts. On August 21 and September 25 [1906], the newspaper *Novoye Vremya*[13] complained bitterly about the activities of the Lettish socialists. The Lettish Social Democratic Party, a branch of the Social Democratic Workers Party of Russia, disclosed a list of police agents. This disclosure was inserted in the party newspaper (circulation: 30,000) and was accompanied by the comment that it was the duty of every honest person to help bring about the liquidation of those spies. The police collaborators were "enemies of the revolution," their property was declared liable to seizure and they themselves were designated for execution. The social democrats have instructed the population to contribute money to the party, but against stamped receipts only. In the latest budget, there was listed among the party's annual receipts totalling 48,000 rubles an item of 5,600 rubles expropriated by the Libau organization for the purchase of weapons. Of course, *Novoye Vremya* is outraged by such "revolutionary legislation" and by this "terror regime."

No one would dare call those actions by the Lettish social democrats anarchism, Blanquism or terrorism. Why? Simply because the armed struggle *clearly* is interrelated with the uprising which took place in December. Such uprisings are bound to reoccur. If Russia is considered as a whole, then this relationship [between armed struggle and armed uprising] is not so clearly noticeable, but it does exist. After all, there is no question but that "partisan" struggle reached its greatest popularity after the December rising. Those actions are related not only to the economic crisis but also to the political crisis. Traditional Russian terrorism was the work of plotting intellectuals. Now, workers or unemployed persons who are members of combat groups usually are leading this struggle.[14] People who like to generalize according to abstract patterns easily may think of anarchism or Blanquism. In the face of an insurrectionist situation as it clearly existed in the Lettish area, such phrases learned by rote obviously are meaningless.

The Lettish example demonstrates that the usual method of analyzing partisan action without regard to the status of the uprising is completely wrong, unscientific and unhistorical. The [concrete] situation must be taken into consideration. The characteristics of the transition periods between large uprisings must be taken into account. The types of struggle which, in a given period, are becoming inevitable should not be criticized with a few clichés such as anarchism, plunder and *Lumpenproletariat*, as is customary among Cadets[15] and the *Novoye Vremya* crowd.

It is said that partisan actions disorganize our work. Let us see to what extent this evaluation is justified, especially with respect to the period after December 1905 and to the areas under martial law and [suffering from] Black Hundreds pogroms. What is it that disorganizes the movement in such an area more: the lack of resistance or the lack of [a well] organized partisan struggle? Compare the situation in Central Russia with that of the Western border regions, such as Poland and Livonia. There is no doubt that in the Western provinces partisan struggles occur far more frequently and have reached a higher stage of development. Contrariwise, there is no doubt that in Central Russia the revolutionary movement is general, and the social democratic movement in particular, is far *more disorganized* [than in the West]. Certainly we would not think of concluding that because of the partisan struggle the Polish and Lettish social democratic movement has suffered from disorganization less [than the movement in Central Russia]. No. The point is merely that the partisan struggle is not responsible for the disorganization of the Russian social democratic workers movement [which occured] during 1906.

In this connection, frequent reference has been made to the peculiarities of national conditions. Such arguments disclose the weakness of the customary objections to partisan struggle. If it is a matter of national conditions, then obviously it is not a matter of anarchism, Blanquism or terrorism, but something else is involved: general Russian or even specifically Russian sins. Analyze this "something else" more *concretely*, gentlemen! You will find that national oppression or national antagonisms explain nothing. These conditions always were present in the Western border regions, yet partisan actions have occurred only in a special historical period. There are many regions where national oppression and antagonisms have been rampant, and yet no partisan struggles are taking place. The fact is that sometimes partisan struggles develop in the absence of any national oppression.[16] A concrete analysis of this question would show that it is not national oppression but the development of the uprising which is decisive. Partisan struggle is an unavoidable form of action at a time when the mass movement has matured to the point of insurrection and when the intervals between the "big battles" of the civil war are becoming shorter.

The movement has not been disorganized by partisan struggles but by the weakness of the party, which does not know how to *take those actions into its own hands.* Consequently, the indictments against partisan warfare, so customary among us Russians, go together with secret, accidental and unorganized partisan actions which, indeed, do disorganize the party. If we do not understand the historical conditions of partisan warfare, then we shall be unable to eliminate its darker sides. In spite of everything, partisan operations occur [because they] are created by powerful economic and political causes. Since we are unable to get rid of those causes, we are unable to prevent this type of struggle. Our complaints about partisan warfare are nothing but complaints about the weakness of our party [which is incapable of] organizing the uprising.

What we said about disorganization also applies to demoralization. Partisan struggle as such does not produce demoralization, which results rather from *disorganization*, undisciplined armed actions and from lack of party leadership. Demoralization, which *unquestionably* has set in, cannot be overcome by disapproving and rejecting the [concept of] partisan struggle. Such censures are by no means sufficient to prevent events which result from profound economic and political causes. It could be objected that, while we may not have the capability of suppressing abnormal and demoralizing happenings, no purpose would be served if the party were to use anomalous and demoralizing tactics. Such a non-Marxist objection would be of a purely liberal-bourgeois character. No Marxist should consider partisan warfare, which is just one of the forms of civil war, as abnormal and demoralizing. Marxists favor class struggle and not social peace. In periods of grave economic and political crisis, the class struggle develops into civil war—that is, into an armed struggle between two parts of the people. In such periods, every Marxist is *obliged* to endorse the cause of civil war. From the Marxist point of view, moral condemnations of civil war are entirely unacceptable.

In situations of civil war, a *combat party* is the ideal type of a proletarian party. This is indisputable. We admit that one may try to prove, and perhaps may be able to prove, the inadvisability of this or that type of struggle at this or that juncture of the civil war. From the point of view of *military expediency*, criticism of the various forms of civil war certainly is justified. We agree that the decisive voice in such questions belongs to those experienced socialist leaders who are familiar with the practical conditions in each locality. But, in the name of Marxist principles, we must insist that civil war be analyzed seriously and that shopworn phrases such as anarchism, Blanquism and terrorism not be thrown into the debate. Senseless partisan actions, such as were indulged in by this or that organization of the PPS[17] in this or that situation, should not be abused for a scare argument against socialist participation in partisan warfare.

One must accept assertions that partisan warfare

disorganizes the [socialist] movement with skepticism. *Every* new form of struggle which involves new dangers and new sacrifices inevitably will "disorganize" organizations unprepared for the new tactics. Our old study groups became disorganized when agitational methods were adopted. Later on, our party committees were disorganized when the party took to demonstrations. In every war, new tactics carry a degree of disorganization into the battle ranks. Yet this is no argument against fighting a war. It merely follows that one must *learn* how to wage war. That is all there is to it.

When I meet social democrats who proudly and self-righteously declare, "we are no anarchists, no thieves, no robbers, we are above [such violent forms of struggle], we reject partisan warfare," then I ask myself: "Do these people understand what they are talking about?" Violent incidents and armed clashes between the Black Hundreds government and the people are happening all over the country. This is inevitable at the present stage of revolution. The population reacts to the attacks by Black Hundreds troopers with armed *coups de main* and ambushes. Because they are spontaneous and unorganized, these counter-attacks may assume inexpedient and *evil* forms. I understand quite well that, due to weakness and lack of preparation by our organization, the party may refrain from assuming, at given places and times, the leadership of such spontaneous actions. I understand that this question must be decided by local practitioners and that the strengthening of weak and unprepared party organizations is not an easy task. But if a social democratic theoretician or writer fails to be saddened by such lack of preparedness and, on the contrary, displays proud self-satisfaction, and conceitedly and enthusiastically repeats slogans on anarchism, Blanquism and terrorism which he memorized in his early youth, then I consider this to be a degradation of the world's most revolutionary doctrine.

It is asserted that partisan actions lower the class-conscious proletariat to the level of drunkards and bums. This is correct. But from this follows only that the party of the proletariat never should consider partisan warfare to be its only or even its chief means of struggle. This particular technique must be integrated with other tactics and be in harmony with the most important methods of combat. Partisan warfare should be enobled by the enlightening and organizing influence of socialism. Without this *last* condition, *all*—clearly all—means of struggle will move the proletariat [which lives] within a bourgeois society close to various non-proletarian strata, whether they stand higher or lower [in social rank].[18] If they are allowed to develop spontaneously, such techniques will lose their effectiveness and their original form and will become prostituted.[19] Strikes which are left to take a spontaneous course degenerate into "alliances," i.e. agreements between business and labor *against* the consumer. Parliament becomes a brothel where gangs of bourgeois politicians are bargaining, wholesale and retail, about "people's freedom," "liberalism," "democracy," republicanism, anticlericalism, socialism and other brands of popular commodities. Newspapers turn into cheap procurers and into tools corrupting the masses and flattering the lowest mob instincts, etc. The socialists know of no universally applicable combat method which would separate the proletariat, as though by a Chinese wall, from those classes of the people which [socially] are situated slighly higher or slightly lower. Socialists use different means at different periods. Those means are chosen in *strict* accordance with ideological and organizational conditions the nature of which must be determined *accurately* [by the Marxian dialectic method].

The bolsheviks[20] have been accused frequently of an unthinking party-oriented [and positive] attitude toward partisan actions. It seems necessary, therefore, to reiterate that the *particular* bolshevik faction[21] which approved partisan warfare defined in its draft [of a social democratic party resolution] the conditions under which armed struggle would be permissible: "Expropriations" of private property are entirely forbidden. "Expropriations" of government property are not recommended, but are *permitted* provided they are accomplished *under party control* and provided the captured money is used for the *purposes of the uprising*. Terrorist partisan acts against representatives of the violent regime and of *active* Black Hundreds groups *are recommended*[22] but are subject to the following restrictions: (1) the popular mood must be taken into account; (2) local conditions under which the workers movement is operating must be considered; (3) care must be taken that no proletarian forces are wasted unnecessarily. The *only* practical difference between the resolution by the unification congress of the [Social Democratic] Party[23] and our draft resolution is that [in the former] "expropriations" of government property were entirely forbidden.

IV

The Russian revolution differs from bourgeois revolutions in Europe in that it displays an immense variety in the methods of struggle. Kautsky predicted this in 1902, at least to a point, when he said that the coming revolution (and he added *perhaps* with the exception of Russia) will not be so much a struggle of the people against the government as a struggle of one part of the people against the other. In Russia we witnessed a broader development of the *second* kind of struggle than during the bourgeois revolutions in the West. The enemies of our revolution have but few followers among the people, but as the fight develops the opponents are getting better and better organized and are gaining support from reactionary groups of the bourgeoisie. Thus, it is natural and unavoidable that in *such* periods, in a period of political general strikes, *the uprising* cannot assume the traditional form of a single blow, limited to a very short time and a very small area.[24] [Under such circumstances], it is natural and unavoidable that the uprising assumes the higher and more complicated form of a protracted civil war enmeshing the entire country—that is, the form of armed struggle by one part of the people against the other. Such a war must be conceived as a series of a few big battles, separated by comparatively long intervals, and a large number of small engagements which take place during these interim periods. If this is so—and it undoubtedly is so—then the task of social democracy is to create organizations most suitable to leading the masses both in the big battles and, so far as practical, in the smaller actions. At a time when the class struggle is developing into civil war, social democrats must consider it their task not only *to participate* in this civil war, but must play the leading role in this conflict. The Social Democratic Party must educate and prepare its organizations in such a way that they will become true belligerents who will not fail to exploit opportunities through which the strengths of the opponent can be sapped.

Unquestionably, this is a difficult task. It cannot be accomplished at once. Similarly, as an entire people is transforming itself in the course of civil war and is learning from the struggle, so our organizations, if they are to fulfill their mission, must be educated and reorganized on the basis of experience.

We do not presume at all to impose on comrades who are carrying on with their practical work any theoretical ideas about tactics, let alone to decide from the vantage point of a desk what role this or that form of partisan struggle should assume during the civil war in Russia. We shall not confuse particular *political orientations* within the social democratic movement with specific partisan actions.[25] But we consider it our task to provide a correct *theoretical* evaluation of the new forms of struggle which life has created.[26] Our business is to fight pitilessly against the clichés and prejudices which are hindering the class-conscious workers from posing a new and difficult question in the right manner and hence from solving it correctly.

Notes

Lenin's text, originally published on October 13, 1906, in No. 5 of his newspaper *Proletari*, is somewhat heavy-handed and occasionally confusing. Fortunately, its main message is clear and penetrating. Hence the editors of *ORBIS* deemed it preferable to print an accurate translation rather than "edit" Lenin's article. For a better understanding of the author's many references to events of the Russian revolution of 1905/1906, the editors have appended footnotes prepared by Stefan T. Possony. In some instances, words were added to make the original text more comprehensible; these words have been placed between brackets. The italics are Lenin's own.

The translation was prepared by Regina Eldor from Volume X of the third edition of *Sochineniya* and from Volume X of the German *Saemtliche Werke*, an authoritative translation prepared by the Lenin Institute in Moscow.

1. The term "partisan war" or "partisan actions" is a euphemism. It does not mean "guerrilla war" in the modern sense but stands for terrorism, holdups and robberies. So-called "expropriations" of money were directed against banks, taxation agencies, post offices, customs houses, railroad stations and similar establishments where large sums of cash were likely to be stored. However, small firms, such as bakeries and village shops, as well as affluent individuals, were also victimized. In many instances, the "expropriations" were planned by professional "fingermen" and executed by expert robbers. Terror was practiced on policemen, soldiers and officials, both in cities and in the rural areas. Operations in the cities were conducted by small "combat groups"; forays in the countryside were sometimes executed by large armed bands which, under the

convenient guise of "partisan warfare," made looting and pillaging a profitable profession. Originally, the social democrats had rejected terrorism, which was a major *modus operandi* of the social revolutionaries. During 1905 and 1906, however, the incidence of terror increased greatly and the bolshevik faction of the Social Democratic Party supported it wholeheartedly. In fact, a large percentage of the bombs used in "partisan warfare" was fabricated in a secret bolshevik laboratory run by Leonid B. Krassin. Most mensheviks were opposed both to terror and "expropriations," but it is interesting that G. V. Plekhanov, the founder of Russian Marxism, favored them, at least for a time. The Russian socialists who opposed terrorism argued that "European means of struggle" be used. They feared that terrorism was harming the reputation of the social democrats and worried about the fact that many, if not most, of the "expropriations" were perpetrated by criminal elements, for purposes of their own personal enrichment.

In reading Lenin's discourse, it should be remembered that, in practical terms, he was advocating alliance between the revolution and crime: Lenin did, in fact, enter into agreements with criminal elements during the partisan warfare period. Later, during World War I, he even recommended a notorious highwayman to the Germans for sabotage operations. (The man received pay but did not commit any acts of sabotage.)

While Lenin was penning his treatise on "partisan war," the terrorist phase of the first Russian revolution was reaching its peak. In October 1906 alone, 121 terror acts, 47 clashes between revolutionaries and the police, and 362 expropriations were reported. [See Boris Souvarine, *Staline, Apercu Historique du Bolchevisme* (Paris: Plon, 1935) p. 92] While it is impossible to draw up exact statistics of the total terror campaign, there is no question that it cost the lives of more than 5,000 policemen and officials. Several millions of rubles were "expropriated" by criminal and revolutionary elements.

2. The terms. "types" or "forms of struggle," and their variations, such as "combat tactics" and "methods of battle," all of which sound awkward in English, denote a key concept in communist conflict doctrine. The term "struggle" is a short notation for "class struggle." Lenin contends that the tactics and techniques of the class struggle must be altered as situations and conditions change. Socialists should have no dogmatic attachment to one particular type of tactic or a particular weapon but should employ those procedures and means which, singly or in combination, are expedient and effective. The point is important since American policy-makers often assume that the communists are wedded to one particular "type of struggle" and that the communists, once they begin to apply one specific method, will continue to do so. Such an interpretation of bolshevik theory

can be reconciled neither with the writings nor the actions of international communism.
3. Lenin alludes here to one of Friedrich Engels' last publications in which, to the chagrin of the radicals, he discussed the difficulties of an armed uprising against a government armed with modern weapons. Engels went so far as to question the usefulness of that revolutionary symbol, the barricade. Lenin also alludes to the gradual shift which at that time was taking place in practically all European socialist parties toward forsaking revolution in favor of evolutionary methods.
4. This refers to a weekly magazine entitled *Bes Zaglavia* (*Without Title*), of which sixteen issues were published between February and May 1906 by S. N. Prokopovich, E. D. Kuskova and others. The editors of this magazine were moderate socialists who believed in democracy. They were friendly to the objectives of the left wing of the Constitutional Democrats (Cadets).
5. Lenin refers to the mass strike movement which began in August 1905 and in October culminated in one of the most complete general strikes of history. It was this strike movement, and particularly the railroad strike, which forced the Russian government to proclaim the so-called "October Manifesto" by which a semi-constitutional regime (Max Weber described it as "sham constitutionalism") was promulgated. Incidentally, these strikes neither were called nor run by the socialist leaders but by the liberal-bourgeois parties, especially the Cadets. Lenin's wording suggests that he was completely aware of this historical fact which, however, he was loath to admit in writing. Lenin and other revolutionaries did not return to Russia until an amnesty, late in October 1905, made it safe for them to do so. Only after the middle-of-the-road parties, whose outstanding demands were met by the "October Manifesto," withdrew from the revolution, did the socialist assume leadership of the revolutionary movement.
6. Alexander Helphand, better known as "Parvus," discussed this vexing problem of barricades as early as 1897—that is, two years after Engels had expressed his doubts. "Parvus" pointed out that barricades, while perhaps no longer militarily useful, could serve as rallying points for the aroused and fighting masses. He considered barricades as a predominantly psychological device suitable for bringing the masses into the streets. Obviously, Lenin, who feared "Parvus" as an intellectually superior competitor, did not want to give him credit for this correct prediction, nor did he want to acknowledge that he now adopted "Parvus' " interpretation of barricade tactics.
7. The "affair of January 22, 1905" is better known as "Bloody Sunday." Lenin's undramatic description of this event, which was the tragic overture to the revolution of 1905, probably is due to the fact that neither the social democrats nor the bolshevik

faction played any significant role in it. The revolution had started without their assistance. The leading revolutionary figure of Bloody Sunday was Father George Gapon, who originally had been involved in "police socialism" and was cooperating with the social revolutionaries in initiating the revolution.

8. The "Black Hundreds" were combat groups set up by parties of the extreme right in order to fight the revolutionaries. They might be considered the ancestors of the Nazi SA and SS, though their organization was not as strong and their membership fluctuated greatly. The Black Hundreds were openly tolerated by the Tsarist police; there is, in fact, the strong possibility that the police itself secretly created these forces. The Black Hundreds rarely, if ever, succeeded in fighting the revolutionaries directly. They were used for anti-semitic pogroms, mostly in poor Jewish districts. The pogroms were launched in the hope that counter-terror ultimately would intimidate the revolutionaries. (It may be added that, according to the official version, the pogroms were "spontaneous.") This hope was based on the mistaken notion, prevalent within the Russian government and the police, that the revolutionary movement was largely the creation of Jewish international circles on whose financial and political support it depended. The assumption was that, if the Jews in Russia were made to pay for the crimes of the revolutionaries, the Jewish world leaders, in order to save their coreligionists, would call off the revolution. The frequency and violence of the pogroms have never been over-rated, and the utility of this anti-revolutionary tactic was very much debated within the Russian government. When it became apparent that the pogroms were totally ineffective in halting the revolution, the Black Hundreds gradually fell into disuse. Their very existence, however, provided the revolutionaries with excellent arguments for their own terror operations.

From 1906 onward, under the premiership of P.A. Stolybin, the revolutionary movement was incapacitated by systematic arrests of revolutionaries, summary executions, exiling to Siberia, and punitive expeditions against partisan bands.

9. The term "armed struggle" is another expression for "partisan action." Lenin had in mind violent actions executed by small groups for secondary objectives such as terrorism and robbery. The term does not denote armed uprising.

10. According to the official legend, the bolsheviks are opposed to terrorism. Lenin's article should dispel any false notions about the bolshevik attitude to political assassinations. Lenin makes it perfectly clear that a true bolshevik never can be against terrorism as a matter of principle: he should oppose terror only if and when murder is inexpedient and ineffective. The bolshevik, by the same token, should favor political assassinations whenever they promise to advance the com-

munist cause.
11. Thus, Lenin admitted that many of the so-called "expropriations" were simply robberies. While Lenin did not openly advocate robberies as a convenient source of income for professional revolutionaries, the "expediency" which he championed was broad enough to include such use of "expropriations."
12. According to the Lenin Institute, Lenin was describing an expropriation which took place on March 26, 1906 at Dushet, near Tiflis, and which was carried out by six men disguised as soldiers of the 263rd Infantry Regiment. The Lenin Institute stated that 315,000 rubles were expropriated. If Lenin's party treasury received only the 200,000 rubles to which he was referring, then 115,000 rubles must have remained in the hands of the "expropriators." Souvarine commented that the robbers were socialist-federalists (that is, they belonged to one of the many splinter groups of the Social Revolutionary Party) and that the bolsheviks got hold of this money "by ruse" (*op. cit.*, p. 91). In other words, true to their Marxian philosophy, they expropriated the expropriators. Note that the action of Dushet was *not* the expropriation in which Stalin participated. Stalin earned his laurels as a bank robber on 26 June, 1907, in Tiflis. In the course of that raid, no less than ten bombs were thrown and 431,000 rubles (or $170,000) seized. The Moscow expropriation was carried out on March 20, 1906, by twenty armed men who attacked a bank, disarmed four guards and took 875,000 rubles, just as Lenin indicated in the text. For a useful discussion of some of these events, the reader is referred to Alexandre Spiridovich, *Histoire de Terrorisme Russe* (Paris: Payot, 1930).
13. *Novoye Vremya* was a leading conservative paper. During 1906, the Lettish revolutionary movement was very well organized and registered some of the more notable successes of the first Russian revolution. The Baltic provinces were the scene of a great deal of partisan action in the modern sense, which it took Russian military forces considerable time to suppress. Socialist ideology contributed only mildly to the Lettish movement's strength: nationalist feelings were a more significant factor. This is one of the first instances of the socialist-nationalist "amalgam" in guerrilla war.
14. Lenin wanted to imply that the partisan actions usually were carried out by authentic "proletarians." There is no evidence to support this statement. The "plotting intellectuals" continued to play a dominant role, and peasants were at least as important in this struggle as workers.
15. This is a reference to the Cadet Party led by P.N. Milyukov.
16. While Lenin's analysis is accurate, he did underrate the importance of the national question during the first Russian revolution. Subsequently he assigned a far higher value to

Revolutionary Guerrilla Warfare:
The Development of Doctrine

nationalism as a revolutionary factor.

17. Lenin was referring to the Polish Socialist Party of which Joseph Pilsudski was the most prominent leader. It is significant that Pilsudski personally led one of the most daring expropriation attacks on a Polish post office himself. Lenin never participated in any of the partisan actions which he was advocating so fervently.
18. This obtuse sentence is of significance only to firm believers in the Marxian doctrine. Lenin wanted to say that some types of struggle would bring the proletariat closer to the middle classes, while others would lead it into closer relationships with the *Lumpenproletariat* and, possibly, with the very poor peasants. His point was that the socialist ideology would preserve the pure class character of the proletarian movement, regardless of the means of struggle employed by it.
19. Lenin presumably meant that if the party loses control over operations, other social forces may be able to exploit the proletarian movement for their purposes.
20. The following paragraph was written by Lenin as a footnote to his article. We have inserted it into the main text to enhance clarity.
21. Not all the bolsheviks were in favor of partisan action.
22. This amounts to Lenin *recommending* terrorism.
23. Lenin was referring to the resolution adopted by the Fourth Congress of the Russian Social Democratic Workers Party at Stockholm during April and May 1906. The difference between Lenin's views and that of the majority of the "unification congress" was considerably greater than indicated here, but Lenin at that time found it necessary to keep his peace with the party, especially since he was not certain of the wholehearted support of his bolsheviks. The Stockholm resolution opposed theft, the expropriation of private funds and of bank accounts, forced contributions, the destruction of public buildings, and railroad sabotage. Lenin succeeded in convincing the congress that it should allow the confiscation of government funds, provided expropriation could be carried out by a revolutionary organization and on its orders. The congress also approved terrorist actions in cases of self-defense.

In September 1906, the Moscow Bolshevik Party Committee issued a resolution which came out far more radically in favor of partisan war. It proclaimed "offensive tactics" to be the only useful tactics. The party was called upon to organize partisan war in cities and villages against the government. The party was to liquidate the most active representatives of the government and to seize money and arms. The resolution suggested that the population at large be invited to support the partisan war. Lenin was in favor of this more radical policy. This article in its entirety is essentially a polemic against the softer resolution of the Stockholm congress.

24. This important sentence refers to uprisings in capital cities. Many revolutionaries believed that the seizure of power could be accomplished by a sudden one-thrust insurrection against the seat of government. Lenin's remark foreshadows the development of Mao Tse-tung's operational doctrine and basically enlarges the concept of uprising into that of civil or guerrilla war.
25. This unclear sentence presumably means that it is wrong to confuse tactics with ideology. Factions of the socialist movement, distinguished from other factions largely by ideological differences, usually had a preference for specific forms of struggle. Conversely, a group specializing in one particular type of combat might be inclined to a correlated ideological orientation. Lenin suggested that the tactics of the revolutionary movement be discussed on their own merit and that ideological questions be discussed in ideological terms.
26. Commenting on Lenin's assertion that the party instead of teaching the masses, is being schooled by them, and that partisan war emerged spontaneously as a riposte to actions by the Black Hundreds, the army and the police, Souvarine said that Lenin's point could be summarized in this fashion: "All that is spontaneous is necessary." This is a paraphrase of a statement by Hegel, all too frequently quoted by Marxists: "All that is real is reasonable." Note the value which Lenin ascribed to spontaneity—a value quite at variance with the subsequent development of the "Leninist--Stalinist doctrine," which claimed to be opposed to spontaneity and placed instead the highest value on organization.

Selections from *Selected Works of Mao Tse-Tung*
abridged by Bruno Shaw

Chapter I: Why Raise the Question of Strategy in Guerrilla War?

In the War of Resistance Against Japan, regular warfare is primary and guerrilla warfare supplementary. . . . Why then raise the question of strategy?

If China were a small country in which the role of guerrilla warfare was only to render direct support over short distance to the campaigns of the regular army, there would, of course, be only tactical problems but no strategic ones. On the other hand, if China were a country as strong as the Soviet Union and the invading enemy could either be quickly expelled . . . or could not occupy extensive areas, then again guerrilla warfare would . . . involve only tactical but not strategic problems.

The question of strategy in guerrilla war does arise, however, in the case of China, which is neither small nor like the Soviet Union, but which is both a large and a weak country. . . . It is in these circumstances that vast areas have come under enemy occupation and that the war has become a protracted one. . . . The protracted nature of the war and its attendant ruthlessness have made it imperative for guerrilla warfare to undertake many unusual tasks; hence such problems as those of the base areas, the development of guerrilla warfare into mobile warfare, and so on. For all these reasons, China's guerrilla warfare against Japan has broken out of the bounds of tactics to knock at the gates of strategy, and it demands examination from the viewpoint of strategy. . . .

Chapter II: The Basic Principle of War Is to Preserve Oneself and Destroy the Enemy

All the guiding principles of military operations grow out of the one basic principle: to strive to the utmost to preserve one's own strength and destroy that of the enemy. In a revolutionary war, this principle is directly linked with basic political principles. . . . In terms of military action this principle means the use of armed force to defend our motherland and to drive out the Japanese invaders. . . . Every war exacts a price, sometimes an extremely high one. Is this not in contradiction with "preserving oneself"? In fact, there is no contradiction at all . . . for such sacrifice is essential not only for destroying the enemy but also for

206   preserving oneself. . . .

Chapter III: Six Specific Problems of Strategy in Guerrilla War Against Japan

Now let us see what policies or principles have to be adopted in guerrilla operations against Japan before we can attain the object of preserving ourselves and destroying the enemy. . . .

The main principles are as follows: (1) The use of initative, flexibility and planning in conducting offensives . . . battles of quick decision within protracted war . . .; (2) co-ordination with regular warfare; (3) the establishment of base areas; (4) the strategic defensive and the strategic offensive; (5) the development of guerrilla warfare into mobile warfare; and (6) correct relationship of command. . . .

Chapter IV: Initiative, Flexibility and Planning In Conducting Offensives Within the Defensive, Battles of Quick Decision Within Protracted War, and Exterior-Line Operations Within Interior-Line Operations

Here the subject may be dealt with under four headings: (1) The relationship between the defensive and the offensive, between protractedness and quick decision, and between the interior and exterior lines; (2) the initiative in all operations; (3) flexible employment of forces; and (4) planning in all operations.

To start with the first. If we take the War of Resistance as a whole, the fact that Japan is a strong country and is attacking while China is a weak country and is defending herself makes our war strategically a defensive and protracted war. . . . It is possible and necessary to use tactical offensives within the strategic defensive, to fight campaigns and battles of, quick decision within a strategically protracted war and to fight campaigns and battles on exterior lines within strategically interior lines. Such is the strategy to be adopted in the War of Resistance as a whole. It holds true both for regular and for guerrilla warfare. . . .

Now let us discuss initiative, flexibility and planning in guerrilla warfare. What is the initiative in guerrilla warfare? In any war, the opponents contend for the initiative, whether on a battlefield, in a battle area, in a war zone or in the whole war, for the initiative means freedom of action for an army. Any army which, losing the initiative, is forced into a passive position and ceases to have freedom of action, faces the danger of defeat or extermination. . . . The question of the initiative is even more vital in guerrilla warfare. For most guerrilla units operate in very difficult circumstances, fighting without a rear, with their own weak forces facing the enemy's strong forces. . . . The initiative is not an innate attribute of genius, but is something an intelligent leader attains through open-minded study and correct appraisal of the objective conditions and through correct military and political dispositions. It follows that the initiative is not ready-made but is something that requires conscious effort. . . .

Next, let us deal with flexibility. Flexibility is a concrete expression of the initiative. The flexible employment of forces is more essential in guerrilla warfare than in regular warfare. A guerrilla commander must understand that the flexibile employment of his forces is the most important means of changing the situation as between the enemy and ourselves and of gaining the initiative. The nature of guerrilla warfare is such that guerrilla forces must be employed flexibly in accordance with the task in hand and with such circumstances as the state of the enemy, the terrain and the local population; and the chief ways of employing the forces are dispersal, concentration and shifting of position. . . . But a commander proves himself wise not just by recognition of the importance of employing his forces flexibly but by skill in dispersing, concentrating or shifting them in good time according to the specific circumstances. . . . Prudent consideration of the circumstances is essential to prevent flexibility from turning into impulsive action.

Lastly, we come to planning. Without planning, victories in guerrilla warfare are impossible. Any idea that guerrilla warfare can be conducted in haphazard fashion indicates either a flippant attitude or ignorance of guerrilla warfare. The operations in a guerrilla zone as a whole, or those of a guerrilla unit or formation, must be preceded by as thorough planning as possible, by preparation in advance for every action. Grasping the situation, setting the tasks, disposing the forces, giving military and political training, securing supplies, putting the equipment in good order, making proper use of people's help, etc.—all these are part of the work of the guerrilla commanders. . . . True, guerrilla conditions do not allow as high a degree of planning as do those of regular warfare. . . . But it is necessary to plan as thoroughly as the objective conditions permit, for it should be understood that fighting the enemy is no joke.

Chapter V: Co-ordination with Regular Warfare

The second problem of strategy in guerrilla warfare is its co-ordination with regular warfare. It is a matter of clarifying the relation between guerrilla and regular warfare on the operational level, in the light of the nature of actual guerrilla operations. . . . Take the case of the guerrilla warfare in the three northeastern provinces. . . . Every enemy soldier the guerrillas kill there, every bullet they make the enemy expend, every enemy soldier they stop from advancing south of the Great Wall, can be reckoned a contribution to the total strength of the resistance. . . .

In addition, guerrilla warfare performs the function of co-ordination with regular warfare in campaigns. . . . If each guerrilla zone or unit goes it alone without giving any attention to co-ordinating with the campaigns of the regular forces, its role in strategic co-ordination will lose a great deal of its significance. . . .

Finally, co-ordination with the regular forces in battles, in actual fighting on the battlefield, is the task of all guerrilla units in the vicinity of an interior-line battlefield. . . . In such cases a guerrilla unit has to perform whatever task it is assigned by the commander of the regular forces. . . . To sit by idly, neither moving nor fighting, or to move about without fighting, would be an intolerable attitude for a guerrilla unit.

Chapter VI: The Establishment of Base Areas

The third problem of strategy in anti-Japanese guerrilla warfare is the establishment of base areas, which is important and essential because of the protracted nature and ruthlessness of the war. . . . With ruthlessness thus added to protractedness, it will be impossible to sustain guerrilla warfare behind the enemy lines without base areas. . . .

Without such strategic bases, there will be nothing to depend on in carrying out any of our strategic tasks or achieving the aim of the war. It is characteristic of guerrilla warfare behind the enemy lines that it is fought without a rear, for the guerrilla forces are severed from the country's general rear. But guerrilla warfare could not last long or grow without base areas. The base areas, indeed, are its rear. . . .

1. The Types of Base Areas

Base areas in anti-Japanese guerrilla warfare are mainly of three types, those in the mountains, those on the plains and those in the river-estuary regions. . . .

Of course, the plains are less suitable than the mountains, but it is by no means impossible to develop guerrilla warfare or establish any base areas there. Indeed, the widespread guerrilla warfare in the plains of Hopei and of northern and northwestern Shantung proves that it is possible to develop guerrilla warfare in the plains. . . .

2. Guerrilla Zones and Base Areas

In guerrilla warfare behind the enemy lines, there is a difference between guerrilla zones and base areas. Areas which are surrounded by the enemy but whose central parts are not occupied or have been recovered, like some countries in the Wutai mountain region . . . are ready-made bases for the convenient use of guerrilla units. . . . Thus the transformation of a guerrilla zone into a base area is an arduous creative process, and its accomplishment depends on the extent to which the enemy is destroyed and the masses are aroused. . . .

As for the big cities, the railway stops and the areas in the plains which are strongly garrisoned by the enemy, guerrilla warfare can only extend to the fringes and not right into these places which have relatively stable puppet regimes. . . .

Mistakes in our leadership or strong enemy pressure may cause a reversal of the state of affairs described above. . . . Such changes are possible, and they deserve special vigilance on the part of guerrilla commanders.

3. Conditions for Establishing Base Areas

The fundamental conditions for establishing a base area are that there should be anti-Japanese armed forces, that these armed forces should be employed to inflict defeats on the enemy and that they should arouse the people to action. . . . If there is no armed force or if the armed force is weak, nothing can be done. This constitutes the first condition.

The second indispensable condition for establishing a base area is that the armed forces should be used in co-ordination with the people to defeat the enemy. . . .

The third indispensable condition for establishing a base area is the use of all our strength, including our armed forces, to arouse the masses for struggle against Japan. . . .

A base area for guerrilla war can be truly established only with the gradual fulfillment of the three basic conditions, that is, only after the anti-Japanese armed forces are built up, the enemy has suffered defeats and the people are aroused.

4. The Consolidation and Expansion of Base Areas

In order to confine the enemy invaders to a few strongholds, that is, to the big cities and along the main communication lines, the guerrillas must do all they can to extend guerrilla warfare from their base areas as widely as possible and hem in all the enemy's strongholds, thus threatening his existence and shaking his morale while expanding the base areas.

Given a protracted war, the problem of consolidating and expanding base areas constantly arises for every guerrilla unit.

5. Forms in Which We and the Enemy Encircle One Another

Taking the War of Resistance as a whole, there is no doubt that we are strategically encircled by the enemy, because he is on the strategic offensive and is operating on exterior lines while we are on the strategic defensive and are operating on interior lines. This is the first form of enemy encirclement. We on our part encircle each of the enemy columns advancing on us along separate routes. . . . This is the first form of our encirclement of the enemy. Next, if we consider the guerrilla bases in the enemy's rear, each area taken singly is surrounded by the enemy on all sides.... This is the second form of enemy encirclement. However, if one considers all the guerrilla base areas together and in their relation to the battle fronts of the regular forces, one can see that we in turn surround a great many enemy forces.... This is the second form of our encirclement of the enemy. Thus there are two forms of encirclement by the enemy forces and two forms of encirclement by our own—rather like a game of *weichi*.[1]

Chapter VII: The Strategic Defensive and The Strategic Offensive in Guerrilla War

The fourth problem of strategy in guerrilla war concerns the strategic defensive and the strategic offensive. This is the problem of how the policy of offensive warfare, which we mentioned in our discussion of the first problem, is to be carried out in practice, when we are on the defensive and when we are on the offensive in our guerrilla warfare against Japan....

1. The Strategic Defensive in Guerrilla War

To wipe out the guerrillas and their base areas, the enemy frequently resorts to converging attacks.... When the enemy launches a converging attack in several columns, the guerrilla policy should be to smash it by counterattack. It can be easily smashed if each advancing enemy column consists of only one unit, whether big or small, has no follow-up units and is unable to station troops along the route of advance, construct blockhouses or build motor roads.... The enemy, though strong, will be weakened by repeated surprise attacks and will often withdraw when he is halfway; the guerrilla units can then make more surprise attacks during the pursuit and weaken him still further....

Should the enemy stay put in our base area, we may reverse the tactics, namely, leave some of our forces in the base area to invest the enemy, while employing the main force to attack the region whence he has come and to step up our activities there, in order to induce him to withdraw and attack our main force; this is the tactic of "Relieving the state of Chao by besieging the State of Wei."[2]

When the enemy retreats, he often burns down the houses in the cities and towns he has occupied and razes the villages along his route, with the purpose of destroying the guerrilla base areas; but in so doing the deprives himself of shelter and food in his next offensive, and the damage recoils upon his own head. This is a concrete illustration of what we mean by one and the same thing having two contradictory aspects. . . . Since we have had the experience of being able to maintain guerrilla warfare during the civil war, there is not the slightest doubt of our greater capacity to do so in a national war. . . .

2. The Strategic Offensive in Guerrilla War

After we have smashed an enemy offensive and before the enemy starts a new offensive, he is on the strategic defensive and we are on the strategic offensive.

At such times our operational policy is not to attack enemy forces which are entrenched in defensive positions and which we are not sure of defeating, but systematically to destroy or drive out the small enemy units and puppet forces in certain areas, which our guerrilla units are strong enough to deal with. . . .

Chapter VIII: The Development of Guerrilla Warfare into Mobile Warfare

To transform guerrilla units waging guerrilla warfare into regular forces waging mobile warfare, two conditions are necessary—an increase in numbers and an improvement in quality. Apart from directly mobilizing the people to join the forces, increased numbers can be attained by amalgamating small units, while better quality depends on steeling the fighters and improving their weapons in the course of the war.

In amalgamating small units, we must, on the one hand, guard against localism, whereby attention is concentrated exclusively on local interests and centralization is impeded, and, on the other, guard against the purely military approach, whereby local interests are brushed aside. . . .

To raise the quality of the guerrilla units it is imperative to raise their political and organizational level and improve their equipment, military technique, tactics and discipline, so that they gradually pattern themselves on the regular forces and shed their guerrilla ways. . . . To accomplish all these tasks requires a prolonged effort, and it cannot be done overnight; but that is the direction in which we must develop.

Chapter IX: The Relationship of Command

The last problem of strategy in guerrilla war against Japan concerns the relationship of command. A correct solution of this problem is one of the prerequisites for the unhampered development of guerrilla warfare.

Since guerrilla units are a lower level of armed organization characterized by dispersed operations, the methods of command in guerrilla warfare do not allow as high a degree of centralization as in regular warfare. . . . However, guerrilla warfare cannot be successfully developed without some centralized command. When extensive regular warfare and extensive guerrilla warfare are going on at the same time, their operations must be properly co-ordinated. . . .

Hence, as opposed both to absolute centralization and to absolute decentralization, the principle of command in guerrilla war should be centralized strategic command and decentralized command in campaigns and battles. . . .

Absence of centralization where it is needed means negligence by the higher levels or usurpation of authority by the lower levels, neither of which can be tolerated in the relationship between higher and lower levels, especially in the military sphere. If decentralization is not affected where it should be, that means monopolization of power by the higher levels and lack of initiative on the part of the lower levels, neither of which can be tolerated in the relationship between higher and lower levels, especially in the command of guerrilla warfare. The above principles constitute the only correct policy for solving the problem of the relationship of command.

This proclamation was written by Mao Tse-tung for the government of the Shensi-Kansu-Ningsia border region and the Rear Headquarters of the Eighth Route Army. Its purpose, said Mao Tse-tung, was to "counter disruptive activities by the Chiang Kai-shek clique."

Revolutionary Guerrilla Warfare:
The Development of Doctrine

On Protracted War
May 1938
A Statement of the Problem

It will soon be July 7, the first anniversary of the Great War of Resistance Against Japan. Rallying in unity, persevering in resistance and persevering in the united front, the forces of the whole nation have been valiantly fighting the enemy for almost a year. . . .

But what actually will be the course of the war? Can we win? Can we win quickly? Many people are talking about a protracted war, but why is it a protracted war? How to carry on a protracted war? Many people are talking about final victory, but why will final victory be ours? How shall we strive for final victory? . . . Defeatist exponents of the theory of national subjugation have come forward to tell people that China will be subjugated. . . . On the other hand, some impetuous friends have come forward to tell people that China will win very quickly without having to exert any great effort. . . . However, most people have not yet grasped what we have been saying. . . . Now things are better; the experience of ten months of war has been quite sufficient to explode the utterly baseless theory of national subjugation and to dissuade our impetuous friends from the theory of quick victory. . . .

A serious study of protracted war is necessary in order to enable every Communist to play a better and greater part in the War of Resistance. . . .

During these ten months of war all kinds of views which are indicative of impetuosity have also appeared. . . . At the outset of the war many people were groundlessly optimistic, underestimating Japan . . . they disagreed with the Eighth Route Army's strategy. . . . After the Taierhchuang[3] victory, some people maintained that the . . . policy of protracted war should be changed. They said such things as, "This campaign marks the last desperate struggle of the enemy."

The question now is: Will China be subjugated? The answer is, No, she will not be subjugated, but will win final victory. Can China win quickly? The answer is, No, she cannot win quickly, and the War of Resistance will be a protracted war.

As early as two years ago, we broadly indicated the main arguments on these questions. On July 16, 1936, five months

Selections from *Selected Works of Mao Tse-Tung*

before the Sian Incident and twelve months before the Lukouchiao Incident, in an interview with the American correspondent, Mr. Edgar Snow,[4] I made a general estimate of the situation. . . . The following excerpts may serve as a reminder:

Question: Under what conditions do you think China can defeat and destroy the forces of Japan?

Answer: Three conditions are required: first, the establishment of an anti-Japanese united front in China; second, the formation of an international anti-Japanese united front; third, the rise of the revolutionary movement of the people in Japan and the Japanese colonies.

Question: How long do you think such a war would last?

Answer: . . . If China's anti-Japanese united front is greatly expanded and effectively organized horizontally and vertically, if the necessary help is given to China by those governments and peoples which recognize the Japanese imperialist menace to their own interests and if revolution comes quickly in Japan, the war will speedily be brought to an end and China will speedily win victory. If these conditions are not realized quickly, the war will be prolonged. . . .

Question: What is your opinion of the probable course of development of such a war, politically and militarily?

Answer: Japan's continental policy is already fixed, and those who think they can halt the Japanese advance by making compromises . . . are indulging in mere fantasy. Moreover, Japan wants to occupy the Philippines, Siam, Indochina, the Malay Peninsula and the Dutch East Indies. . . .

Question: If the war drags on . . . would the Communist Party agree to the negotiation of a peace with Japan and recognize her rule in northeastern China?

Answer: No. Like the people of the whole country, the Chinese Communist Party will not allow Japan to retain an inch of Chinese territory.

Question: What, in your opinion, should be the main strategy and tactics to be followed in this "war of liberation"?

Answer: Our strategy should be to employ our main forces to operate over an extended and fluid front. . . . This means large-scale mobile warfare, and not positional warfare depending exclusively on defense works with deep trenches, high fortresses and successive rows of defensive positions. . . .

Besides employing trained armies to carry on mobile warfare, we must organize great numbers of guerrilla units among the peasants. . . . The Chinese peasants have very great latent power; properly organized and directed, they can keep the Japanese Army busy twenty-four hours a day and worry it to death. . . .

In the course of the war, China will be able to capture many Japanese soldiers and seize many weapons and munitions with which to arm herself; at the same time China will win foreign aid to reinforce the equipment of her troops gradually. . . . The combination of all these and other factors will enable us to make the final and decisive attacks on the fortifications and bases in the Japanese-occupied areas and drive the Japanese forces of aggression out of China. . . .

The Basis of the Problem

Why is the War of Resistance Against Japan a protracted war? Why will the final victory be China's? . . .

The Japanese side. First, Japan is a powerful imperialist country, which ranks first in the East in military, economic and political-organizational power, and is one of the five or six foremost imperialist countries in the world. . . . Secondly, however, the imperialist character of Japan's social economy determines the imperialist character of her war, a war that is retrogressive and barbarous. . . . The reactionary and barbarous character of Japan's war constitutes the primary reason for her inevitable defeat. Thirdly, Japan's war is conducted on the basis of her great military, economic and political-organizational power, but at the same time it rests on an inadequate natural endowment . . . and she cannot stand a long war. . . .

The Chinese side. First, we are a semicolonial and semifeudal country. The Opium War,[5] the Taiping Revolution,[6] the Reform Movement of 1898,[7] the Revolution of 1911,[8] and the Northern Expedition[9]—the revolutionary or reform movements which aimed at extricating China from her semicolonial and semifeudal state—all met with serious setbacks, and China remains a semicolonial and semifeudal country . . . a weak country and manifestly inferior to the enemy in military, economic and

Selections from Selected Works of Mao Tse-Tung

political-organizational power. . . . Secondly, however, China's liberation movement, with its cumulative development over the last hundred years, is now different from that of any previous period. . . . By contrast with Japanese imperialism, which is declining, China is a country rising like the morning sun. . . . China is a very big country with vast territory, rich resources, a large population and plenty of soldiers, and is capable of sustaining a long war. Fourthly and lastly, there is broad international support for China stemming from the progressive and just character of her war, which is again exactly the reverse of the meager support for Japan's unjust cause. . . .

Refutation of the Theory of National Subjugation

The theorists of national subjugation, who see nothing but the contrast between the enemy's strength and our weakness, used to say, "Resistance will mean subjugation," and now they are saying, "The continuance of the war spells subjugation." . . . They can adduce historical instances, such as the destruction of the Sung Dynasty by the Yuan and the destruction on the Ming Dynasty by the Ching, to prove that a small but strong country can vanquish a large but weak one. . . .

What then are the grounds we should advance? . . . The China of today cannot be compared with the China of any other historical period. She is a semicolony and a semifeudal society, and she is consequently considered a weak country. But at the same time, China is historically in her era of progress; this is the primary reason for her ability to defeat Japan. . . .

In the existing international situation, China is not isolated in the war, and this fact too is without precedent in history. In the past, China's wars, and India's too, were wars fought in isolation. . . .

The existence of the Soviet Union is a particularly vital factor in present-day international politics, and the Soviet Union will certainly support China with the greatest enthusiasm;[10] there was nothing like this twenty years ago. . . .

If the subjugationists quote the history of the failure of liberation movements in modern China to prove their assertions first that "resistance will mean subjugation," and then that "the continuance of the war spells subjugation," here again our answer is, "Times are different." . . . These are favorable conditions such as never existed before in any period of our history and that is why the War of Resistance Against Japan, unlike the liberation movements of the past, will not end in failure.

**Compromise or Resistance?
Corruption or Progress?**

. . . The question of compromise has its social roots, and as long as these roots exist the question is bound to arise. But compromise will not avail. To prove the point, again we need only look for substantiation to Japan, China, and the international situation. First take Japan. At the very beginning of the War of Resistance, we estimated that the time would come when an atmosphere conducive to compromise would arise, in other words, that after occupying northern China, Kiangsu and Chekiang, Japan would probably resort to the scheme of inducing China to capitulate. . . . Had China capitulated, every Chinese would have become a slave without a country. . . .

Second, let us take China. There are three factors contributing to China's perseverance in the War of Resistance. In the first place, the Communist Party, which is the reliable force leading the people to resist Japan. Next, the Kuomintang, which depends on Britain and the United States and hence will not capitulate to Japan unless they tell it to. Finally, the other political parties and groups, most of which oppose compromise and support the War of Resistance. . . .

Third, take the international aspect. Except for Japan's allies and certain elements in the upper strata of other capitalist countries, the whole world is in favor of resistance, and not of compromise by China.

. . . Hence we may conclude that the danger of compromise exists but can be overcome. . . .

. . . Every just, revolutionary war is endowed with tremendous power, which can transform many things or clear the way for their transformation. The Sino-Japanese War will transform both China and Japan . . . the old Japan will surely be transformed into a new Japan and the old China into a new China. . . . To say that Japan can also be transformed is to say that the war of aggression by her rulers will end in defeat and may lead to a revolution by the Japanese people. The day of triumph of the Japanese people's revolution will be the day Japan is transformed. All this is closely linked with China's War of Resistance and is a prospect we should take into account.

The Theory of National Subjugation Is Wrong and the Theory of Quick Victory is Likewise Wrong

. . . The subjugationists stress the contradiction between strength and weakness and puff it up until it becomes the basis of their whole argument on the question, neglecting all the other contradictions. . . .

The exponents of quick victory are likewise wrong . . . They presumptuously take the balance of forces at one time and place for the whole situation, as in the old saying, "A leaf before the eye shuts out Mount Tai." In a word, they lack the courage to admit that the enemy is strong while we are weak. . . . These friends have their hearts in the right place, and they, too, are patriots. But while "the gentlemen's aspirations are indeed lofty," their views are wrong, and to act according to them would certainly be to run into a brick wall. . . .

Not that we would not like a quick victory; everybody would be in favor of driving the "devils" out overnight. But we point out that, in the absence of certain definite conditions, quick victory is something that exists only in one's mind and not in objective reality. and that it is a mere illusion, a false theory . . . and we reject the theory of quick victory, which is just idle talk and an effort to get things on the cheap.

Why a Protracted War?

Let us now examine the problem of protracted war. A correct answer to the question "Why a protracted war?" can be arrived at only on the basis of all the fundamental contrasts between China and Japan.

. . . During a certain stage of the war, to a certain degree the enemy will be victorious and we shall suffer defeat. But why is it that at this stage the enemy's victories and our defeats are definitely restricted in degree and cannot be transcended by complete victory or complete defeat? The reason is that, first, from the very beginning the enemy's strength and our weakness have been relative and not absolute; and that, second, our efforts in persevering in the War of Resistance and in the united front have further accentuated this relativeness. . . . On both sides, strength and weakness, superiority and inferiority, have never been absolute. . . . Therefore, in this stage the enemy's victory and our defeat are definitely restricted in degree, and hence the war becomes protracted.

The Three Stages of the Protracted War

Since the Sino-Japanese War is a protracted one and final victory will belong to China, it can reasonably be assumed that this protracted war will pass through three stages. The first stage covers the period of the enemy's strategic offensive and our strategic defensive. The second stage will be the period of the enemy's strategic consolidation and our preparation for the counteroffensive. The third stage will be the period of our strategic counteroffensive and the enemy's strategic retreat. . . .

This first stage has not yet ended. The enemy's design is to occupy Canton, Wuhan and Lanchow and link up these three points. To accomplish this aim the enemy will have to use at least fifty divisions, or about one and a half million men, spend from one and a half to two years, and expend more than 10,000 million *yen*. In penetrating so deeply, he will encounter immense difficulties, which consequences disastrous beyond imagination. . . .

In the second stage, the enemy will attempt to safeguard the occupied areas and to make them his own by the fraudulent method of setting up puppet governments, while plundering the Chinese people to the limit; but again he will be confronted with stubborn guerrilla warfare. . . . The fighting in the second stage will be ruthless, and the country will suffer serious devastation. But the guerrilla warfare will be successful, and if it is well conducted the enemy may be able to retain only about one-third of his occupied territory, with the remaining two-thirds in our hands, and this will constitute a great defeat for the enemy and a great victory for China. . . .

The third stage will be the stage of the counteroffensive to recover our lost territories. Their recovery will depend mainly upon the strength which China has built up in the preceding stage. But China's strength alone will not be sufficient, and we shall also have to rely on the support of international forces and on the changes that will take place inside Japan. . . . Ultimately the enemy will lose and we will win, but we shall have a hard stretch of road to travel.

A War of Jigsaw Pattern

We can say with certainty that the protracted War of Resistance Against Japan will write a splendid page unique in the war history of mankind. One of the special features of this war is the interlocking "jigsaw" pattern which arises from such contradictory factors as the barbarity of Japan and her shortage of troops on the one hand, and the progressiveness of China and the extensiveness of her territory on the other. . . . Its jigsaw pattern manifests itself as follows.

Interior and exterior lines. The anti-Japanese war as a whole is being fought on interior lines; but as far as the relation between the main forces and the guerrilla units is concerned, the former are on the interior lines while the latter are on the exterior lines, presenting a remarkable spectacle of pincers around the enemy. . . .

Encirclement and counterencirclement. Taking the war as a whole, there is no doubt that we are strategically encircled by the enemy because he is on the strategic offensive and operating on exterior lines while we are on the strategic defensive and operating on interior lines. . . .

But our encirclement, like the hand of Buddha, will turn into the Mountain of Five Elements lying athwart the universe, and the modern Sun Wu-kungs [11] —the facist aggressors—will finally be buried underneath it, never to rise again. Therefore, if on the international plane we can create an anti-Japanese front in the Pacific region, with China as one strategic unit, with the Soviet Union and other countries . . . as still another strategic unit, and thus form a gigantic net from which the fascist Wu-kungs can find no escape, then that will be our enemy's day of doom . . . the day of the complete overthrow of Japanese imperialism. . . .

Fighting for Perpetual Peace

The protracted nature of China's anti-Japanese war is inseparably connected with the fight for perpetual peace in China and the whole world. Never has there been a historical period such as the present in which war is so close to perpetual peace. . . . This war, we can foresee, will not save capitalism, but will hasten its collapse. It will be greater in scale and more ruthless than the war of twenty years ago, all nations will inevitably be drawn in, it will drag on for a very long time, mankind will suffer greatly. But, owing to the existence of the Soviet Union and the growing political consciousness of the people of the world, great revolutionary wars will undoubtedly emerge from this war, thus giving it the character of a struggle for perpetual peace. . . . Once man has eliminated capitalism, he will attain the era of perpetual peace, and there will be no more need for war. . . .

History shows that wars are divided into two kinds, just and unjust. All wars that are progressive are just, and all wars that impede progress are unjust. We Communists oppose all unjust wars that impede progress, but we do not oppose progressive just wars. Not only do we Communists not oppose just wars, we actively participate in them. . . .

Our war is sacred and just, it is progressive and its aim is peace. The aim is peace not just in one country but throughout the world, not just temporary but perpetual peace. . . . This is no vain hope, for the whole world is approaching this point in the course of its social and economic development, and provided that the majority of mankind work together, our goal will surely be attained in several decades.

Man's Dynamic Role in War

We have so far explained why the war is a protracted war and why the final victory will be China's, and in the main dealt with what protracted war is and what it is not. Now we shall turn to the question of what to do and what not to do. . . . Let us start with the problem of man's dynamic role. . . .

. . . Protracted war and final victory will not come about without human action. For such action to be effective there must be people who derive ideas, principles or views from the objective facts, and put forward plans, directives, policies, strategies and tactics. Ideas, etc., are subjective, while deeds or actions are the subjective translated into the objective, but both represent the dynamic role peculiar to human beings. We term this kind of dynamic role "man's conscious dynamic role," and it is a characteristic that distinguishes man from all other beings. All ideas based upon and corresponding to objective facts are correct ideas, and all deeds or actions based upon correct ideas are correct actions. . . .

It is a human characteristic to exercise a conscious dynamic role. Man strongly displays this characteristic in war. . . . In seeking victory, those who direct a war cannot overstep the limitations imposed by the objective conditions; within these limitations, however, they can and must play a dynamic role in striving for victory. . . . We do not want any of our commanders in the war to detach himself from the objective conditions and become a blundering hothead, but we decidedly want every commander to become a general who is both bold and sagacious. . . . Swimming in the ocean of war, they must not flounder but make sure of reaching the opposite shore with measured strokes. Strategy and tactics, as the laws for directing war, constitute the art of swimming in the ocean of war.

War and Politics

"War is the continuation of politics." In this sense war is politics and war itself is a political action; since ancient times there has never been a war that did not have a political character. . . . And the tendency among the anti-Japanese armed forces to belittle politics by isolating war from it and advocating the idea of war as an absolute is wrong and should be corrected.

But war has its own particular characteristics and in this sense it cannot be equated with politics in general. "War is the continuation of politics by other . . . means."[12] When politics develops to a certain stage beyond which it cannot proceed by the usual means, war breaks out to sweep the obstacles from the way. . . . But if the obstacle is not completely swept away, the war will have to continue till the aim is fully accomplished. . . . It can therefore be said that politics is war without bloodshed while war is politics with bloodshed. . . . Hence war experience is a particular kind of experience. All who take part in war must rid themselves of their customary ways and accustom themselves to war before they can win victory.

Political Mobilization for the War of Resistance

A national revolutionary war as great as ours cannot be won without extensive and thoroughgoing political mobilization. . . . The mobilization of the common people throughout the country will create a vast sea in which to drown the enemy, create the conditions that will make up for our inferiority in arms and other things, and create the prerequisites for overcoming every difficulty in the war. . . . To wish for victory and yet neglect political mobilization is like wishing to "go south by driving the chariot north," and the result would inevitably be to forfeit victory.

What does political mobilization mean? First, it means telling the army and the people about the political aim of the war. It is necessary for every soldier and civilian to see why the war must be fought and how it concerns him. . . . By word of mouth, by leaflets and bulletins, by newspapers, books and pamplets, through plays and films, through schools, through the mass organizations and through our cadres. . . . We must link the political mobilization for the war with developments in the war and with the life of the soldiers and the people, and make it a continuous movement. This is a matter of immense importance on which our victory in the war primarily depends.

The Object of War

Here we are not dealing with the political aim of the war; the political aim of the War of Resistance Against Japan has been defined above as "To drive out Japanese imperialism and build a new China of freedom and equality." Here we are dealing with the elementary object of war, as "politics with bloodshed," as mutual slaughter by opposing armies. The object of war is specifically "to preserve oneself and destroy the enemy." . . . It should be pointed out that destruction of the enemy is the primary object of war and self-preservation the secondary, because only by destroying the enemy in large numbers can one effectively preserve oneself. . . . In actual warfare the chief role is played by defense much of the time and by attack for the rest of the time, but if war is taken as a whole, attack remains primary. . . . Thus, no technical, tactical, or strategical principles or operations can in any way depart from the object of war, and this object pervades the whole of a war and runs through it from beginning to end. . . .

Offense Within Defense, Quick Decisions Within a Protracted War, Exterior Lines Within Interior Lines

. . . Since Japan is a strong imperialist power and we are a weak semicolonial and semifeudal country, she has adopted the policy of the strategic offensive while we are on the strategic defensive. Japan is trying to execute the strategy of a war of quick decision; we should consciously execute the strategy of protracted war. . . . However, we can make use of our two advantages, namely, our vast territory and large forces, and, instead of stubborn positional warfare, carry on flexible mobile warfare, employing several divisions against one enemy division, several tens of thousands of our men against ten thousand of his, several columns against one of his columns, and suddenly encircling and attacking a single column from the exterior lines of the battlefield. In this way, while the enemy is on exterior lines and on the offensive in strategic operations, he will be forced to fight on interior lines and on the defensive in campaigns and battles. And for us, interior lines and the defensive in strategic operations will be transformed into exterior lines and the offensive in campaigns and battles. . . . We should concentrate a big force under cover beforehand alongside the route which the enemy is sure to take, and while he is on the move, advance suddenly to encircle and attack him before he knows what is happening, and thus quickly conclude the battle. . . .

If we resolutely apply "quick-decision offensive warfare on exterior lines" on a battlefield, we shall not only change the balance of forces on that battlefield, but also gradually change the general situation. . . . After many such battles have been victoriously fought, the general situation between us and the enemy will change. . . . When that happens, these changes, together with other factors on our side and together with the changes inside the enemy camp and a favorable international situation, will turn the overall situation between us and the enemy first into one of parity and then into one of superiority for us. That will be the time for us to launch the counter-offensive and drive the enemy out of the country. . . .

Initiative, Flexibility and Planning

. . . Initiative is inseparable from superiority in capacity to wage war, while passivity is inseparable from inferiority in capacity to wage war. Such superiority or inferiority is the objective basis of initiative or passivity. . . . As for China, though placed in a somewhat passive position strategically because of her inferior strength, she is nevertheless quantitatively superior in territory, population and troops, and also superior in the morale of her people and army and their patriotic hatred of the enemy. . . . Any passivity, however, is a disadvantage, and one must strive hard to shake it off. . . .

Yet war is in no way supernatural, but a mundane process governed by necessity. That is why Sun Wu Tzu's axiom, "Know the enemy and know yourself, and you can fight a hundred battles with no danger of defeat," remains a scientific truth. . . .

The thesis that incorrect subjective direction can change superiority and initiative into inferiority and passivity, and that correct subjective direction can effect a reverse change, becomes all the more convincing when we look at the record of defeats suffered by big and powerful armies and of victories won by small and weak armies. There are many such instances in Chinese and foreign history. Examples in China are the Battle of Chengpu between the states of Tsin and Chu,[13] the Battle of Chengkao between the states of Chu and Han,[14] the battle in which Han Hsin defeated the Chao armies,[15] the Battle of Kunyang between the states of Hsin and Han,[16] the Battle of Kuantu between Yuan Shao and Tsao Tsao,[17] the Battle of Chihpi between the states of Wu and Wei,[18] the Battle of Yiling between the states of Wu and Shu,[19] the Battle of Feishui between the states of Chin and Tsin,[20] etc. Among examples to be found abroad are most of Napoleon's campaigns and the civil war in the Soviet Union after the October Revolution. In all these instances, victory was won by small forces over big and by inferior over superior forces. In every case, the weaker force, pitting local superiority and initiative against the enemy's local inferiority and passivity, first inflicted one sharp defeat on the enemy and then turned on the rest of his forces and smashed them one by one, thus transforming the overall situation into one of superiority and initiative. . . .

To have misconceptions and to be caught unawares may mean to lose superiority and initiative. Hence, deliberately creating misconceptions for the enemy and then springing surprise attacks upon him are two ways—indeed two important means—of achieving superiority and seizing the initiative. What are misconceptions? "To see every bush and tree on Mount Pakung as an enemy soldier" is an example of misconception. And "making a feint to the east but attacking in the west" is a way of creating misconceptions among the enemy. When the mass support is sufficiently good to block the leakage of news, it is often possible by various ruses to succeed in leading the enemy into a morass of wrong judgments and actions so that he loses his superiority and the initiative. The saying "There can never be too much deception in war," means precisely this. . . .

We are not Duke Hsiang of Sung,[21] and have no use for his asinine ethics. In order to achieve victory we must as far as possible make the enemy blind and deaf by sealing his eyes and ears, and drive his commanders to distraction by creating confusion in their minds. . . .

Now let us discuss flexibility. . . . We should know not only how to employ tactics but how to vary them. For flexibility of command the important task is to make changes such as from the offensive to the defensive or from the defensive to the offensive, from advance to retreat or from retreat to advance, from containment to assault or from assault to containment, from encirclement to outflanking or from outflanking to encirclement, and to make such changes properly and in good time according to the circumstances of the troops and terrain on both sides. . . .

Let us now discuss the question of planning. Because of the uncertainty peculiar to war, it is much more difficult to prosecute war according to plan than is the case with other activities. Yet, since "Preparedness ensures success and unpreparedness spells failure," there can be no victory in war without advance planning and preparations. . . . Tactical plans, such as plans for attack or defense by small formations or units, often have to be changed several times a day. . . . A strategic plan based on the overall situation of both belligerents . . . has to be changed when the war moves toward a new stage. . . .

Because of the fluidity of war, some people categorically deny that war plans or policies can be relatively stable, describing such plans or policies as "mechanical." This view is wrong. . . . But one must not deny the need for war plans or policies that are relatively stable over given periods; to negate this is to negate everything, including the war itself, as well as the negator himself. . . .

Mobile Warfare, Guerrilla Warfare And Positional Warfare

A war will take the form of mobile warfare when its content is quick-decision offensive warfare on exterior lines in campaigns and battles within the framework of the strategy of interior lines, protracted war and defense. Mobile warfare is the form in which regular armies wage quick-decision offensive campaigns and battles on exterior lines along extensive fronts and over big areas of operations. . . . Its characteristics are regular armies, superiority of forces in campaigns and battles, the offensive, and fluidity. . . . We must oppose "Only retreat, never advance," which is flightism, and at the same time oppose "Only advance, never retreat," which is desperate recklessness. . . .

We have always advocated the policy of "luring the enemy in deep," precisely because it is the most effective military policy for a weak army strategically on the defensive to employ against a strong army.

Among the forms of warfare in the anti-Japanese war mobile warfare comes first and guerrilla warfare second. . . . Guerrilla warfare does not bring as quick results or as great renown as regular warfare, but "A long road tests a horse's strength and a long task proves a man's heart," and in the course of this long and cruel war, guerrilla warfare will demonstrate its immense power. . . . The principle of the Eighth Route Army is: "Guerrilla warfare is basic, but lose no chance for mobile warfare under favorable conditions." This principle is perfectly correct; the views of its opponents are wrong. . . .

Hence, throughout the War of Resistance China will not adopt positional warfare as primary; the primary or important

Selections from *Selected Works of Mao Tse-Tung*

forms are mobile warfare and guerrilla warfare. These two forms of warfare will afford full play to the art of directing war and to the active role of man—what a piece of good fortune out of our misfortune!

War of Attrition and War of Annihilation

. . . Since there are three forms of warfare, mobile, positional and guerrilla . . . and since they differ in degrees of effectiveness, there arises the broad distinction between war of attrition and war of annihilation. . . . Unless we fight campaigns of annihilation . . . we cannot win time to improve our internal and international situation and alter our unfavorable position. Hence, campaigns of annihilation are the means of attaining the objective of strategic attrition. In this sense war of annihilation *is* war of attrition. It is chiefly by using the method of attrition through annihilation that China can wage protracted war.

But the objective of strategic attrition may also be achieved by campaigns of attrition. Generally speaking, mobile warfare performs the task of annihilation, positional warfare performs the task of attrition, and guerrilla warfare performs both simultaneously. . . .

The strength of the Japanese Army lies not only in its weapons but also in the training of its officers and men—its degree of organization, its self-confidence arising from never having been defeated, its superstitious belief in the Mikado and in supernatural beings, its arrogance, its contempt for the Chinese people and other such characteristics, all of which stem from long years of indoctrination. . . . This is the chief reason why we have taken very few prisoners although we have killed and wounded a great many enemy troops. . . . To destroy these enemy characteristics will be a long process. . . . The chief method of destroying them is to win over the Japanese soldiers politically . . . and, by treating prisoners of war leniently, lead the Japanese soldiers to see the anti-popular character of the aggression committed by the Japanese rulers. . . .

The Possibilities of Exploiting The Enemy's Mistakes

. . . In the ten months of his war of aggression the enemy

has already made many mistakes in strategy and tactics. There are five major ones:

First, piecemeal reinforcement. This is due to the enemy's underestimation of China and also to his shortage of troops. The enemy has always looked down on us. . . . The conclusion the enemy came to was that the Chinese nation is a heap of loose sand. Thus, thinking that China would crumble at a single blow, he mapped out a plan of "quick decision," attempting with very small forces to send us scampering in panic. . . .

Second, absence of a main direction of attack. Before the Taierhchuang campaign, the enemy had divided his forces more or less evenly between northern and central China and had again divided them inside each of these areas. . . .

Third, lack of strategic co-ordination. On the whole, co-ordination exists within the groups of enemy forces in northern China and in central China, but there is glaring lack of co-ordination between the two. . . .

Fourth, failure to grasp strategic opportunities. This failure was conspicuously shown in the enemy's halt after the occupation of Nanking and Taiyuan. . . .

Fifth, encirclement of large, but annihiliation of small, numbers. Before the Taierhchuang campaign . . . many Chinese troops were routed but few were taken prisoner, which shows the stupidity of the enemy command.

. . . However, although much of the enemy's strategic and campaign command is incompetent, there are quite a few excellent points in his battle command, that is, in his unit and small formation tactics, and here we should learn from him.

The Question of Decisive Engagements In the Anti-Japanese War

The question of decisive engagements in the anti-Japanese war should be approached from three aspects: we should resolutely fight a decisive engagement in every campaign or battle in which we are sure of victory; we should avoid a decisive engagement in every campaign or battle in which we are not sure of victory; and we should absolutely avoid a strategically decisive engagement on which the fate of the whole nation is staked. . . .

Are we not afraid of being denounced as "nonresisters"? No, we are not. Not to fight at all but to compromise with the enemy—that is nonresistance, which should not only be denounced but must never be tolerated. We must resolutely fight the War of Resistance, but in order to avoid the enemy's deadly trap, it is absolutely necessary that we should not allow our main forces to be finished off at one blow . . . in brief, it is absolutely necessary to avoid national subjugation. . . .

Is it not self-contradictory to fight heroically first and then abandon territory? Will not our heroic fighters have shed their blood in vain? That is not at all the way questions should be posed. To eat and then to empty your bowels—is this not to eat in vain? To sleep and then to get up—is this not to sleep in vain? Can questions be posed in such a way? I would suppose not. . . . As everybody knows, although in fighting and shedding our blood in order to gain time and prepare the counteroffensive we have had to abandon some territory, in fact we have gained time, we have achieved the objective of annihilating and depleting enemy forces, we have acquired experience in fighting, we have aroused hitherto inactive people and improved our international standing. . . . We are for protracted war and final victory, we are not gamblers who risk everything on a single throw.

The Army and the People Are the Foundation of Victory

Japanese imperialism will never relax in its aggression against and repression of revolutionary China; this is determined by its imperialist nature. If China did not resist, Japan would easily seize all China without firing a single shot, as she did the four northeastern provinces. . . . Now that Japan has launched war against China, so long as she does not suffer a fatal blow from Chinese resistance and still retains sufficient strength, she is bound to attack Southeast Asia or Siberia, or even both. She will do so once war breaks out in Europe; in their wishful calculations, the rulers of Japan have it worked out on a grandiose scale. Of course, it is possible that Japan will have to drop her original plan of invading Siberia and adopt a mainly defensive attitude toward the Soviet Union on account of Soviet strength and of the serious extent to which Japan herself has been weakened by her war against China. But in that case, so far from relaxing her aggression against China she will intensify it,

232 because then the only way left to her will be to gobble up the weak. China's task of persevering in the War of Resistance, the united front and the protracted war will then become all the more weighty, and it will be all the more necessary not to slacken our efforts in the slightest. . . .

The richest source of power to wage war lies in the masses of the people. It is mainly because of the unorganized state of the Chinese masses that Japan dares to bully us. When this defect is remedied, then the Japanese aggressor, like a mad bull crashing into a ring of flames, will be surrounded by hundreds of millions of our people standing upright, the mere sound of their voices will strike terror into him, and he will be burned to death.

Conclusions

What are our conclusions? They are:
"Under what conditions do you think China can defeat and destroy the forces of Japan?" "Three conditions are required: first, the establishment of an anti-Japanese united front in China; second, the formation of an international anti-Japanese united front; third, the rise of the revolutionary movement of the people in Japan and the Japanese colonies. From the standpoint of the Chinese people, the unity of the people of China is the most important of the three conditons." . . .

The existence of serious weaknesses in the War of Resistance may lead to setbacks, retreats, internal splits, betrayals, temporary and partial compromises and other such reverses. Therefore it should be realized that the war will be arduous and protracted. . . .

These are our conclusions. In the eyes of the subjugationists the enemy are supermen and we Chinese are worthless, while in the eyes of the theorists of quick victory we Chinese are supermen and the enemy are worthless. Both are wrong. We take a different view; the War of Resistance Against Japan is a protracted war, and the final victory will be China's. These are our conclusions. . . .

Notes

1. *Weichi* is an old Chinese game in which the two players try to encircle each other's pieces on the board. When a player's pieces are encircled, they are counted as "dead" (captured). But if there is a sufficient number of blank spaces among the

encircled pieces, then the latter are still "alive" (not captured).
2. In 353 B.C., the state of Wei laid seige to Hantan, the capital of the state of Chao. The King of the State of Chi, an ally of Chao, ordered his generals Tien Chi and Sun Pin to aid Chao with their troops. Knowing that the crack forces of Wei had entered Chao and left their own territory weakly garrisoned, General Sun Pin attacked the state of Wei, whose troops withdrew to defend their own country. Taking advantage of their exhaustion, the troops of Chi engaged and routed them at Kueiling (northeast of the present Hotse County in Shantung). The siege of Hantan, capital of Chao, was thus lifted. Since then Chinese strategists have referred to similar tactics as "Relieving the state of Chao by besieging the state of Wei."
3. Taierchuang is a town in southern Shantung where the Chinese Army fought a battle in March 1938 and won its first victory in positional warfare against the Japanese. By pitting 400,000 men against Japan's attacking 70,000 to 80,000, the Chinese defeated the Japanese decisively.
4. Edgar Snow, a native of Missouri, went to the Far East in the mid-1920's, when he was twenty-two. In Shanghai he was associate editor of J.B. Powell's *China Weekly Review*. Later he worked successively for the Chicago *Tribune*, New York *Sun*, New York *Herald Tribune*, and London *Daily Herald*, and during World War II he reported wartime events from Europe and Asia for the *Saturday Evening Post*. He is the author of ten books, including *Red Star Over China* and *The Other Side of the River*.
5. In the early part of the eighteenth century, Britain began exporting increasing quantities of opium to China, importation of which had been prohibited by the Chinese Government. By 1839 the traffic had become so large that Peking sent a special commissioner, Lin Tse-hsu, to Canton to stamp it out. British armed forces subdued the Chinese attempt to assert its sovereign right to prohibit the importation of opium, and the end result was the treaty of Nanking, which provided for the payment of indemnities and the cession of Hong Kong to Britain, and stipulated that Shanghai, Foochow, Amoy, Ningpo and Canton were to be opened to British trade.
6. See note 9, p. 57, in Shaw, *Selected Works of Mao Tse-Tung*.
7. The Reform Movement of 1898 was a short-lived attempt by a budding liberal movment to gain civil reforms under the Empress Dowager Tzu Hsi. It failed and its leader, Tan Szu-tung, was beheaded.
8. China's revolution began in Wuchang, Hupeh, on October 10, 1911 (celebrated as the Double Ten anniversary), with a revolt of government troops against the Manchu regime. Yuan Shih-kai, retired commander of the Manchu Government's armies, was called back into service to defend the dynasty. In January 1912, the provisional government of the Republic of China was set up in Nanking, with Sun Yat Sen as provisional President.

The following month the Manchu "boy-Emperor" was made to abdicate the throne. A deal between the defending and the revolutionary forces resulted in Sun Yat Sen's resigning the presidency and allowing Yuan Shih-kai to be named president in his stead, ending the last imperial Chinese dynasty.

9. The Northern Expedition was a war against the warlords of the northern provinces, a joint Kuomintang-Communist venture, under the command of General Chiang Kai-shek. When its armies reached Hankow, Hupeh, on the Yangtze River in the autumn of 1926, the Communists, under the leadership of Michael Borodin, set up a government entirely dominated by the Communist Party forces. This resulted in a split between the Kuomintang and Communist Party forces, and eventually to war between the two factions.

10. The Soviet Union gave China no support whatever, until seven years later, after Japan had been defeated by two atom bomb attacks, the first on August 6, 1945, on Hiroshima, and the second on August 9, 1945, on Nagasaki. The same day (August 9), the Soviet Union, in compliance with the Yalta Conference agreement signed in February 1945, launched an attack against the Japanese Army in Manchuria.

11. Sun Wu-kung is the monkey king in the sixteenth-century novel *Hsi Yu Chi* [*Pilgrimate to the West*]. He could cover 108,000 *li* (a *li* is about a third of a mile) by turning a somersault. Yet once in the palm of the Buddha, he could not escape from it, however many somersaults he turned. With a flick of his palm, Buddha transformed his fingers into the five-peak Mountain of Five Elements and buried Sun Wu-kung.

12. This definition of war, in fact made by Karl von Clausewitz (1780-1831), is attributed by Mao Tse-tung to V.I. Lenin, *Socialism and War* (Eng. ed., Moscow, 1950), p. 19.

13. Chengpu, in Shantung Province, was the scene of a great battle between the states of Tsin and Chu in 632 B.C. Initially the Chu troops were on top, but the Tsin troops, after retreating 90 *li*, picked the enemy's weak spots and defeated them.

14. Chengkao, in Honan Province, was the scene of battles in 203 B.C. between Liu Pang, King of Han, and Hsiang Yu, King of Chu. When almost defeated, Liu Pang waited until Hisang Yu's troops were in midstream crossing the Szeshui River, then crushed them, and recaptured Chengkao.

15. In 204 B.C. Han Hsin, a general of the state of Han, deployed his much smaller army in pincer formation against the enemy whose troops were then destroyed.

16. Liu Hsiu, founder of the Eastern Han Dynasty, defeated the troops of Wang Mang, Emperor of the Hsin Dynasty, in A.D. 23, by taking advantage of the negligence of Wang Mang's generals, and crushed the enemy troops.

17. Kuantu was in northeast Honan Province, scene of a battle between the armies of Tsao Tsao and Yuan Shao in A.D. 200.

Tsao Tsao, with a meager force, took advantage of his enemy's lack of vigilance, and in a surprise attack set his supplies on fire, throwing the enemy into confusion and then wiping out his main force.

18. In A.D. 208, Chihpi, on the south bank of the Yangtze River in Hupeh Province, was attacked by Tsao Tsao of the state of Wei. The defenders, with only 30,000 men against the attacker's 500,000, set the enemy fleet on fire and crushed his army.
19. Lu Sun, a general of the state of Wu, defeated the far stronger army of Liu Pei, ruler of Shu, at Yiling, near Ichang, Hupeh Province, in A.D. 222. He did so by avoiding battle for seven months until Liu Pei was at his wits' end, and then, taking advantage of a favorable wind, setting fire to Liu Pei's tents and routing his army.
20. In A.D. 383, Hsieh Hsuan, a general of the Eastern Tsin Dynasty, defeated Fu Chien, ruler of the state of Chin, by a simple stratagem. Their armies on opposite sides of the Feishui River in Anhwei Province, Hsieh Hsuan asked Fu Chien to move his troops back from the riverbank, so that he could bring his troops across to do battle. Fu Chien complied, but when he ordered withdrawal, his troops panicked and fled, and Hsieh Hsuan crossed the river and easily defeated those who were left.
21. Duke Hsiang of Sung ruled in the Spring and Autumn Era. In 638 B.C. the state of Sung fought with the powerful state of Chu. One of his officers suggested that as the Chu troops were numerically stronger, it would be well to attack them while they were crossing the river. But the Duke said, "No, a gentleman should never attack one who is unprepared." As a result, he suffered defeat and was himself wounded.

Chapter IV
Modern Revolutionary Guerrilla Warfare in Macro-View: Causes and Contexts

Revolutions are manifestations of tensions within political and social systems. To understand revolutions, therefore, one must first have some insight into the social order. Revolutionary guerrilla warfare usually has disturbed peasant societies, with the peasant base furnishing the revolutionary armies and giving legitimacy to the struggle. This chapter examines the factors conductive to revolutionary guerrilla warfare, especially the nature and structure of peasant society.

Scott G. NcNall and Martha D. Huggins suggest the relationships between "predisposing" factors (the environment) and "precipitating" factors (immediate causes) of revolutionary guerrilla warfare. Focusing primarily on economic problems, national integration and psychological tensions, the authors outline the factors that, singly or in combination, create a revolutionary condition. They also suggest some kinds of short-term change that may spark an uprising where a revolutionary environment exists.

In the next selection G. C. AlRoy asks why the countryside is so often receptive to revolutionary guerrilla warfare. He concludes that the Chinese or Maoist model rather than the Soviet provides the most relevant answers. The author emphasizes "the virtual absence of government in much of the backlands of underdeveloped countries. . . ."

T. Shanin examines the political potentiality of the peasantry, scrutinizing such factors as values, perceptions, and economic pursuits as causes of the breakdown of peasant society. Although peasants are normally viewed as a class, they do not necessarily act as a monolithic entity. Nevertheless, peasant society is relatively self-contained and isolated from the rest of society. In this context, according to the author, repression can stimulate a viable opposition under the control of professional rebels.

The essay by Eric Wolf begins where the previous selections stop. Using arguments similar to those of Shanin, the author seeks the nature of peasant rebellions in the 20th century. Wolf observes that peasants are normally passive: "the peasant is slow to rise." Only such trauma as modernization can break down this passivity. Confronted by demographic and ecological crises, the peasantry undergoes a crisis of authority. The state comes to be perceived as an antagonist.

The selection by James Scott analyzes peasant society in

238 Southeast Asia. Reviewing several models of association and conflict, Scott concludes that political action does not depend on horizontal ties or primordial sentiments, but on patron-client relationships based on reciprocal needs and services. The durability of such quid-pro-quo relationships depends upon the maintenance of traditional power bases; if these are weakened, the relationships are undermined. Revolutionary cadres may then hope to eliminate and supplant the older patrons.

Guerrilla Warfare: Predisposing and Precipitating Factors
by Scott G. McNall and Martha Huggins

Revolutions do not begin with a particular event but owe their inception to deep social, political, economic, and psychological tensions within the social system. Once these conditions are created, any event can act as a catalyst, sparking the revolution. The purpose of this paper is to examine predisposing factors of revolutionary guerrilla warfare, which create a revolutionary environment, and their relationship to precipitating factors, which act as a catalyst.

Predisposing Conditions

Revolutionary war is mass based, class conscious, makes use of organized violence, and assumes the political system and its leaders to be illegitimate. If those in power allow the legitimacy of the system to diminish, revolutionary conditions will develop and a political vacuum will allow the growth of revolutionary organizations. The factors that create these conditions span the range of the political, social, economic, and psychological. We identify a number major of precipitating influences.

1. Level of Development

A variety of theories relate levels of development with the potential for revolution. Davies (1962), for example, has argued that there is a J-shaped curve that can be applied to development and revolution: if development accelerates, causing expectations to increase, followed by a decline in development rate while expectations remain high, then the potential for revolution is high. But in fact the potential for conflict is related to both the level of development and the stage of development. Thus, as societies move from traditional, to transitional, and finally to the modern stage, there is a direct relationship between increase in per capita income and domestic violence, up to the modern stage. There is a curvilinear relationship between level and stage of development and level of violence. Russett concludes

> that underdeveloped nations must expect a fairly high level of civil unrest for some time, and that very poor states should probably expect an increase, not a decrease, in domestic violence during the next few decades.

Russet's explanation of why nations past the traditional-transitional phase experience a decrease in domestic violence is that as

> the economic sources of discontent diminish, the ordinary nonviolent processes of government become more accessible and effective in satisfying demands, and the government itself becomes better able to control its citizens and to prevent them from resorting to violence with any hope of success (Cited in Bwy, 1968: 24, 26).

But predisposing economic conditions may result from a lack of progress reflected in prolonged depression or stagnation. Strikes are not usually directly aimed at altering the balance of power in a country, or at changing the basic institutions, but at obtaining improved wages and working conditions. The kind of response that concerns us is revolutionary violence, stemming from economic despair, whose purpose is to alter the nature of the social system. In the case of anomic or particularistic violence the problem is seen as an aberration in the system; in the case of organized revolutionary violence the problem is seen as the system itself. When economic despair produces a belief that the problems are due to the nature of the institutions, the potential for revolutionary war increases.

2. Rate of Change

Tanter and Midlarsky have argued that there is a relationship between achievement, aspirations, expectations, and level of violence. Adapting Davies' argument, they note that a people's aspirations are related to the rate of economic growth, or of achievement. There is a fairly even balance between achievement and aspirations as long as there are no reversals in the actual achievement, such as might occur with a depression. Achievements can be political and cultural as well as economic. In Tanter and Midlarsky's framework:

> Expectations represent a change in outlook caused primarily by an *immediate* decrease in the production of social commodities. Aspirations are more in the nature of a hope and an optimism generated by *long-term* past performance.

The distance between the two concepts [revolutionary gap] may be seen as a measure of the potentiality for the occurrence of a violent revolution (Tanter and Midlarsky, 1967: 271).

They also hypothesize a relationship between rate of increase in the gross national product and the sharpness of the reversal prior to the revolution. Their examination of domestic violence in several countries tends to bear them out, if one looks only at those nations still in the traditional-transitional phase of development.

It should be emphasized that the breaks experienced need not be negative ones. We can recall Durkheim's discussion of the relationship between "fortunate crises" and suicide rates. A "fortunate" crisis affected the suicide rate in the same manner that an economic disaster did (Durkheim, 1951: 243). An economic break creates conditions of anomie and normlessness. What we look for as a precondition, then, is any rapid change, in either direction, of a country's economic fortunes.

3. Distribution of Land (Wealth)

Land reform is often part of the program of revolutionaries, especially in countries where there is a "dispossessed" group of peasants who are "alienated from the fruits of their labor." Tanter and Midlarsky again furnish evidence for the analysis of this predisposing condition. There is a relationship between land inequity and domestic violence (Tanter and Midlarsky, 1967: 277).

Hector Bejar, in discussing the potential for revolution in Peru, notes that the inequity of land distribution is one of the major preconditions for revolutionary activity.

There is an enormous concentration of land among a very few owners. One percent of all agricultural and cattle raising units occupies 75 percent of the land involved in agriculture; 0.1 percent of the total number of landowners holds 60.9 percent of the land which is being utilized. Of the 17 million hectares of arable soil, 10 million are part of 1,000 great landholdings, and only 1,933,000 are in the hands of peasant communities (Bejar, 1970: 25).

The distribution of wealth in general, as well as of land in particular, may entail comparable revolutionary possibilities.

4. Minority Groups

The suppression of, or the failure to integrate, minority groups in the larger society contributes to the potential for revolutionary warfare. (There appears to be no necessary relationship between this and development.) The reasons are basic:

a. If a group is not integrated, there is little or no difficulty on their part in defining the political system and/or its leaders as illegitimate.

b. An unintegrated group possesses, by definition, an alternative ideology.

It is not simply that a minority group can take over the country, or secede, as the Ibo attempted to do in Nigeria. Rather, they may support revolutionaries who appeal to them, or remain neutral as other groups struggle for control. Minority groups may, indeed, be "used" by both guerrillas and counter-insurgents. A case in point is the Montagnards in Vietnam.

In some cases the natives of a country are treated as an oppressed minority, even though they are the majority of the population. For example:

> The Algerian population, both Arab and Berber, was kept in a state of docility by the semi-educated marabouts, or holy men who functioned under French encouragement, and who, as they were invariably susceptible to French influence, were also known as Beni Oui Oui or 'Yes-men'. France ruled as far as was possible, especially in the countryside, through the 'caids', or French-appointed tribal leaders, who were also regarded as Beni Oui Oui. At municipal and local level there was a lack of Muslim representation, where the Europeans were easily able to out-vote them. Muslims, even though they had achieved French citizenship, did not have the same rights as the Europeans (O'Ballance, 1967: 26).

5. Foreign Presence

The presence of a foreign power can provide a focus for

discontent. It may thus stimulate a revolutionary ideology and provoke nationalism. A foreign presence may occur through dominant business interest, as in a number of "third world" countries. A foreign power may be seen as supporting an unpopular regime, or it may *be* the regime. Horowitz has discussed the problem in Latin America.

> Revolutionary sentiments are not simply a consequence of underdevelopment, but more properly a result of polarization between developed and underdeveloped sectors. And these polarities confront each other in Latin America not exclusively as a class question, but as a question in international stratification. The large-scale foreign-corporation, along with the large-scale penetration of foreign military-bureaucratic forces. produces the same effect in the "periphery" as it does at the "cosmopolitan center": it serves to absorb foreign wealth and soak up foreign labor power (Horowitz, 1969: 18).

The impact may also divide sectors of the native population from one another and intensify class feelings.

6. External War

Problems which existed or were latent before an external war may be exacerbated or brought to the surface by it. Many armed insurrections have broken out after an occupying force left a country. Often, resistance groups have continued their battle against quisling or pre-war regimes, for example, in Greece at the end of World War II. Here there was no one contending faction whose legitimacy was clearly accepted by the majority of the populace: the sense of political community was absent.

External war can fragment a society, even a victorious one, without foreign occupation, if it weakens the ability of the state to maintain the position and privileges of dominant classes and groups. Barnet suggests, in fact, that external war is the most usual precipitant cause of revolutionary war.

> In most of the revolutions of the past fifty years war has served that function. World War I completed the process of demoralization and exhaustion of the czarist bureaucracy,

leaving its short-lived successor, the Kerensky government, prey to an almost bloodless Bolshevik coup. The Japanese invasion disrupted the Kuomintang's weak political control over China and made Mao's victory possible. In the Philippines, Vietnam, Greece, Malaya, Indonesia, Yugoslavia, and elsewhere foreign invasion undermined the control of the old elites and opened the process to new forces (Barnet, 1968: 51).

7. Suppression and the Failure of Peaceful Change

As Lanternari has pointed out, the suppression of nativist religious movements in developing African countries often forced these groups to go underground and become stronger. The emotional appeal and the discipline of dissident groups may be intensified. Too, suppression may drive together opposition factions which otherwise could not cooperate.

Che Guevara (1962) has argued that before guerrilla warfare can be successful peaceful means of change must be exhausted, or the populace must come to believe that the possibilities have been exhausted. If a group can show that it has been blocked illegitimately from its rightful role, the revolutionary potential increases. Of course, a group may deliberately make demands which a central government cannot or will not meet in order to place upon the regime the onus of illegitimacy. Insurrection itself may be a means to discredit the government: the Tumpamaros tried to use a policy of urban terror to discredit the government of Uruguay by showing it could not maintain order. Guerrillas may cause the government to suspend the civil rights of people not directly involved in insurgent activities, which in turn may convert apathy into opposition and support for guerrilla activities. The most recent example of this is in Northern Ireland: here IRA attacks led to the policy of internment, which increased and solidified Catholic support for the IRA.

8. Alternative Ideology and the Intellectuals

We will adopt a loose definition of ideology to make its impact more clear. Simply, we define ideology as a set of attitudes, values, and beliefs about the way things are and ought to be. An alternative ideology is a challenge to traditional definitions and perspectives. A variety of ideologies, e.g., religious, political, social, may challenge existing ideologies.

Sociologists have argued the causal importance of ideologies since Weber's seminal discussion of the role of the Protestant ethic in the development of capitalism. The point, often missed, was not that there is something inherent about Protestantism *per se* which creates the capitalist, but that Protestantism offered a break with traditional views of the world. Protestantism represented a new ideology. Similarly, Lanternari (1965) points out that Christian missionary efforts in Africa were the source of some of the nationalistic movements that developed there. Protestantism offered a new sense of personal worth and challenged colonial as well as traditional values.

Perhaps the clearest statement of how Protestantism produces revolutionaries has been given by Michael Walzer (1965), who argues that Puritanism and Calvinism were political responses to the disorder and chaos of the sixteenth and seventeenth centuries. Attacking the older politico-religious order, they were alternative ideologies. The "saints" of the new religions became spokesmen for new ways of life or new socio-political systems; they were progenitors of the modern revolutionary, who in turn also fulfills the role of prophet.

A few men who challenge the system are sometimes sufficient to foster a revolutionary ideology. This is why, in the modern revolutionary's view, simply holding out is so important. As Guevara has shown (1962), the insurgent has an important initial advantage: he need not win, all he must do is to prevent defeat. It may be added that revolutionary martyrs may succeed posthumously. The revolutionary leader, then, acts as a charismatic prophet around whom dissident elements in the population can focus their interests. The guerrilla as prophet need not be a socially marginal individual, as Peter Berger (1963) has pointed out in his analysis of Weber's notion of charisma. Nor should it be assumed automatically that the guerrilla leader represents a movement in opposition to the basic values of society. He may attempt to reaffirm basic values.

It is often the function of the intellectuals in the movement to create ideology where it is lacking.

> Now Chief, Cervantes pursued, I took a fancy to you the first time I laid eyes on you and I like you more and more . . . because I realize what you are worth. . . . You do not yet realize your lofty noble function. You are a modest man without ambitions, you do not wish to realize the exceedingly important role you are destined to play in the revolution. It is not true that you took up arms simply because of Senor Monico. You are under arms to protest against the evils of all the *caciques* who are overrunning the whole nation. We are the elements of a social movement which will not rest until it has enlarged the destinies of our motherland. We are the tools Destiny makes use of to reclaim the sacred rights of the people. . . . What moves us is what men call ideals; our action is what men call fighting for a principle. . . .
>
> Hey, there . . . Macias called, pull down two more beers (Azuela, 1929).

Often the political philosophy develops out of the activity of the insurgents. Debray, for one (1967: 56), has argued that the role of armed insurrection is to politically educate the masses: "the most successful form of political propaganda is successful military action."

Economic development may furnish an important alternative ideology. The promise of new consumer products, jobs, and education for one's children is a powerful force in mobilizing the people of a country. It is not necessary for the vision of the good life to be generated within the country. With tourism and the mass media bringing to people in all countries the image of life in affluent societies, it is possible for discontent to be generated from without. Tannenbaum has indicated that much Latin American discontent can be traced to contact with American labor and with ideas and practices that clash with the feudal system (Tannenbaum, 1960: 167).

Barnet summarizes some of the key revolutionary functions of Communist ideology:

> . . . the communist theorists proclaim the inevitability of revolution, which is no small boost to the morale of a ragged band of rebels hiding in the jungles or the hills. Communist doctrine also offers a dogma through which one can identify

friends and enemies and plan a strategy of revolution. Moreover, it offers a plausible vision of the good society. Finally, there is really no other body of theory to which to turn (Barnet, 1968: 55).

There must, then, be some *cause*, whether political, social, economic or ethnic. It must appeal to the population and the insurgents for an extended period of time and take them through times of defeat as well as success. There can be changes of emphasis in the cause as the revolution progresses. For example, a group may begin with a reform cause and shift to one that includes land reform and socialism, as may have occurred in Castro's Cuba. But a cause must not be so abstract as to lose the attention of the masses. Bejar, in discussing the failure of guerrilla activities in Peru in 1965, makes this clear.

The ideal proclaimed by the guerrillas necessarily appeared remote to the peasants, who were interested above all in their concrete and even local demands. While the guerrillas advocated social revolution, the peasants wanted more tangible things, the realization of small demands that the revolutionaries were not always successful in incorporating into their program—in spite of the fact that these demands are the means for raising the people to a higher level of consciousness. The guerrilla's program was much more complicated and their goals much more distant (Bejar, 1970: 116).

Some writers have claimed that there must be a "revolutionary cadre" of intellectuals to articulate the class interests of and politically educate the masses.

. . . revolutions are invariably preceded by the "transfer of allegiance" of a society's intellectuals and the development by them of a new political "myth." If intellectuals have any obvious social "functions," . . . they are surely these: to socialize the members of a society outside of the deomestic context, in schools and adult learning situations; to reinforce and rationalize attitudes acquired in all social contexts; and to provide meaning to life and guidelines to behavior by means of conscious doctrines where events have robbed men of their less conscious bearings (Eckstein, 1965: 150).

Transitional societies which are educationally progressive but economically underdeveloped may be unable to use the talents of educated people. Hyde describes how this condition contributed to revolution in Sarawak, where young Chinese pursued a traditional cultural yearning for education.

> . . . the student who had worked hard and long for his exams all too frequently found that graduation still did not qualify him to do anything much more than find a living within the Chinese community itself. The best that the majority could hope for was that they would become assistants in Chinese sundries shops or clerks working with Chinese firms (Hyde, 1969: 66).

In general, men do not question the existing and dominating ideologies of their social system unless there is a severe dysjuncture in their lives and in their society. As Karl Mannheim has pointed out,

> As long . . . as the traditions of one's national and local group remain unbroken, one remains so attached to its customary ways of thinking that the ways of thinking which are perceived in other groups are regarded as curiosities, errors, ambiguities, or heresies. At this stage one does not doubt either the correctness or the unity and uniformity of thought in general (Mannheim, 1968: 7).

For Mannheim it was rapid vertical mobility that caused people to become aware of their ideologies and evaluate them. Thus, economic crises or rapid economic development would be a factor in causing people to question traditional definitions. This is one reason why, in traditional-transitional societies, rapid development is coupled with a high rate of ideological disruption. So, too, could internal war itself be a factor in causing people to question their ideologies. It might be possible for a guerrilla group through assassination, terror and sabotage to change the social position and the life-chances of enough people to make them re-evaluate existing ideologies.

In *The Ruling Class*, Mosca argues that one of the causes of a revolution is the "distance" between the elites and the masses (Mosca, 1965: 204-207). If the masses believe the elite have the same basic values they do, then the chances of revolt are less. DeGrazia gives an example: the Polish lords of the Middle Ages were ruthless in their exploitation of the peasant, but the peasants did not rebel so long as the nobles "lived among them, spoke their language, swore the same oaths, ate the same kind of food, wore the same style of clothes, exhibited the same manners or lack of them, had the same rustic superstitions" (cited in Eckstein, 1965: 146). Revolution did occur, according to DeGrazia, when the lords took on French manners and actually behaved in a more humane fashion to the peasants.

But other analysts have noted that division and disaffection may be crucial within as well as between classes. In *The Russian Revolution*, Trotsky argued that revolution required three things: (1) the political consciousness of a revolutionary class; (2) discontent in the intermediate layers, or middle class; and (3) loss of morale in the ruling class (Trotsky, 1919: 311). If elites are torn by dissent, disruption and intrigue, the task of the revolutionary is easier. Just as the alienated intellectual challenges the notion of community and signals the loss of common values, so does a disorganized or disaffected elite. It may be vital to note that intellectuals tend to come from intermediate or higher social strata, and so, thus, does revolutionary leadership.

These predisposing factors create the possibility of internal warfare. It is obvious that the more of them that operate strongly together, the more probable becomes a revolutionary condition.

Precipitating Factors

Predisposing conditions are necessary but not sufficient for revolution. This is similar to the distinction that Lieberson and Silverman make between the precipitating and underlying conditions of race riots (1965). The specific events which precipitate it are not the causes of a riot or of a revolution. There are many comparable situations in which revolutions sometimes do and sometimes do not break out.

Inconsistencies of policy or leadership often trigger unrest. A government may change its strategy or tactics of repression, increasing or decreasing it. Either can trigger revolution, or raise the level of participation in a revolution already under way. In Vietnam, American bombing of peasant villages, destruction of rice paddies, machine-gunning of water buffalo, and other destructive measures increased active support and sympathy for the Viet Cong. The widespread use of torture by the French during the Algerian revolution also raised the level of repression and aroused public opposition.

If a government shifts from a high to a low level of repression, it is not at all certain that the response would be support or acquiescence. First, such a change would probably reflect a state of crisis or malaise, and would represent a concession. But it is unlikely that a group seeking significant social or political change would stop at a simple change in level of repression, which might suggest government weakness. Rather, there would be increased demand for further reforms. A population hardened to severe repression is not as likely as one unaccustomed to it to fear the use of violence as a means to achieve political ends. Some countries have an endemic history of violence by those in power against the people; turning the tables would come easy. At the same time, signs of impending change may stiffen the determination of adherents of the status quo. Events in Soviet-dominated East Europe in the 1950's, following the death of Stalin, well illustrate these possibilities. The removal or death of a repressive leader has opened the door to unrest in many instances.

Levine (1959: 422-423), in a discussion of the rise of anti-European violence in Africa, argues persuasively that

> (a) in a territory where the government has pursued a relatively consistent policy favoring African self-government, there will be little anti-European violence; (b) where the government has pursued a consistently repressive policy toward African self-rule, there will also be little anti-European violence; (c) where the government has pursued inconsistent policies toward African political autonomy, there will be a good deal of anti-European violence.

References

Azuela, Mariano. *The Underdogs*. New York: Brentano's, 1929.

Barnet, Richard J. *Intervention and Revolution*. New York: World, 1968.

Guerrilla Warfare: Predisposing and Precipitating Factors

Bejar, Hector. *Peru 1965.* New York: Monthly Review Press, 1970.

Berger, Peter L. "Charisma and Religious Innovation: The Social Location of Israelite Prophecy." *American Sociological Review,* 1963.

Blasier, Cole. "Studies of Social Revolution: Origins in Mexico, Bolivia, and Cuba." *Latin American Research Review,* 1967.

Bwy, D. P. "Political Instability in Latin America: The Cross-Cultural Test of a Causal Model." *Latin American Research Review,* 1968.

Campbell, Arthur. *Guerrillas: A History and Analysis.* New York: John Day, 1968.

Clausewitz, Karl von. *On War.* New York: Modern Library, 1943.

Davies, James C. "Toward a Theory of Revolution." *American Sociological Review,* 1962.

Debray, Regis, *Revolution in the Revolution?* New York: Grove Press, 1967.

Durkheim, Emile. *Suicide.* New York: Free Press, 1951.

Eckstein, Harry. "On the Etiology of Internal Wars." *History and Theory,* 1965.

Gott, Richard. *Guerrilla Movements in Latin America.* London: Thomas Nelson, 1970.

Guevara, Ernesto "Che". *Che Guevara on Guerrilla Warfare.* New York: Praeger, 1962.

Horowitz, Irving Louis, Josue de Castro and John Gerassi. *Latin American Radicalism.* New York: Vintage Books, 1969.

Hyde, Douglas. *The Roots of Guerrilla Warfare.* London: Bodley Head, 1968.

Knorr, Klaus. "Unconventional Warfare: Strategy and Tactics in Internal Political Strife." *Annals of the American Academy,* 1962.

Lanternari, Vittorio. *Religions of the Oppressed.* New York: New American Library, 1965.

LeVine, Robert A. "Anti-European Violence in Africa: A Comparative Analysis." *Journal of Conflict Resolution*, 1959.

Lieberson, Stanley and Arnold R. Silverman. "The Precipitants and Underlying Conditions of Race Riots." *American Sociological Review,* 1965.

Mannheim, Karl. *Ideology and Utopia.* New York: Harcourt, Brace, World, 1963.

Marcuse, Herbert. *One Dimensional Man.* London: Routledge and Kegan Paul, 1964.

Mosca, Gaetano. *The Ruling Class.* New York: McGraw-Hill, 1965.

O'Ballance, Edgar. *The Algerian Insurrection, 1954-62.* Hamden: Archon, 1967.

Oppenheimer, Martin. *The Urban Guerrilla.* Chicago: Quadrangle, 1969.

Reik, Miriam. "Ireland: Religious War—or Class Struggle?" *Saturday Review,* 1972.

Roszak, Theodore. *The Making of a Counter Culture.* New York: Anchor Books, 1969.

Rummel, R. J. "Dimensions of Conflict Behavior Within Nations, 1946-59." *Journal of Conflict Resolution,* 1966.

Tannenbaum, Frank. "On Political Stability." *Political Science Quarterly,* 1960.

Tanter, Raymond. "Dimensions of Conflict Behavior Within and Between Nations, 1958-60." *Journal of Conflict Resolution,* 1966.

Tanter, Raymond and Manus Midlarsky. "A Theory of Revolution." *Journal of Conflict Resolution,* 1967.

Trotsky, Leon. *The History of the Russian Revolution.* London:

British Socialist Party, 1919.

Walzer, Michael. *The Revolution of the Saints.* Cambridge: Harvard University Press, 1965.

This essay is adapted and abridged from a paper given at the annual meeting of the American Sociological Association, 1971.

Insurgency in the Countryside Of Underdeveloped Countries
by G.C. Alroy

Who said history never repeats itself? It apparently just has in the world of Communism—and with a vengeance. For Marx, who had turned his teacher Hegel upside down when fashioning his own conception of world revolution, has now suffered the same fate at the hands of his Chinese disciples. Just as he had replaced the struggle of nations in Hegel's historical dialectic with the struggle of social classes, so the Chinese have replaced the urban proletariat in his blueprint for global revolution with the peasantry, the class for whose insurrectionary quality Marx had only contempt. This heresy has been affirmed by a startling statement from Peking issued on the twentieth anniversary of Japan's surrender in World War II. Flying in the face of all orthodox Marxist dogma, the statement (as reported in the New York *Times*) proclaimed that "the peasants constitute the main force of the national-democratic revolution against the imperialists and their lackeys. . . . The countryside, and the countryside alone, can provide the revolutionary bases from which the revolution can go forward to final victory." The statement emphasized that "Mao Tse-tung's theory of establishing revolutionary base areas in the rural districts and encircling the cities from the countryside is of outstanding and universal importance for the present revolutionary struggles of all the oppressed people, and particularly for the revolutionary struggles of the oppressed nations and peoples in Asia, Africa, and Latin America against imperialism and its lackeys." And, on a global scale, "if North America and Western Europe can be called 'the cities of the world,' [and] Asia, Africa, and Latin America constitute 'the rural areas of the world' . . . the contemporary revolution also presents a picture of the encirclement of cities by the rural areas."

While this heretical statement must have puzzled some, the fact is that it merely dramatizes a familiar trend in red revolutionism, one at least as old as the asserted peasant character of the Communist Chinese insurgency some thirty years ago. True, the trend has been resisted by the more orthodox Russian Communists. For instance, in his *Conversations with Stalin*, Milovan Djilas tells that in 1944 the men in the Kremlin rejected his interpretation of the Titoist Revolution, in which he had stressed the new revolutionary role of the peasantry. (Perhaps the Soviets already sensed in that novelty a threat to their leadership of international Communism.) But the spectacular successes of the Chinese Revolution and similar insurgencies have given the heresy an unprecedented momentum in the years since the Second World War. The popular writings of Jean-Paul Sartre and others have glorified and romanticized the peasant insurgents in the "non-white" world; these *Damnés de la terre*, as Franz Fanon eulogized them, have been depicted ever more pure and pristine as the last hope of revolutions betrayed by the corrupt men of the cities. After the success of the Castro Revolution, which even non-Communists termed a "peasant revolution," Guevara not only publicly affirmed the validity of Mao's rural strategy for all Latin America, but also heaped scorn on the vaunted revolutionism of Marx's urban proletariat. The cities were not merely tactically disadvantageous to armed insurgents, they even sapped the will to fight. For, according to Guevara, "The ideological influence of the cities inhibits the guerrilla struggle by raising hopes for peacefully organized mass struggle."

Indeed, if nothing succeeds like success, even in Marxist dogma, the Chinese heretics would seem to stand on some firm ground. The successes of rural-based insurgencies in our age are not only spectacular in themselves, not only impressive in comparison with the fortunes of non-rural insurgencies, but downright upsetting in their implications for conventional thinking about revolution in general. With few notable exceptions, as in the Palestine terror, incumbent regimes appear remarkably capable of controlling urban insurgencies. (Successful non-rural insurgencies have, by and large, not been urban either, but almost purely military.) This is true even where they seem to express profound popular discontent, as in Budapest in 1956. East Berlin and Poznan tell the same sad tale, though the human sacrifice there was not entirely futile, since some later liberalization was won. But could any regime ever rouse more desperate hatred than that of the ghetto fighters or Bor-Komorowski's Poles for the Nazis? Yet there was Warsaw, twice drowned in blood! Actually, the most sustained, most successful insurgency against even these ruthless (and efficient) rulers was waged in one of the most backward rural areas in Europe, in Yugoslavia. And even if the Yugoslavs did not stand alone—for the Allies aided them—the fact of backwardness may have had much to do with their exceptional triumph.

It would be reassuring if at least the rural-based insurgencies

Insurgency in the Countryside
of Underdeveloped Societies

supported our comfortable moral convictions. But by and large they do not. Consider the rural-based insurgencies, at least long-sustained when not successful, in the "transitional" countries.[1] These were against regimes that could not possibly have been hated more than the Nazis and were often genuinely progressive. Consider too that it took the Soviet Union itself over twenty years to pacify vast areas in Central Asia. And surely the challenge of the Arab Middle Ages to the twentieth century in the Yemeni backlands cannot be dismissed by those inclined, like most of us, to rate the progressivism of incumbent governments inversely by the strength of insurgencies against them.

Strange though it may seem, especially in view of the legendary inertness and reactionary instincts of the peasantry, the countryside of the developing world does indeed seem to possess the remarkable revolutionary potential claimed by the Chinese heretics, though not to the extravagant degree claimed. But why? One would imagine that the rural setting offers some advantage to insurgents and that great difficulties of terrain give even less than popular insurgents a fighting chance against even ruthless incumbents. But why should insurgents find the countryside of underdeveloped countries particularly propitious for their aims?

II

One principal reason is the appalling performance ability of governments in the rural areas of underdeveloped countries. While the Chinese statement refers to this fact, many of us are unaware that the typical effectiveness of regimes in those areas of the world is not merely low for the purpose of ambitious socio-economic reforms, but frequently insufficient to even maintain a semblance of order. Occasionally reports from rural areas in Asia, Africa, and Latin America speak of the actual absence of the national government. What this may mean in even a relatively stable country has recently been indicated by Karl von Vorys in his *Political Development in Pakistan;* he found that government effectiveness in simple law enforcement in vast border areas approximates zero. It takes a little time for such a fact to sink into our consciousness.

Yet it may be better *not* to be aware of this poor performance ability at all than to take it in the sense in which the

Chinese Communists do. Not only does their historic statement confuse matters by speaking of the inability of "imperialist" conquerors fully to control the countryside—for this may apply to such cases as colonial incumbents—but it distorts when it holds that the insurgent rural peoples resist a hated government because they are fighting for a better one. The bare truth of the matter is that these peoples typically resist *all* national, modern governments—including the one which is nominally their own. And their national government is typically not even sufficiently present among them to really exist.

Under these conditions, not only do insurgents in a very real sense rise against a government that is not there, but also against one not capable of getting there fast. For the infrastructures in communication and transportation taken for granted in economically developed societies typically do not exist in underdeveloped ones. What else could underdevelopment mean more poignantly than the utter inability of government to deploy its forces effectively in its own territory? It is ironic that challenged regimes should be derided so often as unrepresentative of the mass of their people, when their basic weakness is precisely their being too much like their technically backward people. Totalitarian regimes in such societies actually illustrate the very opposite of this; they remain vastly unrepresentative in this sense for some time, and their development is devoted to their machineries of control and coercion. This is, in effect, their first, almost intuitive, reaction upon taking power. Then they safely take the peasants' land away.

Moreover, whatever governmental presence does exist in rural areas of underdeveloped countries is typically inefficient, even when serious advances in public administration have been made at the center. After all, the same deficiencies which obstruct governmental control over the rural populace also handicap control over the outlying officials themselves. The governmental presence thus very often amounts to arbitrary and fitful rule by corrupt potentates, a notorious condition in which even the lowly gendarme may outrage the peasantry almost at will.

Besides appearing to rise against a government that is not there, insurgents in the backlands of underdeveloped countries in a very real sense often do not initiate the act of insurrection either. The fact is that these areas are often in a state of permanent insurrection anyway. Modern guerrillas usually appear

Insurgency in the Countryside
of Underdeveloped Societies

where rural brigands always have been, and usually still are. The modern guerrillas follow an ancient pattern, for larger insurrection is almost invariably epidemic where rural brigandage is endemic. This was so since time immemorial: Wat Tyler coincided with the fabled Robin Hood in England; the France of the *coquins, cotreux, tuchins,* and other bandits was the France of the great *jacqueries;* the Germany of Schinderhannes, the fabled brigand hero, was the Germany of the Great Peasant War; in Eastern Europe, the time of the Ruthenian brigand Dovbush was the time of one of the greatest waves of rural risings in history; and the Andalusia of *bandolerismo* was the Andalusia of the most severe rural risings in Spanish history before the Civil War. Actually, who can tell brigand from rebel? Were Emiliano Zapata and Pancho Villa Mexican brigands or agrarian revolutionaries? But why bother? We know that Castro, a truly radical revolutionary, assiduously and successfully cultivated the Robin Hood image for himself. More than that, we know that Mao's Red Army actually absorbed whole bands of traditional Chinese brigands in its ranks for years. Why should not other revolutionaries elsewhere again do the same?

Indeed, the other principal reason why the rural areas in underdeveloped countries are so propitious for revolution is the kind of peasantry residing in them. One cannot emphasize too much the backward condition of these men, since the conventional association of radical revolutionaries with social progress may easily obscure it. The fact of the matter is that these backward peoples offer many substantial advantages for even progressive insurgents. They are manipulable in ways difficult for us to understand and to an extent unsuspected by most of us. In the prevailing state of lawlessness in which such peasants live, even the modern guerrilla may appear in traditionally defined roles at least as familiar as the nominal government itself. Perhaps we should elaborate this a little.

The virtual absence of government in much of the backlands of underdeveloped countries simply means that the traditional parochial loyalties of family, clan, or tribe dominate. These peoples live by and large in essentially feudal or pre-feudal settings, within variously effective traditionalist forms of governance. To again quote von Vorys on backward areas of Pakistan:

Tribal laws administered by the tribal hierarchies remain determinant. It is a case in point that the government does not interfere with the manufacture and carrying of arms in these areas and is rather helpless when rail lines are taken up by the tribesmen to serve as the chief supply of iron for the weapons. Nor does it try to interfere with the settlement of disputes. The kinship group adjudicates such matters, or in the case of interfamily discord the issue is resolved by the tribal council (*jirga*). Although time and again the tribes have agreed formally to refrain from harboring fugitives from justice, the fact is that very few such persons are turned over to the authorities. More often, when such a fugitive requests sanctuary according to *Nanawatai,* he is hospitably entertained and protected from arrest.

Under typical conditions, even modern guerrillas may appear as any other group of outlaws; and while this is not an unmixed blessing—for peasants have fought and betrayed the Robin Hoods as often as they have sheltered them—the boon to the insurgents is extraordinary. The guerrillas appear to follow an established tradition of rebellion; they fight the traditional foe; they incorporate traditional bands; they even take men and taxes as legitimately, or illegitimately, as the nominal government—if they can. However, since the modern guerrillas are not *just* another group of outlaws, but typically possess skills of large-scale organization, efficient and honest cadres, and other resources which the nominal governments often cannot match, they can tax and conscript. And often they do.

Yet, eating their cake and having it too, the socio-radical insurgents may harness the forces of reform which may exist among the peasants as well as exploit their traditionalism. Just because backwardness works for the insurgents, they are not deprived of the rewards of antitraditionalism. The land question, sometimes an acute one in transitional societies, is an obvious illustration. Although the conventional explanation, that the peasants fight for land, may all too crudely oversimplify the manner and motive of their involvement in revolutionary warfare, the revolutionaries' espousal of agrarian reform has surely not hurt them either. The craving for land, where it exists, may help the revolutionaries most with the great mass of the passive peasants, those among whom they must "swim like fish in water." As for the others—the more mobile, more daring, more exceptional peasants who actually join the insurgents—they are likely to be men who really flee the land rather than fight for it for themselves; who break from traditionalism, who perhaps want modernity, who see it in the radical insurgents, and are attracted to them.

III

However impressive the advantages for insurgency in the rural areas of transitional societies may be, it is imperative to point out that they do not necessarily assure its success. Notwithstanding the sanguine language of the Chinese Communist leaders, other factors affect the outcome of insurgencies; otherwise all the Third World would permanently be on fire. International alignments, geographic contiguity, availability of resources, and other, often sobering, factors have much to do with whether or not a commitment to launch revolutionary warfare is made. Nor are the Chinese Communists by themselves capable of making such commitments around the world, even if they wanted to. Other men must undertake the hardship and risks of such warfare.

But it is equally important to see that it is objective conditions of backwardness on which revolutionary warfare thrives. Without this understanding, the so-called battle for the hearts and minds of men may be just an empty slogan—not only when it is invoked by the insurgents and their apologists, but also by their opponents. The battle is really between exploiting unglamourous backwardness and overcoming it. To a considerable extent, the guerrillas' alleged great closeness to the people thus appears to be true, but in a sense which the apologists of revolutionary wars surely never intended. On the other hand, how fundamentally futile appears the conventional conception of counterinsurgency! It is ironic, though perhaps understandable, that these doctrines should address themselves mostly to the guerrilla stage of revolutionary wars in mainly military terms, when it is precisely this phenomenon which would appear to be practically endemic to the rural areas in transitional societies. Moreover, while guerrillaism is not invincible in every case, as in Malaya, the Soviet Union, Cuba under Castro, and the Philippines, revolutionary war may be even less vulnerable in that particular stage than in the next one, the shift to conventional warfare, actually the stage which counterinsurgency doctrines usually gave up as lost.

Nor are the conditions of backwardness we have examined quite the same as those normally envisaged by advocates of socio-economic reforms against the threat of insurgency. While their reforms seem geared chiefly to the fulfillment of popular aspirations and social justice, we have been impressed not only with the absence of conditions in which these concepts could mean what most of us think, but even more with the truly crucial importance of organization, manipulation, control-ability, in the rural areas of the underdeveloped world. And this, unlike popular aspirations, chiefly concerns cadres, elites, structured small groups of men, rather than the large masses we often hear about.

It goes without saying that the substantially developed societies, at which the reformist approach ultimately aims, would indeed be immune to the threat of guerrillaism, just as claimed. But what good does it do? True, some time in the distant future all societies will have reached that blessed state; but there is neither time enough, nor sufficient governmental ability in the regimes concerned, even if there were enough will, to effect that transition as a stabilizing factor. Actually, intensive development will itself most certainly constitute a serious, further destabilizing factor.

Just the same, the work of development has to be done. And in the long critical meantime, the regimes concerned may find the odds favoring the pursuit of control and effective integration of their peoples, by organization and other means, even more than producing those popular satisfactions we normally associate with economic development—ownership of land, liberty, material abundance. We may not like the sound of this, but have we a better explanation for the remarkably good record of totalitarian regimes against insurgencies? All claims of their apologists notwithstanding, these regimes are not exactly high in popularity—effective, yes; popular, hardly. Actually, one tends to see them as much more popular than they are precisely because they have so much staying power. Is it not possible—indeed probable—that we tend to see totalitarian insurgents also as more popular for the same reason? Precisely because organization is so awfully scarce in the rural areas of the underdeveloped world, even a little bit goes a long way toward seriously challenging a government, thus further enhancing the impressive staying power of so many insurgencies in such areas.

We may wonder, finally, how insurgents can generate organizational prowess, honest cadres, disciplined ranks, and strongly committed elites where others lack them. After all, they operate in the same deprived setting as the governments they challenge. The answer seems to be their ideological fervor, which serves now in the same way in which religious fervor has often served in the traditional societies of the past—as a substitute for the material underpinning of effective organization. It is easy to see that the insurgent groups on whom Peking counts most possess infinitely more of this vital *ersatz* than many regimes in the world can even hope to have. But just because the cards are stacked against many regimes in the underdeveloped world, there is no good reason for adding gratuitous insult to their injury, their great vulnerability to insurgency, by assuming that its sheer existence is proof of their innate depravation; nor is intelligence served when revolutionary successes in the rural areas of underdeveloped countries are, as so often happens, explained in terms which have little to do with the real world.

Notes

1. Algeria, Angola, Bolivia, Burma, China, Colombia, the Congo, Cuba, India, Indonesia, Iraq, Malaya, Mozambique, Peru, the Philippines, the Sudan, Venezuela, Vietnam, among others.

The Peasantry as a Political Factor
by T. Shanin

Peasants are the majority of mankind. For all but comparatively few countries, 'the people' (as opposed to 'the nation') still denotes 'the peasants'; the specific 'national culture' closely corresponds to peasant culture; 'the army' means young peasants in uniform, armed and officered by men different from themselves. And yet one has to be reminded of it.

'It is a commonplace to say that agrarian history as such is neglected—the fact is too obvious to be denied'[1]—this holds true for many branches of social science as far as the countryside is concerned. The decade which has elapsed since this passage was written had not much improved the situation, apart from several notable exceptions in the fields of anthropology and history in the last few years. Indeed in the growing flood of social science publications, the few existing rural studies have almost been submerged. But reality seems to confute this solipsism of 'civilized mind'. Day by day the peasants make the economists sigh, the politicians sweat, and the strategists swear, defeating their plans and prophecies all over the world—Moscow and Washington, Peking and Delhi, Cuba and Algeria, Congo and Vietnam.

Even more striking than the neglect of study, are the emotional undertones and diversity of opinion which shroud this subject.'Mitrany's[2] 400 pages bring together but a fraction of the views expressed. Writers, scientists and politicians have all contributed to the discussion, in which the image of the peasant has swung from that of an angelic rustic humanist to a greedy, pig-headed brute. For example in Russia in the same period the peasantry has been held to be the 'real autocrat of Russia'[3] and 'non-existent historically speaking'.[4] This kind of verbal contest did not make reality much clearer. The peasantry went its own way quite oblivious of being an intellectual nuisance.

The emotional tension of ambiguous contempt or utopian praise, the allegory replacing definition as well as the acute shortcomings in our conceptual grasp of peasantry are all strongly felt in Western intellectual tradition. The neglect of the subject is but a symptom of this. It calls for serious study in the sociology of knowledge into the 'eidos' of intellectual image makers when dealing with 'the class that represents the barbarism within civilization'.[5] The peasantry as an 'undecipherable hieroglyphic to the understanding of the civilised'[6] seems to be determined by a conglomeration of factors, of which one stands out as crucial. The real peasantry does not fit well into any of our concepts of contemporary society. This 'maddening' peasant quality seems to lie at the roots of the problems of research in this field.

In this paper we shall start by trying to define the *differencia specifica* of the peasantry—the uniqueness by which the peasantry may be defined and selected. The definition of peasantry will necessarily consist of both static and dynamic elements, either one of which is insufficient on its own. From this starting point we shall proceed to the problem of peasantry as a part of society, and then to the patterns of political influence of this entity. In dealing with this subject other approaches are feasible and indeed needed. The translation of rich, complex reality into a verbal form of fewer dimensions makes many approaches possible and valid, subject to a recognition of the limitations involved.

'Peasant society and culture has something generic about it. It is a kind of arrangement of humanity with some similarities all over the world.'[7] In this way Redfield summarises a wide comparison of peasants in different periods and countries. Peasantry appears to be 'a type without localization—not a typical anthropologist's community'.[8]

The peasantry consist of small producers on land who, with the help of simple equipment and the labour of their families, produce mainly for their own consumption, and for the fulfilment of their duties to the holders of political and economic power.

(i) The relation to land and the specific character of agricultural production, lies at the roots of the specific features of the peasant family farm.

The produce from the farm fulfils the basic consumption needs of the peasant family, and allows the peasant relative independence from other producers and the market. This makes for the great stability of peasant households, which in crises are able to maintain their existence by increased effort, lowering their own consumption and partially withdrawing from market relations.

The mainly agricultural production puts limits on the density and concentration of population and influences human intercourse. Furthermore it determines the cycle of peasant labour activities and life. Nature introduces an element of interference beyond human control, with which all the peasant community is faced.

The holding of land by being 'a necessary and generally sufficient condition to enter the occupation'[9] acts together with other factors as an entrance ticket into the peasantry. Moreover the position in the hierarchy of peasant sub-groups is to a great extent defined by the amount of land held.[10]

In peasant households, land appears as a traditionally defined and stable holding, which does not necessarily imply legal ownership. We shall define land property as socially accepted rights of holding and utilizing land—rights which are apart from those gained by labour and capital input. These rights are expressed by the possibility of transferring them, at least temporarily.

Land property in a wide sense appears on the one hand as the customarily defined peasant family holding, and on the other hand as politically formalised, legal ownership. In actual fact, the legal ownership of peasant land may lie with the peasant himself, the commune, the landlord or the state, the land becoming therefore a private plot, communal property or a customary lease-holding. 'Landlords are not needed to establish the fact of peasantry'.[11] Their acquisition of part of the peasants' produce, and even their political and administrative interference generally failed to break the basic features of the peasant/land interrelation.

(ii) The family farm is the basic unit of peasant ownership, production, consumption and social life. The individual, the family and the farm, appear as one indivisible whole. 'The identification of interest of family and farm holding seems to be a typical characteristic of the traditional peasant family'.[12] The farm takes the dual form of a production and consumption unit. The balance of consumption needs, family labour available and farm potentialities, strongly influences peasant activities. The profit and accumulation motives rarely appear in their pure and simple form, which makes the neat conceptual models of maximisation of income in a market society of most doubtful applicability to peasant economy.[13] The new rapidly developing patterns of industrialising society 'are found outside agriculture which still remains the domain of the familistic model'.[14]

Peasant property is at least *de facto* family property. The head of the family appears as 'the manager rather than proprietor of family land',[15] and his position 'has rather the character of management of common family property'.[16] These two descriptions given by different scholars about the peasantry of two different countries show striking similarities. Whatever the imposed national legal structure, peasants seem to act within this social frame.

The family's social structure determines the division of labour, status and social prestige. Moreover, 'The family is the production team of the farm and the position in the family defines the duties to the farm, the functions and rights attached. The rhythm of the farm defines the rhythm of family life.'[17]

The prestige and position of an individual in peasant society is basically determined by two factors, as is his own self-evaluation and image. These factors are firstly the status of the family he is born into, and secondly his position within this family. His position within the family once again does not depend primarily on personal ability, but on his progression through certain basic ascribed positions, i.e. childhood; partial maturity before marriage; the period after marriage but before full independence; independence which may be gained either by leaving the family farm and establishing his own, or by becoming head of the family farm on his parents' death or retirement; and finally the period of his own retirement.[18] Therefore marriage is 'an absolute postulate'.[19] Family interest directs the choice—and an unmarried man (even a farm-owner) 'arouses unfavourable astonishment' and 'does not count',[20] not being able to carry out farm activities to the full. The main definition of the family lies in the full participation in the farm unit, the hard core of which consists of a married couple or polygamous group and their offspring. The family to the Russian peasant at the beginning of the twentieth century is generally 'the people who eat from the same dish', and to the French peasant in the same period 'the people who are locked behind the same lock'.[21] Family solidarity provides the basic framework for mutual help, control and socialisation. The individualistic element of personal feelings submits to the formalised frame of accepted family role behaviour. Forming the basic nucleus of peasant society, the family farm determines peasant everyday action, interrelation and values. Together with the mainly natural economy, it makes for the segmentation of peasant society into small units with a remarkable degree of self-sufficiency and ability to withstand economic crises and market pressures.

The relative uniformity of peasant family patterns in different societies and periods diminishes when we move to the level of the village community. At this level the national uniqueness of political organization and social value structure becomes increasingly influential.

(iii) The fundamental importance of occupation in defining men's social position, role and personality, is well known, though poorly studied. Galeski, however, in his previously quoted book, *Peasants and Farming as an Occupation*, concerns himself with this problem in both an analytical and empirical manner. The ambiguity in the occupational definition of the farmer's trade, seems to stem from its unique character. Apart from its family structure, the necessary ties to land and the relatively high independence from the market, its uniqueness lies in its being a peculiar exceptionally wide set of interrelated functions carried out on a rather unspecialised level. Although many of the tasks fulfilled by the peasant are also carried out by other occupational groups, the uniqueness of the peasant's work is the combination of functions performed. This leads to many special characteristics of every day peasant life, as well as its resistance to industrialisation. The process of growing specialisation in the countryside leads to the development of a rural, non-farming population. Simultaneously the farmer's function is progressively narrowed and becomes more professionalised as the farm develops into an enterprise. The peasant becomes a farmer. However, the tasks which cannot be easily divided into a few repetitive actions and mechanised, for example livestock management, still remain largely his special province.

These features of farming determine the process of socialisation and occupational education of the young as one which is highly diffused, personal, informal and conducted mainly within the framework of the family.

(iv) The peasantry is a pre-industrial social entity which carries into contemporary society specific, different and older elements of social interrelation, economics, policy and culture. This will be elaborated in the following section, but at this stage we should like to stress that we refer not only to the 'relicts of the way of production which already belongs to the past',[22] not only to belated development, but to a specific development.[23]

Most of the existing definitions of peasantry are covered by that proposed above. One definition, however, stands aside from the others, that of A.L. Kroeber[24] taken over by R. Redfield,[25] which approaches the peasantry as 'a part society, with part culture'. This will be taken into account, according to the adopted line of reasoning, in the section on the interrelation of peasantry with society as a whole.

A concept of a social stratum cannot be limited to any empircally chosen group.[26] The relation between theoretical models and operational definitions used consciously or unconsciously in empirical studies seems to be one of the most difficult problems of modern sociology.'[27]
Yet the importance and validity of conceptual definition of peasantry for any kind of research in this field seems to us beyond doubt.

The peasantry appears not only as a distinctive social group, but as a general pattern of social life, which defines a stage in the development of human society. 'The peasantry is a way of living', says Fei[28] in his classical description of Chinese society. This general pattern of social life makes its appearance as a sector of earlier tribal, mainly nomadic, society, becomes decisive and marks a historically distinctive period, that of a society of small producers and then gradually sinks as a section within industrial society. The appearance of the small-producer pattern of life is marked by the major change referred to as 'agricultural revolution'.[29] This stage created the basis for stable settlement, land division and a revolutionary rise in productivity which brought with it the possibility of a comparatively stable surplus allowing for annual fluctuations in yield. Production has become determined to an increasing extent by the labour utilised.[30]

Property relations and nuclear units appear as the major indices of economic and social life, and may be used to delineate the society of small producers.

Property relations barely exist in tribal-nomadic society.[31] They appear in their wide sense discussed above, in the small producers' society, and become fully, legally formalised in the capitalistic, industrial one. The kinship group is the basis of social relations in nomadic, tribal society, and remains so in the narrower familism of a small producers' society. The individual in his own right 'does not count', he is but a part of the family whole. The town and market-centered industrialising society, however, broke down this frame. The individual becomes the basic nuclear unit of society, free to interact in the new, huge and complex social hierarchies and structures. Therefore the prevalence of family units of production and family property may well demarcate the social pattern of small producers and the historical periods in question.

The society of small producers shows a distinctive 'cultural pattern',[32] features of which persist in the peasantry of industrialising societies.[33] The basically 'social, rather than economic' way of reasoning, 'the lack of calculation' (i.e. of the maximising of income as the only aim determinant) were widely documented by Thomas and Znaniecki and stressed by every keen student of peasant life.[34] A great deal has been said about the irrational behaviour of the peasants as far as land,[35] loans,[36] 'fair-prices'[37] and income[38] are concerned. Peasant logic seemed to be changeable and subjective,[39] with elements of what can be called pre-Socratic thought, by which two contradictory opinions may be held simultaneously. What remains sometimes overlooked is the fact that the peasants' 'exposed stupidity' is evidence, not necessarily of an absence of thought, but rather of a frame of reference and pattern of thought peculiar to the group, and serving their needs well.[40]

This point is borne out increasingly by recent studies. R. E. F. Smith has already pointed out the cyclical rather than linear concept of time held by Russian peasants, which is clearly linked to their productive life.[41] Pitt-Rivers defines the main features of a closed community as habitual personal contact, wide endogamy, homogeneity of values, emphasis on strict conformity, intense group solidarity, marked ideological egalitarianism, etc., and this may serve as a generalisation of recent anthropological research into specific peasant cultures.[42] The clash of this particular culture and its gradual weakening to the external and new *Weltanschauung* of the industrialising and 'civilised' world, is an important element of social history.

The village structure, to a much greater extent than the family farm, presents us with features unique to a specific country and period.[43] In the context of the village community or peasant commune, the peasant reaches the level of nearly complete self-sufficiency. The appropriation and division of land, marriage, sociability, religious needs are generally taken care of at the village level. The common interests of communal rights as well as productive actions which need the participation of more than one family provides for co-operation generally coupled with some grass roots democracy. All this makes the word *mir*, i.e. 'the world' and 'peace', used by the Russian peasants to refer to their village commune, a significant description of its function. The village is the peasant's world. The small producers' society consisted of innumerable village segments generally under the spell and suppression of alien, political hierarchies.

The small producers' society falls historically in the intermediate period between the nomadic tribal and the industrialising societies. The word intermediate sometimes tends to be used interchangeably with 'unstable,' 'temporary' and even 'not important to look at'. However, the small producers' pattern of society proved as lasting as and no less stable than any other historical pattern. The society based on organic, cyclical, non-structured dynamics, with the subsistence family farm as its basic nucleus proved exceptional inner stability all over the world. Indeed one does not need Wittfogel's hydraulic despotism to explain the striking examples of arrested structural change collected in his book.[44] The basic social nucleus of the family subsistence farm with its cyclical and organic stability seems to be far more of a common element in all the societies quoted than their 'hydraulic' features. Furthermore, a surplus-exhausting, highly centralised state, bears, if any, the potentialities of structural change, by the introduction of a powerful external pressure on the world of natural economy and cyclical stability.

The peasant backbone in the small producers' society dissolves under the influence of the rise of a market and town-centered economy, and the consequent industrialisation. An analysis of the development of the economic surplus is needed for the understanding of this process.[45] The development of agriculture provided the basis for industrialisation, but the farms themselves remained to a great extent apart from this new social framework.[46]

The producing and trading town introduces general social patterns alien to the world of small producers. The impersonal, warfare-like, profit-centered market relations are here at the root of human relationships. The man freed from the ties and protection of the family, becomes an individual participant in mass society, structured by huge bureaucratic hierarchies. The accumulation of anonymous capital determines economic growth. The general claim of efficiency and achievement provides the core of the social value system.

By the advantages of capital concentration, rise in productivity, spread of education, political weight and population growth, the urban society rapidly overtakes the countryside, and becomes the main determinant of social and economic change. The peasants' small producers' world becomes a mere segment of a world very differently structured. Moreover, whilst still preserving elements of uniqueness, the countryside develops a special relation with the town, which becomes increasingly decisive in its own development. The town's lead is felt by the increasing influence of market relations, the draining off of surplus labour and capital, the professionalisation of agriculture, the spread of mass culture and production and the evidence of 'social disorganisation'. [47]

The approach to the countryside's development is a town-centered society, as simply belated and not different, proved wrong but persistent. In fact, we may clearly see three parallel patterns of spontaneous development of the countryside:

i) The competition of large-scale, capital intensive, mechanised agriculture, gradually destroys the small farms. The concentration of land-ownership is followed by growing concentration of production. Agriculture, being fully taken over by an industrial method of production, becomes 'merely a branch of industry'.[48] This development is apparent in the large farms of the United States, North Italy and Central France,[49] as well as in some of the Soviet *sovkhozy*. Yet the special features of the farming occupation create difficulties for its division into simple, repetitive actions, i.e. its full automation. This, together with the competitive strength of the family farm unit and the fact that synthetic foods are relatively unimportant, halt the processes of the 'food factory' becoming the main method of food production.

ii) The town-centered society makes for the development of the peasants into a professional stratum of farmers. The poorer villagers are increasingly drained from the countryside by the expanding urban areas. The same happens to the peasant entrepreneurs and to part of the economic surplus. The middle peasants, relying on the advantages of the family production unit and the increasing cooperative movement, fight successfully for their place in the market society. These unique features of the development of the farmer stratum has already been pointed out by Marx[50] and described as *the* only way for the peasantry to develop by O. Bauer.[51] The latest studies of Polish and German sociologists have shown the growth of a new stratum of worker-peasants who supplement their agricultural, mainly subsistence, production by hiring out their labour.

This pattern of development of the peasantry into a cohesive, increasingly narrowing and professionalised occupational group of farmers, is clearly seen in most parts of North-West Europe. Although increasingly tied to the industrialising society, farming still maintains some of its peculiar elements.

The socialist states which allow for the activity of small producers in the countryside, providing them with the necessary aid, and curbing capitalist development (Russia in the NEP period, contemporary Poland and Jugoslavia) bring the above pattern to its clearest expression.

iii) The third pattern of development appears mainly in the so-called under-developed societies. The population explosion, the developing market relations and the industrial competition with traditional handicrafts breaks up the cyclical equilibrium of society. The slow industrialisation is not able to drain the countryside of its excess labour nor to provide sufficient capital accumulation. The potential surplus is swept away by growing consumption needs. In the small producers framework this development is not expressed by increasing unemployment, but by 'hidden' under-employment, excess population in the countryside, lowering of average income per head and increasing misery.[52]

iv) As distinct from these three spontaneous trends of development, the increasing strength of the modern state and the wish of the revolutionary elites to tackle the problem of development within the framework of socialist, collectivistic thinking makes for the appearance of state organised collectivisation of agriculture. This pattern is qualitatively different from the spontaneous trends, by being a conscious plan put into operation by a political hierarchy. The evaluation of its success in

The Peasantry as a Political Factor

any of the different forms taken would seem to be premature. Yet in the Soviet Union, where the earliest attempts were made, the specific elements of peasant life, the strength of the peasant farm plot and the unique ability of farming to defeat town-designed plans was proved to quite a surprising degree.

The difficulties of achieving a conceptual grasp of the peasantry have been clearly felt in discussions about the place of the peasantry in society. Even people starting from similar theoretical assumptions reach opposite conclusions. The peasantry is a class to Stalin,[53] a petit bourgeois mass to Kritsman,[54] and not a class but a notion, to Plekhanov.[55]

This, clearly, is partly due to differences in definition. Ossowski[56] elaborates the three different ways the concept of social class is used by Marx, and many other conceptual sub-divisions of society have been applied by different writers. Different analytical aims, inter-related with different concepts of society made for this diversity. Yet the type of conceptual sub-division of society chosen is crucial to any interpretation of social structure, functioning and change.

The main European sociological tradition[57] of conceptual sub-division of contemporary society stems from Marxian class analysis. Social class is approached as a unity of interest, expressed in group sub-culture, feeling of belongingness and common action, and shaped by conflict relation with other classes. Society is structured by this dialectic composition of inter-class conflict and unity.

If we take the criteria for defining class as being the distribution of power,[58] the control of means of production,[59] or the organisation of production,[60] the peasantry in an industrialising society will fall into either a huge amorphous group of 'ruled', or into an even more amorphous group of middle-classes.[61] The peasantry as a qualitatively distinct entity disappears. This leads the majority of Marxist social scientists to approach the peasantry as a disappearing remainder of pre-capitalist society—as 'not existing historically speaking'. Yet when a major part of the population remains outside the concept of society as a whole, the definition seems to be sadly inadequate, even if the consolation of looking into the future is offered. Unfulfilled predictions seem to be the inevitable result of such a model.

Max Weber's modification of the Marxist concept of class

puts market relations at the roots of class definition.[62] 'Class situation is, in this sense, ultimately market situation.'[63] 'Class situations are further differentiated, on the one hand, according to the kind of property that is usable for returns, and, on the other hand, according to the kind of services which can be offered in the market.'[64] For Weber therefore, 'owners of warehouses' and 'owners of shares', for example, constitute social classes, as much as do industrial workers and peasants. The shortcomings of an unlimited analytical division of society into small sub-groups, when approaching social reality has already been indicated by Marx in his unfinished manuscript on social class.[65]

In history, the peasantry has acted politically many times as a class-like social entity. Moreover, the peasantry of industrialising society has proved its ability for cohesive political action, not only when facing traditional land-owners in belated pre-capitalist battles. For their common interest has also driven the peasants into political conflicts with large capitalist landowners, with various groups of townsmen and with the modern state.

The polarisation of the countryside in an industrialising society into capitalist owners and rural proletariat as predicted by Marxists, was arrested by the urban drain of capital and labour, as well as by the specific features of a peasant family farm economy. The widely accepted image of the countryside being rapidly split by the inevitable economic trends of polarisation, slowed down to some extent by non-economic factors, proved over-simplified. Economic counter-trends seem to act in the opposite direction, greatly influencing the final result. Furthermore the elements of specific culture, consciousness and 'the meaning attached'[66] to the class position proved most persistent. All this made peasant cohesiveness as the potential basis of political class formation much stronger than the predictions of Russian Marxists or American strategists would lead us to believe.

On the other hand the basic division of peasants into small local segments, the diversity and vagueness of political aims, considerably weakens their political impact. Hence, how far peasantry may be regarded as a class is not a clear-cut problem, but rather should be seen as a question of degree and historical period. If we posed an imaginary scale or continuum, we could say that the peasantry would appear as a social entity of comparatively low 'classness', which rises in crisis situations.

But the peasantry's special features as a social group are not merely quantitative. Marx's classical description of the duality of peasant social character, (on the one hand it is a class, on the other it is not)[67] leaves the riddle unsolved. In so far as the peasantry is not a class, what is it—granting its qualitative existence?

A class position is basically a social interrelation—a conflict inter-relation with other classes and groups. Outside this inter-relation a class ceases to exist. Yet 'because the farmer's produce is essential, and at the lowest level, sufficient for human existence, the labour of a farmer is necessary for the existence of a society, but the existence of a society as a whole, is not to the same extent necessary for the existence of a farmer.'[68] The peasants prove this by withdrawing from the market in crisis situations, and indeed sometimes use this ability consciously or unconsciously as a tool of political influence.

The main duality of the peasant's position in society is in their being on the one hand a social class (of low 'classness') and on the other 'a different world', bearing the elements of a distinctive pattern of social relation—a highly self-sufficient society in itself.

The peasantry is the social phenomenon in which the Marxist tradition of class analysis meets the main conceptual dichotomy of non-Marxist sociological thinking: Maine's brotherhood—economic competition; de Coulanges familistic-individualistic; Tönnies' Gemeinschaft—Gesellchaft and Durkheim's mechanic (segmentarian)—organic societies.[69] This unique duality of 'class' and 'society' accounts for conceptual difficulties, yet may well serve as a qualitative definition of the peasantry especially when delineating it from the wide amorphous groups of 'middle classes', 'exploited masses' or 'remainders of feudalism'.

As already stated, A. L. Kroeber advanced a definition of peasants as 'constituting part societies with part cultures, definitely rural, yet live in a relation to a market town . . . lack the isolation, political autonomy and self-sufficiency of a tribal population, yet their local units maintain much of their old identity, integration and attachment to soil.'[70] Redfield elaborates Kroeber's point and adds 'there is no peasantry before the first city.'[71]

The anthropological approach by which the extent of

cultural self-sufficiency is used as an index of social development seems to be valid. Moreover, research centered around the problem of the development from tribal to a small producer's society will necessarily stress different factors than that centered round the development from a small producer's society into an industrialising one. However, Redfield's definition of peasantry seems to be too narrow, and his definition of tribal society too all-inclusive. Settlers in many countries, cut off from towns, far from noblemen and even generally out of the reach of the state and its tax-collectors can hardly be labelled tribal. Yet these groups share the main features common to peasants. Moreover, they seem to prove the peasantry's self-sufficiency, its ability to exist out of the spell of noblemen and the town. The social/political significance of this seems to have neccessitated the structure of power relations of pre-capitalistic society. For this very self-sufficiency made political suppression necessary to the rulers.

The political impact of the peasantry was generally marked by its basic socio-political weaknesses. The vertical segmentation into local communities, clans and groups, the differentiation of interest within the communities themselves, the difficulties in the crystallisation of nation-wide symbols and aims and in the rise of leadership and organisation made for what we have called low 'classness'. Technological backwardness, especially in the fields of communication and military action, brought to nothing many political attempts. Yet the peasantry had its socio-political points of strength in being the main food producer, its dispersion in rural areas and its numerical size. Its 'monopoly' of food production proved of crucial importance in times of crisis. The spread of the countryside could become a stronghold. Numerical strength may tip the balance.

Yet in the long run the basic weakness of the peasantry prevails. The peasantry proved no match for the smaller, closely-knit, well-organised, technically superior groups, and was time and time again double-crossed or suppressed politically and militarily. However, granting all this, the peasantry's existence and action cannot be ignored as politically sterile and therefore without significance. For not only victors and rulers determine political reality.

The spread of industrialisation and mass culture gives the peasantry new possibilities of communication and cultural cohesiveness. Yet at the same time it lowers the importance of the countryside in the national production, curbs by international trade the 'food-monopoly', stimulates polarisation, raises the technological advantages of government's power of suppression. Once again the line of development seems to weaken the peasants' political influence.

However, the peasants' chances of influencing the political sphere increases sharply in times of national crises. When the non-peasant social forces clash, rulers split or foreign powers attack, the peasantry's attitude and action may well prove decisive. Whether this potentiality becomes reality is mainly dependent upon the peasants' ability to act in unison with or without organisation. This in turn is dependent upon the cohesiveness of the peasantry, its economic social and cultural similarities and interaction, and their reflection on the ideological sphere.

A comparison between the peasantry's political and military action in pre-industrial and in contemporary society remains to be done. We shall, however, concentrate on the latter.

The patterns of peasant political action and influence are determined by its character as a social entity.

1. *Independent class action*, as described by Marxist class theory. In this pattern, a social class crystallises in conflict, creates its nation-wide organization, works out its ideology, aims and symbols and produces its leaders from within. This form of political action seems to be fairly typical of the main social classes. However, for contemporary peasantry this pattern of political action is the least frequent. The 'Green movement' in Eastern Europe, the peasant unions in Russia—1905, and China—1926, the Zapata movement in Mexico and all their counterparts in the rest of the world have to be studied comparatively to understand the mechanics of this pattern of peasant action.[72]

2. *Guided political action*, in which the class is moved by an external uniting power elite. This pattern of action may become especially important as far as the peasantry is concerned. The conservative cyclical stability of both the farm and the village, and its political implication may generally be overcome only by a severe crisis, coupled by an external factor of a sweeping political and emotional force. This external organiser of the peasantry may be found in millennial movements, secret societies, Russian cossacks, Mexican *sinarquistas*, French Bonapartism or Mao's

people's army, and provides the peasantry with the missing factor of unity on a wide scale. The common element found in all these very different movements is a closely-knit group of activists, with its own momentum, specific structure of organisation, aims and leadership, to whom the peasantry is an object to be led and/or manipulated. The peasantry in this case may be 'used' (consciously tricked into action alien to its own interests) or 'led to achieve its own aims'.[73] Yet the definition of aims remains in the hands of qualitatively distinct leaders. The peasants' interests and attitudes are only one of the factors to be taken into account. As stated by Marx, referring to the French peasantry in the mid-nineteenth century, 'they are consequently incapable of enforcing their class interest in their own name whether through a parliament or through a convention. They cannot represent themselves, they must be represented. Their representative must at the same time appear as their master . . .'[74] The only thing to be objected to in this statement is absolutism, which has been refuted by later events.

The low-'classness' of peasantry, makes the study of peasant movements especially illuminating for the sociological analysis of the leading elites. Owing to the lower self-consciousness of the peasantry and hence its weaker influence on its own leaders, the elite group dynamics seem to appear in a purer form. Moreover, it leads us to look into the problem of class-like masses (i.e. social groups acting temporarily as class entities without bearing all the features of a class) and into their place in political development (e.g. the Russian soldiers in 1917--18).

3. *The fully spontaneous, amorphous political action.* This pattern seems to be highly typical of peasant impact on political reality, and may take one of two forms:

(a) Local riots appearing as 'sudden' short outbursts of accumulated disappointment and rebellious feeling. Generally easily overcome by the suppressive centralised power, these riots may act as a check on central policy (indicating necessary changes). When interrelated with crises in other areas and spheres, they may develop into nation-wide movements capable of determining major political changes.

(b) The second form is peasant passivity. The conceptual grasp of passivity as a factor of dynamics is not easy. Yet the spontaneous lowering of production activities by Russian peasantry in 1920 proved strong enough to defeat a government which was victorious

The Peasantry as a Political Factor

in war against powerful enemies. Enormous numbers of governmental decrees and orders have been defeated all over the world by the peasantry's spontaneous, stubborn and silent nonfulfillment. Furthermore, the influence of peasant conservative apathy has many times proved decisive for the victory of the establishment over the revolutionaries. That nonresistance is a special peasant contribution to politics, elaborated and sophisticated by L. Tolstoy and Gandhi, has been suggested by R. E. F. Smith and points in the same direction. The interrelation between the basic features of peasant society and passive resistance seems to be evident.

In the study of the political life of a society, military action has a place of special importance. Clausewitz' remark that war is an extension of politics by other means holds true, not only in the interrelation between countries. This leads to the need for a special consideration of the army and guerrilla warfare as frames of contemporary peasant political action.

The modern army of mass enlistment is one of the few nationwide organisations in which the peasantry actively participates. The segmentation of the peasantry is thereby broken. By cultural intercourse and intermixture, if not by indoctrination, the peasant-soldier is taught to think in wide national, and not village-limited terms. He is taught organisation, complex co-operative action, co-ordination, modern techniques and military skills. The army provides him with a hierarchial institution through which he may rise as a leader and be trained for this position. Even where some national organisations are represented on a rural level, the army generally provides the peasant with the framework for the most active participation.

The increase in peasant ability to act politically, when put in the army, is generally successfully curbed by rigid discipline and by control from non-peasant officers. Yet in a time of crisis, this suppression disappears and the peasant army, its attitudes, action and refusal to act may become decisive. Moreover, the experience gained in army service acts as an important influence in the villages. The ex-serviceman by his new experience tends to become a leader and a channel through which outside influences reach the villagers. In attempting to organise politically, the peasants frequently refer back to their army experience. The Russian *Tamanskaya armiya* and the 'Green army' of the Black

Sea, the F.L.N., the Chinese 'people's militia', the Zapata and Villa armies in Mexico served not only as military but as the main political organization, a kind of party in arms.[75]

The army, as this kind of organisation, may bear the marks of both the first and second patterns of political action, i.e. the peasantry as a class for itself, or as a 'guided' socio-political entity.

During the last decade partisan warfare, by its success, has been moved into the center of public attention. American strategists approach guerrilla warfare as a specific military technique to be taught by smart sergeants along with saluting and target practice. Their remarkable failure in both guerrilla and anti-guerrilla warfare is the best comment on this approach.

Guerrilla warfare is the most suitable form of expressing peasant militancy. Its record seems to be as old as the peasantry itself. Innumerable rebels, brigands and outlaws appear in popular memory as well as in the real history of every people. The ability of the amorphous guerrilla 'army' to disperse in times of need into the sympathetic peasant mass and the expanses of the countryside, its ability to utilise various degrees of peasant militancy and friendly passivity, its ability to survive without outside supplies and the adequacy of primitive military techniques, may make the guerrillas unbeatable by modern military methods.

Yet the essentially peasant character of guerrilla warfare provides not only its strength but also its weakness: segmentarism, lack of a crystallised ideology and aims, lack of stable membership. These essential weaknesses may be overcome by an injection of a hard core of professional rebels, making the revolt into guided political action. The professional rebels' nation-wide ideological and organisational cohesiveness, their stability and zeal and their ability to work out a long-term strategy enables them to unite the peasantry, transforming its revolt into a successful revolution. Yet the main key to the understanding of guerrilla warfare has to be sought not in the marvels of the rebels' organisation, but in their interrelation with the peasanty, not only in the military techniques of the few, but in the sociology of the masses.[76]

There are subjective determinants of military action generally labelled 'morale', the resistance of which to quantitative analysis does not limit their importance in the shaping of reality. Peasant revolts all over the world display common cultural features which, in all their complexity, seem to be better grasped by the synthetic expressions of the arts than by the analytical tools of the social sciences. The leader-hero, the legends which surround him, and his personal charisma, take to a large extent the place of ideology and organisation as a uniting factor. The picturesque image of the young peasant rebel challenges the mundane nature of every day peasant life. The childish display of exhibitionism described by Znaniecki[77] is typical of the peasant's attempts to establish his own personality when breaking out of his rigid family ties, explains much of the spirit of peasant fighters. All these features influence the general character of peasant units as a fighting force, together with the specific values and self-image of the leading elites.

The main stream of contemporary sociology has by-passed the traditional peasantry. Rural sociology has been localised in, and financed by, rich, industrial societies, and consequently centered upon the problem of how to promote the farming minorities into fully productive and wealthy members of 'civilised society'. Few sociologists have elevated the peasantry from the footnote to the page.

Yet if historical and social significance were the criteria for subjects of study, we would be virtually swept by publications dealing with peasantry. Innumerable problems of contemporary world political and economic development lead us back to the subject of the peasantry, and its understanding and misunderstanding by policy makers. To take but one example, the history of the Soviet Union was to a great extent determined by the sequence of the ruling party's evaluation—prediction—policy—policy's unexpected results, time and time again, in 1918, 1920, 1927–1929, etc.[78] and we may continue in this vein until 1964. Countless other examples could be cited in Africa, Asia, Latin America, etc., etc.

Only a combination of both conceptual and factual research by different disciplines and approaches, may overcome the astonishing short-comings of our knowledge of the peasantry and the methodological difficulties involved. Limping along main roads achieves more than marching along side roads.[79]

Notes

1. F. Dovring: *Land and Labour in Europe* 1900-1950, 1956, p.5.
2. D. Mitrany: *Marx Against Peasants,* 1951, dealing with Marxist as well as populist ideology.
   See also S.H. Franklin: Reflection on the Peasantry, in *Pacific Viewpoint* III No. 1.
3. V. Chernov—as quoted by J. Maynard: *The Russian Peasant,* 1962, p 97.
4. G. Plekhanov on the Russian peasantry.
5. K. Marx: *Capital,* Vol. III, Hamburg, 1890, p. 348.
6. K. Marx and F. Engels: *Selected Works,* 1950, Vol. I, p. 159.
7. R. Redfield: *Peasant Society and Culture,* 1956, p. 25.
8. *Ibid.* pp. 23-25.
9. B. Galeski: *Chlopi i Zawod Rolnika,* Warszawa, 1963, p. 48.
10. *Ibid.* p. 47: 'A rise within the professional group of farmers is traditionally achieved by enlargement of the land holding, by a rise from the position of owner of a small farm to the position of an owner of a bigger one, and the description "good farmer" is generally attached in the view of the village to all owners of the biggest farms without exception and is not linked to the real professional skill or effectiveness of their work.'
11. Redfield: *op. cit.,* p. 28.
12. Galeski: *op. cit.,* p. 140.
13. The proofs of that statement cannot be brought 'marginally' and the reader is referred to the studies by Znaniecki, Galeski or Chayanov and his group (though this last seemed clearly to overstate his own case).
14. Galeski: *op. cit.,* p. 57.
15. W. I. Thomas and F. Znaniecki: *The Polish Peasant,* 1958, p. 92.
16. B. Mukhin: *Obychnyi poryadok nasledovaniya,* 1888, p. 62.
17. *Ibid.* p. 140.
18. Thomas and Znaniecki: *op. cit.,* p. 93. See a very similar analysis in A. Vasil'chikov: *Zemeledel'e i zemvladenie,* 1876, Vol. 2, p. 21.
19. Thomas and Znaniecki: *op. cit.* p. 107.
20. *Ibid.* p. 107.
21. A. Chayanov: *Organizatsiya krest'yanskogo khozyaistva,* 1925, p. 21.
22. K. Marx: *Selected Works, op. cit.,* Vol. II, p. 303.
23. Galeski: *op. cit.*
24. In Marxist studies appearing generally as the definition of 'small' or 'middle' or 'parcela' peasant to be delineated from the capitalist farmers.
25. A. L. Kroeber: *Anthropology,* 1923, p. 284. See also R. Redfield: *The Primitive World and its Transformation,* 1953.
26. For analysis see Galeski: *op. cit.,* pp. 16-19.
27. *Ibid.* p. 19.

28. R. Bendix and S. M. Lipset: *Class, Status and Power*, 1953, p. 32. 'The peasantry, the key toward understanding of China, is a way of living, a complex of formal organisation, individual behaviour and social attitudes closely knit together for the purpose of husbanding land with simple tools and human labour.'
29. See for example V. Gordon Childe: *Social Evolution*, 1951.
30. See E. Mandel: *Traité d'économie marxiste*, Jerusalem, 1964, pp. 33-36, 41.
31. Except fighting strangers at the tribal hunting territories.
32. Cultural patterns being defined for that purpose as 'the lens of mankind through which men see; the medium by which they interpret and report what they see'. C. Wright Mills, *Power Politics and People*, 1962, p. 406.
33. Cf. Redfield: *Peasant Society and Culture, op. cit.*
34. Even Marx mentions the rural societies in which the laws of labour are never developed and therefore the 'general economic laws of society' do not work.
35. Thomas and Zananiecki: *op. cit.*, p. 173.
36. *Ibid.* p. 161.
37. *Ibid.* p. 169.
38. *Ibid.* p. 166.
39. See for example Mukhin: *op. cit.* p. 311, stating that the peasants' court or meeting tends to decide property disagreements 'according to men', i.e. in accordance with the personality of the people involved rather than the proofs presented.
40. See for example the Polish sociologists' studies of the prestige determinants of peasant economic action or Chayanov's proofs of the 'economically "rational" renting of land when the cost of rent is higher than the additional profit gained, a sensible thing in conditions of wasted surplus labour.
41. R. E. F. Smith: 'A Model of Production and Consumption on the Russian Farm', University of Birmingham, Centre for Russian and East European Studies, Discussion Paper RC/D 1, 1964, p. 11.
42. See J. Pitt-Rivers: *The Closed Community and Its Friends*, 1957. For a summary of contemporary anthropological research see C. Geertz in *Biannual Review of Anthropology*, 1961, pp. 1-41.
43. For tabulation see S. Eisenstadt: *The Political System of Empires*, pp. 34-35 and the supplementary tables.
44. K. Witfogel: *Oriental Despotism*, 1963.
45. *The Transition from Feudalism to Capitalism—a Symposium*, 1963; Mandel: *op. cit.*
46. Mandel: *op. cit.*, p. 173. According to Mandel's evidence even in the present day U.S.A. 1,250,000 small farms show a mainly natural economy.
47. Thomas and Znaniecki: *op. cit.*, p. 1122.
48. K. Marx: *A Contribution to the critique of political economy*, 1904, p. 303.

49. See the reports from the international socialist symposium on the European peasantry as published in *Al Hamishmar*, Tel-Aviv 1964.
50. K. Marx: *Capital*, Vol. III.
   The moral of this story, which may also be deduced from other observations of agriculture, is that the capitalist system works against a rational agriculture, or that a rational agriculture is irreconcilable with the capitalist system, although technical improvements in agriculture are promoted by capitalism. But under this system, agriculture needs either the hands of the self-employing small farmer, or the control of associated producers.'
51. O. Bauer: *Bor'ba za zemlyu*, 1926, p. 203.
52. See for an example *Pourquoi les travailleurs abandonnent la terre*, Geneva, 1960, pp. 138, 144, which reports on India as follows:
   During the years 1941-51 the natural growth of labour force in the countryside was absorbed in Agriculture   70.3%
   in Services   28.3%
   in Industry   1.4%
   During the years 1931-51 the share of workers engaged in agriculture rose from 71 to 74 % of the working population and in 1952 74 % of peasant families held less than 2 ha. land and 1/3 is reported as landless.
   For an elaboration of the mechanism of such social processes see I. H. Boeke: *Economics and Economic Policy in Dual Societies*, 1953.
53. I. Stalin: *Problems of Leninism*, 1945, p. 510.
54. See the introduction to A. Gaister: *Rassloenie sovetskoi derevni*, 1928, p. XIII.
55. G. Plekhanov.
56. See S. Ossowski: *Class Structure in the Social Consciousness*, 1963.
57. We shall not touch upon the hierarchial status groups and American studies of social stratification which are irrelevant to our study.
58. See R. Dahrendorf: *Class and Class Conflict in Industrial Society*, 1959.
59. Marx: *Selected Works*, op. cit. p. 33 and V. I. Lenin: *Collected Works*, (5th ed.) Vol. 39, p. 15.
60. See works by Bogdanov, Makhaiski, etc. and may be traced to St. Simon.
61. See B. Semenov: *Problema Klassov . . .*, 1959.
62. For exposition of all aspects see J. Rex: *Key Problems of Sociological Theory*, 1961.
63. G. H. Gerth and C. Wright Mills: *From Max Weber*, 1961, p. 182.
64. Bendix: *op. cit.*, p. 64.
65. Marx: *Capital*, Vol. III, 1909, pp. 1031-1032.

66. See Rex: *op. cit.*, p. 138.
67. Marx: *Selected Works, op. cit.*, p. 303.
    'In so far as millions of families live under economic conditions of existence that separate their mode of life, their interests and their culture from those of other classes, and put then in hostile opposition to the latter, they form a class. In so far as there is merely a local interconnection among these small-holding peasants, and the identity of their interests begets no community, no national bond and no political organisation among them, they do not form a class.'
68. Galeski: *op. cit.*, p. 49.
69. See R. Redfield: *The Little Community*, 1955, pp. 139-143.
70. Kroeber: *op. cit.*, p. 284.
71. R. Redfield: *The Primitive World*, 1953, p. 31.
72. For an important insight into the influence of the stratification of peasantry on political action, see H. Alavi: Peasantry and Revolution, *The Socialist Register*, 1965.
73. For presentation of this dichotomy see for example: V. Lenin: *What is to be done?* as compared with G. Sorel: *Reflections on Violence*.
74. For example Marx: *Selected Works, op. cit.*, p. 303.
75. One of its forms is described by Marx in the *Communist Manifesto* when speaking about the early stage of bourgeois class organisation as: 'an armed and self-governing association in the mediaeval commune'. See Marx: *Selected Works, op. cit.*, p. 34.
76. Inroads into research on this subject were made by E. Hobsbawn: *Primitive Rebels* 1959, and Vietnam and the Dynamics of Guerrilla Warfare in *New Left Review*, 33.
    See also A. Gilly and others in *Monthly Review*, Vol. 17, on guerrilla activities in Latin America.
77. Thomas and Znaniecki: *op. cit.*, p. 103.
78. See for example E. H. Carr in *Proceedings of the British Academy*, XLIX, 1963, pp. 90-91.
79. This paper was primarily prepared for an inter-disciplinary conference on peasantry, sponsored by the Centre for Russian and East European Studies of the University of Birmingham. The generalisations attempted here have been based on preliminary studies of Russian, Chinese, Polish and Mexican societies, with the consequent severe limitation of excluding many important areas, especially the densely populated South Asia.

On Peasant Rebellion
by Eric R. Wolf

Six major social and political upheavals, fought with peasant support, have shaken the world of the twentieth century: the Mexican revolution of 1910, the Russian revolutions of 1905 and 1917, the Chinese revolution which metamorphosed through various phases from 1921 onwards, the Vietnamese revolution which has its roots in the Second World War, the Algerian rebellion of 1954 and the Cuban revolution of 1958. All of these were to some extent based on the participation of rural populations. It is to the analysis of this participation that the present article directs its attention.

Romantics to the contrary, it is not easy for a peasantry to engage in sustained rebellion. Peasants are especially handicapped in passing from passive recognition of wrongs to political participation as a means for setting them right. First, a peasant's work is more often done alone, on his own land, than in conjunction with his fellows. Moreover, all peasants are to some extent competitors, for available resources within the community as for sources of credit from without. Secondly, the tyranny of work weighs heavily upon peasants: their life is geared to an annual routine and to planning for the year to come. Momentary alterations of routine threaten their ability to take up the routine later. Thirdly, control of land enables them, more often than not, to retreat into subsistence production should adverse conditions affect their market crop. Fourthly, ties of extended kinship and mutual aid within the community may cushion the shocks of dislocation. Fifthly, peasants' interests—especially among poor peasants — often cross-cut class alignments. Rich and poor peasant may be kinfolk, or a peasant may be at one and the same time owner, renter, share-cropper, labourer for his neighbours and seasonal hand on a near-by plantation. Each different involvement aligns him differently with his fellows and with the outside world. Finally, past exclusion of the peasant from participation in decision-making beyond the bamboo hedge of his village deprives him all too often of the knowledge needed to articulate his interests with appropriate forms of action. Hence peasants are often merely passive spectators of political struggles or long for the sudden advert of a millennium, without specifying for themselves and their neighbours the many rungs on the staircase to heaven.

If it is true that peasants are slow to rise, then peasant participation in the great rebellions of the twentieth century must obey some special factors which exacerbated the peasant condition. We will not understand that condition unless we keep in mind constantly that it has suffered greatly under the impact of three great crises: the demographic crisis, the ecological crisis and the crisis in power and authority. The demographic crisis is most easily depicted in bare figures, though its root causes remain ill understood. It may well be that its ultimate causes lie less in the reduction of mortality through spreading medical care, than in the diffusion of American food crops throughout the world which provided an existential minimum for numerous agricultural populations. Yet the bare numbers suffice to indicate the seriousness of the demographic problem. Mexico had a population of 5.8 million at the beginning of the nineteenth century; in 1910—at the outbreak of the revolution—it had 16.5 million. European Russia had a population of 20 million in 1725; at the turn of the twentieth century it had 87 million. China rumbered 265 million in 1775, 430 million in 1850 and close to 600 million at the time of the revolution. Viet-Nam is estimated to have sustained a population of between 6 and 14 million in 1920; it had 30.5 million inhabitants in 1962. Algeria had an indigenous population of 10.5 million in 1963, representing a fourfold increase since the beginnings of French occupation in the first part of the nineteenth century. Cuba had 550,000 inhabitants in 1800; by 1953 it had 5.8 million. Population increases alone and by themselves would have placed a serious strain on inherited cultural arrangements.

The ecological crisis is in part related to the sheer increase in numbers; yet it is also in an important measure independent of it. Population increases of the magnitude just mentioned coincided with a period in history in which land and other resources were increasingly converted into commodities—in the capitalist sense of that word. As commodities they were subjected to the demands of a market which bore only a very indirect relation to the needs of the rural populations subjected to it. Where, in the past, market behaviour had been largely subsidiary to the existential problems of subsistence, now existence and its problems became subsidiary to the market. The alienation of peasant resources proceeded directly through outright seizure or through coercive purchase, as in Mexico, Algeria and Cuba; or it took the form—especially in China and Viet-Nam—of stepped-up capitalization of rent which resulted in the transfer of

resources from those unable to keep up to those able to pay. In
addition, capitalist mobilization of resources was reinforced
through the pressure of taxation, of demands for redemption
payments and through the increased needs for industrially
produced commodities on the part of the peasantry itself. All
together, however, these various pressures disrupted the
precarious ecological balance of peasant society. Where the
peasant had required a certain combination of resources to effect
an adequate living, the separate and differential mobilization of
these resources broke that ecological nexus. This is perhaps best
seen in Russia where successive land reforms threatened continued peasant access to pasture, forest and ploughland. Yet it is
equally evident in cases where commercialization threatened
peasant access to communal lands (Algeria, Mexico, Viet-Nam),
to unclaimed land (Cuba, Mexico), to public granaries (Algeria,
China), or where it threatened the balance between pastoral and
settled populations (Algeria). At the same time as commercialization disrupted rural life, moreover, it also created new
and unsettled ecological niches in industry. Disruptive change in
the rural area went hand in hand with the opening up of incipient but uncertain opportunities for numerous ex-industrial
peasants. Many of these retained formal ties with their home
villages (Algeria, China, Russia); others migrated between
country and industry in continuous turnover (especially Viet-
Nam). Increased instability in the rural area was thus accompanied by a still unstable commitment to industrial work.

Finally, both the demographic and the ecological crisis
converged in the crisis of authority. The development of the
market produced a rapid circulation of the *elite*, in which the
manipulators of the new 'free-floating resources'—labour bosses,
merchants, industrial *enterpreneurs*—challenged the inherited
power of the controllers of fixed social resources, the tribal chief,
the mandarin, the landed nobleman.[1] Undisputed and stable
claims thus yielded to unstable and disputed claims. This rivalry
between primarily political and primarily economic power-
holders contained its own dialectic. The imposition of the market
mechanism entailed a diminution of social responsibilities for the
affected population: the economic *entrepreneur* did not concern
himself with the social cost of his activities; the traditional
power-holder was often too limited in his power to offer
assistance or subject to co-optation by his successful rivals. The

advent of the market thus not merely produced a crisis in peasant ecology; it deranged the numerous middle-level ties between centre and hinterland, between the urban and the rural sectors. Commercialization disrupted the hinterland; at the very same time it also lessened the ability of power-holders to perceive and predict changes in the rural area. The result was an ever-widening gap between the rulers and the ruled. That such a course is not inevitable is perhaps demonstrated by Barrington Moore,[2] who showed how traditional feudal forms were utilized in both Germany and Japan to prevent the formation of such a gap in power and communication during the crucial period of transition to a commercial and industrial order. Where this was not accomplished—precisely where an administrative militarized feudalism was absent—the continued widening of the power gap invited the formation of a counter-*elite* which could challange both a disruptive leadership based on the operation of the market and the impotent heirs of traditional power, while forging a new consensus through communication with the peasantry. Such a counter-*elite* is most frequently made up of members of provincial *elites*, relegated to the margins of commercial mobilization and political office; of officials or professionals who stand midway between the rural area and the centre and are caught in the contradictions between the two; and of intellectuals who have access to a system of symbols which can guide the interaction between leadership and rural area.

Sustained mobilization of the peasantry is, however, no easy task. Such an effort will not find its allies in a rural mass which is completely subject to the imperious demands of necessity. Peasants cannot rebel successfully in a situation of complete impotence; the powerless are easy victims. Therefore only a peasantry in possession of some tactical control over its own resources can provide a secure basis for on-going political leverage. Power, as Richard Adams[3] has said, refers ultimately 'to an actual physical control that one party may have with respect to another. The reason that most relationships are not reduced to physical struggles is that parties to them can make rational decisions based on their estimates of tactical power and other factors. Power is usually exercised, therefore, through the common recognition by two parties of the tactical control each has, and through rational decision by one to do what the other wants. Each estimates his own tactical control, compares it to the other, and decides he may or may not be superior'.

The poor peasant or the landless labourer who depends on a landlord for the largest part of his livelihood, or the totality of it, has no tactical power: he is completely within the power domain of his employer, without sufficient resources of his own to serve him usefully in the power struggle. Poor peasants, and landless labourers, therefore, are unlikely to pursue the course of rebellion, unless they are able to rely on some external power to challenge the power which constrains them. Such external power is represented in the Mexican case by the action of the Constitutionalist army in Yucatan, which liberated the peons from debt bondage 'from above'; by the collapse of the Russian army in 1917 and the reflux of the peasant soldiery, arms in hand, into the villages; by the creation of the Chinese Red Army as an instrument designed to break up landlord power in the villages. Where such external power is present the poor peasant and landless labourer have latitude of movement; where it is absent, they are under near-complete constraint. The rich peasant, in turn, is unlikely to embark on the course of rebellion. As employer of the labour of others, as money-lender, as notable co-opted by the State machine, he exercises local power in alliance with external power-holders. His power domain with the village is derivative; it depends on the maintenance of the domains of these power-holders outside the village. Only when an external force, such as the Chinese Red Army, proves capable of destroying these other superior power domains, will the rich peasant lend his support to an uprising.

There are only two components of the peasantry which possess sufficient internal leverage to enter into sustained rebellion. These are (a) a landowning middle peasantry or (b) a peasantry located in a peripheral area outside the domains of landlord control. Middle peasantry refers to a peasant population which has secure access to land of its own and cultivates it with family labour. Where these middle-peasant holdings lie within the power domain of a superior, possession of their own resources provides their holders with the minimal tactical freedom required to challenge their overlord. The same, however, holds for a peasantry, poor or 'middle', whose settlements are only under marginal control from the outside. Here landholdings may be insufficient for the support of the peasant household; but subsidiary activities such as causal labour, smuggling, live-stock raising—not under the direct constraint of an external power domain—supplement land in sufficient quality to grant the peasantry some latitude of movement. We mark the existence of such a tactically mobile peasantry: in the villages of Morelos in Mexico; in the communes by the central agricultural regions of Russia; in the northern bastion established by the Chinese Communists after the Long March; as a basis for rebellion in Vietnam; among the *fellaheen* of Algeria; and among the squatters of Oriente Province in Cuba.

Yet this recruitment of a 'tactically mobile peasantry' among the middle peasants and the 'free' peasants of peripheral areas poses a curious paradox. This is also the peasantry in whom anthropologists and rural sociologists have tended to see the main bearers of peasant tradition. If our account is correct, then—strange to say—it is precisely this culturally conservative stratum which is the most instrumental in dynamiting the peasant social order. This paradox dissolves, however, when we consider that it is also the middle peasant who is relatively the most vulnerable to economic changes wrought by commercialism, while his social relations remain encased within the traditional design. His is a balancing act in which his equilibrium is continuously threatened by population growth; by the encroachment of rival landlords; by the loss of rights to grazing, forest and water: by falling prices and unfavorable conditions of the market: by interest payments and foreclosures. Moreover, it is precisely this stratum which most depends on traditional social relations of kin and mutual aid between neighbours; middle peasants suffer most when these are abrogated, just as they are least able to withstand the depredations of tax collectors or landlords.

Finally—and this is again paradoxical—middle peasants are also the most exposed to influences from the developing proletariat. The poor peasant or landless labourer, in going to the city or the factory, also usually cuts his tie with the land. The middle peasant, however, stays on the land and sends his children to work in town; he is caught in a situation in which one part of the family retains a footing in agriculture, while the other undergoes 'the training of the cities'.[4] This makes the middle peasant a transmitter also of urban unrest and political ideas. The point bears elaboration. It is probably not so much the growth of an industrial proletariat as such which produces revolutionary activity, as the development of an industrial work force still closely geared to life in the villages.

Thus it is the very attempt of the middle and free peasant to remain traditional which makes him revolutionary.

If we now follow through the hypothesis that it is middle peasants and poor but 'free' peasants, not constrained by any power domain, who constitutes the pivotal groupings for peasant uprisings, then it follows that any factor which serves to increase the latitude granted by that tactical mobility reinforces their revolutionary potential. One of these factors is peripheral location with regard to the centre of State control. In fact, frontier areas quite often show a tendency to rebel against the central authorities, regardless of whether they are inhabited by peasants or not. South China has constituted a hearth of rebellion within the Chinese State, partly because it was first a frontier area in the southward march of the Han people, and later because it provided the main zone of contact between Western and Chinese civilization. The Mexican north has similarly been a zone of dissidence from the centre in Mexico City, partly because its economy was based on mining and cattle-raising rather than maize agriculture, partly because it was open to influences from the United States to the north. In the Chinese south it was dissident gentry with a peasant following which frequently made trouble for the centre; in the Mexican north it was provincial business men, ranchers and cowboys. Yet where there exists a poor peasantry located in such a peripheral area beyond the normal control of the central power, the tactical mobility of such a peasantry is 'doubled' by its location. This has been the case with Morelos, in Mexico; Nghe An province in Viet-Nam; kabylia in Algeria; and Oriente in Cuba. The tactical effectiveness of such areas is 'tripled' if they also contain defensible mountainous redoubts: this has been true of Morelos, Kabylia and Oriente. The effect is 'quadrupled' where the population of these redoubts differs ethnically or linguistically from the surrounding population. Thus we find that the villagers of Morelos were Nahuatl-speakers, the inhabitants of Kabylia Berber-speakers. Oriente province showed no linguistic differences from the Spanish spoken in Cuba, but it did contain a significant Afro-Cuban element. Ethnic distinctions enhance the solidarity of the rebels; possession of a special linguistic code provides for an autonomous system of communication.

It is important, however, to recognize that separation from

the State or the surrounding populace need not only be physical or cultural. The Russian and the Mexican cases both demonstrate that it is possible to develop a solid enclave population of peasantry through State reliance on a combination of communal autonomy with the provision of community services to the State. The organization of the peasantry into self-administering communes with stipulated responsibilities to State and landlords created in both cases veritable fortresses of peasant tradition within the body of the country itself. Held fast by the surrounding structure, they acted as sizzling pressure-cookers of unrest which, at the moment of explosion, vented their forces outward to secure more living-space for their customary corporate way of life. Thus we can add a further multiplier effect to the others just cited. The presence of any one of these will raise the peasant potential for rebellion.

But what of the transition from peasant rebellion to revolution, from a movement aimed at the redress of wrongs, to the attempted overthrow of society itself? Marxists in general have long argued that peasants without outside leadership cannot make a revolution; and our case material would bear them out. Where the peasantry has successfully rebelled against the established order—under its own leaders—it was sometimes able to reshape the social structure of the country-side closer to its heart's desires; but it did not lay hold of the State, of the cities which house the centres of control, of the strategic non-agricultural resources of the society. Zapata stayed in his Morelos; the 'folk migration' of Pancho Villa simply receded after the defeat at Torreon; The Ukrainian rebel Nestor Makhno stopped short of the cities; and the Russian peasants of the Central Agricultural Region simply burrowed more deeply into their local communes. Thus a peasant rebellion which takes place in a complex society already caught up in commercialization and industrialization tends to be self-limiting, and hence anachronistic.

The peasant Utopia is the free village, untrammelled by tax collectors, labour recruiters, large landowners, officials. Ruled over, but never ruling, peasants also lack any acquaintance with the operation of the State as a complex machinery, experiencing it only as a 'cold monster'. Against this hostile force, they had learned, even their traditional power-holders provided but a weak shield, though they were on occasion willing to defend them if it proved to their own interest. Thus, for peasants, the State is a negative quantity, an evil, to be replaced in short shrift by their own 'home-made' social order. That order, they believe, can run without the State; hence peasants in rebellion are natural anarchists.

Often this political perspective is reinforced still further by a wider ideological vision. The peasant's experience tends to be dualistic, in that he is caught between his understanding of how the world ought properly to be ordered and the realities of a mundane existence, beset by disorder. Against this disorder, the peasant has always set his dreams of deliverance, the vision of a *mahdi* who would deliver the world from tyranny, of a Son of Heaven who would truly embody the mandate of Heaven, of a 'white' Tsar as against the 'black' Tsar of the disordered present.[5] Under conditions of modern dislocation, the disordered present is all too frequently experienced as world order reversed, and hence evil. The dualism of the past easily fuses with the dualism of the present. The true order is yet to come, whether through miraculous intervention, through rebellion, or both. Peasant anarchism and an apocalyptic vision of the world, together, provide the ideological fuel that drives the rebellious peasantry.

The peasant rebellions of the twentieth century are no longer simple responses to local problems, if indeed they ever were. They are but the parochial reactions to major social dislocation, set in motion by overwhelming societal change. The spread of the market has torn men up by their roots, and shaken them loose from the social relationships into which they were born. Industrialization and expanded communication have given rise to new social clusters, as yet unsure of their own social positions and interests, but forced by the very imbalance of their lives to seek a new adjustment. Traditional political authority has eroded or collapsed; new contenders for power are seeking new constituencies for entry into the vacant political arena. Thus when the peasant protagonist lights the torch of rebellion, the edifice of society is already smouldering and ready to take fire. When the battle is over, the structure will not be the same.

No cultural system—no complex of economy, society, polity and ideology—is ever static; all of its component parts are in constant change. Yet persists. If they begin to exceed these limits, however, or if other components are suddenly introduced from outside, the system will be thrown out of kilter. The parts of the system are rendered inconsistent with each other; the system grows incoherent. Men in such a situation are caught painfully between various old solutions to problems which have suddenly shifted shape and meaning, and new solutions to problems they often cannot comprehend. Since incoherence rarely appears all at once, in all parts of the system, they may for some time follow now one alternative, now another and contradictory one; but in the end a beach, a major disjuncture will make its appearance somewhere in the system.[6] A peasant uprising under such circumstances, for any of the reasons we have sketched, can—without conscious intent—bring the entire society to the state of collapse.

Notes

1. S. N. Eisenstadt, *Modernization: Protest and Change*, Englewood Cliffs, Prentice-Hall, 1966.
2. Barrington Moore, Jr., *Social Origins of Dictatorship and Democracy*, Boston, Beacon Press, 1966.
3. Richard N. Adams, 'Power and Power Domains', *American Latina*, Year 9, 1966, p. 3-21.
4. Germaine Tillion, *France and Algeria: Complementary Enemies*, p. 120-1, New York, Knopf, 1961.
5. Emanuel Sarkisyanz, *Russland and der Messianismus des Orients: Sendangsbewusstsin and politischer Chiliasmus des Ostens*, Tubingen, J. C. B. Mohr, 1955.
6. Godfrey and Monica Wilson, *The Analysis of Social Change*, Cambridge, Cambridge University Press, 1945.

Patron-Client Politics and Change in Southeast Asia
by James Scott

The analysis presented here is an effort to elaborate the patron-client model of association, developed largely by anthropologists, and to demonstrate its applicability to political action in Southeast Asia. Inasmuch as patron-client structures are not unique to Southeast Asia but are much in evidence particularly in Latin America, in Africa and in the less developed portions of Europe, the analysis may possibly have more general value for an understanding of politics in less developed nations.

Western political scientists trying to come to grips with political experience in the Third World have by and large relied on either (or both) of two models of association and conflict. One model is the horizontal, class model of conflict represented most notably by Marxist thought. It has had some value in explaining conflict within the more modern sector of colonial nations and in analyzing special cases in which rural social change has been so cataclysmic as to grind out a dispossessed, revolutionary agrarian mass. By and large, however, its overall value is dubious in the typical nonindustrial situation where most political groupings cut vertically across class lines and where even nominally class-based organizations like trade unions operate within parochial boundaries of ethnicity or religion or are simply personal vehicles. In a wider sense, too, the fact that class categories are not prominent in either oral or written political discourse in the Third World damages their *a priori* explanatory value.

The second model, and one which comes much closer to matching the "real" categories subjectively used by the people being studied, emphasizes primordial sentiments (such as ethnicity, language, and religion), rather than horizontal class ties. Being more reflective of self-identification, the primordial model naturally helps to explain the tension and conflict that increasingly occurred as these isolated, ascriptive groups came into contact and competed for power. Like the class model, however—although less well developed theoretically—the primordial model is largely a conflict model and is of great value in analyzing hostilities between more or less corporate and ascriptive cultural groupings.[1] Important as such conflict has been, it hardly begins to exhaust the political patterns of Southeast Asia and Africa, let alone Latin America. If we are to account, say, for intra-ethnic politics or for patterns of cooperation and coalition building *among* primordial groups, then the primordial model cannot provide us with much analytical leverage.

The need to develop a conceptual structure that would help explain political activity that does not depend solely on horizontal or primordial sentiments is readily apparent in Southeast Asia.[2] In the Philippines, for example, class analysis can help us understand the recurrent agrarian movements in Central Luzon (e.g., Sakdalistas and Huks) among desperate tenants and plantation laborers; but it is of little help in explaining how Magsaysay succeeded in weaning many rebels away from the Huks, or, more important, in analyzing the normal patterns of political competition between Philippine parties. In Thailand, primordial demands may help us discern the basis of dissident movements in North and Northeast Thailand, but neither primordialism nor class analysis explains the intricate pattern of the personal factions and coalitions that are at the center of oligarchic Thai politics. The almost perpetual conflicts between the central Burman state and its separatist hill peoples and minorities are indeed primordial, communal issues, but communalism is of no use in accounting for the intra-Burman struggles between factions within the Anti-Fascist People's Freedom League (AFPFL) or, later, within the military regime. Ethnicity and class do carry us far in explaining racial hostilities and intra-Chinese conflict in Malaya, but they are less helpful when it comes to intra-Maylay politics or to interracial cooperation at the top of the Alliance party.[3]

As these examples indicate, when we leave the realm of class conflict or communalism, we are likely to find ourselves in the realm of informal power groups, leadership-centered cliques and factions, and a whole panoply of more or less instrumental ties that characterize much of the political process in Southeast Asia. The structure and dynamics of such seemingly *ad hoc* groupings can, I believe, be best understood from the perspective of patron-client relations. The basic pattern is an informal cluster consisting of a power figure who is in a position to give security, inducements, or both, and his personal followers who, in return for such benefits, contribute their loyalty and personal assistance to the patron's designs. Such vertical patterns of patron-client linkages represent an important structural principle of Southeast Asian politics.

Until recently the use of patron-client analysis has been the province of anthropologists, who found it particularly useful in penetrating behind the often misleading formal arrangements in small local communities where interpersonal power relations were salient. Terms which are related to patron-client structures in the anthropological literature—including "clientelism," "dyadic contract," "personal network," "action-set"—reflect an attempt on the part of anthropologists to come to grips with the mosaic of nonprimordial divisions. Informal though such networks are, they are built, they are maintained, and they interact in ways that will permit generalization.

Although patron-client analysis provides a solid basis for comprehending the structure and dynamics of nonprimodial cleavages at the local level, its value is not limited to village studies. Nominally modern institutions such as bureaucracies and political parties in Southeast Asia are often thoroughly penetrated by informal patron-client networks that undermine the formal structure of authority. If we are to grasp why a bureaucrat's authority is likely to depend more on his personal following and extrabureaucratic connections than on his formal post, or why political parties seem more like *ad hoc* assemblages of notables together with their entourages than arenas in which established interests are aggregated, we must rely heavily on patron-client analysis. The dynamics of personal alliance networks are as crucial in the day-to-day realities of national institutions as in local politics; the main difference is simply that such networks are more elaborately disguised by formal facades in modern institutions.

In what follows, I attempt to clarify what patron-client ties are, how they affect political life, and how they may be applied to the dynamics of Southeast Asian politics. After 1) defining the nature of the patron-client link and distinguishing it from other social ties, the paper 2) discriminates among different varieties of patron-client bonds and thereby establishes some important dimensions of variation, and 3) examines both the survival of and the transformations in patron-client links in Southeast Asia since colonialism and the impact of major social changes (such as the growth of markets, the expanded role of the state, and so forth) on the content of these ties.

I. The Nature of Patron-Client Ties

The Basis and Operation of Personal Exchange. While the actual use of the terms "patron" and "client" is largely confined to the

Mediterranean and Latin American areas, comparable relationships can be found in most cultures and are most strikingly present in preindustrial nations. *The patron-client relationship—an exchange relationship between roles—may be defined as a special case of dyadic (two-person) ties involving a largely instrumental friendship in which an individual of higher socioeconomic status (patron) uses his own influence and resources to provide protection or benefits, or both, for a person of lower status (client) who, for his part, reciprocates by offering general support and assistance, including personal services, to the patron.*[4]

In the reciprocity demanded by the relationship each partner provides a service that is valued by the other. Although the balance of benefits may heavily favor the patron, some reciprocity is involved, and it is this quality which, as Powell notes, distinguishes patron-client dyads from relationships of pure coercion or formal authority that also may link individuals of different status.[5] A patron may have some coercive power and he may also hold an official position of authority. But if the force or authority at his command are alone sufficient to ensure the compliance of another, he has no need of patron-client ties which require some reciprocity. Typically, then, the patron operates in a context in which community norms and sanctions and the need for clients require at least a minimum of bargaining and reciprocity; the power imbalance is not so great as to permit a pure command relationship.

Three additional distinguishing features of patron-client links, implied by the definition, merit brief elaboration: their basis in inequality, their face-to-face character, and their diffuse flexibility. All three factors are most apparent in the ties between a high-status landlord and each of his tenants or sharecroppers in a traditional agrarian economy—a relationship that serves, in a sense, as the prototype of patron-client ties.[6]

First, there is an *imbalance in exchange between the two partners which expresses and reflects the disparity in their relative wealth, power, and status*. A client, in this sense, is someone who has entered an unequal exchange relation in which he is unable to reciprocate fully. A debt of obligation binds him to the patron.[7] How does this imbalance in reciprocity arise? It is based, as Peter Blau has shown in his work, *Exchange and Power in Social Life*,[8] on the fact that the patron often is in a position to supply unilaterally goods and services which the potential client and his family need for their survival and well being. A locally dominant landlord, for example, is frequently the major source of protection, of security, of employment, of access to arable land or to education, and of food in bad times. Such services could hardly be more vital, and hence the demand for them tends to be highly inelastic; that is, an increase in their effective cost will not diminish demand proportionately. Being a monopolist, or at least an oligopolist, for critical needs, the patron is in an ideal position to demand compliance from those who wish to share in these scarce commodities.

Faced with someone who can supply or deprive him of basic wants, the potential client in theory has just four alternatives to becoming the patron's subject.[9] First he may reciprocate with a service that the patron needs badly enough to restore the balance of exchange. In special cases of religious, medical, or martial skills such reciprocation may be possible, but the resources of the client, given his position in the stratification, are normally inadequate to reestablish an equilibrium. A potential client may also try to secure the needed services elsewhere. If the need for clients is especially great, and if there is stiff competition among patron-suppliers, the cost of patron-controlled services will be less [10] In most agrarian settings, substantial local autonomy tends to favor the growth of local power monopolies by officials or landed gentry. A third possibility is that clients may coerce the patron into providing services. Although the eventuality that his clients might turn on him may prompt a patron to meet at least the minimum normative standards of exchange,[11] the patron's local power and the absence of autonomous organization among his clients make this unlikely. Finally, clients can theoretically do without a patron's services altogether. This alternative is remote, given the patron's control over vital services such as protection, land, and employment.

Affiliating with a patron is neither a purely coerced decision nor is it the result of unrestricted choice. Exactly where a particular patron-client dyad falls on the continuum depends on the four factors mentioned. If the client has highly valued services to reciprocate with, if he can choose among competing patrons, if force is available to him, or if he can manage without the patron's help—then the balance will be more nearly equal. But if, as is generally the case, the client has few coercive or exchange resources to bring to bear against a monopolist-patron whose services he desperately needs, the dyad is more nearly a coercive one.[12]

The degree of compliance a client gives his patron is a direct function of the degree of imbalance in the exchange relationship—of how dependent the client is on his patron's services. An imbalance thus creates a sense of debt or obligation on the client's part *so long as it meets his basic subsistence needs* and represents, for the patron, a " 'store of value'—social credit that . . . (the patron) can draw on to obtain advantages at a later time." [13] The patron's domination of needed services, enabling him to build up savings of deference and compliance which enhance his status, and represents a capacity for mobilizing a group of supporters when he cares to. The larger a patron's clientele and the more dependent on him they are, the greater his latent capacity to organize group action. In the typical agrarian patron-client setting this capacity to mobilize a following is crucial in the competition among patrons for regional preeminence. As Blau describes the general situation,

> The high-status members furnish instrumental assistance to the low-status ones in exchange for their respect and compliance, which help the high-status members in their competition for a dominant position in the group.[14]

A second distinguishing feature of the patron-client dyad is the *face-to-face*, personal quality of the relationship. The continuing pattern of reciprocity that establishes and solidifies a patron-client bond often creates trust and affection between the partners. When a client needs a small loan or someone to intercede for him with the authorities, he knows he can rely on his patron; the patron knows, in turn, that "his men" will assist him in his designs when he needs them.[15] Furthermore, the mutual expectations of the partners are backed by community values and ritual.

In most contexts the affection and obligation invested in this tie between nonrelatives is expressed by the use of terms of address between partners that are normally reserved for close kin. The tradition of choosing godparents in Catholic nations is often used by a family to create a fictive kinship tie with a patron—the godfather thereby becoming like a brother to the parents.[16]

Whether the model of obligation established is father-son, uncle-nephew, or elder-younger brother, the intention is similar: to establish as firm a bond of affection and loyalty as that between close relatives. Thus while a patron and client are very definitely alive to the instrumental benefits of their assocation, it is not simply a neutral link of mutual advantage. On the contrary, it is often a durable bond of genuine mutual devotion that can survive severe testing.

The face-to-face quality of the patron-client dyad, as well as the size of the patron's resource base, limits the number of direct active ties a single patron can have.[17] Even with vast resources, the personal contact and friendship built into the link make it highly unlikely that an active clientele could exceed, say, one hundred persons. The total following of a given patron may be much larger than this, but normally all except 20-30 clients would be linked to the patron through intermediaries. Since we are dealing with positive emotional ties (the ratio of "calculation" to affection may of course vary), a leader and his immediate entourage will be comparatively small.

The third distinctive quality of patron-client ties, one that reflects the affection involved, is that they are *diffuse, "whole-person" relationships rather than explicit, impersonal-contract bonds*. A landlord may, for example, have a client who is connected to him by tenancy, friendship, past exchanges of services, the past tie of the client's father to his father, and ritual coparenthood. Such a strong "multiplex" relation, as Adrian Mayer terms it,[18] covers a wide range of potential exchanges. The patron may very well ask the client's help in preparing a wedding, in winning an election campaign, or in finding out what his local rivals are up to; the client may approach the patron for help in paying his son's tuition, in filling out government forms, or in getting food or medicine when he falls on bad times. The link, then, is a very flexible one in which the needs and resources of the partners, and hence the nature of the exchange, may vary widely over time. Unlike explicit contractual relations, the very diffuseness of the patron-client linkage contributes to its survival even during rapid social change—it tends to persist so long as the two partners have something to offer one another.[19] Just as two brothers may assist each other in a host of ways, patron-client partners have a relationship that may also be invoked for almost any purpose;

the chief differences are the greater calculation of benefits and the inequality that typifies patron-client exchange.

The Distinctiveness of the Patron. The role of patron ought to be distinguished from such role designations as "broker," "middleman," or "boss" with which it is sometimes confounded. Acting as a "broker" or "middleman"—terms which I shall use interchangeably—means serving as an intermediary to arrange an exchange or transfer between two parties who are not in direct contact. The role of middleman, then, involves a three party exchange in which the middleman functions as an agent and does not himself control the thing transferred. A patron, by contrast, is part of a two-person exchange and operates with resources he himself owns or directly controls.[20] Finally, the terms "middleman" and "broker" do not specify the relative status of the actor to others in the transaction, while a patron is by definition of superior rank to his client.

Important as this distinction is, it is easily lost sight of for two reasons. First, it is not always a simple task to determine if someone personally controls the resources he uses to advance himself. What of the case in which a civil servant distributed the subordinate posts in his jurisdiction to create an entourage? Here it would seem that he was acting as a patron, inasmuch as the jobs he gave out were meant as *personal* gifts from the store of scarce values he controlled and were intended to create a feeling of personal debt and obligation among recipients. The social assessment of the nature of the gift is thus crucial. If we were to find, on the other hand, that the civil servant was viewed as someone who had acted as an agent of jobseekers and put them in touch with a politician who controlled the jobs, then he would be acting as a broker. It is only natural that many an ambitious public official will seek to misrepresent acts of brokerage or simple adherence to the rules as personal acts of patronage, thereby building his following.[21] To the extent that he succeeds in representing his act as a personal act of generosity, he will call forth that sense of personal obligation that will bind his subordinates to him as clients.[22]

A second potential source of confusion in this distinction is that the terms designate roles and not persons, and thus it is quite possible for a single individual to act both as a broker and a patron. Such a role combination is not only possible, but is empirically quite common. When a local landowning patron, for example, becomes the head of his village's political party he is likely to become the middleman between many villagers and the resources controlled by higher party officials. In this case he may have clients for whom he also serves as broker. The diffuse claims of the patron-client tie actually make it normal for the patron to act as a broker for his clients when they must deal with powerful third parties—much as the patron saint in folk Catholicism who directly helps his devotees while also acting as their broker with the Lord.[23] If on the other hand, the political party simply gives the local patron direct control of its programs and grants in the area, it thereby enhances his resources for becoming a patron on a larger scale and eliminates the need for brokerage.

Patrons ought finally to be differentiated from other partly related terms for leadership such as "boss," "caudillo," or "cacique." "Boss" is a designation at once vague and richly connotative. Although a boss may often function as a patron, the term itself implies (a) that he is the most powerful man in the arena and (b) that his power rests more on the inducements and sanctions at his disposal than on affection or status. As distinct from a patron who may or may not be the supreme local leader and whose leadership rests at least partly on rank and affection, the boss is a secular leader *par excellence* who depends almost entirely on palpable inducements and threats to move people. As we shall show later, a settled agrarian environment with a recognized status hierarchy is a typical setting for leadership by patrons, while a more mobile, egalitarian environment is a typical setting for the rise of bosses. The final two terms "caudillo" and "cacique" are most commonly used in Latin America to designate the regional—often rural—bosses. Again the implication is that coercion is a main pillar of power, and in the case of the *caudillo*, a personal following is common.[24] A "cacique" or "caudillo" may act as patron to a number of clients but he typically relies too heavily on force and lacks the traditional legitimacy to function mainly as a patron. At best, a *cacique* or *caudillo* may, like a boss, be a marginal special case of a patron, providing that a portion of his following is beneath him socially and bound to him in party by affective ties. Over time, however, a metamorphosis may occur. Just as the successful brewery owner of late 18th century England might well anticipate a peerage for his son, the *cacique* who today imposes his rule by force may do well enough to set his son up as a landowner, whose high status and legitimacy strengthens his role as patron.

Patron and Clients as Distinctive Groupings. To this point, the discussion has centered on the nature of the single link between patron and client. If we are to broaden the analysis to include the larger structures that are related by the joining of many such links, a few new terms must be introduced. First, when we speak of a patron's immediate following—those clients who are directly tied to him—we will refer to a patron-client *cluster*. A second term, enlarging on the cluster but still focusing on one person and his vertical links is the patron-client *pyramid*. This is simply a vertical extension downward of the cluster in which linkages are introduced beyond the first-order.[25] Below are typical representations of such links.

Patron-Client Cluster   **Patron-Client Pyramid**

Although vertical ties are our main concern, we will occasionally want to analyze *horizontal dyadic ties*, say, between two patrons of comparable standing who have made an alliance. Such alliances often form the basis of *factional systems* in local politics. Finally, patron-client *networks* are not ego-focused but refer to the overall pattern of patron-client linkages (plus horizontal patron alliances) joining the actors in a given area or community.

Patron-client clusters are one of a number of ways in which people who are not close kin come to be associated. Most alternative forms of association involve organizing around categorical ties, both traditional—such as ethnicity, religion, or caste—and modern—such as occupation or class—which produce groups that are fundamentally different in structure and dynamics. The special character of patron-client clusters stems, I believe, from the fact that, unlike categorically-based organizations, such clusters a) have a basis of *membership that is specific to each link*,[26] and b) are *based on individual ties to a leader* rather than on shared characteristics or horizontal ties among followers.

Some other important distinctions between categorical and patron-client groupings follow from these particular principles of organization. Here I rely heavily on Carl Landé's more elaborate comparisons between dyadic followings and categorical groups.[27]

1. *Members' Goals:* Clients have particularistic goals which depend on their personal ties to the leader, whereas categorical group members have common goals that derive from shared characteristics which distinguish them from members of other such groups.
2. *Autonomy of Leadership:* A patron has wide autonomy in making alliances and policy decisions as long as he provides for the basic material welfare of his clients, whereas the leader of a categorical group must generally respect the *collective* interest of the group he leads.
2. *Stability of Group:* A patron-client cluster, being based on particularistic vertical links, is highly dependent on its leader's skills and tends to flourish or disintegrate depending on the resources of the leader and the satisfaction of individual client demands. A categorical group, by contrast, is rooted more firmly in horizontally shared qualities and is thus less dependent for its survival on the quality of its leadership and more durable in its pursuit of broader, collective (often policy) interests.
4. *Composition of Group:* Patron-client clusters, because of the way they are created, are likely to be more heterogeneous in class composition than categorical groups which are based on some distinctive quality which members share. *By definition,* patron-client pyramids join people of different status rankings while categorical groups *may or may not* be homogeneous in status.
5. *Corporateness of Group:* In a real sense a patron-client cluster is not a group at all but rather an "action-set" that exists because of the vertical links to a common leader—links which the leader may activate in whole or

in part.[28] Followers are commonly not linked directly to one another and may, in fact, be unknown to each other. An organized categorical group, by contrast, is likely to have horizontal links that join members together so that it is possible to talk of a group existence independent of the leader.

Although this listing is not exhaustive, it does illustrate the special character of patron-client networks. Bearing in mind the generic qualities of these ties, we now turn to the range of variation within the genus.

II. Variation in Patron-Client Ties

One could potentially make almost limitless distinctions among patron-client relationships. The dimensions of variation considered here are selected because they seem particularly relevant to our analytical goal of assessing the central changes in such ties within Southeast Asia. Similar distinctions should be germane to the analysis of other preindustrial nations as well.

The Resource Base of Patronage

A potential patron assembles clients on the basis of his ability to assist them. For his investment of assets, the patron expects a return in human resources—in the form of the strength of obligation and the number of clients obligated to him. The resource base or nature of the assets a patron has at his disposal can vary widely. One useful basis for distinguishing among resources is the directness with which they are controlled. Patrons may, in this sense, rely on a) their own knowledge and skills, b) direct control of personal real property, or c) indirect control of the property or authority of others (often the public). The resources of *skill and knowledge* are most recognizable in the roles of lawyer, doctor, *literatus*, local military chief, teacher—religious or secular. Those equipped with these skills control scarce resources that can enhance the social status, health, or material well being or another. Inasmuch as such resources rest on knowledge, they are less perishable than more material sources—although the *time* of the expert *is* limited—and can be used again and again without being diminished. Such resources are relatively, but not entirely, secure. In the case of lawyers and literati, for example, the exchange price of their services depends respectively on the continued existence of a court system and the veneration of a particular literary tradition, both of which are subject to change. The value of a local military chief's protection is similarly vulnerable to devaluation once the nation state has established local law and order.

Reliance on *direct control of real property* is a second common means of building a clientele. Traditionally, the typical patron controlled scarce land. Those he permitted to farm it as sharecroppers or tenants became permanently obligated to him for providing the means of their subsistence. Any businessman is in a similar position; as the owner of a tobacco factory, a rice mill, or a small store he is able to obligate many of those of lower status whom he employs, to whom he extends credit, or with whom he does business. This kind of resource, in general, is more perishable than personal skills. A landlord has only so many arable acres, a businessman only so many jobs, a shopkeeper so much ready cash, and each must carefully invest those resources to bring the maximum return. Like any real property, moreover, private real property is subject to seizure or restrictions on its use.

A third resource base available to the potential patron is what might be called *indirect, office-based property*. Here we refer to patrons who build a clientele on the strength of their freedom to dispense rewards placed in their trust by some third party (parties). A village headman who uses his authority over the distribution of communal land to the poor or the distribution of *corvée* labor and taxation burdens in order to extend his *personal* clientele would be a typical example of traditional office-based patronship. One can classify similarly office-holders in colonial or contemporary settings whose discretionary powers over employment, promotion, assistance, welfare, licensing, permits, and other scarce values can serve as the basis of a network of personally obligated followers. Politicians and administrators who exploit their office in this way to reward clients while violating the formal norms of public conduct are, of course, acting corruptly. Finally, we should add private-sector office-holders in the private sector such as plantation managers, purchasing agents, and hiring bosses, who may also use their discretionary authority to nurture a clientele.

Indirect, office-based property is least secure in many respects, as its availability depends on continuity in a position that is ultimately given or withdrawn by third parties. A landlord will usually retain his local base whereas an office holder is likely to be swept out by a new victor at the polls or simply by a power struggle within the ruling group. In spite of the risks involved, these posts are attractive because the resources connected with many of them are far greater than those which an individual can amass directly.

The categories of resources just discussed are not mutually exclusive. It is common, for example, for a patron to have a client who is obligated to him by being a tenant on his land and also by having secured an agricultural loan through his patron's chairmanship of the ruling party's local branch. The resources that cement a dyadic tie may thus be multiple—it is often a question of deciding which is the predominant resource. Much the same analysis can be made of a patron-client cluster or network, since a patron may have clients who are bound to him by quite different resources, and it is often important to determine what the main resource is that holds the cluster together.

Resource Base of Clientage.

As the other member of a reciprocating pair, the client is called upon to provide assistance and services when the patron requires them. The variation in the nature of such assistance is another means of distinguishing one patron-client dyad from another. Here one might want to differentiate: (1) *labor services and economic support,* as provided by a rent-paying tenant or employee, (2) *military or fighting duties,* such as those performed by members of a bandit group for their chief, and (3) *political services* such as canvassing or otherwise acting as an agent of a politician. Within the "political service" category one may wish to separate electoral services from nonelectoral political help. I should add here that the term "clients" can refer to those who are in the middle of a patron-client pyramid—being a client to someone higher up and a patron to those below. In this case, a superior patron will be interested in his client's potential services, but those services will include the size, skills, assets, and status of the client's own subordinate following.

Just as a patron-client dyad can be distinguished by the main resource base of clientage so can a patron-client cluster be categorized by the modal pattern of client services for the cluster or pyramid as a whole.

Balance of Affective and Instrumental Ties.

By definition instrumental ties play a major role in the patron-client dyad. It is nonetheless possible to classify such dyads by the extent to which affective bonds are also involved in the relationship. At one end of this continuum one might place patron-client bonds which, in addition to their instrumental character, are reinforced by affective links growing, say, from the patron and the client having been schoolmates, coming from the same village, being distant relatives, or simply from mutual love. Comparable affective rewards may also spring from the exchange of deference on the one hand and *noblesse oblige* on the other in a settled agrarian status network—rewards that have value beyond the material exchanges they often involve.[29] At the other end of the spectrum lies a dyadic tie much closer to an almost neutral exchange of goods and services. The more purely coercive the relationship is and the less traditional legitimacy it has, the more likely that affective bonds will be minimal.

This distinction has obvious analytical value. If we were to look at a patron's entire following, we would be able to classify each vertical bond according to the ratio of affective to instrumental rewards involved. One could, of course, do the same for horizontal alliances. Using this criterion we could identify a set of followers among whom the ratio of affective to instrumental ties was relatively high, reflecting perhaps distant kinship, old village or neighborhood ties, or comparable bonds. The loyalty of this set of followers would be less dependent upon a continued flow of material benefits, simply because their loyalty is partly based on nonmaterial exchanges. As we move beyond this partly affective following to a patron's other supporters, the weight of instrumental, usually material, ties becomes relatively more important. The nature of a man's following—the balance of affective to instrumental ties obligating his clients to him—can tell us something about its stability under different conditions. When a patron increases his material resource base, it is his instrumental following that will tend to grow rapidly, and when he is in decline, that same following will shrink rapidly, as clients look for a more promising leader. The degree of dependence on material incentives within a following is, in principle, a quality one could measure by establishing how much more than their present material rewards a rival patron would have to offer to detach a given number of another's clients.

The affective-instrumental distinction just made leads to a similar, but not identical, distinction between the *core* and *periphery* of a man's following. These categories actually are distributed along a continuum; at the *periphery* of a man's following are those clients who are relatively easy to detach while at the *core* are followers who are more firmly bound to him. The periphery is composed of clients bound largely by instrumental rewards, while the core is composed of clients linked by strong affective ties, *as well as* clients who are attracted to a patron by such strong instrumental ties that they seem unbreakable.[30] This amounts, in effect, to a distinction between a man's virtually irreducible following and his more or less fluctuating, "fair-weather" following. Patrons can then be differentiated by the size of their core-following relative to their peripheral-following. A landlord or a businessman will generally have a sizable core group composed both of his friends, kin, etc., and of his tenants of employees. This nucleus is his initial following; his clientele may grow larger, but it is unlikely to contract further than this durable core. A politician or bureaucrat, on the other hand, unless he is privately wealthy, is likely to have a comparatively smaller core group composed mostly of those with whom he has strong affective ties and, hence, a relatively large proportion of "fair-weather" clients. The blows of fortune such a politician or administrator suffers are more likely to be instantly and fully reflected in a reduction of the size of his clientele, which is largely a calculating one. Politicians, and bureaucrats, because they have smaller core followings and because they can, through their office, often tap vast resources, are apt to have meteoric qualities as patrons; the landholder, by contrast, is likely to cast a steadier, if dimmer, light.

Balance of Voluntarism and Coercion

There are obvious and important differences in the degree of coercion involved in a patron-client bond. At one end are the clients with virtually no choice but to follow the patron who directly controls their means of subsistence. Here one might place a tenant whose landlord provides his physical security, his land, his implements and seed, in a society where land is scarce and insecurity rife. Nearer the middle of this continuum would perhaps be the bonds between independent smallholders who depend on a landlord for the milling and marketing of their crops, for small loans, and for assistance with the police and administration. Such bonds are still based on inequality, but the client, because he has some bargaining power, is not simply putty in his patron's hands. Finally, let us assume that an electoral system has given clients a new resource and has spurred competition among patrons for followings that can swing the election to them. In this case the inequality in bargaining power is further reduced, and the client emerges as more nearly an independent political actor whose demands will receive a full hearing from his patron.

In general, the oppression of the client is greater when the patron's services are vital, when he exercises a monopoly over their distribution, and when he has little need for clients himself. The freedom of the client is enhanced most when there are many patrons whose services are not vital and who compete with one another to assemble a large clientele—say for electoral purposes.

The greater the coercive power of the patron vis-à-vis his client, the fewer rewards he must supply to retain him. A patron in a strong position is more likely to employ *sanctions*—threats to punish the client or to withdraw benefits he currently enjoys—whereas a relatively weaker patron is more likely to offer inducements—promises to reward a client with benefits he does not now enjoy.[31] In each instance, superior control over resources is used to gain the compliance of followers, but the use of sanctions indicates a higher order of power than the use of inducements.

Assessment of the coercive balance and of the ratio of sanctions to inducements can be made not only for a dyad but also for a patron-client cluster or pyramid. The cluster of a local baron with a private army may be held intact by a mix of deference and sanctions, while a campaigning politician may build a cluster simply with favors if he has no coercive power or traditional legitimacy. Each cluster or pyramid has its special vulnerability. The coercive cluster will be jeopardized by a breach of the patron's local power monopoly, and a cluster based on inducements will be in danger if its leader's income or access to public funds is cut off.

Durability Over Time.

Patron-client dyads may be rather ephemeral, or they may persist for long periods.[32] In a traditional setting they are likely to last until one of the partners has died. Knowing how durable such ties are can also tell us something about the structure of competition over time. Where dyads are persistent they tend to produce persistent factional structures with some continuity in personnel over time, at least stable clusters or pyramids that may recombine in a variety of ways but are constructed from the same components. Where dyads are fragile, personal alignments may undergo an almost total reordering within a decade.

Since patron-client clusters are based ultimately on power relations, they will endure best in a stable setting that preserves existing power positions. A particular patron will thus retain his clients as long as he continues to dominate the supply of services they need. A patron is also likely to keep his followers if the scope of reciprocity that binds them is greater. That is, the more of the client's vital needs a patron can meet (i.e., if he can supply not only land and security but also influence with the administration, help in arranging mortgages or schooling, and so forth), the greater the tendency for the tie to be invoked frequently and to endure over long periods. Compared with patrons who can provide only legal services, only financial help, or only educational advantages, the *multiplex* bond between patron and client is a solid linkage that serves many needs; since it is more of a whole-person tie, it will be called into action often.

Homogeneity of Following.

A patron may have a heterogeneous set of followers drawn from all walks of life, or he may have a following composed, say, of only his poor sharecroppers or only clerical subordinates in his office.[33] The proportion of a man's supporters who share social characteristics and the salience of those social characteristics to them constitute a measure of how homogeneous a following is. Since a patron, by definition, occupies a higher social station than his clients, the greater the homogeneity in a following, the greater the latent shared interests among followers that might threaten the relationship. When a landed patron whose clients are his tenants, for example, sells off what had been common pasture land, all his tenants are equally affected. Their shared situation and the common experiences it provides create a potential for horizontal ties, whereas a heterogeneous clientele lacks this potential.

Field Variables.

Occasionally, we will want to describe and contrast configurations of patron-client clusters within a political arena rather than dealing with a single cluster or dyad. Four particularly useful distinctions in this respect are a) the degree of monopoly over local resources by a single patron, b) the degree of monopoly over links to other structures by a single patron, c) the *density* of patron-client linkages in the population,[34] and d) the extent of differentiation between different pyramids and clusters. The first two variables are self-explanatory and measure the degree of dominance exercised by a patron over local and supralocal resources. "Density" refers to the proportion of a given population that is a part of the patron-client network. In some situations, for example, a large part of the lower classes may not actually have any vertical links of clientage to a patron. To gauge accurately the explanatory power of patron-client politics in a political field requires that we know for how much of the population such ties are effective. Finally, the degree of differentiation among clusters is a means of discerning whether one cluster looks pretty much like the next one or whether many clusters are socioeconomically distinct. In the classical feudal situation, the pyramidal structure of one lord's small domain was similar to that of his neighbor—the social structure of the landscape resembled a repetitive wallpaper pattern—and competition was thus between almost identical units. In other circumstances, pyramids may be differentiated by predominant occupation, by institutional affiliation, and so forth, so that the seeds of a distinctive and perhaps durable *interest* have been sown.

III. Survival and Development of Patron-Client Ties in Southeast Asia

A. Conditions for Survival

As units of political structure, patron-client clusters not only typify both local and national politics in Southeast Asia, they are also as characteristic of the area's contemporary politics as of its traditional politics. In one sense, the "style" of the patron-client link, regardless of its context, is distinctively traditional. It is particularistic where (following Parsons) modern links are universal; it is diffuse and informal where modern ties are specific or contractual; and it produces vertically-integrated groups with shifting interests rather than horizontally-integrated groups with durable interests. Despite their traditional style, however, patron-client clusters both serve as mechanisms for bringing together individuals who are not kinsmen and as the building-blocks for elaborate networks of vertical integration. They cannot, therefore, be merely dismissed as vestigial remains of archaic structures but must be analyzed as a type of social bond that may be dominant in some contexts and marginal in others.

In my view, most of traditional and contemporary Southeast Asia has met three necessary conditions for the continued vitality of patron-client structures: (1) *the persistence of marked inequalities in the control of wealth, status, and power which have been accepted (until recently) as more or less legitimate;* (2) *the relative absence of firm, impersonal guarantees of physical security, status and position, or wealth,* and (3) *the inability of the kinship unit to serve as an effective vehicle for personal security or advancement.*

The first condition is more or less self-evident. A client affiliates with a patron by virtue of the patron's superior access to important goods and services. This inequality is an expression of a stratification system which serves as the basis for vertical exchange. Classically in Southeast Asia, the patron has depended more on the local organization of force and access to office as the sinews of his leadership than upon hereditary status or land ownership. Inequalities were thus marked, but elite circulation tended to be comparatively high. With the penetration of colonial government and commercialization of the economy, land ownership made its appearance (especially in the Philippines and Vietnam) as a major basis of patronage. At the same time access to colonial office replaced to some extent victory in the previously more fluid local power contests as the criterion for local patronage. Although land ownership and bureaucratic office have remained two significant bases of patronship in postcolonial Southeast Asia, they have been joined—and sometimes eclipsed as patronage resources—by office in political parties or military rank.

If inequities in access to vital goods were alone sufficient to promote the expansion of patron-client ties, such structures would predominate almost everywhere. A second, and more significant, condition of patron-client politics is the absence of institutional guarantees for an individual's security, status, or wealth. Where consensus has produced an institutionalized means of indirect exchange—one that is legally based, uniformly enforced, and effective—impersonal contractual arrangements tend to usurp the place of personal reciprocity. A patron-client dyad, by contrast, is a *personal* security mechanism and is resorted to most often when personal security is frequently in jeopardy and when impersonal social controls are unreliable. In this context, direct personal ties based on reciprocity substitute for law, shared values, and strong institutions. As Eric Wolf has noted, "The clearest gain from such a (patron-client) relation . . . is in situations where public law cannot guarantee adequate protection against breaches of non-kin contracts." [35]

It is important to recognize the unenviable situation of the typical client in less developed nations. Since he lives in an environment of scarcity, competition for wealth and power is seen as a zero-sum contest in which his losses are another's gain and vice-versa.[36] His very survival is constantly threatened by the caprice of nature and by social forces beyond his control. In such an environment, where subsistence needs are paramount and physical security uncertain, a modicum of protection and insurance can often be gained only by depending on a superior who undertakes personally to provide for his own clients. Operating with such a slim margin, the client prefers to minimize his losses—at the cost of his independence—rather than to maximize his gains by taking risks he cannot afford. When one's physical security and means of livelihood are problematic, and when recourse to law is unavailable or unreliable, the social value of a personal defender is maximized.

The growth of strong, institutional orders that reduce the need for personal alliances was a rare occurence—the Roman and Chinese imperial orders being the most notable exceptions—until the 19th and 20th centuries, when modern nation-states developed the technical means to impose their will throughout their territory. Before that, however, the existence of a fair degree of local autonomy was inevitable, given the limited power available to most traditional kingdoms. The greater that autonomy, or what might be called the *localization* of power, the more decisive patron-client linkages were likely to be. In settings as diverse as much of Latin America, feudal Europe, and precolonial Southeast Asia, the localization of power was pervasive and gave rise to networks of patron-client bonds. From time to time in South East Asia a centralizing kingdom managed to extend its power over wide areas, but seldom for very long or with a uniform system of authority. A typical Southeast Asian kingdom's authority weakened steadily with increasing distance from the capital city. Beyond the immediate environs of the court, the ruler was normally reduced to choosing which of a number of competing petty chiefs with local power bases he would prefer to back.[37] Such chiefs retained their own personal following; their relationship to the ruler was one of bargaining as well as deference; and they might back a rival claimant to the throne or simply defy demands of the court when they were dissatisfied with their patron's behavior. Thus, the political structure of traditional Southeast Asia favored the growth of patron-client links, inasmuch as it was necessary for peasants to accommodate themselves to the continuing reality of autonomous personal authority at almost all levels.

The localization of power is in many senses as striking a characteristic of contemporary as of traditional Southeast Asia. As Huntington aptly expressed it, "The most important political distinction among countries concerns not their form of government but their degree of government"[38] Many of the outlying areas of Southeast Asian nations, particularly the upland regions of slash-and-burn agriculturalists, are only intermittently subject to central government control and continue to operate with much autonomy. By far the most important manifestation of the localization of power, however, has occurred within the very bureaucratic and political institutions that are associated with a central state. The modern institutional framework is a relatively recent import in Southeast Asia; it finds minimal support from indigenous social values and receives only sporadic legal enforcement. With the exception of North Vietnam and Singapore, where a portion of the intelligentsia with modernizing ideologies and popular backing have taken power, these new institutions do not command wide loyalty and must therefore fight for survival in a hostile environment. The net effect of this fragile institutional order is to promote the growth of personal spheres of influence within ministries, administrative agencies, and parties. Sometimes the vertical links are strong (e.g., Thailand) and sometimes a high degree of decentralization or "sub-infeudation" occurs (as in parliamentary Burma from 1955 to 1958). In either case, what replaces the institution are elaborate networks of personal patron-client ties that carry on more or less traditional factional struggles rather than operate as agents of a hierarchical organization. Patron-client politics are thus as much a characteristic of faction-ridden central institutions as of the geographical periphery in these nations.

The third condition under which patron-client bonds remain prominent relates directly to the capacity of such ties to foster cooperation among nonkin. As a mechanism for protection or for advancement, patron-client dyads will flourish when kinship bonds alone become inadequate for these purposes.

Although kinship bonds are seldom completely adequate as structures of protection and advancement even in the simplest societies, they may perform these functions well enough to minimize the need for nonkin structures. Such is the case among small isolated bands of hunters and gatherers, among self-sufficient, corporate lineages and within corporate villages.[39] None of these conditions, however, is particularly applicable to Southeast Asian societies. The highland areas are inhabited by poorly integrated minorities but only rarely are these minorities so isolated as to lack economic and political ties with the larger society. Corporate lineages, outside traditional Vietnam, are uncommon in low-land Southeast Asia where bilateral kinship systems lead to overlapping kindreds rather than mutually exclusive lineages. Finally, corporate village structures (except in Java and perhaps Vietnam's Red River delta) are not typical of Southeast Asia. The scope for nonkin ties in general, and patron-client links in particular, has thus been quite wide throughout the region.

Even when government did not impinge much on their activities, villagers in traditional Southeast Asia still had need of extrakin and extravillage contacts. They needed to secure marriage partners, to assure themselves protection and contacts for the limited but vital trade carried on between villages, and finally to establish an outside alliance in case a village quarrel forced them to seek land and employment elsewhere.[40] If vertical dyadic ties were of some value in the traditional context, they assumed a more decisive role in the colonial and postcolonial periods. First, the commercialization of the economy and the growth of markets enhanced the value of cooperative arrangements among nonkin. Both corporate kin groups and corporate village structures had depended on a certain level of economic autarchy for their vitality—an autarchy which colonial economic policy quickly eroded. These corporate structures (where they existed) tended "to lose their monopoly over resources and personnel in situations where land and labor became free commodities."[41] As the communal land controlled by the village dwindled, as outsiders came increasingly to own land in the village, and as villagers increasingly worked for nonkin, the value of patron-client links increased for all concerned.

In traditional Southeast Asia, as in feudal Europe, then, the inability of kindreds to provide adequate protection and security fostered the growth of patron-client structures. The limited effectiveness of kindreds as units of cooperation and security was further reduced by the new structures and uncertainties of the colonial economy. Within this new economy, the goals of wealth, protection, power, and status could not be realized without outside links to nonkin (and often nonvillagers), and the establishment of these links, for the most part, followed the patron-client model.

The relative decline in the protective capacity of kindreds (which, given the absence of strong, predictable new institutions would widen the scope of patron-client ties) accelerated the *political* transformation of the colonial and postcolonial period as well. Both administration and electoral politics created new political units that did not generally coincide with the kindred or with the traditional village. As Gallin has shown for electoral politics in Taiwan, the political vitality of the corporate lineage is sapped by changes in the governing unit which no longer permit a single lineage to dominate. The lineage thus loses much of the material basis of its previous solidarity, and new dyadic ties become the means by which winning coalitions are built in the new unit.[42] Political consolidation, like economic consolidation, beyond minimal kin and village units can thus enlarge the potential role for such nonkin structures as patron-client clusters.

Considering all three criteria, Southeast Asian states, like most traditional nations, satisfy most conditions for the survival of patron-client structures as a common means of cooperation. First, the disparities in power and status that form the basis of this kind of exchange have, if anything, become more marked since the colonial period. Second, nonkin structures of cooperation have always been important in the complex societies of Southeast Asia and have become more significant because of the new economic and political dependencies introduced by colonialism and nationhood. Finally, with the possible exceptions of Singapore and North Vietnam, the nations of Southeast Asia have not developed strong modern institutions which would begin to undermine purely personal alliance systems with impersonal guarantees and loyalties.

At this point in the argument, it is essential to show how patron-client structures, as one form of vertical cleavage, coexist with communalism, another form of vertical cleavage, in Southeast Asia. If loyalty to an ethnic or religious group is particularly strong it will mean that the only possible partners in most patron-client dyads will be other members of the same community. Since the community is a categorical group which excludes some possible dyadic partnerships, it represents a different form of cleavage from patron-client links. Vertical dyadic bonds can, nonetheless, coexist with communal cleavage in at least three ways: 1) as intercommunal patron-client ties above corporate communities; 2) as intercommunal patron-client ties above factionalized communities, and, 3) as intracommunal patron-client structures. First, when communal groups do deal corporately with the outside—as quite a few small, highland tribes do in Southeast Asia—we may get patron-client ties that join their leaders, as clients, to regional or national leaders. If two distinct corporate communities were linked through their leaders who were clients of an outside leader, the structure could look like that in Figure A.

Figure A

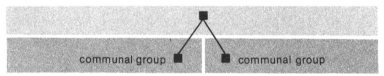

More often in Southeast Asia, however, the second situation prevails in which a number of *patrons with separate followings within the same communal group* compete for the most advantageous links to the outside. A simple representation of this pattern is presented in Figure B, in which two patrons are linked

Figure B

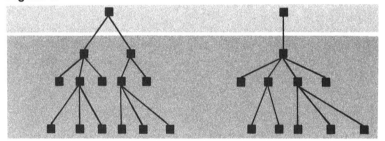

as clients to an outside patron and thereby have established a working alliance against a third patron in the same communal group who is linked to a different outside leader. Here, the communal group is rent by factionalism and has multiple ties to the outside world.[43] The vertical links outside the communal group, however, are likely to be somewhat weaker or more tentative than links within the community. This is so because all competing subordinate patrons and their clientele fall within a communal unit which shares a potentially strong interest; if the communal group as a whole were threatened, the shared parochial links would serve as the basis for a unity that might supersede any exterior patron-client links. The situation described in Figure B is only likely to arise, then, if there are no salient collective threats to the communal group as a whole.

The mixture of communalism and patron-client structures portrayed in Figures A and B focuses on the extracommunal patron-client links that achieve a measure, however weak, of intercommunal integration. A third mixture of communal and dyadic association focuses instead on intracommunal politics alone. This would be represented by just the boxed portion of Figure B, which indicates that, even if communal conflict is widespread, it may well be that the *intra*communal politics of each contending group is best described by the patron-client model.

The salience of communal feeling, especially in Malaysia, Burma, and Laos, but also in Indonesia and Vietnam makes such mixtures of communalism and patron-client politics common. Except at the apex of the political structure where a leader may have leaders of smaller communal groups as his clients, most patrons have followings that are almost exclusively drawn from their own community. Intercommunal integration tends to take place near the apex of the political structure with the base of each communal pyramid remaining largely separate. The links that represent this integration tend, moreover, to be fragile and to disintegrate in the face of a community wide threat. Both communalism and patron-client links share the political stage, but patron-client structures are most prominent in periods of peace and stability. In addition, the process of politics *within* each communal group—in effect holding communal affiliation constant—is usually best analyzed along patron-client lines. In nations such as Thailand, the Philippines, and Cambodia, which are comparatively homogeneous culturally, there are few communal barriers to the proliferation of patron-client linkages. Thus, the patron-client model can be applied to those nations in its "pure" form since communal affiliation is not important in creating discontinuous patron-client networks.

B. The Transformation of Traditional Patron-Client Ties

1. **The General Trend.** The typical patron in traditional Southeast Asia was a petty local leader. Unlike the representative of a corporate kin group or a corporate village structure (rare outside Vietnam and Java, respectively), the local patron owed his local leadership to his personal skills, his wealth, and occasionally to his connections with regional leaders—all of

which enhanced his capacity to build a personal following. The fortunes of such petty leaders waxed or waned depending on the continuing availability of resources and spoils which served to knit together a following. Perhaps the most striking feature of local patron leadership in Southeast Asia was its fluidity and instability, which contributed to a relatively high rate of local elite circulation. In contrast to India, where hereditary office-holding and landholding provided somewhat greater continuity, the typical local leader in Southeast Asia had put together many of the necessary resources of wealth, force, connections, and status on his own and could probably only promise his son a slight advantage in the next round. Two important reasons for this oscillation in local power are a) the weakness of the central state, which lacked either the force or durability to sustain and guarantee the continuation of local power elites, and b) the relative ease with which clients in a slash-and-burn economy could, if dissatisfied simply move to another area, thus undermining their ex-patron's basis of power.[44]

Patron-client systems have survived—even flourished—in both colonial and postindependence Southeast Asia. There have been important changes, however. New resources for patronage, such as party connections, development programs, nationalized enterprises, and bureaucratic power have been created. Patron-client structures are now more closely linked to the national level with jobs, cash, and petty favors flowing down the network, and votes or support flowing upward. In the midst of this change, old style patrons still thrive. Highland leaders, for example, still operate in a personal capacity as patron/brokers for their people with the lowland leaders. Landowners in the Philippines and elsewhere have used their traditional control of land and the tenants who farm it to win positions of local or regional party leadership. Whatever the particular form they take, patron-client networks still function as the main basis of alliance systems among nonkin throughout Southeast Asia.

The nature of patron-client bonds within Southeast Asia has varied sharply from one period to the next and from one location to another. Different resources have risen or plummeted in value as a basis of patronage depending upon the nature of the political system:

Patron Client Politics and Change in Southeast Asia

Secular Trends in Patron-Client Ties

Quality	Traditional	Contemporary
1. Duration of bond	more persistent	less persistent
2. Scope of exchange	multiplex	[increasingly] simplex
3. Resource base	local, personal	external links, office based
4. Affective/instrumental balance	higher ratio of affective to instrumental ties	lower ratio of affective to instrumental ties
5. Local resource control	more local monopoly	less local monopoly
6. Differentiation between clusters	less differentiation	more differentiation
7. Density of coverage	greater density	less density

The capacity to mobilize an armed following was particularly valuable in the precolonial era; access to colonial office was a surer basis of patronage than armed force in the colonial period; and the ability to win electoral contests often became the central resource with the advent of independence. Not only have resource bases proved mercurial over time, but the nature of patron-client ties in the indirectly ruled highland areas has remained substantially different from lowland patterns. Amidst this variety and change, it is nevertheless possible to discern a number of secular trends in the character of patron-client bonds. Such trends are far more pronounced in some areas than others, but they do represent directions of change that are important for our analysis.

(1) In comparison with more bureaucratic empires, patron-client bonds in precolonial Southeast Asia were not, as I have pointed out, markedly persistent. With the quickening of social change brought about by the commercialization of the economy and the penetration of the colonial state into local affairs, however, a patron's resource base became even more vulnerable to the actions of outside forces over which he had little or no control. It was an ingenious patron indeed who could survive the creation of the colonial state, the export boom, the depression of the 1930s, the Japanese occupation, and independence with his resources and clientele intact. The major exception to this trend was the colonial period in indirectly ruled areas where colonial military and financial backing of traditional rulers, if anything, brought a stability—or stagnation—to political systems that had been more chaotic. Elsewhere, patron-client links tended to become more fragile and less persistent.

(2) With the differentiation of the economy and its effects on the social structure, the scope of exchange between patron and client tended to narrow somewhat. Where traditional patrons could generally serve as all purpose protectors, the newer patron's effectiveness tended to be more specialized in areas such as political influence, modern sector employment, or administrative influence. Although patron-client ties remained flexible and personal, the more limited capacities of the patron tended to make relationships less comprehensive and hence less stable.[45]

(3) The traditional patron for the most part operated with personally controlled local resources. One effect of the colonial period—and independence as well—was to increase radically the importance of external resources for local patronage. A following based on purely local office or landholding was seldom sufficient to sustain a patron in a new environment where schools, agricultural services, regional banks, and public employment represented competing sources of patronage. The growing role of outside resources, in most cases, thus led to competition among patrons, each of whom recruited followings with the particular resources at his command.[46] In addition, since those who controlled the new resources were generally officeholders subject to transfers or political changes at the center, the new patrons were less secure than older patrons and probably more inclined to maximize their gains over the short run.

(4) Because the new patron-client ties were weaker and less comprehensive, and because the new patrons were often from outside the local community, the instrumental nature of the exchange became more prominent. A relationship that had always involved some calculations of advantage lost some of its traditional legitimacy and grew more profane. Patron-client exchanges became more monetized, calculations more explicit, and concern centered more on the rate of return from the relationship rather than on its durability. This trend meant that newer patron-client clusters were likely to have a comparatively large "fair-weather" periphery, a comparatively small core-following, and a less "constant" patron as well.

(5) The breakdown of local patron monopolies follows logically from most of the changes we have already discussed. Where one local landowner or traditional leader had once dominated he now faced competitors who might be local ad-

ministrators of state welfare in loan programs, teachers in new secular schools, a local trader or businessman, or the resident manager of a foreign-owned plantation. Factional strife which reflects this competition was most common in villages where socioeconomic change and government penetration had been far-reaching, and less common in more traditional areas.[47]

(6) As differentiation occurred within the local societies, they gave rise to patron-client clusters that were distinct. A bureaucrat might have a following primarily within his agency, a businessman among his laborers, and a landowner among his tenants. This process of differentiation among clusters provided the potential basis for durable group interests inasmuch as many clusters now had an institutional distinctiveness.

(7) While the changes we have examined may have assisted the vertical integration of patron-client pyramids, they tended to reduce the universality of coverage. That is, more and more people in the new market towns and cities, on plantations, and on small plots they rented from absentee landlords were no longer attached—or were very weakly attached—to patrons. These new elements of the population varied greatly in their interests and their levels of organization, but, in any event, they fell outside the older patron-client network.

While some long run trends in patron-client ties seem clear, it is difficult to say anything about the balance between voluntarism and coercion over time. On the one hand, changes in the economy have made clients less autonomous and more dependent on patrons for protection against a fall in world prices, for cash advances before the harvest, and so forth. Also contributing to a decline in the client's bargaining position is the imported legal system of property guarantees which allow a wealthy man, if he so chooses, to resist pressures for redistribution that operated in a traditional setting. On the other hand, the breakdown of local patronly monopolies and the exchange resources that electoral systems often place in the hands of clients work in the opposite direction. Given these contradictory tendencies, one can draw the tentative conclusion that patron coerciveness has declined only where extralocal recources *and* competitive elections are common and has elsewhere either increased or remained the same.

In general, patron-client ties have tended to become more instrumental, less comprehensive, and hence less resilient. They still represent diffuse personal bonds of affection when compared to the impersonal, contractual ties of the marketplace, but the direction of change is eroding their more traditional characteristics. Even this supple traditional protective mechanism has had to pay a certain price to survive in the midst of a nation-state with a commercialized economy. The durability and legitimacy of the patron-client tie was best served when all of a client's dependencies were focused on a single patron. But as Godfrey and Monica Wilson have shown, this situation is less and less likely since the process of modernization tends to create multiple dependencies—each of less intensity—rather than concentrating dependence on one person.[48] The slowly weakening comprehensiveness of the link is, ultimately, what undermines its sanctity and legitimacy for the client.

2. The Dynamics of the Transformation.

The engine behind the shift in patron-client ties was largely provided by the penetration of the local arena by an intrusive national economy and national political system. This penetration wrought two major changes that transformed patron-client links: a) during the colonial period, especially, it impaired the effectiveness of local redistributive pressures and b) particularly after independence it "nationalized" access to patronly resources, thus creating new bases of patronage and devaluating old ones.

Traditional peasant societies, operating in an economy of great scarcity in which one family's gain is another's loss, have generally developed a variety of social control mechanisms that guarantee a measure of security to each family and temper the centrifugal forces generated by the struggle for subsistence.[49] These mechanisms commonly involve forcing anyone who has accumulated considerable wealth to redistribute a portion of it. A wealthy man is pressed to assume expensive ceremonial offices, to make large religious contributions, to give loans and donations, and so forth. He trades his wealth for prestige, and, by so providing for at least the minimum well-being of others, he becomes a legitimate patron with a personal entourage of those obligated to him.

The central fact about these redistributive mechanisms, however, is that they operate by virtue of a local power situation. That is, the wealthy man in a peasant village can seldom rely on outside force or law to protect him; instead, his wealth and position are ultimately validated by the legitimacy he acquires in the local community. Unless a wealthy individual can persuade most of the community that his wealth is no threat to them or can win enough personal allies to sustain his position, he is in danger. Colonialism, however, broke the relative autonomy of the local arena and hence weakened many of the community's redistributive pressures. Supported in effect by the power of the colonial regime to enforce its notion of law, the patron could increasingly ignore local levelling pressures. If he lost much of the social approval he previously enjoyed, he had gained an outside ally with the power to guarantee his local position. The colonial power situation thus offered the *older* patron new leverage in the local arena—leverage which was further strengthened by the growing complexity of colonial society. As Blau has explained,

> Social approval has a less pervasive significance as a restraining force in complex societies than in simpler ones, because the multiplicity of groups and the possible mobility between them in complex societies allows deviants of nearly all sorts to escape from the impact of community disapproval by finding a sub-group of likeminded persons. . . .[50]

Absentee landlords, the new urban wealthy, and minority communities (who were relatively impervious to local social approval so long as they had colonial backing) were *new* elements in colonial society which could escape patronly obligations. The colonial system thus tended to allow existing patrons greater latitude for exploitation while producing a class of wealthy nonpatrons.

If the intrusion of external power could strengthen the hand of an existing patron, it could also create a resource base for the rise of new patrons. The activities of the colonial regime included the hiring, firing, and promotion of public employees, the dispensing of contracts, and the granting of licenses and permits, all of which could be used to create a personal following. With independence, not only did local leaders take over responsibility for all these decisions, but the scope of government activity and regulation was generally expanded into new areas such as community development. The survival or demise of a local patron often depended, as Geertz has shown, on how successful he was in tapping these new bases of power.[51]

Except for the rare local patrons—especially in indirectly ruled areas—who were able to monopolize these external resources, the new situation produced more competition and mobility among patrons. Many potential clients quickly discovered that their needs were best served by a patron who had access to the institutions which controlled the use of these external resources. In any local context this shift could be measured by the rise of new patrons who were wholly or partly based in these new structures. Studying the incorporation of Sardinia into the Italian nation-state in the 20th century, Alex Weingrod has documented the growing importance of such externally based patrons.[52] The proportion of outsiders asked to be godparents, for example, increased dramatically from 1920 to 1960, and patrons with links to the ruling party and the state bureaucracy had increased their followings at the expense of traditional landholders. A similar process has occurred in Southeast Asia as the integration of villages into a national economy and political system tended to produce a number of more specialized local patrons who often became factional leaders.

Most of the transformations in patron-client bonds that we have been discussing apply with greatest force to the directly ruled, lowland areas of Southeast Asia where the colonial impact was both swift and far-reaching, and where colonial officials more thoroughly replaced indigenous leaders. In the indirectly ruled areas—such as highland Burma, the Unfederated States of Malaya, most of Indonesia's Outer Islands, Cambodia, Laos (and perhaps Thailand belongs here as a limiting case of indirect colonial influence)—these generalizations must be qualified. To ease the financial and administrative burden of colonial rule in these areas, the colonizers generally kept local rulers in place and used them as agents. Since these were by and large peripheral areas of marginal commercial interest, the pace of economic change tended to be slower as well.

The effects of this policy on patrons, in contrast to the directly ruled regions, were twofold. First, local patron/leaders tended to be strengthened by colonial backing and the new powers given to them. What had probably been a fairly unstable and minimal chieftaincy now became a local regime stabilized and extended by the colonial power. Secondly, the sanction of colonial authority permitted many such leaders to broaden the resource base of their authority.[53] It is true of course that a local patron's new source of strength entailed some threat to his legitimacy, but since the colonial regime demanded little beyond the maintenance of law and order in those areas, it was seldom crippling. On the other hand, the annoited patron now had the means to eclipse his rival patrons. He not only had his traditional authority and the discretionary administrative powers given him by the colonial regime, but he could use his power to purchase land, control local trade, and act as the commissioned agent of private firms. Frequently, then, the local ruler gained a new lease on political life as the dominant local figure owing to his wealth, his administrative power, and a measure of traditional legitimacy. Given the slower pace of economic change and state penetration in these regions, the local leader had to contend less with new competitors who flourished amidst such changes. The strength many patrons achieved under indirect rule is nowhere more apparent than in postcolonial elections in which they often could deliver most of their region's vote.

3. Electoral Politics and Patron-Client Ties.

Most Southeast Asian states have had functioning electoral systems at one time since their independence. Although only the Philippines retains a parliamentary system, the electoral studies that do exist can tell us much about the effects of party competition on patron-client bonds and, beyond that, highlight some of the unstable features of patron-client democracies.

The dynamics of electoral competition transformed patron-client relations in at least four important ways: (1) it improved the client's bargaining position with a patron by adding to his resources; (2) it promoted the vertical integration of patron-client structures from the hamlet level to the central government; (3) it led to the creation of new patron-client pyramids and the politicization of old ones; and (4) it contributed to the survival of opposition patron-client pyramids at the local level.

First, with popular elections the client gained a new political resource, since the mere giving or withholding of his vote affected the fortunes of aspirants for office. Nor were voters slow to realize that this resource could be turned to good account. Even someone with no other services of value to offer a patron found that the votes of his immediate family were often sufficient to secure the continuous assistance of a local politician. This pattern could be found throughout Southeast Asia in electoral situations but is most striking in the Philippines, where most patron-client ties are centered around landholding and elections. The Filipino politician, as Wurfel points out, does favors *individually* rather than collectively because he wishes to create a personal obligation of clientship.[54] The voter, for his part, asks that his patron/politician favor him because of a personal obligation to reciprocate.

In one sense, popular elections can be seen as a reestablishment of the redistributive mechanisms of the traditional setting. Once again a patron's position becomes somewhat more dependent on the social approval of his community—a social approval that is now backed by the power to defeat him or his candidate at the polls. Unable to depend on outright coercion, and faced with competitors, the electoral patron knows he must (unless his local economic power is decisive) generally offer his clients better terms than his rivals if he hopes to maintain his local power.

Second, nationwide elections make it necessary for a national party to establish a network of links extending down to the local level. For the most part a party does this by taking advantage of existing patron-client clusters and incorporating them into its structure. The competitive struggle of Indonesian parties to forge such links in Java during the 1950s is apparent in the accounts Feith and Geertz give of electoral campaigns.[55] They agree that effective campaigning in the village took the form of activating and politicizing preexisting personal links rather than mass meetings or policy stands. The campaign, as Feith says

> ... a race for a foothold in these villages ... a foothold involving allegiance of as many as possible of their influential people. Here the first step was to secure the support of those whose authority was accepted by the village prominents. Thus the parties struggled with one another for influence with the bupatis, the wedanas, and the tjamats ... with the local military commanders and the heads of local offices of Religion, Information, and Mass Education, kijaijis, ulamas, heads of clans, old guerrilla leaders. ... [56]

As elsewhere in Southeast Asia, a party succeeded best at the polls by securing the adhesion of the important local patrons, who would deliver their clients as a matter of course. Working on voters individually or by class affiliation made little sense when most of the electorate was divided into patron-client clusters. The affiliation of patrons was often gained by making them candidates, by promising them jobs or other patronage, or even by cash payments. It is clear, however, that, in comparison with those parties who had to create new links to village leaders, parties such as the Nahdatul Ulama which could rely on bonds that antedated the election, were in a stronger position. Nash's account of the 1960 election in upper Burma reveals a similar pattern of patron mobilization. When a local patron was approached to join U Nu's faction of the AFPFL on the promise of later patronage, he was able to get thirty-nine others—his relatives and those who owed him money or for whom he had done favors, i.e., his clients—to join as well.[57]

The nature of the new exchange relationship that gives vitality to this patron-client pyramid is similar in most electoral systems. The local patrons and their clients provide votes at election time, hopefully carrying the village, while the party undertakes to help its local adherents (through their patron) with jobs, help in dealing with the bureaucracy, providing public works, and so forth. Since the winning party can generally offer more support to local allies than the opposition can, local patrons are likely to display a "bandwagon effect," switching allegiance to a probable winner. In addition, the party's need for a powerful local base is likely to lead to a certain localization of power. In return for delivering local votes for its list, the party is likely to give its local patron a wide discretion in administrative and development decisions affecting the locality. Thus many local patrons are able to entrench themselves further as dominant figures.

A third consequence of elections for the patron-client structure is to promote the expansion of patron-client ties and the politicization of existing bonds. Knowing that an electoral victory is important, a local patron with a modest following will probably try to obligate more clients to him in order to strengthen his electoral position. Patrons who have previously been politically inactive would "immediately convert their private power such as control over sharecroppers, debtors, kinsmen, heighbors, etc., into public political power in the form of votes."[58] Given these tendencies, the patron-client structures in a given community are most evident immediately before an election, especially a hotly contested one, when the contestants attempt to activate any links that might advance their cause.

A final point about the impact of elections on patron-client structures is that they tend to heighten factionalism and unless one cohesive party completely dominates, to promote the survival of local opposition factions. In most traditional settings, patron rivalry was largely limited to the local arena so as not to invite external intervention. An electoral system, by contrast, creates rival national or regional parties which need allies at the local level. A weak faction that might previously have been forced to compose its differences with a dominant faction, can now appeal for external support. Many of these external allies are able to provide their local adherents with patronage, cash, or other favors so as to maintain a local foothold. The net effect of electoral competition is thus to exacerbate many of the latent factional differences among patron-client clusters and occasionally to buttress weak patrons whose position would otherwise have disintegrated.

The effects we have attributed to elections can be compared to the situation in Thailand, where elections have only rarely been any more than a device to legitimate self-selected rulers. There the local client's vote is not important enough to materially improve his bargaining position with a patron, and the vertical integration of patron-client clusters had not gone very far beyond the central institutions of the bureaucracy and armed forces. In Thai villages unlike the electoral settings of Java in the 1950s, or the Philippines, many patron-client clusters are of purely local significance and are not highly politicized. Local factional conflict, as a result, is much less striking in Thailand than where competitive elections have helped to subsidize it.

To this point we have focused on the general influence of elections on patron-client structures. Depending on the region and the party in question, however, there has been a noteworthy variation in the connection between party and patron-client structures. The essential distinction is one between a party that has created its own network of patron-client linkages from the center and a party that relies on preexisting patron-client bonds and merely incorporates them into its organization. This corresponds to René Lemarchand's distinction between types of party machines in Africa.

> In situations where micro-level clientelistic structures provide the essential linkages between the party machine and the masses . . . the machine is superimposed upon, and in some ways tributary of, the clientelistic subsystem. A distinction must therefore be drawn between the more orthodox types of machine . . . in which patronage becomes the essential source of cohesion, and what one might call the neo-traditional machine, in which exchange processes between the center and periphery are mediated by, and contingent upon, the operation of traditional forms of clientelism.[59]

As Lemarchand adds, the "orthodox" machine is more dependent on material inducements since its linkages are of more recent origin and hence more instrumental. "Neo-traditional" machines, by contrast can rely somewhat more heavily on established patterns of deference, though they must bargain with nearly autonomous local patrons. This distinction has validity in Southeast Asia as well, both in accounting for the different styles of politics in the indirectly ruled, more traditional areas as compared to the directly ruled, heavily commercialized areas and also in explaining the structural differences between the more traditionalist as compared to more modern parties. In the indirectly ruled areas, political parties confronted fairly stable constellations of local patron power which would have been difficult to destroy. It was simpler to come to terms with local leaders rather than to try to circumvent them, even though this accommodation tended to divide the party into a coalition of political fiefdoms. In directly ruled areas where competition among patrons and heavily instrumental ties were more common, a party had a greater opportunity to create new linkages but at perhaps greater expense in favors and patronage. Throughout Southeast Asia all parties had to adapt themselves to these differences in social structure in different regions of the country: the PNI of Indonesia operated differently in Central Java than in the Outer Islands; U Nu's AFPFL faction could not win the support of the hill tribes in the same way they won the vote of the lowland Burmese, and the Nationalists in the

Philippines campaigned differently in central Luzon than in Mindanao.

There were also systematic differences between the neo-traditionalist and modernist parties within a given area. In Java, for example, the Nahadatul Ulama incorporated more traditional patron-client ties than did the PNI, which could rely more on its access to material rewards and local administration. On the East Coast of Malaya, the Islamic PMIP appealed most to the more traditional peasantry while the ruling UMNO concentrated on the town population and those dependent on federal funds. One would also guess that the Ba Swe faction of the AFPFL had a more instrumental base in towns and the modern sector, while U Nu's faction was more frequently based on existing patron-client links in the countryside. Simply knowing the kind of patron-client bonds a party had created or incorporated could reveal a great deal about the party's cohesiveness, the nature of its local base, and the extent of its reliance on material inducements.

4. The Inflationary Character of "Patron-Client Democracy."

The introduction of competitive elections in Southeast Asia increased the pressures on regimes for the downward distribution of tangible benefits. In return for votes flowing up the vertical chain of patron-client structures, each patron depended upon the downward distribution of patronage in the form of administrative favors, land grants, public employment, and so on, in order to keep his own pyramid of followers intact. Elections, by themselves, had shifted the balance of exchange so that it favored

Distributive Pressures of Elections

A. Strongest when	B. Weakest when
1. Elections determine powerholders	Elections have marginal significance
2. Regime is weak, unstable regime	Regime is strong, stable regime
3. Socioeconomic change is extensive (direct rule, lowland areas)	Socioeconomic change is less extensive (indirect rule, highland areas)
4. Party is modernist, secular'	Party is traditionalist, religious

Examples of Strong Distributive Pressure (A): Indonesia (until 1955), Burma (until 1959 at least), the Philippines.

Examples of Weaker Distributive Pressures (B): Thailand, and perhaps Malaysia (until 1967).

the client somewhat more than before. The consequence of this shift in exchange terms was a greater flow of material benefits toward the base of the patron-client network.

The strength of the downward distributive pressures generated by electoral procedures in Southeast Asia depended primarily on four variables which are stated in contrasts.

Each of the above variables related to the strength of incentives impelling a party to maximize its clientele and the degree to which that clientele will depend on concrete material incentives rather than ties of affection or deference. The first and obvious requirement for distributive pressures is that elections be important in the selection of an elite. Secondly, a shaky regime or party will be in a less advantageous position to resist client demands than a strong one, since an election is an all or nothing affair and uncertainty over the outcome will raise costs; when the race is close, the party in its recruitment efforts knows that the marginal value of the extra dollar, patronage job, or development grant is all the greater. That is one reason why distributive pressures were greater in 1955 in Indonesia when an election fraught with uncertainty would determine which parties would form a coalition, than in Malaysia in 1964 when the question was not whether the Alliance would win but whether or not it would take two thirds of the seats. The importance of social change (item three) is based on the observation that patron-client ties in less traditional areas typically require the patron to deliver more in the way of instrumental, material rewards. The maintenance of a loyal patron-client network in a traditional area where deference is strong, will, I assume, cost a party somewhat less in material rewards and favors than will a network of the same size in a built-up area where traditional patron-client bonds have eroded. Finally, a neo-traditionalist party such as the UMNO in Malaysia can, in part, rely upon the traditional legitimacy of many of its leaders while a party of "new men" such as the PNI in Indonesia or the AFPFL in Burma has to rely more often on highly instrumental ties. Thus a weak party led by "new men" and relying on votes from among an uprooted population is likely to develop a patchwork patron-client structure that is very expensive to maintain. It is indicative of just how much financial backing such structures require that only a ruling party with access to the public till can generally afford the construction costs.[60]

The distributive pressures experienced by such regimes manifest themselves in familiar ways. Government budgets, and of course deficits, swell quickly with expenditures on education, growing public employment, community development projects, agricultural loans, and so forth. Particularly since votes in Southeast Asia are to be found in the countryside, one would expect that regimes with strong electoral pressures would spend more in the grass-roots rural areas than would regimes without such pressures. Given such pressures, local expenditure is also arranged as much as possible so that benefits can be distributed individually since that arrangement is more appropriate to patron-client exchange patterns. Even with pork-barrel programs a local party leader will claim personal responsibility for the gift and personally help distribute whatever employment or subcontracting it includes. The capacity of the regime to keep its network intact and win elections depends on its capacity to provide rewards for the lower tiers of its structure at a constant or even expanding rate.

A regime that is dependent on its particularistic distributive capacity is also unlikely to solve its financial dilemmas either by structural reform or by tapping new sources of revenue. Most conceivable structural reforms, such as land redistribution, would strike at the resource base of many patrons and are thus unacceptable to parties whose *policy* interests coincide with the desires of its dominant patrons. Such regimes also have a most difficult time raising revenue from internal taxation. A rise in direct taxation would threaten their base of support; and, in fact, they are notorious for the undercollection of revenues due them, since favors to their clients often take the form of either leaving them off local tax rolls or ignoring debts they owe the government. The Burmese peasants connected to U Nu's faction of the AFPFL, for example, were almost universally in default on agricultural loans they had received as party supporters. They assumed the loan was a gift for clientship and knew that a government dependent on their votes could scarcely press matters.

If this analysis is correct, regimes under intense distributive pressures will characteristically resort to budget deficits, especially in election years, to finance their networks of adherents. Their reliance on heavily instrumental and highly monetized patron-client ties will also make it difficult for them to avoid running down their foreign exchange reserves to maintain their strength at the polls. The division of expenditures within the budgets of such regimes should also reveal a heavy emphasis on distributive expenses at the local level. Empirical studies of budget distribution, budget deficits, and foreign exchange expenditures over time in parliamentary Burma, Indonesia, and the Philippines, when compared with similar statistics for nonparliamentary periods in these same countries, or say, with statistics from Thailand and Malaysia should confirm this prediction.[61]

Democratic regimes which must cater to the strong distributive pressures generated by their electoral clientele are thus particularly vulnerable to the vagaries of world prices for primary products on which their budgets depend. As long as the economy expanded and world prices were buoyant, they could afford the costs in public jobs, pork-barrel projects and loan programs to solidify and expand their huge patron-client network. But a stagnating economy or declining world prices threatened the entire structure they had pieced together, since it relied so heavily on material inducements and relatively little on affective ties. In this context it may be that the collapse of Korean war-boom prices for primary exports was the crucial blow to democracy in Indonesia and Burma. The Philippines may narrowly have escaped a similar fate by virtue of their longer and more legitimate democratic tradition, as well as by their not having suffered as proportionately large a loss in foreign exchange. Malaysia was less vulnerable, since she had just become independent and her strong government faced only moderate distributive pressures, while the Thai military elite was even less reliant on its distributive performance. The political stability or instability of parliamentary forms in these nations in the late 1950s was thus strongly affected by the strength of distributive pressures fostered by these political systems in the mid 1950s.

Notes

1. Two influential anthropologists who employ this mode of analysis are: Clifford Geertz ("The Integrative Revolution," in Geertz, ed., *Old Societies and New States* [New York: The Free Press of Glencoe, 1963]); and Max Gluckman (*Custom and Conflict in Africa* [Oxford: Basil and Blackwell, 1963]).
2. A number of political studies of Southeast Asia have dealt with factionalism or patron-client ties. The most outstanding is Carl

Landé's *Leaders, Factions, and Parties: the Structure of Philippine Politics*, Monograph No. 6 (New Haven: Yale University—Southeast Asia Studies, 1964). For the Thai political system, Fred W. Riggs' *Thailand: The Modernization of a Bureaucratic Polity* (Honolulu: East-West Center Press, 1966) and David A. Wilson's *Politics in Thailand* (Ithaca: Cornell University Press, 1962) pursue a similar line of analysis; and for Burma, see Lucian W. Pye, *Politics, Personality, and Nation-Building: Burma's Search for Identity* (New Haven: Yale University Press, 1962). Some notable attempts to do comparable studies outside Southeast Asia are: Colin Leys, *Politicians and Policies: An Essay on Politics in Acholi Uganda 1962-1965* (Nairobi: East Africa Publishing House, 1967); Myron Weiner, *Party-Building in a New Nation: The Indian National Congress* (Chicago: University of Chicago Press, 1967); Paul R. Brass, *Factional Politics in an Indian State* (Berkeley and Los Angeles: University of California Press, 1965); Frederick G. Bailey, *Politics and Social Change: Orissa in 1959* (Berkeley and Los Angeles: Universtiy of California Press, 1963).

3. Class as well as ethnicity is relevant to Malay-Chinese conflict, since the different economic structure of each community places them in conflict. Many a rural Malay experiences the Chinese not only as pork-eating infidels but as middlemen, money lenders, shopkeepers, etc.—as the cutting edge of the capitalist penetration of the countryside.

4. There is an extensive anthropological literature dealing with patron-client bonds which I have relied on in constructing this definition. Some of the most useful sources are: George M. Foster, "The Dyadic Contract in Tzintzuntzan: Patron-Client Relationship," *American Anthropologist*, 65 (1963), 1280-1294; Eric Wolf, "Kinship, Friendship, and Patron-Client Relations," in Michael Banton, ed., *The Social Anthropology of Complex Societies*, Association of Applied Social Anthropology Monograph #4 (London: Tavistock Publications, 1966) pp. 1-22; J. Campbell, *Honour, Family, and Patronage* (Oxford: Clarendon Press, 1964); John Duncan Powell, "Peasant Society . . . ," p. 412, Carl Landé, *Leaders, Factions and Parties . . . ,"* Alex Weingrod, "Patrons, Patronage, and Political Parties," *Comparative Studies in Society and History*, 10 (July, 1968), pp. 1142-1158.

5. Powell, "Peasant Society and Clientelist Politics," *American Political Science Review*, 64 (June, 1970), 412.

6. Another comparable model, of course, is the lord-vassal link of high feudalism, except in this relationship the mutual rights and obligations were of an almost formal, contractual nature. Most patron-client ties we will discuss involve tacit, even diffuse standards of reciprocity. Cf. Ruston Coulborn, ed., *Feudalism in History* (Princeton: Princeton University Press, 1956).

7. In most communities this sense of obligation is a strong moral force, backed by informal community sanctions that help bind the client to the patron. A good account of how such feelings of debt reinforce social bonds in the Philippines is Frank Lynch's description of *utang na loob* in *Four Readings in Philippine Values*, Institute of Philippine Culture Papers, No. 2 (Quezon City: Aleneo de Manila Press, 1964).
8. Peter M. Blau, *Exchange and Power in Social Life* (New York: Wiley, 1964), pp. 21-22. Blau's discussion of unbalanced exchange and the disparities in power and deference such imbalance fosters is directly relevant to the basis of patron-client relationships.
9. These general alternatives are deduced by Blau (p. 118) and are intended to be exhaustive.
10. Later, we will examine certain conditions under which this may actually occur.
11. There is little doubt that this last resort usually acts as a brake on oppression. The proximate causes for many peasant uprisings in medieval Europe during hard times often involved revocation of small rights granted serfs by their lords—e.g., gleaning rights, use of the commons for pasturage, hunting and fishing privileges, reduction of dues in bad crop years—rights which offered a margin of security. Such revolts, even though they generally failed, served as an object lesson to neighboring patrons. Cf. Friedrich Engels, *The Peasant War in Germany* (New York: International Publishers, 1966); Norman Cohn, *The Pursuit of the Millennium* (New York: Harper, 1961); and E. B. Hobsbawm, *Primitive Rebels* (New York: Norton, 1959).
12. Blau, *Exchange and Power in Social Life*, pp. 119-120, makes this point somewhat differently: "The degree of dependence of individuals on a person who supplies valued services is a function of the difference between their value and that of the second best alternative open to them." The patron may, of course, be dependent himself on having a large number of clients, but his dependence upon any one client is much less than the dependence of any one client upon him. In this sense the total dependence of patron and client are similar, but almost all the client's dependence is focused on one individual, whereas the patron's dependence is thinly spread (like that of an insurance company—Blau, p. 137) across many clients. Cf. Godfrey and Monica Wilson, *The Analysis of Social Change, Based on Observations in Central Africa* (Cambridge: The University Press, 1945), pp. 28, 40.
13. Blau, p. 269.
14. Blau, p. 127.
15. The classic analysis of the functions of gift-giving (prestation) in creating alliances, demonstrating superiority, and renewing obligations, is Marcel Mauss, *The Gift: Forms and Functions of*

Exchange in Archaic Societies (Glencoe, Illinois: The Free Press of Glencoe, 1954).
16. See Sidney Minta and Eric Wolf, "An Analysis of Ritual Co-Parenthood (Compadrazgo)," *Southwestern Journal of Anthropology* 6 (Winter, 1948) pp. 425-437.
17. Carl Landé, "Networks and Groups in Southeast Asia: Some Observations on the Group Theory of Politics," Unpublished manuscript (March, 1970), p. 6.
18. Adrian C. Mayer, "The Significance of Quasi-Groups in the Study of Complex Societies," in Michael Banton, ed., *The Social Anthropology of Complex Societies,* pp. 97-122. Mayer would call a short-term, contractual interaction that was limited in scope a *simplex* tie.
19. In another sense the patron-client dyad is fragile. Since it is a diffuse, noncontractual bond, each partner is continually on guard against the possibility that the other will make excessive demands on him, thus exploiting the friendship. A patron may, for example, prefer to hire an outsider for an important job because he can then contractually insist that the work be of top quality. With a client, it would be a delicate matter to criticize the work. As in friendship, "the diffuseness of the [patron-client] obligation places a corresponding demand for self-restraint on the parties if the relationship is to be maintained." William A. Gamson, *Power and Discontent* (Homewood, Illinois: Dorsey, 1968), p. 167.
20. A *broker* does in a real sense *have a resource: namely, connections.* That is, the broker's power—his capacity to help people—is predicated on his ties with third parties.
21. U.S. Congressmen spend a good portion of their time trying to seize personal credit for decisions which benefit their constituents whether or not they had anything to do with the decision—as broker or patron. For similar reasons, cabinet ministers in Malaysia and elsewhere have travelled about the country with government checks in hand, making grants to mosques, temples, and charitable groups in a way that will dramatize the largesse as an act of personal patronage. Every government decision that benefits someone represents an opportunity for someone to use that act to enlarge the circle of those personally obligated to him.
22. And it naturally follows that in underdeveloped countries, where the patrimonial view of office is especially strong, a public post could be a client-creating resource.
23. Foster, "The Dyadic Contract in Tzintzuntzan," pp. 1280-1294.
24. For good descriptions of both types of leadership, see Eric R. Wolf and E. C. Hansen, "Caudillo Politics: A Structural Analysis," *Comparative Studies in Society and History,* 9, 2 (January, 1967), 168-179 and Paul Friedrich, "The Legitimacy of a Cacique," in Marc J. Swartz, ed., *Local Level Politics:*

Social and Cultural Perspectives (Chicago: Aldine 1968), pp. 243-269.
25. The terms "cluster," "network," and "first" and "second" orders are adapted from a somewhat similar usage by J. A. Barnes, "Networks and Political Processes," in Swartz, ed., pp.
26. Mayer, "The Significance of Quasi-Groups . . ." p. 109.
27. Landé, "Networks and Groups . . ." (unpublished manuscript), pp. 6-12.
28. Mayer, p. 110.
29. There is no contradiction, I believe, in holding that a patron-client link originates in a power relationship and also holding that genuine affective ties reinforce that link. Affective ties often help legitimate a relationship that is rooted in inequality. For an argument that, in contrast, begins with the assumption that some cultures engender a psychological need for dependence, see Dominique O. Mannoni, *Prospero and Caliban: The Psychology of Colonization* (New York: Praeger, 1964).
30. F. G. Bailey uses the terms "core" and "support" in much the same fashion: see his "Parapolitical Systems," in Swartz, ed., *Local-Level Politics*, pp. 281-294.
31. Here I follow the distinctions made in Blau, *Exchange and Power in Social Life*, pp. 115-118. Other power theorists have made the same distinction.
32. Both J. A. Barnes, "Networks and Political Process," in Swartz, pp. 107-130, and Powell, *Peasant Society* . . . , p. 413, discuss this variable.
33. This variable thus relates not to a dyad but to the following in a cluster or pyramid.
34. My use of the term is adapted from Barnes, p. 117.
35. Wolf, "Kinship, Friendship, and Patron-Client Relations . . . ," p. 10.
36. In this connection, see my *Political Ideology in Malaysia* (New Haven: Yale University Press, 1968), chapter 6; and for zero-sum conceptions among peasants, see George M. Foster, "Peasant Society and the Image of Limited Good," *American Anthropologist*, 65 (April, 1965), 293-315.
37. See, for example, Edmund R. Leach, "The Frontiers of Burma," *Comparative Studies in Society and History*, 3 (October, 1960), 49-68.
38. Samuel P. Huntington, *Political Order in Changing Societies* (New Haven: Yale University Press, 1968), p. 3.
39. Corporate villages are included here since they generally stress shared kinship links to a common ancestor. Part of the corporate character of the Javanese village was perhaps further reinforced as a consequence of the collective exactions required by Dutch colonial policy. "Sanctioned reciprocity" is probably a better term for village structures in Java and Tokin than "corporate."
40. Richard Downs,"A Kelantanese Village of Malaya," in Julian H.

Steward, ed., *Contemporary Change in Traditional Societies,* Vol. II: *Asian Rural Societies* (Champaign-Urbana, Illinois: University of Illinois Press, 1967), p. 147.

41. Wolf, "Kinship, Friendship, and Patron-Client Relations . . . ," p. 5. For a brilliant account of the same process in England, see Karl Polanyi, *The Great Transformation* (Boston: Beacon Press, 1957).

42. Bernard Gallin, "Political Factionalism and Its Impact on Chinese Village School Organization in Taiwan," in Swartz, ed., pp. 377-400.

43. A combination of situations one and two would occur when the tacit rules within a communal group allowed patron-client conflict but forbade the losing or weaker patrons within the communal group from maintaining ties to outside leaders.

44. See, for example, Edmund R. Leach, *The Political Systems of Highland Burma* (Cambridge: Harvard University Press, 1954), and J. M. Gullick, *Indigenous Political Systems of Western Malaya,* London School of Economics Monographs on Social Anthropology #17 (London: University of London/Athlone Press, 1958).

45. Again, indirectly ruled areas were often exceptions in that local rulers tended to take on new powers under the colonial regime and thus became more comprehensive patrons than in the past.

46. For Malaysia, M. G. Swift, *Malay Peasant Society in Jelebu* (London: University of London, 1965), pp. 158-160 captures this shift in local power. A general treatment of such changes is contained in Ralph W. Nicholas, "Factions: A Comparative Analysis," in M. Banton, general ed., *Political Systems and the Distribution of Power,* Association of Applied Social Anthropology Monograph #2 (London: Tavistock Publications, 1965), pp. 21-61.

47. In his study of politics in an Indonesian town, Clifford Geertz has shown that the more traditional hamlets were more likely to be united under a particular leader than were hamlets which had changed more; *The Social History of an Indonesian Town* (Cambridge: M.I.T. Press, 1965), Chapter 6. This finding is corroborated by Feith's study of the 1955 Indonesian elections; Herbert Feith, *The Indonesian Elections of 1955,* Interim Report Series, Modern Indonesia Project (Ithaca: Cornell University, 1961), pp. 28-30. A comparative study of two Burmese villages also supports this conclusion: cf. Manning Nash, *The Golden Road to Modernity* (New York: Wiley, 1965). In this context, directly ruled lowland areas tended to develop factional competition among different patrons, while less directly ruled areas (especially highland areas) more frequently retained some unity behind a single patron who remained their broker with the outside world.

48. G. and M. Wilson, *The Analysis of Social Change,* pp. 28, 40.

49. For a description of such mechanisms see Clifford Geertz, *Agricultural Involution* (Berkeley: University of California Press, 1963); George M. Foster, "Peasant Society and the Image of Limited Good," *American Anthropologist* 67, 2 (April, 1965), pp. 293-315; Swift, *Malay Peasant Society* . . ., and Mary B. Hollnsteiner, "Social Control and Filipino Personality," *Symposium on the Filipino Personality* (Malcati: Psychological Association of the Philippines, 1965), p. 24.
50. Blau, *Exchange and Power in Social Life*, p. 114.
51. Geertz shows how local leaders often managed to become agents of the local sugar mills—buying crops, renting land, and recruiting labor and thereby enlarging their power in the community. *The Social History* . . ., p. 57.
52. Alex Weingrod, "Patrons, Patronage, and Political Parties," pp. 388, 397.
53. Similar assessments of the effects of indirect rule can be found in M. G. Swift, pp. 148-149; Harry J. Benda and John Bastin, *A History of Modern Southeast Asia* (Englewood Cliffs, N.J.: Prentice-Hall, 1968), pp. 75-122. The best accounts of the pattern, however, come from India. Cf. Bailey, *Politics and Social Change*, Chapter 4, and Paul R. Brass, *Factional Politics in an Indian State* (Los Angeles and Berkeley: University of California Press, 1965), Chapter 4.
54. David Wurfel, "The Philippines," in *Comparative Studies in Political Finance: A Symposium, Journal of Politics*, 25 (November, 1963), 757-773.
55. Feith, *The Indonesian Elections of 1955;* Geertz, *The Social History of an Indonesian Town.*
56. Feith, p. 79.
57. Manning Nash, "Party Building in Upper Burma," *Asian Survey*, 3 (April, 1963), pp. 196-202.
58. Nicholas, "Factions . . .," p. 45.
59. René Lemarchand, "Political Clientelism and Ethnicity: Competing Solidarities in Nation-Building," *American Political Science Review*, 66 (March, 1972).
60. Adrian C. Mayer seems to have this distinction in mind, in his study of an Indian town comparing the "hard" campaign of the Jan Sangh, which relied on durable social ties and tried to prevent defections, and the local Congress Party, which ran a "soft" campaign of short-term links by promising favors and benefits to intermediaries. See Mayer, "The Significance of Quasi-Groups in the Study of Complex Societies," p. 106.
61. Many of these data are not in a form that permits easy comparisons. Although budget deficit and foreign exchange figures seem to fit this pattern, statistical confirmation will have to await further research.

**Chapter V
Modern Revolutionary Guerrilla Warfare:
Some Micro-Views**

The difficulties in formulating a general theory of revolution stem primarily from the differences in the complex contexts within which revolutions occur. Models and generalizations are essential for the study of revolution, but may not account for all the variations of this phenomenon: one must study each revolution within its own context. The selections in this chapter furnish concrete examples to make possible some comparison of models and generalizations to the realities of actual situations.

The selection by Chalmers Johnson examines the relevance of Maoist theory in the context of the Vietnamese revolutionary struggles against the French and later against South Vietnam and the United States. In Vietnam the Viet Minh and the Viet Cong followed the Maoist strategy but incorporated modifications based on unique national characteristics and developing situations. Focusing on the Maoist concepts of "mass line," "self-reliance," and "protracted war," the author notes that the Viet Minh significantly modified the protracted war concept in their decisive victory over the French at Dien Bien Phu. Later the Viet Cong adopted a strategy based on what they believed was the psychological vulnerability of the American position in Vietnam. Subordinating the mass line and protracted war, the Viet Cong long strove for decisive victories in order to convince American opinion of the hopelessness of engaging in a counter-revolutionary war. Nevertheless, Johnson concludes that the Vietnamese struggles remain closely wedded to the Maoist strategy in that civilian populations were associated with long-term revolutionary military activities under party control. The underlying strategy was opportunism in exploiting social grievances to defeat the enemy.

The essay by Sam C. Sarkesian examines the political and social patterns in Malaya that led to the "Emergency" of 1948. The author outlines the historical development of the communal society which thwarted the Malayan Communist Party's attempt to create a united front against the British. The racial antagonisms fostered by communalism motivated the Malays to take sides with the British against the revolutionaries. While the Malayan insurgents attempted to follow Maoist principles, they failed to overcome the fatal weaknesses of their position. The author concludes that communal political and social loyalties prevented the maturing of the Maoist revolutionary initiatives and doomed the revolution from the outset.

The next two selections focus on Che Guevara's principles and practice. J. Moreno studies the Guevara model in light of the Cuban and Bolivian experiences. Using Guevara's Bolivian diaries as his primary source, Moreno emphasizes the "foco" as his major theoretical contribution; a highly trained group of guerrilla fighters who are "not only the vanguard but also the political and military center of the revolution." Moreno compares this concept with Maoist guerrilla base doctrine.

Edward Friedman's critique of Moreno's assessment argues that Guevara's "foco" reflects a basic misunderstanding of Maoist doctrine. The base areas are not rigid in concept or practice, the author asserts; the Maoist model rests or recognition that each revolution has its own characteristics, reflecting the peculiarities of its political and social context. Friedman argues that Guevara's defenders have missed the point of his failure in Bolivia: at the root of revolutionary guerrilla warfare is peasant support and mobilization. Friedman reiterates that a specific approach successful in one country should not be presumed automatically applicable to another.

The final selections are examinations of struggles in Portuguese Africa, i.e. Angola, Mozambique and Guinea-Bissau, with a study of that in Guinea-Bissau by Lars Rudebeck and an assessment of the war in Mozambique by Eduardo Mondlane, the assassinated leader of the revolutionary movement there. These studies identify one of the most serious problems of these movements, internal rivalries and divisiveness. The movement in Angola has been especially hampered by these problems. The exception so far has been in Guinea-Bissau, where Amilcar Cabral was the accepted leader and the revolution has been most successful. Here Cabral's assassination early in 1973 was perhaps designed to foster internal difficulties.

External considerations, especially the attitudes of neighboring countries, have played an important role in determining the character of these wars. For example, Guinea-Bissau is completely surrounded (apart from the seacoast) by Black African states which support the revolution. Here the revolutionary organization attained a classic three-tiered military structure: local militia; a regular guerrilla force; and a regular army. Although the movements in all three territories are trying to develop a guerrilla army effective enough to conduct a war of attrition, it appears that only in Guinea-Bissau have they been able to hold a significant territory and achieve a widely effective peasant political mobilization. In each territory the revolution has developed into a peasant-urban protracted war similar to the Maoist model.

The coup d'etat in Portugal in April 1974 and the emergence of a revolutionary regime has perhaps radically altered the situation in Portuguese Africa.

The Third Generation of Guerrilla Warfare
by Chalmers Johnson

During the past thirty years Asian Communist parties have helped instigate and have pursued five major guerrilla conflicts against Asian non-Communist, Western-allied, or Western colonial regimes. Of these five struggles — China, 1937-49; Malaya, 1948-58; the Philippines, 1946-54; the Viet Minh Movement, 1956-54; and the Viet Cong Movement, 1959-present — two, Malaya and the Philippines, made no important contribution, except perhaps a negative one, to the developing body of Asian Communist theory on the strategy of revolution. The Malayan and Philippine conflicts must be left out of account because their Communist leaders did not utilize directly the heritage of Mao Tse-tung's revolution and because both were defeated. But the other three revolutionary wars — the Chinese Communist, the Franco-Vietnamese and the Vietnamese Communist — all show signs of mutual influence on the progressive development of what might be called an emerging Asian Communist "tradition" of revolutionary theorizing. Of perhaps greater interest, the leaders of the third of these three successive generations of Asian Communist revolutionaries have based their struggle on a strategy that is influenced by, but significantly different from, that of their revolutionary grandfathers.

"Guerrilla warfare," in the Asian Communist context, is an element of a concrete revolutionary strategy — that is, of a plan of coordinated behavior intended to bring a revolutionary association to power in a given social system. Our concern in the present discussion is neither with the sources of social disequilibrium that may have mobilized Asian revolutionaries (the "roots of revolution") nor with the particular ideological "future cultures" that Asian Communist revolutionaries hope to create. We are concerned not with why some Asian Communists are revolutionaries nor with what they propose to build on the ruins of the cultures that bred them, but with *how* they think they can defeat their enemies in order to begin their social rebuilding.

The Maoist Tradition:
Military Power Through Political Means

Mao Tse-tung has been the primary strategic innovator in the Asian Communist revolutionary tradition. He began thinking about how to achieve a revolutionary victory after the Leninist strategic tradition had been decisively defeated in Asia — that is, after Chiang Kai-shek's 1927 defeat of the Chinese Communist Party and the Collapse of the Comintern-inspired united front. Although not all Asian Communists learned the lesson of 1927 as well as Mao (*viz.*: the strategy pursued by the Indonesian Communist Party, prior to its destruction in late 1965, under circumstances somewhat similar to those in China during 1927), Mao and later leaders inspired by him concluded that henceforth Communist revolutions could only take the form of revolutionary wars. Whether they knew it or not, they were responding to Trotsky's earlier gloss on Lenin's strategy of revolution: you can succeed without the army, but you cannot succeed against it. Mao's first axiom became that, in order to make a revolution in China (later generalized to include all of the allegedly neocolonial nations and territories of the world), the Communist Party had to build an army and use it to defeat its enemies in military combat.

Despite many differing interpretations by foreign observers of the Chinese Communist revolution, the Chinese Communists themselves interpret it in exclusively military terms. For example, Lin Piao, Peking's Minister of Defense and "Chairman Mao's closest comrade in arms," has said:

> Comrade Mao Tse-tung's theory of and policies for people's war have creatively enriched and developed Marxism-Leninism...The special feature of the Chinese revolution was armed revolution against armed counter-revolution. The main form of struggle was war and the main form of organization was the army which was under the absolute leadership of the Chinese Communist Party, while all the other forms of organization and struggle led by our Party were coordinated, directly or indirectly, with the war.

Since Mao does not admit of an alternative road to victory save the battlefield, the goal of all revolutionary Communist activity must be to prepare for success in battle: "The seizure of power by armed force, the settlement of the issue by war, is the central task and the highest form of revolution. This Marxist-Leninist principle of revolution holds good universally, for China and for all other countries." In short, "Every communist must understand this truth: political power grows out of the barrel of a gun."[1]

Chiang Kai-shek had demonstrated to Mao and the small band of military leaders and peasant rebels who fled to the countryside with him after the debacle of 1927 that a threatened regime defended by a professional army and willing to use that army against Communist rebels could thwart Leninist strategy. Mao therefore oriented his own revolutionary strategy, following upon his axiomatic definition of revolution as revolutionary war, to the solution of two problems: 1) the building of a Communist army, and 2) the development of tactics whereby an objectively weaker military force could defeat an objectively superior military force. His grand strategy, in its most fundamental military sense, was to obtain a decisive intelligence advantage over his otherwise invincible foes. If he could obtain near-perfect intelligence concerning his enemy's strength and movements — and at the same time deny the enemy such intelligence about his own forces — he could begin to correct the material and professional imbalance between the two antagonists. With intelligence, he could introduce the element of surprise, set ambushes, concentrate superior numbers at any selected point, choose the time and place of fighting, avoid all evenly-matched or unfavorable engagements, escape mopping-up campaigns, and contribute to the demoralization of the enemy's rank and file.

The fundamental dynamic problem of this strategy is how to create the web of intelligence upon which it is based. Mao found an answer to this problem in the Leninst doctrine of the united front. However, whereas Lenin saw the united front as a tactic for legitimizing his party's activities, helping to position it within the target regime for purposes of carrying out a *coup d'etat*, Mao conceived of a united front with the mass of the population so that the people could serve as the source of his army's manpower and as its intelligence-collecting network. In order to obtain this desired level of cooperation from the population, Mao argued, the revolutionary directorate had to discover some issue salient among the masses which the party could champion; the party thereby gains access to the people's sympathies and is able to organize the masses for guerrilla warfare.

As revolutionary strategist with his own basic values and

goals thoroughly compartmentalized, Mao recognized the need to exploit opportunistically *any* political issue that might bring about the desired level of mass organization. Most of the inner-party struggles during the revolution concerned Mao's efforts to get his idea of the flexible use of a united front accepted by his party colleagues. Throughout the shifting alliances and political developments of the 1927-49 period, Mao remained committed to the goal of army-building, and he fought against those party members who confused the aims of this or that particular united front effort with the ultimate goal of a military victory by the Communist Party.

As early as 1930, Mao had identified the following issues as potentially exploitable bases for his guerrilla-oriented united front: 1) the contradictions between imperialism and the Chinese nation, and among imperialists themselves; 2) the contradictions within the counter-revolutionary ruling cliques; 3) the contradictions between rulers and the broad masses of the ruled; 4) the contradictions between the landlords and the peasantry; 5) the contradictions between the bourgeoisie and the working class; 6) the contradictions between warlords and their troops; and 7) the contradictions between the counter-revolutionary regime and the intellectuals and students.[2] At the outset of his guerrilla revolution, Mao chose number four — the contradictions inherent in the agrarian situation — as the most immediately promising basis for the united front, and he experimented with several different formulations of Red Army land policy in trying to exploit this issue. But he never forgot the other contradictions and always stood ready to shift his efforts to one or another in light of changing political realities.

From 1931, with the heightening of Japanese pressure on China, and particularly after 1935, when Mao became formal head of the Communist Party, he switched the party's united front policy to contradiction number one — that between imperialism and the Chinese nation. It was on this basis, in the context of the Japanese invasion of China, that Mao obtained the secure mass organization that ultimately allowed him to defeat the Kuomintang.

Mao's strategy is multifaceted and complex. There are, however, three aspects of it that are particularly relevant to a discussion of his successors' variations: first, what the Chinese Communists call the "mass line;" second, Mao's emphasis on "self-reliance;" and third, his three-phase periodization of the "protracted war."

The "mass line," despite its increasing ideological ossification in China after 1958, was one of Mao's key organizational insights; and during the revolutionary period, when it was relatively faithfully pursued, the mass line helped keep Mao's party free from the ideological dogmatism that has so often isolated Marxist-Leninist parties from popular political support. According to the mass line, the party must eschew a doctrinarie application of the Marxist-Leninist theory of social contradictions in formulating its policies. Instead, the party must conduct on-the-spot investigations into the social and political problems agitated the masses, and it must then interpret these findings in terms of Communist ideology, taking care not to run ahead of popular consciousness. By insisting on a rigid adherence to the mass line, Mao was able to forge a genuinely popular mass movement based on true sources of mass discontent; he thereby avoided the analytical and political sterility of so many Marxist movements when confronted with, for example, mass nationalism. Mao's development and adherence to the mass line is closely connected with the second important facet of his strategy, his emphasis on self-reliance.

Mao has preached self-reliance to successive generations of guerrillas partly because he himself won a military victory without any of the contemporary advantages enjoyed by Communist guerrillas, such as privileged santuaries, external sources of arms, or international champions seeking to obtain recognition or belligerent status for Communist forces. Lin Piao has argued:

> Revolution or people's war in any country is the business of the masses in that country and should be carried out primarily by their own efforts; there is no other way...If one does not operate by one's own efforts, does not independently ponder and solve the problems of the revolution in one's own country and does not rely on the strength of the masses, but leans wholly on foreign aid — even though this be aid from socialist countries which persist in revolution — no victory can be won, or be consolidated even if it is won.[3]

During his own struggle for power, Mao was forced to stress

self-reliance because external aid was never made available to him. However, later Communist guerrillas have used international political and military aid and the coordination of strategies in ways that Mao never dreamed of.

The third facet of Mao's strategy, protracted war, like the mass line and self-reliance, is linked to his lack of external assistance and to his assessment of the initial superiority of the enemy's armed forces. Mao wrote his famous and influential tract, *Lun Ch'ih-chiu Chan* ("On Protracted War"), during the first year after the Japanese invasion. Because of Japan's overwhelming power vis-a-vis China, Mao argued, it would take a long time for the Chinese guerrillas to equalize the military imbalance between the two; however, so long as his followers were willing to commit themselves to a long war of attrition, given the favourable conditions for a guerrilla united front created by the invasion, victory would ultimately be theirs. "Will China perish?" he asked. "No. She will have her final victory. Can China win a quick victory? No, this must be a protracted war. Is this conclustion correct? I think it is so."[4]

In order to make his promise of eventual victory credible to party members during the dark days of May June, 1938, Mao had to come up with an analysis that pointed to an ultimate Chinese triumph and at the same time offered practical guidelines during the early periods when victory was not in sight. He accomplished this by means of his famous three-stage periodization of the war: "the period of the enemy's strategical offensive, the period of the enemy's strategical defense and of our preparations for counter-offensive, and the period of our strategical counter-offensive." For each period, Mao set forth various combinations of tactics to be pursued, and although he was never too precise about this, the thrust of his analysis was clear: stages one and two were to be periods of guerrilla warfare and attrition, while stage three would be period of victorious "strategic assault."

Mao's *On Protracted War* has become his most famous revolutionary military tract, and his three-stage blueprint is common knowledge today among Communist guerrillas throughout the world. This is an anomaly. In China, Mao's third stage never arrived during the period when Mao's forces were engaging a foreign foe, the Japanese army; and Lin Piao, in his 1965 reassertion of Maoist theory, refers only in passing to the three stages, placing much greater stress on the development of rural guerrilla bases than on the types of military operations that are appropriate to each stage of the war. Lin Piao accurately emphasizes that the importance of the Anti-Japanese War period was that it allowed the Communist Party to champion nationalistic resistance to the invader and thereby gain its first military significant mass following. *On Protracted War* itself should be read as a morale builder for the hard-pressed Chinese forces during the early phases of the war, and as Mao's realistic appraisal of the Japanese Army's strength and staying power. The Chinese Communist revolutionary victory is significant not because of any particular set of precepts advanced by Mao during its course (especially ones so abstract that they could not influence behavior significantly) but because it is the clearest example to date of a successful, internally derived, mass-based, militarily oriented strategy of revolution.

The Viet Minh Legacy:
Decisive Victory Shortens the Protracted War

Probably the single, most lasting contribution that Mao Tsetung has made to the Asian Communist revolutionary tradition is his definition of revolution as revolutionary war. The Indochinese Communists who organized the Viet Minh movement against the French colonialists faced a comparatively easier task than Mao did, and one in which various nonviolent political options conceivably could have brought success. However, with the French resort to armed counter-revolution and with the precedent of Communist military victory emerging in China at precisely the same time, the Vietnamese movement for political independence took the form of a revolutionary colonial war. In retrospect, Asian Communist revolutionary theorists have tended to see this development as inevitable, as part of a Marxist law, proved by the Chinese and Viet Minh cases. One reason for the outbreak and exacerbation of the later Viet Cong conflict is that the Vietnamese Communists of the 1960's have never envisioned, nor are they organized for, any type of electoral or other nonviolent resoultion of fundamental political problems (other than one that would ratify their own military victory).

The Viet Minh movement occupies a transitional position in the development of Asian Communist revolutionary strategy. It was explicitly influenced by Mao's victory, and Vietnamese

Communist revolutionaries have analyzed their activities using concepts first formulated by Mao. For example, General Nguyen Giap asserts:

> 1949 saw the brilliant triumph of the Chinese Revolution and the birth of the People's Republic of China. This great historic event which altered events in Asia and the world, exerted a considerably influence on the war of liberation of the Vietnamese people. Viet Nam was no longer in the grip of enemy encirclement, and was henceforth geographically linked to the socialist bloc.[5]

Giap compliments Mao by using the term "people's war, people's army," and throughout his book of that title he employs such Maoist concepts as the three stages of the protracted war, self-reliance, and the basing of guerrilla warfare on popular support — regardless of whether these concepts are relevant to Giap's own struggle.

The Viet Minh revolution and the Chinese Communist revolution are actually far from similar. Guerrilla warfare in its relatively pure Chinese form is a strategy of poverty — both politically and militarily. The political operations that must precede the opening of guerrilla warfare, including adherence to united front principles such as the mass line, are dictated by the need to build a truly popular supporting structure, one that can sustain carefully controlled and slowly escalated military attacks on the target regime's forces. It is a strategy of military operations tailored to remain just under the threshold where professional armed forces could easily decimate the nascent revolutionary army. This strategy is delicate, painful, and slow in both its organizational preparation and actual implementation; no revolutionary party would voluntarily choose it if more direct alternatives were available.

In their struggle with the French occupiers, the Viet Minh Communists chose rather unthinkingly to follow Chinese precedents, just as they did a few years later with regard to land reform; but their revolutionary task was not as difficult as Mao's, and they enjoyed immensely greater military supplies than did the Chinese revolutionaries. The Viet Minh revolution was not a civil war but a colonial war of independence. Mass line investigations of popular grievances and trial-and-error testing of political programs were unnecessary because the revolutionaries could assume virtually universal anti-French sentiment and because the Viet Minh army was not totally dependent on the population for military support. The Viet Minh revolution contributed to the Asian Communist tradition of political violence, but it also began the reorientation of the tradition away from pure guerrilla warfare.

There is significant evidence of these Viet Minh modifications in guerrilla strategy. For example, the Viet Minh leaders considered the arrival in 1950 of Chinese Communist forces on Vietnam's borders more important than the continuing need to project a nationalistic image to the indigenous population: In November 1945, in response to united front demands, the Indochinese Communist Party had "dissolved" itself; however, on March 3, 1951, the Communists abandoned the "Viet Minh" facade (Viet Minh is an acronym meaning "League for the Independence of Vietnam") and emerged as the Vietnam Lao Dong (Workers) Party. By doing this they allied themselves more closely with the Socialist bloc, which was then supplying them with significant amounts of arms and technical assistance, but they also tended to turn the nationalistic struggle into a civil war. Even though the Lao Dong Party continued to fight against the French, many Vietnamese concluded that Communist Party "nationalism" did not have room for *all* persons of Vietnamese nationality. After 1954, some 800,000 to 1,000,000 Vietnamese migrated to the south, while only some 30,000 to 100,000 went northward. Although it was probably unforeseen at the time, the creation of the Lao Dong Party contributed directly to the division of the country three years later and to the reopening of the civil war a decade later. Viet Minh revolutionary strategy placed more immediate value on its international alliances than it did on its popular infrastructure, and this constituted a departure from the guerrilla strategy of the Chinese Communists.

Another difference lay in the types of relationships the Viet Minh army maintained with the mass of the population. As in China, the revolutionary army was raised from peasant villages, whose inhabitants supported the struggle against the French, and the Vietnamese villages similarly supplied the revolutionary army with intelligence, logistic support, labor, guerrillas, militia and refuge. However, because the revolutionary army was not as dependent on the population as the Chinese Communist forces had been, the Viet Minh cadres began to use, on a small scale at this time, a tactic eschewed by the Chinese as counterproductive and contrary to the basic logic of guerrilla warfare — namely, terrorism. Although the elimination of traitors is to be expected in any revolutionary war, Vietnamese Communists displayed a mush greater willingness to use terror against even potential traitors than had the Chinese, and they never showed the Chinese skill in developing methods of rehabilitation or reform of domestic opponents or wavering elements. Terrorism in this context served less to win the active support of the population than to raise the cost to the population of supporting the revolutionaries' enemies.

With regard to self-reliance, the Viet Minh revolutionaries did all of their own fighting, but they enjoyed secure sanctuaries across the Chinese border and a supply of military equipment unprecedented in the Chinese case until 1945 at the earliest. The key dates in the development of Viet Minh strategy are 1950 and 1953. In 1950 the Chinese Communists arrived on the Vietnamese borders, and in 1953 the Korean truce released large amounts of Chinese equipment for transfer to Vietnam. Chinese aid, consisting primarily of U.S. arms captured either from the Kuomintang armies or from U.S. forces in Korea, included ammunition, light arms, ant-aircraft guns, heavy mortars, 105 mm guns, and a few trucks of Russian manufacture. This equipment greatly enhanced the military capabilities of the Viet Minh army, but it also tended to give the army a more professional military orientation than had obtained in the Chinese People's Liberation Army. For example, the North Vietnamese army constituted an identifiable interest group within the postrevolutionary regime from the outset.

The enhanced military capability of the Viet Minh forces made possible the spectacular victory of positional warfare at Dien Bien Phu — the Viet Minh's most significant, and certainly most controversial, contributions to the evolving strategy of Asian Communist "people's war." In *On Protracted War*, Mao cautioned against "decisive battles on which are staked the destiny of the nation." He wrote:

> The rash advocates of quick victory cannot endure the arduous course of the protracted war. They want a quick

victory, and whenever conditions turn slightly for the better, they clamor for a strategic war of decision. If their wish were carried out, the entire resistance would be jeopardized, thus sacrificing our protracted war and falling into the vicious trap of the enemy.

Officers of most counter-guerrilla armies would agree with Mao; what they pray for most of all is a formal battle in which the rebels will "stand and fight."

The battle of Dien Bien Phu, contrary to Mao's advice, is an example of a patiently prepared, carefully launched, thoroughly successful assault against an enemy that according to some was overextended and according to others simply made a strategic mistake. Although the defeat was not militarily "decisive" with regard to the enemy's entire war-making potential, it carried with it such a powerful psychological punch that it produced valuable reverberatory effects on the overall war-making *commitment* of the defending forces.

Vo Nguyen Giap himself is both candid and cautious in discussing the gamble he took in waging the battle of Dien Bien Phu:

> Early in 1954, while the enemy was feverishly making preparations for his offensive against our free territory in the Fifth Zone, our plan was to concentrate big forces to attack on the Western Highlands, which was an important strategic position where the enemy was relatively exposed...Such was the essence of the strategic direction of the Dien Bien Phu campaign and of the Winter-Spring campaign as a whole...Its main object was the destruction of enemy manpower. It took full advantage of the contradictions in which the enemy was involved and developed to the utmost the spirit of active offensive of the revolutionary army.[6]

The battle conceivably could have gone the other way, probably with equal if not greater impact on the morale of the revolutionary forces. But it did not. Although "its main object was the destruction of enemy manpower, it took full advantage of the contradictions in which the enemy was involved," and herein lies a source of later modification of Maoist doctrine and a justification for trying shortcuts in Mao's long, arduous timetable.

The Viet Minh revolution partook formally of the basic aspects of the Maoist approach to the problems of revolution. The Viet Minh Communists accepted the necessity of wagering a revolutionary war; they utilized guerrilla methods to overcome the enemy's professional military advantages; and they oriented their activities to the military defeat of the enemy's armed forces. At the same time, they did not follow Mao's strategy closely. As late as 1965, Lin Piao was arguing:

> In order to win a people's war, it is imperative to build the broadest possible united front and formulate a series of policies which will ensure the fullest mobilization of the basic masses as well as the unity of all the forces that can be unified...To rely on the peasants, build rural base areas, and use the countryside to encircle and finally capture the cities — such was the way to victory in the Chinese revolution.[7]

The Viet Minh did some of these things, although less thoroughly than had the Chinese, and they did other things that the Chinese thought risky and in violation of theory. By producing the withdrawal of French forces, however, they too gained a victory — without being forced to annihilate the French Army. The Viet Minh cut short the protracted war and thereby set their own successor generation to rethinking the problems of overall revolutionary strategy and to experimenting with a new definition of revolutionary "victory."

The Viet Cong Synthesis: Political Power Through Military Means

On February 10, 1962, Viet Cong guerrillas threw four hand grenades into a crowded village theater near Can Tho, South Vietnam. A total of 108 persons were killed or injured, including 24 women and children. On September 12, 1963, Miss Vo Thi Lo, 26, a school teacher in An Phuoc village, Kien Hoa province, was found near the village with her throat cut. She had been kidnapped by the Viet Cong three days earlier. In June 1965, Saigon officials reported that 224 rural officials had been assassinated or kidnapped during the month, doubling the rate of April and May. On January 17, 1966, the Viet Cong in Kien Tuong province detonated a mine under a highway bus, killing 26 Vietnamese civilians, 7 of them children. These types of revolutionary activities, undertaken systematically and over a period of several years, differentiate radically the Viet Cong strategy of revolution from that pursued by Mao Tse-tung.

On May 13, 1959, the Central Committee of the Lao Dong Party, meeting in Hanoi, declared that the time had come to begin the task of "liberating the South...to struggle heroically and perseveringly to smash the Southern regime." In accordance with the basic orientation of Asia Communist revolutionary strategy, the Lao Dong Party determined that this task would require a revolutionary guerrilla war, and accordingly it began to build the political platform on which guerrilla military operations could be sustained. Basing its united front appeal on propaganda directed against the policies of the Ngo Dinh Diem regime, on December 20, 1960 it created a united front organization, the National Liberation Front of South Vietnam (NLF), and it began both to mobilize the approximately 10,000 Viet Minh left behind in the South in 1954 and to infiltrate into the South guerrilla cadres of Southern origin. NLF political workers undertook to organize peasant villages into a guerrilla infrastructure.

This approach appears familiar, but there was a fundamental difference between it and earlier precedents. Ten years of development in military technology, particularly in the realm of air power and the use of helicopters, had made a purely military rationale for guerrilla warfare seem hopelessly unrealistic and time-consuming. The Vietnamese Communists' former enemy, the French army, was itself proving at the time in Algeria that military answers to Mao's strategy could be developed. Moreover, the danger of American intervention against the rebels or against Hanoi itself was clearly recognized; the NLF served both as a focal point for building an indigenous united front in the South and as a lightning rod protecting Hanoi from the obvious danger of direct retaliation.

Other alternatives to a purely military approach were sought. "Contradictions" within the Saigon government suggested that the Viet Cong could isolate it politically and demoralize or win over its army without actually being forced to reverse the rebel-government military imbalance and defeat the defenders' armed forces in a Maoist "third-stage" positional war. The concept of three stages was retained in Viet Cong strategic planning, but the first two stages were reoriented away from steadily increasing guerrilla attrition and toward a political transformation of the target system through violent means. As Douglas Pike observes:

> Not military but sociopsychological considerations took precedence. Military activities and other forms of violence were conceived as means of contributing to the sociopolitical struggle...The two hundred to five hundred "guerrilla incidents" per week that went on in Vietnam week after week and month after month for five years had no purpose in themselves — and indeed when viewed in themselves often made no sense — except to serve the political struggle movement. Thus the primary purpose of the violence program was to make possible the political struggle movement.[8]

Selective terrorism not only terrorizes people; it atomizes them and causes them to make individual calculations of the relative costs and benefits in the short run of particular courses of behavior. In this sense it is somewhat comparable to the extortion rackets aimed at poorly protected, poorly socialized, ghetto dwellers and shopkeepers in American cities. The difference in Vietnam is that the Viet Cong not only institutionalized this type of daily cost-benefit analysis among the peasants of the South — which resulted in the peasants' growing disbelief in *any* policies promulgated by Saigon — but it also sought to legitimatize its activities by playing on impulses toward national unification, family and ethnic loyalties, religious differences, and hopes for peace and stability.

Viet Cong strategy, prior to the 1965 American intervention in force, differed from both Chinese and Viet Minh strategy in that military activity prepared the war for a political showdown, rather than political activity preparing the way for a military showdown (there never was a political showdown between the Communists and the Kuomintang in China). The Viet Cong assaults began in the villages and worked their way up toward district and provincial governmental levels, often enhanced by inept reactions from the central government but, in any case, carried out with a relentlessness not found in either Chinese or Viet Minh practice.

The Third Generation of Guerrilla Warfare

Because of differences in their basic conceptions of revolutionary strategy, the Viet Cong revolution also differed in detail from the Chinese and Viet Minh cases. There was no mass line investigation of popular grievances, except on a tactical basis to take timely advantage of particular developments. The Communist revolutionary effort, having been initiated from outside the embattled territory, was not dependent upon a critical level of political grievance for its sustenance. Similarly, the Viet Cong revolution was not self-reliant, either in terms of its equipment, core manpower or strategic inspiration. Equally significant, its front organization enjoyed international coordinated efforts by Communist nations and organizations to advance its claim to a popular political existence separate from the Saigon government and allegedly preferred by the people of South Vietnam.

Since 1963, Viet Cong strategy has been undergoing various transformations in response to the changing reactions and capacities of the enemy. Although it is impossible to comment on this subject in any definitive way, we can indicate the lines along which Viet Cong thinking appears to have developed. Having defined the goal of the struggle in the South as a political rather than a military victory, the Viet Cong at first sought to make credible by violence the inability of the Siagon government to govern. In response to a growing American presence, they increased their military activity — but this time with the additional intention of exploiting "contradictions" in the American position. (There is no evidence that the Vietnamese Communists ever concluded that they alone could defeat the United States through guerrilla warfare over any realistic time period.)

The "contradictions" that the Viet Cong perceive in the American position include: 1) the falling out between the U.S. and some of its allies over the war in Vietnam; 2) the inability of the U.S. to match its extraordinary technical superiority to Viet Cong maneuverability and military intelligence; 3) America's domestic aversion to the war, producing opportunities for any ruling administration's political rivals (the Viet Cong do not generally predict a "collapse" of the American home front, but they do see the war as perhaps engendering in the U.S. an advantageous political competition — one that might bring to power a popular leader, such as General de Gaulle vis-a-vis Algeria, who would end the war on terms favorable to the rebels);

4) unintended consequences of the American style of large-scale operations in Vietnam, such as corruption in the distribution of American aid, which weaken the already attenuated stability and legitimacy of the Saigon government; and 5) the possible generation of a true, anti-American, nationalist reaction among the people of South Vietnam as a result of the large number of Americans sent there.[9]

Like General Giap's "taking full advantage of the contradictions in which the enemy was involved" at the time of the battle of Dien Bien Phu, the Viet Cong revolutionaries (who of course include Giap) have tailored their military activities to exploit these perceived weaknesses. On the one hand, the Viet Cong have tried to create the impression that they are determined to wage a protracted (if need be, endless) war; on the other hand, they have aimed for "decisive victories" on the model of Dien Bien Phu (e.g., the Tet offensive of 1968). Both lines are intended to damage psychologically the U.S. commitment to the war, thus generating a negative cost-benefit analysis among American political leaders and thereby producing a victory more like that obtained by the Viet Minh than by the Chinese Communists. The essence of the strategy is not to defeat the U.S. and allied forces militarily but to convince the Americans through the use of violence, both pervasively and at selected points, that their position is hopeless.

Viet Cong revolutionary guerrilla strategy has thus come a long way from the doctrines of Mao Tse-tung. In a sense, Viet Cong strategy is a theory of neither "revolutionary" nor "guerrilla" war. It is not revolutionary to the extent that it created social grievances rather than responding to them, and it loses its guerrilla characteristics to the extent that it no longer depends upon an overtly nonbelligerent but covertly engaged population to provide its army with an overwhelming intelligence advantage. Nevertheless, the strategies of Mao, the Viet Minh, and the Viet Cong are linked by their willingness to associate civilian populations with military activities and by their relatively undoctrinaire, opportunistic readiness to exploit any social "contradiction" in order to bring about the violent defeat of the enemy.

Notes

1. Quotations from Lin and Mao are from Lin Piao, *Long Live the Victory of People's War* (Peking: Foreign Languages Press, 1965), pp. 3, 26, 44.
2. Mao Tse-tung, "A Single Spark Can Start a Prairie Fire" (January 1930, *Selected Sorks* (New York: International Publishers, 1954), I, pp. 119-21.
3. Lin Piao, *op. cit.*, pp. 38, 41-42.
4. As translated in American Consulate General, Hong Kong, *Current Background*, No. 249 (July 8, 1953), p. 23.
5. Vo Nguyen Giap, *People's War, People's Army* (Hanoi: Foreign Languages Publishing House, 1961), p. 22.
6. *Ibid.*, pp. 162, 165.
7. Lin Piao, *op. cit.*, pp. 11, 21.
8. Douglas Pike, *Viet Cong* (Cambridge: M.I.T. Press, 1966), pp. 32, 99. Pike first used the term "third generation of revolutionary guerrilla warfare" in this book, p. 36.
9. For evidence from Viet Cong sources on these perceived contradictions, see Melvin Gurtov, "Hanoi on War and Peace," RAND Corporation document P-3696, December 1967 (to be published in the forthcoming book edited by John Boettiger, *Vietnam and American Diplomacy*).

The Malayan Emergency—
The Roots of Insurgency
by Sam C. Sarkesian

Revolutionary guerrilla warfare is primarily a political conflict supported by unconventional military means. It originates from basic social grievances, and cannot be successful unless based on the particular characteristics of a given society. The structure and tensions of Malayan society provided an ideal base upon which to build a revolutionary movement. Yet the very same conditions prevented the revolutionary guerrilla movement from developing into a national movement and, it may reasonably be argued, caused the downfall of the insurgency they stimulated.

Leadership (in this case the Malayan Communist party) is a key ingredient of revolution. The morale, organization and skills of the revolutionary cadre are crucial considerations in determining the conduct and outcome of a revolutionary war. Other important factors are the general demographic, economic and geographic profiles. Population distribution, literacy, urbanization, and welfare influence the direction and strategy of the insurgency. Geography influences these factors and provides the given natural setting within which revolutionary tactics and strategy develop. In Malaya, population and geographic characteristics created a disadvantage for the insurgency, which the British were quick to exploit.

The Malayan insurgency is one of the few in recent times in which a colonial power was able to defeat an insurgency movement. This uniqueness enhances its importance as a case study of revolutionary and counter-revolutionary guerrilla warfare. The Malayan political system provides a classic example of the importance of non-military factors.

The Plural Society

Malaysia is a "plural society." There are Indians, Pakistanis, aborigines, and European; but it is the communal nature of the Chinese and Malay social systems that chiefly shape the internal dynamics of Malaysia. The distinctions, and the historical tensions between these two groups were important in the Emergency in Malaya.[1]

The Malays consider themselves the original and only "legitimate" inhabitants of Malaya. The Chinese began to arrive in the early 1800's after the establishment of Singapore, where the rapid growth of ocean trade required a large labor force. Many

Chinese merchants also came to take advantage of increased opportunities. Singapore became, and remains, predominantly Chinese.[2] The Chinese labor influx also reached the west coast of Malaya, drawn from the 1850's onward by the development of tin mining and, later, rubber plantations. Secret societies, characteristic of mainland China, accompanied the immigrants. These societies became engulfed in bloody factional conflicts over control of mine labor and recruitment of members.[3] The inability of the Malays to control these conflicts, combined with the demands of British and Chinese commercial interests, prompted the British to intervene and impose "advisors" upon some of the local Sultanates. This advisory system evolved into a colonial administration based on "direct and indirect" rule; generally, the west coast states were formed into the Federated States of Malaya, while the east coast states were loosely organized into the Unfederated States of Malaya.[4]

The peculiarities of the Chinese immigration tended to create special conditions of some importance during the later Emergency. The Chinese immigrants usually arrived as members of groups or as a result of connections with compatriots already in Malaya. This provided a degree of stability and continuity in a strange land with strange customs.

> Thus the Chinese immigrants sought security in a strange land by joining their traditional informal associations or by clinging to the leadership of individual labor contractors. . . . they continued to believe that the course of wisdom was to avoid the government, and that security and riches were to be found in precisely those places where the government was least conspicuous. This meant that they were not particularly disturbed about joining associations which operated on the outer fringe of legal respectability.[5]

This distrust of government and acceptance of organizations not "legal" were manifested by surrendered insurgents during the Emergency.

> The fact that most of the SEP's (Surrendered Enemy Personnel-Chinese) thought of the government as remote from them was in large part a function of life in a plural society composed of distinct racial communities. For the SEP's the

barriers between the races were very real.⁶

Distrust for the government, which was oriented toward the Malays, was buttressed by Chinese racial attitudes.

The SEP's (Surrendered Enemy Personnel) left little doubt of their strong sense of racial superiority to Malays and Indians and at times of their downright contempt for them. In their efforts to minimize their relations with Malays and Indians and to preserve their identity as Chinese, the SEP's perceived communism to be of value. Communism, as a result of both its record in Malaya and its history in China, was considered by the vast majority of the SEP's to be a movement of and for the Chinese.⁷

Economic life reinforced communal distinctions. Chinese businessmen were the middlemen between producers and Malayan consumers. The Chinese tended to live in the cities. Those who did not were located in the tin and rubber belts or operated market gardens close to urban areas.⁸

Although the vast majority arrived as laborers, many soon found in the stability and order of Malaya opportunities for demonstrating their entrepreneurial talents. Becoming the largest element first in the urban centers, and then in the country as a whole, they came to dominate the retail and commercial life of Malaya.⁹

Linguistic, religious, social, and political considerations set Chinese, Malays and Indians apart from each other.

The three major domiciled communities in Malaya contrast sharply in physical appearance, language, religion, and mores. Few can speak, and still fewer are literate in each other's languages; market Malay is a lingua franca for many whereas English serves the same purpose for some. Intermarriage is unusual, and groups within communities tend to follow traditional economic pursuits.¹⁰

Each race developed an autonomous and distinctive society with its own values and perceptions. Chinese success in the economic sphere, in which wealth was correlated to status, was in sharp contrast to the unassertive if not bucolic Malay mores and values. The basic distinctions and friction between the two cultures remain to this day. [11]

The growing domination of the Chinese led the British to promulgate laws protecting the Malays from their inroads in land tenure and ownership, the professions and citizenship. For example, the Civil Service was not open to non-Malays until the 1950's;[12] the Malay Reservation Enactments of 1913 and 1933 made "rice growing a Malay monopoly by denying lands suitable to the growing of this crop to other communities."[13] Chinese discontent also was a reaction to the official policy of bolstering Malaya importance and loyalty.

> . . . the maintenance of the position, authority, and prestige of the Malay Rulers must always be a cardinal point in British policy; and the encouragement of indirect rule will probably prove the greatest safeguard against the political submersion of the Malays which would result from the development of popular government on western lines. For, in such a government the Malays would be hopelessly outnumbered by the other races owing to the great influx of immigrants that has taken place in Malay during the last few years. [14]

But British policy succeeded in gaining the trust and affection of the Malays. This attitude generally survived, although shaken by the Japanese Occupation, through the period of the Emergency. [15]

The role of the Indians and Pakistanis in the Emergency was minor. The Malayan Communist Party (MCP) did not make serious inroads in these groups, primarily because of the circumstances in which a majority of them came into Malaya. The Indians were hired to work the rubber plantations. I .ter they also became middlemen and moneylenders. In contrast to the Chinese, Indians were generally recruited by government authorities, which maintained an interest in their welfare. The Indian laborers were not obliged to work for private contractors, but remained under government supervision.

Thus, the Indian immigrants arrived with a far more positive orientation toward the government than was the case with the Chinese. They tended to look to legal authority for guidance and advice and to the government for protection from their informal leaders and compatriots.[16]

Indians of middle class who were fluent in English were often employed in government offices. The Indians also had had experience with British government in their own homeland and appreciated their opportunities in legal political activities.[17]

In contrast, the Chinese were little involved in government or politics. Their sense of loyalty, dependence and obligation was toward the family, not the state. Thus the plural society stimulated loyalties peculiar to each race.

Geography and Demography

Geography was a major determinant in the Emergency. Eighty percent of the land is covered by dense jungle. Ninety percent of the population live in a cleared strip 10 miles wide and 500 miles long along the Western coast. Much of the interior is virtually uninhabited.[18] Most of the rubber plantations and tin mines are also located in this cleared strip. The tin mines and rubber plantations were the preserve of Chinese and Indians. It was in this area that the war was fought, since the major goal of the insurgents was the control or neutralization of this economically productive area and population.

In the initial stages, however, the insurgents attempted to establish themselves in the squatter areas located on the fringes of the jungle.

> In 1948, however, at least half the rural Chinese population did not live in the villages at all. It was mainly on the 423,000 squatters, still living in the jungle fringes to which so many had fled during the Japanese occupation, that the Communist jungle army in 1948--50 depended for supplies and information.[19]

It has been estimated that 100,000 aborigines live in the jungle areas. They had been untouched by civilization or government until the insurgents were forced back into the jungles. Only after they were employed to grow food and serve as helpers for the insurgents did the British make any effort to bring government to the aborigines.[20]

Other geographic factors influenced the course of the Emergency. Malaya is a long peninsula; its only border, over 300 miles, is with Thailand on the north. Its greatest width is 200 miles (the average width is much less) and its length is 480 miles. The shoreline is approximately 1,200 miles. Consequently, Britain's navy and Thai cooperation prevented supplies from reaching the insurgents from abroad. However, it was on the remote and jungled border that the insurgents operated most freely. One author has written:

> I studied an immense map in the police station. Almost every foot of it was green, to denote ubiquitous jungle, and it was covered with thousands of curving and zigzagging black lines, to denote streams. In such country there can be no frontier posts and no effective frontier patrols; and although the Siamese authorities do their best to help, and allow British patrols to operate inside Siam, there can be no major operations, as in Malaya itself. The Siamese Government could probably not afford a resettlement policy in New Villages, even if it wished to do so.[21]

A mountain chain down the center of the peninsula "divides Malaya into two unequal parts, the larger to the east of the range; the smaller and more populous constitutes the west coastal area of settlement to the west of it."[22] This range extends about 300 miles the length of Malaya. A secondary range lies to the east. The insurgency was confined largely to the west coast. It is interesting to note that in Malaysia today there exists a regional distinction between the more advanced and developed west coast as contrasted to the less developed and mainly traditional east coast settlements.

The Malayan Communist Party

The Malayan Communist Party was formed in 1930 as an outgrowth of the South Seas Communist Party, which was established in Singapore in 1928.[23] Although there were attempts to introduce Communism to the Malays through culturally related Indonesian channels, it appears that it first came to Malaya as part of the general ideological attitudes of some of the mass of Chinese immigrants.[24] Communism made little headway amongst the Malays.

> There are many reasons for the cool reception that the Malays have given the overtures of Communism. The hold of tradition is still strong among most Malays and it has enabled them to find great personal contentment in a simple pattern of life. Over the centuries they have developed adequate methods for resolving localized social conflicts and thus have felt little need of, or desire for, complicated political activities. For those Malays who have left the kampong, the break from their village ways has not been extremely harsh or impersonal. For the vast majority, urban life has meant employment in service industries and in the government. The better educated have been able to look to careers in the civil service and government-supported schools, and for the rest the customary practice has been to enlist in the police and military. Of all the possible careers in a modern society, these are perhaps the most capable of providing a sense of personal security and the least likely to encourage attitudes favorable to Communism.[25]

The Chinese immigrants, it has been seen, were not hesitant about joining groups which were at or beyond the fringe of legality. Many of the Chinese immigrants composed a plantation-industrial wage labor force. It has also been suggested that a minority of teachers in the Chinese schools were also attracted to Communism:

> Many of them were aspiring intellectuals who had failed to find positions in China and who had come to Malaya with embittered radical views. The Chinese groups in Malaya were prosperous enough to give them employment in small independent schools, in night schools, or as private tutors.[26]

But the MCP did not win a large following, even among the a-political Chinese, many of whom were loyal to the anti-Communist Kuomintang regime of Chiang Kai-shek which governed the home country. The MCP changed to the tactic of infiltrating existing Chinese organizations, with some success. In

Modern Revolutionary Guerrilla Warfare:
Some Micro-Views

the middle 1930's the MCP reorganized and became somewhat more effective. It developed and controlled many labor groups and in 1935 organized its spearhead, the General Labor Union, which conducted a series of strikes in the late 1930's. However, improving economic conditions following the worldwide depression of 1929 blunted their effectiveness. [27]

In China, after the Japanese invasion of 1937, a truce was called between the Kuomintang and the Communist Party to form a united front against the foreign enemy. In Malaya the MCP was thus able to lead "anti-Japanese" groups among sections of the Chinese population, including Kuomintang followers, which had no communist sympathies. By the end of 1939 the MCP was an experienced underground organization able to foment a series of violent anti-British outbreaks (as well as strikes). Early in 1940 the Kuomintang group, realizing that it was being used by the MCP, withdrew its support and denounced the Communists. [28]

By the time of the Japanese invasion of Malaya in 1942 the MCP had established cells in the urban areas as well as the countryside. The MCP established the Malayan People's Anti-Japanese Army (MPAJA) shortly before the fall of Singapore. An ancillary organization, the Malayan People's Anti-Japanese Union (MPAJU), was also established for political work, intelligence, supply, and recruiting. These organizations also enlisted many Malays. [29]

To the British, the MCP was the only organization capable of conducting guerrilla warfare against the Japanese. They were undoubtedly aware of the ultimate designs of the MCP, but under the pressure of the war the British took a calculated risk and supported the MCP with supplies, training, and direction. But during this period of cooperation the MCP was planning for the future. [30] The anti-Japanese achievements of the MPAJA were slight. Nevertheless, "when the war ended, about 7,000 guerrillas came out of the jungle fully convinced that it was their might which had defeated the enemy, and they were welcomed by large elements of the civil population as heroes." [31]

Several weeks of political vacuum intervened from the Japanese surrender in 1945 to the return of British troops. The flight of the British four years earlier had created administrative disruption. The Japanese policy of dividing the population, repressing the Chinese and stimulating Malayan and Indian nationalism, subverted British pre-war policies. Japanese policy had also caused economic disintegration.

> This was particularly so in the urban areas where trade and industry were to all intents and purposes obliterated by Japanese occupation, and unemployment became rampant. Trade with the surrounding countries, particularly Thailand and Burma, which supplied a high proportion of Malay's staple food—rice, was to all intents and purposes cut off. A food crisis rapidly emerged. Under such circumstances there was a natural tendency for the people to engage in food production, mainly by "squatting" upon any suitable land and setting up their small vegetable gardens there.

To alleviate the crisis the Japanese forced scores of thousands out of urban areas into the countryside as "squatters" to raise food. In these areas, populated almost exclusively by Chinese, and elsewhere, there was no functioning governmental structure.

After the surrender of the Japanese the MCP emerged from the jungle and established a quasi-government in many parts of the peninsula.[32] Initially welcomed by Malayans, the MCP soon lost some of the good will it had gained. Its military arm, the MPAJA, conducted mass trials by "people's courts" of alleged collaborators and traitors.

> Striking against not only Chinese who opposed them but also Malays and Indians, the Communists introduced a wave of fear into a country which had hoped to realize peace and security after surviving the terrors of the Japanese occupation. Since large numbers of Malays and Indians had worked for the Japanese and the majority of these peoples had not openly opposed the conquerors, the MPAJA's announced objective of punishing and even killing all who had assisted the recent enemy produced serious racial tensions. Fear of the MPAJA became fear of the Chinese, and just as the MPAJA casually confused "collaboration" with opposition to Communism, many Malays lost sight of the distinction between Chinese who were Communists and Chinese in general. In some areas shocking racial riots erupted. . . . [33]

However, terrorism did not take place in all MPAJA controlled areas. The MPAJA "kept good discipline in some areas, initiated terrorism in others, but at no time did they attempt to extend their control over the whole peninsula." [34]

After the return of the British in 1945 the MCP was allowed to operate relatively freely. Arrangements were made to demobilize the MPAJA and turn in weapons to the British administration. However the MCP, evidently intent on carrying out a policy of "national liberation", maintained an effective armed element.

> The stocks of buried arms accumulated by the MCP remained, and the "secret" section of the MPAJA remained ready for swift mobilization. It is estimated that, at the end of 1945, this secret section numbered about 4,000. Moreover, the Party ensured the continuation of its control over all former MPAJA members by setting up MPAJA Ex-Comrades Associations. The MCP itself now functioned legally but its true leaders nevertheless remained underground.

Old Comrades Associations covered the country, and in turn established a shadow military organization. Similarly, political and social organization were not neglected. [35]

Initially a "moderate," Soviet-directed strategy was followed, emphasizing labor agitation and organization to establish a shadow control of the foreign-controlled mining and plantation sectors. [36] Then, beginning in 1947 (as the cold war intensified), "their objective was to paralyze the vitally important rubber and tin industries by disrupting their labor forces as a prelude to the overthrow of the government." The massive effort in the labor movement was supplemented by popular front organizations and youth movements in the schools. [37]

In 1947 Communist plans ripened.

> The MCP and its numerous front organizations staged public rallies, demonstrations and strikes day after day with apparent impunity, in contrast to the strict prohibition of all political organizations and activities by the colonial government before World War II. To the simple-minded rural folk, the act of running away from the enemy from which the colonial government was supposed to protect

them, had forfeited the colonial government any further authority to govern, and the MCP was not slow in encouraging this line of thinking among the people.[38]

The following table indicates the extent of unrest:[39]

Year	Number of Strikes Malaya	Singapore	Man Days Lost Malaya	Singapore
1947	291	45	696,036	492,708
1948	181	20	370,464	128,657
1949	29	3	5,390	6,618
1950	48	1	37,067	4,692
1951	58	4	41,365	20,640

The British reacted by promulgating regulations prohibiting anyone with less than three years' membership from holding office in a union. Anyone convicted of a variety of crimes was also barred from office. Finally, all federations were outlawed that were not composed of allied trades. These regulations had the effect of excluding practically all the professional communists and reduced the MCP influence in the trade unions to insignificance. In constructive policy, the year 1948 inaugurated the Federation Agreement of February which provided significant concessions to Malayan demands for political progress toward independence.[40]

The failure of labor and civil strife, Soviet influence at a time of intensifying cold war, and the threat of stability caused by economic recovery and progress toward Malayan self-government combined to plunge the MCP toward a more desperate strategy.[41]

Unless the party could show constant progress toward the goal of revolution, the situation could only get worse. For the party to sustain its power position, no slackening in the efforts of the rank and file could be permitted. The MCP had gained much, but at the expense of asking much of its supporters. To ask for more would be impossible without tangible evidence of progress, and all that had been achieved could be lost if routine replaced extraordinary effort. The returns of exhortation were diminishing.[42]

The result was a campaign of terrorism, assassination, and sabotage, followed by guerrilla warfare.

Insurgent Organizational Structure

The Malayan insurgents generally followed orthodox Communist organizational practice with interlocking links between Party, armed forces and front organizations at all levels, although there were some differences between the various state organizations. The attempt to adapt organization to the peculiarities of each of the states within the Federation was only partially successful because of the predominantly Chinese character of the MCP, which also diminished the "united front" appeal.

The organization of the Party and the Malayan Races Liberation Army (MRLA) followed a dual and parallel hierarchical structure from top to bottom. The apex of the Party and the movement was the Central Committee, which served as the executive Committee for the Politburo and the Military High Command. The Regional Bureaus also occupied a relatively high place in the hierarchy.[43]

The Party encountered serious difficulty in maintaining control over its subordinate agencies. A British intelligence report stated that "M.C.P. communications, though organized with efficiency and maximum security, particularly at the higher levels, remain a very slow and tedious process." Because of the nature of the terrain and dependence on human couriers, it was difficult for the Politburo to coordinate. Committee members at lower levels "are in fact virtual dictators within their respective spheres of operation provided they keep within the defined limits of Central and State policy."[44]

At the day-to-day operational level the District Committees, after the first year of the war, directed political, military, and Min Yuen activity. The Min Yuen, "the People's Movement," provided the political and economic support for the MRLA and the Party, furnishing supplies, intelligence and recruits. (During the Japanese occupation the Malayan People's Anti-Japanese Union had performed this role.) The District Committees were directly under State or Regional Committees and controlled the local branches. In turn, the state and Regional Committees were under the control of the Central Committee.

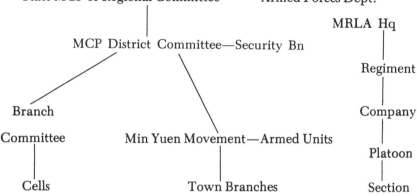

MCP Organization[46]

The District Committee has assumed a major significance in the Party structure. It is responsible for the organization and control of the Min Yuen (people's Movement) which penetrates labour, provides the Army with vital food supplies, information and recruits and which, through the agency of its armed (civil) units, is able to keep the "struggle" alive even while regular M.R.L.A. formations lie dormant. The members of one type of armed civil unit composed mainly of District, Area and Branch Party members and Min Yuen executives, live in the jungle in close proximity to the squatter area in which they operate. They carry out normal Min Yuen activities, eliminate "traitors" and perform simple acts of sabotage which, when necessary, the assistance of the M.R.L.A. Members of other lesser formations continue to pursue their normal livelihood, but occasionally operate as armed ancillary units. M.R.L.A. formations, on the other hand, have no contact with the people and are employed almost exclusively for the more hazardous tasks of attacking the Security Forces, Police Stations, guarded estates and other major targets.[45]

In addition to the non-combat organizations, there were armed units in the Min Yuen, normally organized on an area basis and under control of the District Committees. Some of these armed units closely paralleled organizations and functions of the MRLA.[47]

In 1947, it was estimated there were about 12,500 full time MCP members, 12,000 of whom were Chinese and the rest Indians (with only 35 Malays). The MRLA was organized into twelve regiments (subdivided into companies and sections), each with an assigned geographic area of operations, totalling about 4,000. Over 90 percent of the MRLA were of Chinese descent. The remainder consisted of Malays (primarily in the state of Pahang), some Indians, and a few Japanese deserters. A great majority of the Chinese were from the labor and squatter classes, the remainder from educated and student classes and the professions. These people generally filled the middle and higher executive posts.[48]

This organizational structure reveals a basic, fatal weakness. Organized into geographical units near compact settlement areas, hampered by geography as well as the enemy in communications and movement, the insurgents lacked speed and flexibility—basic requirements of guerrilla warfare. Moreover, the inability to attract significant non-Chinese support simplified the government's problems of identification and interdiction as it limited popular support. The counterrevolutionary forces were able to isolate ethnic and geographic communities from each other, spatially and socially, and deny the revolutionaries access to physical and human resources. The Malays joined with the British, making the insurgency manageable. When the British moved to grant independence to Malaya they reduced the impact of revolutionary appeals. Even under these relatively favorable conditions the British had to struggle for twelve years before the Emergency could be declared ended in 1960. (In 1974 remnants of the insurgents still operated along the Malaya-Thailand border.)[49] The experience taught a clear lesson for both revolutionary and counterrevolutionary: strategy and tactics must be rooted in the political and social systems of the particular society if there is to be any hope of success.

Notes

1. The term Malaya signifies the country of Malaya proper during the Emergency as contrasted to Malaysia, which now includes Malaya proper and the islands of Eastern Malaysia but not Singapore, which is a sovereign nation. The term Malay refers specifically to the Malay race. Malayans refers to all of the peoples inhabiting Malaya. For an excellent discussion of the Chinese community in Malaya from the tin mining days to after the Emergency see Victor Purcell, *The Chinese in Southeast Asia* (London, 1966), pp. 223-356.
2. Bela C. Maday, et al, *Area Handbook for Malaysia and Singapore* (Washington, 1964), p. 48. There were 5,000 people, mostly Chinese, in Singapore in 1819.
3. Purcell, p. 272.
4. Lennox A. Mills, *Malaya: A Political and Economic Appraisal* (Minneapolis, 1958), pp. 3-12.
5. Lucian Pye, *Guerrilla Communism in Malaya* (Princeton, 1956), p. 53.
6. *Ibid.*, p. 107.
7. *Ibid.*, p. 207.
8. Maday, p. 27. It has been estimated that by 1947 the Chinese were 68 percent of the urban and 45 percent of the total population. T. E. Smith, *Population Growth in Malaya* (London, 1952); C. A. Fisher, *Southeast Asia* (New York, 1964), p. 634. The approximate numbers were: Malays, 2,500,000; Chinese, 2,600,000; Indians and Pakistanis, 600,000; total (including others), 5,850,000. The Chinese were a majority on the island of Singapore, a minority on the peninsula.
9. Pye, p. 54.
10. J. Norman Parmer, "Malaysia" in George M. Kahin (ed), *Government and Politics of Southeast Asia* (Ithaca, 1964), pp. 315-316.
11. Maday, p. 177.
12. Vernon Bartlett, *Report from Malaya* (New York, 1955), p. 108.
13. U.S. Department of State, Office of Intelligence Research, *The Problem of Agrarian Reform in British Malaysia* (Washington, 1951), p. 7. This publication has an excellent account of the problems of preserving Malay predominance over the Chinese in rural areas since World War II.
14. John Bastin and Robin W. Winks, *Malaysia: Selected Historical Readings* (New York, 1966) pp. 351-352.
15. *Ibid.*, p. 351.
16. Pye, p. 56.

17. Maday, p. 129.
18. Ibid., p. 9. See also Richard L. Clutterbuck, *The Long, Long War: Counter-insurgency in Malaya and Vietnam* (New York, 1966), p. 45. "Because the rubber and tin are worked almost entirely by the Chinese and Indians, and because the Malays' agricultural kampongs (villages) are generally set apart, most of the villages astride the road through the rubber and tin areas are wholly Indian or Chinese. It was around these Chinese villages that the war was fought."
19. Clutterbuck, p. 46.
20. Bartlett, p. 47. See also Richard Miers, *Shoot to Kill* (London, 1949) for an account of the aborigines' role in the Emergency.
21. Ibid., pp. 74-75.
22. Maday, p. 177.
23. Good brief surveys may be found in J. H. Brimmell, *A Short History of the Malayan Communist Party* (Singapore, 1956), and *Communism in South East Asia* (New York, 1959) by the same author. Also see Purcell; Pye; and Harry Miller, *The Communist Menace in Malaya* (New York, 1955).
24. See Pye, p. 51; "It would be hard to determine the exact time when the Chinese first began to bring ideas about Communism to Malaya, since they arrived in the motley baggage of the immigrants, who remained relatively isolated from the general society."
25. Ibid., p. 49.
26. Ibid., p. 54.
27. Ibid., pp. 58-62.
28. Reference Division, Central Office of Information, *The Fight Against Communist Terrorism in Malaya* (London, no date), p. 6; Miller, p. 62f; Pye, p. 63f; C. C. Too, *Some Salient Features in the Experience in Defeating Communism in Malaya, with Particular Regard to the Method of New Villages* (unpublished manuscript, 1966), p. 2. Maday, p. 656, states that by 1939 there were 5,000 MCP members.
29. Waller, p. 4.
30. F. Spencer Chapman, *The Jungle is Neutral* (London, 1952); P.B.G. Waller, *Notes on the Malayan Emergency: Strategies and Organization of the Opposing Forces* (Menlo Park, 1968), p. 5. Miller, p. 62f. Chapman is excellent on jungle war.
31. Pye, p. 69. During the three years and eight months of occupation the MPAJA inflicted only a few hundred casualties on the Japanese. In contrast, the MCP claimed it executed during this period 2,542 "traitors," mainly Chinese who opposed it. See also Maday, p. 657.
32. Too, p. 2f.
33. Pye, p. 71.
34. Maday, p. 658. See also Reference Division, p. 7.
35. Waller, p. 6. See also Pye, pp. 71-73; Clutterbuck, p. 22; Miller, p. 60.

36. Maday, pp. 529f, 658. Apparently the MCP was split over two policies proposed for use at the end of the war: one, called the "Chinese line", favored an immediate attempt to seize total Power; the other, the "moderate policy," favored a return to labor agitation and organization as recommended or directed by the Soviets. Since the Secretary General of the MCP, Lai Tek, favored the "moderate" course, it was adopted. See also Too, p. 5.
37. Waller, p. 6. See also Clutterbuck, p. 25f, who suggests the possibility of a coup d'état.
38. Too, p. 4.
39. Adapted from Maday, p. 556. See also Miller, p. 74.
40. Clutterbuck, p. 28; Miller, p. 77. Miller writes (p. 74): "But adding to the restiveness among the population caused by industrial strife was the ugly birth of banditry, much of which really could not be laid at the doors of the "secret army" of the Communist Party, as this had not yet decided to give itself shooting practice. The majority of the bandit gangs who plundered and killed, mostly in Perak, Kedah, and Pahang, comprised men of the Malayan Overseas Chinese Self Defense Army (M.O.C.S.D.A.), which had been formed by the Chinese Nationalists for the express purpose of fighting the Communists—in a World War III! They were vicious groups, but many men later surrendered and in 1948 formed the nucleus of Chinese jungle squads in Perak which were sent out on the track of the Communist terrorist."
41. Clutterbuck, p. 28f; Waller, p. 6, Commissioner of Police, Federation of Malaya, *A Paper on the Situation in Malaya*, 2 November 1950, p. 6; Pye, pp. 83f; Miller, p. 76; Roderick Dhu Renick, Jr., *The Emergency Regulations of Malaya: Background, Organization, Administration, and Use as a Socializing Technique* (unpublished Masters' thesis, Tulane University, 1964), p. 22.
42. Pye, pp. 81-82. Pages 79-82 provide a summary of internal Party problems during this period.
43. See Waller, p. 16f, for a detailed discussion of the various political organizations.
44. Federal War Council, Joint Intelligence Advisory Committee, *The Potential of the Malayan Communist Party*, 1950, Federation of Malaya, p. 10.
45. Federal War Council, *Potential*, p. 10. Italics appear in original document. See also Waller, p. 6.
46. Fred Barton, *Salient Operational Aspects of Paramilitary Warfare in Three Asian Areas* (Operations Research Office, Johns Hopkins University, April, 1953), p. 40. See also Pye, p. 22 and Waller, p. 17.
47. Federal War Council, *Potential*, p. 7. "In some areas, the MRLA has had to detach considerable numbers of armed personnel to form a nucleus of the Armed Units. It appears that District

Committees now are required to recruit and expand Armed Units from their own resources. In some areas there are considerable programmes of expansion. The numerical strength of units is in all probability limited principally by the availability of weapons."

48. Ibid., p. 5; Waller, pp. 10-11, 18-19; Barton, p. 27. Waller also states that since only "30 percent of the MRLA were Party members—although most were Communist-inclined, the MCP found it necessary to construct a Party organization within the MRLA itself for control and political education of the members and to recruit new members for the Party."

49. For the defeat of the revolution see further above, chapter I.

Che Guevara on Guerrilla Warfare: Doctrine, Practice, and Evaluation
by J. Moreno

Most theories of revolution seem to agree that certain preconditions must be met if a revolutionary situation is to arise.[1] The peculiar contribution of Ernesto Che Guevara to understanding revolutions is that according to him such preconditions can be created [1:4].[2] Few men in the world today would be better qualified than Guevara to sustain such theory with empirical evidence from his own participation in revolutions. After his experience in Guatamala, Cuba, the Congo and, perhaps, other parts of the world, Guevara was considered, by friends and enemies alike, as one of the world's top-ranking guerrilla fighters of the twentieth century. By the time he started a new daring experiment in Bolivia he was reckoned as one of the most articulate theorists in the field. The death of Guevara in the mountains of Bolivia on October 9, 1967, brought commotion and mixed feelings to his enemies and admirers the world over. His enemies were exultant, first because the dreaded Guevara was dead, and second because his death was 'clear evidence' that his theories were wrong. His admirers were sad because he was dead, but were also elated because the puzzle of his disappearance had been unraveled and because his death at the hands of Bolivian rangers trained by a U.S. military mission, far from being 'evidence' that his theories were wrong, was 'evidence' that they were right. In order to decide whether Guevara's theory of revolution was right or wrong a series of value judgments needs to be made and the final outcome will, naturally, be in line with these evaluative presuppositions. It is beyond the scope of this essay to enter such discussion. It is my intention to investigate, as dispassionately as I can, to what extent the empirical evidence of the Cuban and the Bolivian experiments proves or disproves the major generalizations of Guevara's theory of revolution. It is with mixed feelings that I approach the subject, first because it is quite difficult to enter such a study without taking sides, and second because of the scarcity of documents available to the writer. It goes without saying that whatever findings are presented in this paper are subject to further clarification and correction when more material with evidence relevant to this subject is made available to the public.

Modern Revolutionary Guerrilla Warfare:
Some Micro-Views

1. The Theory of the 'Foco'

The main contribution of Guevara to a theory of revolution is that the necessary conditions to a revolutionary situation can be created through the emergence in rural areas of highly trained guerrilla fighters organized into a highly cohesive group called the 'foco'.[3] Most theories of revolution, Marxist and non-Marxist alike, have basically agreed that certain objective and subjective conditions are to be met before a revolutionary situation can develop.[4] Guevara also is aware that certain minimum preconditions are needed to kindle the first spark.[5] He admits that certain grievances must exist which society is unable to redress through the ordinary mechanisms of tension management. He agrees that all other legal avenues to solve the conflict must be exhausted before violence can be used. Finally he also suggests that while the government enjoys legitimacy (through honest constitutional elections) or an appearance of it (through rigged elections) the use of violence will not be seen as appropriate. Three preconditions, therefore, are minimally required, according to Guevara, before a revolutionary situation can be developed through the use of the guerrilla *foco* [1: 4]:

1. A lack of legitimacy by the incumbent elite to govern the country.
2. Existence of tensions that cannot be redressed by regular channels.
3. All legal avenues to change the situation are perceived as closed.

The presence of these preconditions does not guarantee the success of the guerrilla *foco*. They make up the socio-political frame within which the *foco* can become a catalyst for a full fledged revolutionary situation. Whether or not the *foco* will function as a catalyst will depend on other factors. Our task here is precisely to determine which, according to Guevara, are the factors that make a guerrilla *foco* truly operative and successful.

A. What Kind of Men Make up the Foco

The guerrilla fighter, according to Guevara, must be a social reformer, a man who fully dedicates himself to destroy an unjust social order to replace it with something new [1: 30]. The ideal type of guerrilla is seen by the Argentine physician as a man who leads an ascetic life with impeccable morality with strict self control and who seeks to introduce social reforms through personal example [1: 40, 65, 67]. In order to achieve such dedication to a cause, the guerrilla must be ideologically motivated. Social justice and freedom are suggested as basic tenets of such an ideology [1: 35].[6]

The life in the guerrilla band is a long and painful process of learning. Not because a man has a rifle, a back pack, sleeps on a hammock and is hunted by the police, can he be called a guerrilla. All such characteristics are shared by bandits. According to Guevara, life in the guerrilla *foco* gives the fighter both an opportunity to become a revolutionary (which is the highest rank of the human species), and to become a man. Those who are not able to reach either stage, are advised to give up [7: 196].

B. Composition and Organization of the Foco

The *foco* is basically made up at first, of some 25 to 35 men under the politico-military leadership of a man in charge of the whole operation. The *foco* has supporters and sympathizers in the city, but will not receive orders from any organized group or party stationed in the urban centers. It operates as the vanguard of a popular army. The *foco* will establish close relations with the peasants of the area in which it operates, but at no time will it sacrifice the mobility and safety of the guerrilla band for the sake of a village or territory.

The *foco* is not an end in itself, but only a means to create the revolutionary situation. However, because the *foco* is necessary to speed up the revolutionary process, its survival seems to be at times an end in itself [2: 1]. This is the case particularly in the first stages when the *foco* is still establishing itself in an area hardly known to the guerrillas and when the support of the peasants has not been obtained. Other functions of the *foco*, such as attacking the enemy, protecting the peasants or implementing agrarian reform will never be undertaken if by these the very existence of the *foco* is jeopardized.

The social composition of the *foco* should closely reflect that of the population of the area in which it operates. This means that a high percentage of the membership should be peasants. The guerrilla band has to use the environment as a powerful ally against an enemy far superior in manpower and equipment. Climbing hills and running through jungles often becomes too hard a task for the city dwellers. Under these circumstances the peasant often qualifies as a better soldier [1:34]. The peasant class, however, because of its backwardness and isolation cannot provide the leadership of the *foco*. Students, workers and intellectuals are called for this task, but the peasants will provide the bulk of the liberating army [3].

C. Functions of the Foco

The emergence of a revolutionary *foco* represents an open challenge both to the legitimacy of the government and to its exclusive right to use force to maintain stability, which in turn tends to generate legitimacy. The challenge to its legitimacy could be easily dismissed if it were not for the open confrontation presented by the small band to the so far unchallenged mechanisms of repression, particularly the armed forces. As the repressive ability of the government decreases, the credibility of the threat to its legitimacy presented by the *foco* increases. Consequently the primary function of the *foco* is to minimize, neutralize and exterminate the ability of the government to curb opposition and maintain stability. The armed forces thus become the primary target of the guerrilla band.

Against an enemy far superior in number and equipment, Guevara recommends the use of constant mobility, constant vigilance and constant wariness [2: 13]. Moral and psychological rather than physical extermination of the enemy is sought in planning ambushes and in selecting tactics and strategies. Only at the end of the war, when the armed forces are entirely demoralized and the *foco* has grown to the full size of a rebel army, is it possible to enter into regular, large-scale battle [3; 2: 15].

As the credibility of the potential and actual threat of the guerrilla band to the *status quo* forces increases, the *foco* begins to operate as an integrative center of attraction. On the one hand police repression against government opposition will grow in the cities to cut off supply and communication lines to the rebels, to retaliate against sabotage and/or purely 'to get even' for the government losses in guerrilla ambushes. As tension, fear and lawlessness mount in the urban centers, avenues of redress and accommodation are perceived as entirely closed. Two choices are open for those in the opposition: either to go into exile or to the mountains.[7]

On the other hand, the peasants of the area where the *foco* is in operation, who often in the past were robbed and terrorized by the regular army, begin now to realize that a common cause exists that places them and the guerrillas on the same side.[8] At this point the peasants start cooperating with the guerrillas, furnishing them with much needed knowledge of the environment (roads, paths, shortcuts, location of villages, rivers, etc.) and with information about deployment of army troops, ambushes, etc. Essential foodstuffs are also provided by the peasants. Most important of all, the peasants begin to join the guerrilla group, giving the *foco* a true local character [5: 72, 81]. Again the regular army retaliates against the peasants for helping the guerrillas by burning their huts and destroying their crops. The behavior of the army, so much in contrast with that of the rebels, further enrages the hatred of the peasants who now begin to seek protection and justice with the rebels.

Thus, by drawing support from the urban and rural population, the *foco* begins to play an increasing function in the mobilization of the masses against the existing government. The men in the mountains are no longer seen as one of the many groups that make up the opposition. They begin to be seen as a coherent, highly effective group that poses a challenge to the government. They begin to be seen as the vanguard of an armed struggle of the urban and peasant masses against the repressive methods of a government which is illegitimate and unpopular. Guerrilla warfare becomes, according to Guevara, a people's war, a mass struggle, the vanguard of which is the *foco* [2: 2].

D. The Foco as Vanguard and Center of the Revolution

If the revolutionary *foco* is to succeed in creating the subjective conditions for a revolutionary situation, it is clear that it must coordinate its actions with those of others also opposing the established order. If a small band is to succeed not in a *coup d'état*, but in arousing the populace into a mass popular struggle, it is evident that many other groups and organizations are to be involved in mobilizing support both in urban and rural areas.

Such support is essential particularly in the early stages of the formation of the *foco*, during which time its catalytic function is not fully in operation. Because of the relative weakness of the *foco* in this take-off stage, dependence on city leaders for weapons, logistic support and recruits is greater than at any other time. It would be misleading, however, to interpret this dependence as subordination of the *foco* to the city leaders.

In classical approaches to guerrilla warfare, including those of Mao Tse-tung, Giap and Ho Chi Minh, it is taken for granted that guerrilla war is fought by a group of highly dedicated men, who receive precise orders from the party, whose leadership remains aloof from the battlefront. In other words, the guerrilla band is the vanguard, but is not the center of the revolution. The approach of Guevara is vastly different and seems to be most appropriate for Latin American context [2: 10-11].[9]

According to Guevara the guerrilla *foco* is not only the vanguard but also the political and military center of the revolution. Other political forces of the opposition may gather around this nucleus, but initiative, power of decision making, direction of the struggle, tactics and strategies, will entirely and exclusively depend on the leadership of the *foco*. The rationale behind this centralization of powers in the hands of the guerrillas is not only tactical but strategic. It is Guevara's assumption that the *foco* is basically a microcosm of the revolution that is beginning to take place. While for those outside the *foco* it operates as a catalyst of revolution helping to create the conditions for the mass uprising, for those inside the *foco* itself it operates as a school of true revolutionists. And in fact, while for others the revolution will not start until the actual seizure of power by the vanguard, for those participating in the life of the *foco* the revolution has indeed started from the time the *foco* was constituted. Likewise, as the revolution at a later stage produces the ideology, leadership, tactics and strategies that seem to fit best the revolutionary process at each particular point without accepting orders or directives from outside itself, in the same manner, the *foco* cannot subordinate itself to the directions of anybody or any group which is outside the center of action. All other urban and rural groups should accept the political and military leadership emanating from the unified command of the *foco*.[10]

II. The Practice of the Revolutionary Foco

An attempt will be made in the second part of this paper to put together whatever empirical evidence is available to substantiate Guevara's theory of the *foco* as the catalyst of a revolutionary situation. Such evidence is collected from Guevara's own writings and speeches, particularly from his diaries of the Cuban and Bolivian campaigns. In both instances Guevara describes his personal participation in setting up a revolutionary *foco* and furnishes us with observations about the behavior of the guerrillas, the organization of the group and its relation to the outside world.

As a methodological note I would like to point out that both diaries, particularly the one from Bolivia, could be used as a unique collection of field notes of a keen and highly disciplined observer. The Cuban memoirs were based on field notes taken by Guevara and were published by him some time after the campaign. The Bolivian Diary presents the unedited field notes themselves. Very little work has been done by trained sociologists as participant-observers in revolutionary situations. My own work as participant observer in the 1965 uprising in the Dominican Republic is one of the few studies of this kind.[11] Until more research of this kind is conducted, the best sociologists can do is to study the field notes of other participants.

Such observations can provide the sociologist with data for a crude qualitative analysis of the sources of strain and conflict in the group, of the degree of solidarity, morale and ideological commitment required of the individual if he is to participate fully in group life under constant tension. By classifying somewhat more systematically such observations or by gathering information and remarks made by Guevara on events, situations and individuals, an insightful researcher could succeed in building classifications or typologies that could be of great use for further study of such an unconventional topic as guerrilla warfare.

In the following pages I will attempt to show to what extent the experiences collected in Guevara's diaries substantiate the major tenets of the theory of the *foco*, or to what extent it only helps to understand them. Or perhaps, instead of talking about substantiating theories, I should talk in terms of the extent to which the empirical experiences indicate that Guevara's doctrine was an adequate guide for guerrilla activity as a catalyst for the revolutionary situation.[12]

I will be using the evidence furnished by Guevara in both diaries to illustrate each of the four points discussed in the first part of this paper. For methodological reasons I will draw from the Cuban Diary first and then from the Bolivian experiment. It seems to me that the Cuban experiment was closer than the Bolivian to what Guevara had in mind as an ideal type of a revolutionary *foco* going through its natural stages of development. In fact, I believe that the model for the doctrine of the *foco* was formulated in Guevara's mind, at least in part, upon confrontation with the Cuban experience. I also believe that in the Bolivian experiment Guevara was seeking both the implementation of a doctrine and the verification of some theoretical generalizations.

A. What Kind of Men Made up the Foco

In Cuba, after the fiasco of *Alegria de Pio*, where the men who landed with Fidel Castro were decimated, those who managed to escape gathered again at the ridge of the Sierra Maestra mountains. By the end of December 1956 there were approximately twenty guerrillas. Soon some peasants began to join the guerrillas and by February 1957, after the first two battles were won, some new recruits arrived from the city of Manzanillo. Although the small group of original combatants seems to have been quite cohesive and enjoyed a high degree of morale, Guevara pointed out that shortly after new recruits arrived the size of the troop had to be cut down, for there was a group of men with very low morale [5: 36-40].

From the start discipline was strictly enforced; three crimes were punishable by death: insubordination, desertion and defeatism [5: 36, 130]. Traitors, informers and bandits were summarily tried and executed [6: 178-84]. Still, in May 1957 Guevara remarked that desertions were not uncommon due to the physical and moral inability of the men to endure the hardships of the guerrilla life. The struggle against the lack of physical, ideological and moral preparation of the men was a daily one; the results were not always encouraging [5: 88].

The Cuban guerrillas evidently saw themselves as social reformers in the implementation of revolutionary justice against some members of the armed forces who had terrorized the peasants [5: 25]. On the other hand, students and professionals of middle-class extraction now living in contact with poverty and deprivation began to feel solidarity with the peasants. A makeshift clinic was set up by Guevara and other technical services were provided by the guerrillas. The middle-class members of the *foco*, through contact with the peasants, began to see the need for agrarian reform [5: 81].

In Bolivia, the hard core of the *foco* was made up of some seventeen Cubans that Guevara brought with him. To this group three other men from Peru were added. About another twenty-nine men from Bolivia joined the *foco* at one time or other [7; 108-9, 158]. The field notes show remarkable difference between the Cuban and Bolivian groups. The Cubans, most of them veterans of the Cuban struggle, were disciplined, highly trained, ideologically motivated, fully aware of the international dimension of the struggle and displayed a high degree of group solidarity [7: 29, 87, 170, 189, 196].[13]

The recruitment of Bolivians was Guevara's chief concern from the start. After the opposition of the Bolivian Communist party to the guerrilla *foco* became manifest, the recruitment of new fighters slowed down. The new recruits lacked the training, discipline and morale of the Cubans and often showed antagonism to them [7: 43, 65, 67, 71, 117, 170, 196]. This was to be expected since the Cubans were veterans in guerrilla warfare, while the Bolivians had no previous experience. By the end of August, however, some of the Bolivians had already shown leadership abilities. But a month later some of the best Bolivian fighters were killed. Making the summary of the last month of the life of the *foco*, Guevara was able to write: 'the morale of the group is quite good and I only have doubts about Willy who might try to desert' [7: 209, 230, 232].

The Bolivian Diary shows dramatically how the first task of the emerging *foco* was to train the men into a school of discipline, self control and endurance facing physical and psychological hardships. Hunger, thirst, lack of sleep, physical exhaustion are part of the everyday routine when the guerrillas are not attacking or being chased by the enemy. On at least two occasions Guevara criticized his own lack of self control. Two months before his death he wrote, 'I feel like a human scrap . . . and on some occasions I have lost self control' [7: 196, 205-8].

B. Composition and Organization of the Foco

In Cuba approximately twenty men escaped the first attack of Batista's forces after the landing of the *Gramma* in December 1956. Crescencio Pérez, a peasant of the area, helped the men to find the way to the mountains. Fidel Castro occupied his leadership position from the beginning. With the cooperation of Crescencio Pérez and other peasants, communications for logistic support were established with Celia Sánchez, Frank País and other leaders in urban centers [5: 20-2]. Less than two months after the landing, the *foco* began to operate, staging a surprise attack on a tiny outpost of the army at *La Plata*. This military victory was highly significant as a symbol: it showed the world that the rebels were still alive and it showed the rebels that they could beat the army.

At this time the rebels could not count on the support of the peasants in any large scale. At least half a dozen peasants, however, had joined the rebels and were playing an increasingly important role not only as guides, but also in getting supplies for the group. [5: 30, 46-7]. By April 1957, five months after the landing, Guevara reports that peasants made the mobile vanguard of their column [5: 72-3]. Communications with the cities were established, some city-leaders had come to discuss the situation with Fidel, and finally new city-recruits and ammunition started to arrive from the cities. Two U.S. journalists, Herbert Matthews and Robert Taber, visited the rebels about this time [5: 51, 63, 85]. In May, Fidel reorganized the rebel forces and after discharging those who had not adapted themselves to the new style of life, still had about 125 men [5: 90]. In July, Fidel divided his forces into two columns and Guevara was appointed to lead the second column made up of some 75 men. The two columns enjoyed tactical independence, but were under Fidel's unified command. Soon after that, two more columns were created as offshoots of Fidel's column to widen the operational sphere of the *foco*. By this time the *foco* had emerged from the stage of self-preservation into a new stage of expansion. Guevara characterizes this period as one of balance of power between the government forces and the guerrillas. Both were able to defend their positions, but were not able to make inroads into the other's territory [6: 199].

The social composition of the Bolivian *foco* was mentioned

Che Guevara on Guerrilla Warfare:
Doctrine, Practice, and Evaluation

earlier: 18 Cubans, 29 Bolivians, and 3 Peruvians. Little is known of the social extraction of these men: there were three physicians besides Guevara, two students, some miners, and one Indian of Aymará extraction. There is reason to believe that men such as Chino, Bigote, Coco and Inti Peredo and others were middle class in origin. From the apparent lack of adaptation to the physical environment, it seems clear that most of the men, if not all, were from the capital or other cities [7: 207-8, 235, etc.].

The recruitment took place in the first five months. Despite the fact that Mario Monje and the Communist party had clearly refused to cooperate under the leadership of Guevara, a network of contacts with Cuba, Peru, Brazil and Argentina was established to channel support for the *foco* [7: 53-5, 62, 65, 73, 87]. By the end of the second month, however, Guevara's diary begins to show increasing concern for the lack of participation of Bolivians in the new *foco*. At the end of the sixth month, Guevara pointed out ominously that several casualties (and one desertion) had taken place, but not a single new recruit had been incorporated [7: 71, 108, 131].

A serious problem that Guevara had to face in Bolivia, but not in Cuba, stemmed from the heterogeneity of the group that existed in Bolivia. The division between Cubans and Bolivians was rather sharp, and the latter resented from the start the role played by the former in the leadership of the *foco*. Guevara used every opportunity to ease tensions and solve conflict, stressing the international character of the struggle, the communality of the interests and goals and the solidarity of a fight against common enemies [7: 43, 53, 61, 68, 85, 117]. There is evidence that Guevara was very fond of some of the Cubans who came with him from Havana. But he was well aware of the need to develop leadership among the Bolivians and gave them every opportunity to use their abilities [7: 71, 127, 166, 209].

Perhaps the most serious blow to Guevara's plan to develop a *foco* in the Bolivian mountains was the negative response of the peasants to the call to join the struggle. Month after month, in summing up the major points of the period, Guevara painfully realizes the absolute lack of peasant participation. In the last summary before his death, Guevara wrote that the peasants not only did not help, but had become informers for the army [7: 131, 152, 170, 232]. This lack of cooperation contrasts sharply with the generous participation of the Cuban peasants described by Guevara as early as April 1957 [5: 72].

The stage of development reached by the Bolivian *foco* at the end of the first year of struggle was quite different from that attained by its Cuban counterpart. By the end of the first year the Bolivian *foco* had not succeeded in taking off from the stage of training and self-preservation. However, if the internal structure of the Bolivian and Cuban *focos* is compared in terms of cohesion, leadership and morale, no basic difference seems to exist between the two. Despite the frequent criticism Guevara made about lack of discipline, of self control and of ability to endure hardhsips, he agreed that the group was steadily growing in morale, solidarity and fighting capability [7: 87, 91, 95, 99, 125, 152]. At some point at the end of June, Guevara noted that the morale of the whole group was high: 'all Cubans are excellent fights and there are only three Bolivians who are still lazy' [7: 170].

C. Functions of the Foco

As mentioned earlier, the attack on the tiny army outpost at La Plata constituted a symbolic victory for Castro's forces in Cuba. Both the illegal government of Batista and the Cuban public had to face the facts: a group of rebels had taken over an army outpost in the mountains. It was an open challenge not only to Batista's claim of legitimacy which nobody recognized, but a threat to his monopoly on the use of force which everybody reluctantly accepted. Four months after the battle at La Plata the first major attack of the rebels took place at the barracks of El Uvero. The meaning of this battle was not purely symbolic since the army lost fourteen men, nineteen were wounded and the other fourteen were captured by the rebels. As a consequence of this attack the army withdrew its forces from a number of outposts that could not easily be protected without major reinforcements [5: 101].

As the repressive ability of Batista's army against the *foco* decreased, the striking power and range of action of the *foco* increased. The credibility of the threat increased with the opening of a new front, and other groups emerged to challenge the forces of the government: on March 13 a group of university students attacked the presidential palace in an attempt to kill Batista; in May an anti-government expedition landed in Oriente province; in June the Student Directorate started a new front in the Escambray mountains; in September an uprising took place at the Naval Base of Cienfuegos [6: 198]. All these attempts failed to overthrow Batista but succeeded in spreading a generalized belief of dissatisfaction and in giving credit to the *foco* as the only credible challenge to the forces of the government.

With increased police brutality in the cities and the assassination of civilian leaders, such as those of Pelayo Cuervo and Frank País, many turned their hopes toward the guerrilla *foco* as the best expression of popular struggle against an oppressive government [6: 202]. However, not all recruits arriving in the Sierra Maestra made good guerrilla fighters. In several passages Guevara pointed out that many of the new recruits were slow in adapting themselves to the hardships of guerrilla living, and that some had to be discharged, while still others deserted the guerrilla band. It is also clear, however, that the rebel band kept growing, since, according to Guevara, by early July their number had risen to about 200 men [6: 128, 131, 133]. By this time, he believed, a qualitative change had taken place in the Sierra Maestra and a truly liberated zone began to exist. Government forces had withdrawn from the area, and both peasants and landowners began to recognize the rebel forces as the true government of the region by paying taxes to and obtaining protection from the rebel forces [6: 178-84; 192-5].

In Bolivia, from the seizure of power by General Barrientos in a *coup d'état* in November 1964 no serious threat to his stay in office had been posed by any opposition group, apart from the miners' strike of May and September 1965. Barrientos' stern reaction to crush the opposition of the miners was supported by the army and by some leaders of the peasant leagues. U.S. foreign aid in the form of economic loans and military missions to Bolivia seem to have also helped Barrientos' position, if not to legitimize his claim to govern, at least to strengthen his monopoly in the use of force.[14]

The arrival of Che Guevara in Bolivia in the early days of November 1966 was a well calculated attempt to produce a credible threat to the legitimacy of Barrientos' power and to his monopoly in the use of force. Four months after establishing the guerrilla *foco*, the first encounter with the army took place. The results were highly favorable to the rebels: seven soldiers dead, fourteen prisoners including two officers, and four wounded.

Weapons and ammunition were also captured. From this moment the existence of the guerrilla *foco* was known all over the country through official broadcasts of the government and through clandestine communiques of the rebels under the name of The National Liberation Army of Bolivia [7: 243-7].

There were twelve or fifteen encounters between the army and the guerrillas before the one that took place at the *Quebrada del Yuro* where Guevara was wounded and taken prisoner. In all these battles, with the exception of one on September 26, the results were quite favorable to the rebels. The army lost from 35 to 40 men; some 25 were wounded and around 40 were made prisoners by the rebels. The losses for the rebels were quantitatively small, although some were qualitatively significant. Before the end of August Guevara had lost only eight men in combat [7: 127, 166, 189, 228]. To these casualties suffered by the small rebel band, one has to add the loss of approximately twelve men who made up the rearguard of the guerrillas and who got stranded from the rest of the group shortly after the first encounter with the army [7: 129, 131, 135]. The rearguard was made up of nearly one-third of the guerrillas, under the command of two Cubans. Despite efforts made by Guevara to find them, contact was never re-established. The guerrilla *foco* was thus divided prematurely, which considerably weakened its mobility and striking capability. On August 31 the Bolivian army eliminated the rearguard. At the end of every month in summing up the major events of the period in his diary, Guevara always refers dramatically to the failure to establish contact with the rearguard as one of the negative factors affecting their plans. To these very sensitive losses, one should finally add the desertion of a few who at different times gave up the fight.[15]

Putting all these losses together, the guerrilla *foco* was reduced from 45 men in March to 29 in April, 25 in May, 24 in June, 22 in July, 20 in August, and 15 in September. When Guevara was ambushed by the army on October 8, there were sixteen men with him, of whom three at least were wounded. Guevara dramatically contrasted the decrease in the number of fighters with the number of army troops actually engaged in the hunt. Two days before his capture Guevara noted in his diary that two divisions of the army were engaged in the operation with at least 1,800 men hunting the seventeen guerrillas [7: 235, 237].

It was not, however, the loss of men that made the Bolivian *foco* fail. It was not even the effectiveness of the rangers trained by U.S. instructors. Men were also lost in the Cuban fight, and there is no reason to believe that the Bolivian army was any better than its Cuban counterpart. In evaluating every month the operations conducted by the army, Guevara pointed out that the Bolivian army was ineffective and utterly incapable. Only in September did he note that the army had improved in efficiency [7: 152, 170, 191, 232]. Daniel James claims that the first rangers trained by the U.S. military mission completed their training around the middle of August. The efficiency of these troops was soon felt in the defeat of the rearguard of the *foco* on August 31 and in the battles of September and October that finally crushed the guerrillas [8: 55-6]. There were two other factors that doomed the Bolivian *foco* to fail: first, the lack of communication and recruitment from the urban centers, and second, the lack of response from the peasants. From April to October in summing up and in evaluating the events of the month, Guevara ominously singled out these two factors as the most important functions the *foco* was failing to fulfill [7: 131, 152, 170, 232]. Lines of communication with the capital city of La Paz, with Argentina, Cuba and other Latin American countries were cut soon after the first encounter with the army [7: 132]. No logistic support, medicine or ammunition was received by the rebel band in more than six months. Still worse, no one came from the cities to reinforce them. Support in the cities for the emerging *foco* never materialized, and Guevara expressed his feelings that if some 50 to 100 could be recruited in the cities, it would not matter if the actual number of combatants was limited to 10 or 25 [7: 170, 191]. The last messages received by Guevara were dated in May and June 1967; promises of reinforcements were made. The reinforcements never arrived [7: 258].

It was quite clear to Guevara that the cooperation of the peasants was necessary if the *foco* was to succeed in Bolivia. The question of time was an important one. It would be unrealistic to think that by merely setting up the *foco* the guerrilla band could count on the unconditional support of the peasants. It was expected that in Bolivia the peasants might react with mistrust at first, but it was hoped that later they would slowly start siding with and giving protection to the rebel band. This second stage was never reached. Every month from April to September,

Guevara painfully repeated the same note: 'we have not succeeded in developing peasant support.' In the summary of September, Guevara pointed out that, 'the peasants are not helping us in any way: on the contrary they are becoming informers for the army [7: 131, 152, 189, 232].

D. The Foco as Vanguard and Center of the Revolution

In his Cuban diaries, Guevara points out that as early as February 1957 signs of disagreement were found between the *llano*-leaders and the guerrillas of the 26 of July Movement [5: 50]. The isolation and lack of communication between the guerrilla fighters and their supporters in the cities led to some strategic and ideological differences which had to be eliminated if the *foco* was to operate from the mountains as a true catalyst of a revolutionary situation. In order to iron out such differences a meeting was arranged between the guerrillas and the urban leaders, and as a consequence new reinforcements were sent to the mountains [5: 61].

In the early part of 1957 several opposition groups had made serious and dramatic attempts to overthrow Batista. On July 12, Castro published the *Manifesto de la Sierra*, in which he advocated the formation of a united front against Batista. In October the most important groups of the opposition met in Miami and drew up a program of action against Batista. Felipe Pazos, former president of the Cuban National Bank, represented the 26 of July Movement at the meeting and signed the so-called *Miami Pact*. Castro from the mountains soon condemned Pazos for signing such a document without consultation with the leaders and fighters in the Sierra and denounced the document as not acceptable to them.

Out of this dissension between the *llano* and *sierra* leaders within the 26 of July Movement, it became apparent that in no circumstance would the former impose their will upon the latter. Indeed a united front and a unified leadership were required to mobilize the masses and to break the structure of the established order. But this unified leadership had to be centered around those men who precisely constituted a credible threat to the dictator. By this time, Fidel Castro and his companions had spent a full year in the *sierras*, and Batista's forces had proven entirely unable to eliminate them. Consequently they had the right to lead others into the struggle [6: 211-27].

As the *foco* consolidated its position, it became the center of attraction for many in the opposition both in the cities and rural areas. When the *Pact of Caracas* was signed by most opposition leaders on July 20, 1958, the demands of the *foco* were accepted by all. From this moment it was clear to all that the *foco* was not only the vanguard of the revolution, but was also its *center*. When in January 1959 the *sierra* leaders came down to the cities and refused to accept a military junta or any other form of compromise with the government, everybody accepted their leadership. Nobody could have dared, at this point, dispute their right to conduct the revolution the way they wanted.

There are some who still argue that it was not Castro and his guerrilla band who brought about the downfall of Batista.[16] They point out that the middle class was alienated, the rich did not feel secure, some generals betrayed Batista, bishops wrote pastoral letters, U.S. embargoes were enforced, the army disintegrated and that the whole political structure collapsed. In my opinion, all these are manifestations of a revolutionary situation, which was typical not only of Batista's Cuba, but of many other Latin American countries. By setting up a guerrilla *foco*, Castro and his men affected the revolutionary situation in two ways: first, they helped accelerate and spread the process of social disintegration of the old structure to the whole system, and second, they made people aware of such a situation. Awareness is only a first step in the process of mobilization for action. When the masses decided to act, they followed the lead of those who had been effectively active. The leaders of the *foco* were now to become the leaders of a mass struggle: the revolution was about to start.

As for Bolivia, it is quite possible that long before arriving there Guevara had asked himself how the Communist Party would react to his decision to start a guerrilla *foco* in Bolivia. He knew only too well that the MR-13 in Guatemala and the FALN in Venezuela had run into serious opposition to their guerrilla activities from the official Communist parties in those countries. He also knew that bringing some Cubans with him to start a war of liberation for Bolivia would certainly produce ill feelings among some Bolivians whose cooperation he needed. He was convinced, on the other hand, of the international character of his mission and that in this continent objective conditions were present that would guarantee the emergence of a revolutionary situation. 'Obviously, in the countries where this condition existed, it would be criminal not to act to seize power' [2: 4].

Although the plan of Guevara had continental ambitions, the struggle had to materialize first on Bolivian soil and with Bolivian help because, according to Guevara and his associates, the country enjoyed the best objective conditions for a revolution and the best geographic location to start a continental struggle [8: 280, 285, 287]. It was therefore essential for Guevara's plan that the *foco* be made up of Bolivians fighting for their own national liberation front [7: 247]. Twenty days after the *foco* was started, Guevara received information that a leader from Peru was planning to send him twenty recruits. Guevara was not pleased with the news because it meant that the *foco* would become internationalized before contacts could be established with the official Communist Party in Bolivia whose secretary general, Monje, was out of the country. Guevara discussed the matter with the Bolivian members of the *foco* who seemed to react favorably to the plan. At the end of the first month, Guevara mentioned that his main interest at this point was to increase the number of Bolivians to twenty and showed concern with the reaction of the Communist Party to the emergence of the *foco* [7: 36, 37]

There is little doubt that Guevara foresaw the possibility of conflict between the *foco* and the Communist Party. He made it clear to the members of the party who had joined the *foco* that it was up to them to decide whether they would break with the official party line in adopting another course of action. He also made it clear to them that no compromise could be accepted with regard to the idea of a unified leadership in the revolution [7: 44].

Mario Monje, secretary general of the Communist Party, arrived in the mountains on December 31, 1966, to see Guevara. The discussion between Guevara and Monje soon was centered around the sensitive issue of the leadership of the *foco*. Monje insisted that as secretary general of the party it was his right and duty to be the leader of the revolution, at least for the time that it would be confined to Bolivian soil. He showed readiness to resign his position in the city and come to lead the guerrillas. Guevara seems to have been ready to compromise in all other issues, but not in the one of a unified leadership under his direction. At this point an impasse was reached. Monje was allowed to discuss the matter with the members of the party who were in the *foco*, and none of them sided with him. The next day he left for La Paz. Guevara addressed his followers and emphasized that the *foco* was willing to cooperate with anybody, but explained that no compromise could be accepted that would jeopardize its leadership [7: 54].

On two more occasions Mario Monje was mentioned in Guevara's field notes. He was blamed for dissuading recruits to join the *foco* and for sending Fidel Castro inaccurate information about the situation. On the last day of January Guevara accused Monje of double-crossing the rebels. The Communist Party, according to Guevara, was now taking up arms against the *foco*. He was confident, however, that the very honest and committed would eventually side with him.[17] The party was mentioned three other times in Guevara's notes, which show that he was hopeful that a splinter group within the party led by Kolle and 'Paulino' would come to their support. The much needed help of recruits and medical supplies never arrived [7: 152, 170, 189, 285].

The diary of Pombo, one of Guevara's lieutenants who arrived in Bolivia late in July 1966 to do the goundwork before Guevara's arrival, furnishes information which helps to explain why the *llano* leaders, and particularly the Communist Party, failed to support the emerging *foco*. The first difficulty encountered by Guevara's associates in preparing the struggle was the apathy of the Bolivians to any action [8: 261]. Whether this apathy was a reaction against the Cuban guerrillas who were running things in Bolivia was not indicated in the diary. The notes suggest, however, that these same people were convinced that objective conditions were present to start the struggle in Bolivia [8: 262, 266, 280]. The second factor was divisionism between factions and leaders who not only disagreed with the Cubans but also among themselves in matters of strategies and tactics. Divisions existed within the Communist Party (Monje-Zamora-Kolle), between the party and other Marxist groups (Guevara-Sánchez) and even between the Central Committee and the secretary general of the party [8: 260-9]. The third factor to obstruct the creation of a guerrilla *foco* as the center was ideological. Monje and other communist leaders thought that conditions were ripe for the struggle. The uprising, however, should take the form of a *coup d'état*, not of mass popular struggle. Only if this failed should they plan to start guerrilla war [8: 256-7, 263, 265, 267]. Naturally this ideological position was an open negation of Guevara's theory of the *foco* and could never be reconciled with it.

After the all too apparent break with the communists was completed the only hope for mustering support for the *foco* in the cities were the miners and the students, both groups with a long revolutionary tradition in Bolivia. As the *foco* became known after its military victories over the army from April to August it was expected that support from such groups would increase. In June 1967 the miners went on strike against the government. Barrientos attributed the strike to Guevara's work, although apparently direct links between him and the miners did not exist [7: 169]. Guevara, however, took the opportunity to write a message to the miners exhorting them to make a common front with the *foco* by joining the guerrillas [7: 248]. By then, however, Barrientos had sent the army into the mines and some 87 miners had been killed [7: 165].

When a few months later, Loyola, the young teacher who was one of Guevara's contacts in La Paz, was put in jail a number of teachers and students struck to protest the action of the government. Guevara found in the strike signs of solidarity between the teachers and the *foco* [7: 224]. Although neither this incident, nor that of the miners materialized into any serious threat to the government, the *foco* was indeed beginning to function as a catalyst to a revolutionary situation. It had taken much more time than in Cuba where after one year in operation the *foco* had become the center of the revolution. Guevara and his men knew that wide differences existed between the Cuban and the Bolivian cases [8: 273-4]. They had foreseen the possibility of a prolonged struggle that might take from seven to ten years to achieve final success [8: 263]. It was also Guevara's opinion that their struggle was continental in character and that it would be unrealistic to think of a revolution succeeding in Bolivia alone without at least having a revolution in a coastal country if not in all of Latin America. 'If that doesn't happen', Guevara concluded the day he arrived, 'this revolution will be smothered' [8: 285].

Conclusion

An attempt was made in this paper to compare Guevara's theory of guerrilla warfare against the empirical evidence furnished in his diaries to the Cuban and Bolivian campaigns. It was suggested that it would be more appropriate to talk about a doctrine and its application rather than of a theory and its empirical substantation.

The paper suggests that Guevara was basically in agreement with most theories of revolution that certain objective and subjective preconditions are to be present before a revolutionary situation can emerge. Guevara's special contribution to the theory of revolution is that such conditions can be created by the revolutionary *foco*. To use one of Guevara's favorite analogies, the *foco* is the midwife of revolution [2: 6]. Posing a credible threat to a government seen as illegitimate is one of the primary functions of the *foco*. By attacking the army and other mechanisms of repression the *foco* challenges the efficiency of the government to curb opposition. As the inefficiency of the government increases the opposition forces are polarized toward the *foco*, which at this point begins to operate like a vanguard and center of the revolution.

In comparing the practice of the *foco* in the Cuban and the Bolivian campaigns the paper did not find any basic differences between the two in the internal structure and functions of the *foco*. The same types of men, the same kind of training, the same style of leadership existed in both experiments. Group morale, integration and ideological commitment was high in both experiences. The two *focos* went through a similar stage of consolidation before they went into action attacking the armed forces. Both in like manner posed a serious threat to the ability of the government to curb opposition. For the first nine months their attacks on the army were highly successful. These successes of both *focos* in establishing themselves and in posing a threat to the government indicate that Guevara's doctrine for guerrilla activity was adequate in both cases.

Sharp differences between the two cases were found in *external* circumstances, the control of which lay beyond the immediate reach of the *foco* in the stage of consolidation. There was a negative response of the Bolivian peasants and an apathy or even opposition of the *llano* leaders toward the emerging *foco*. By the end of the first year of struggle the Cuban *foco* had become the vanguard and center of the revolution with full peasant support, and the *llano* leaders were well aware of the symbolic and actual strength of the *sierra* leaders and in most cases were ready to follow their lead. By the end of one year in Bolivia, the *llano* leaders were openly undermining any possibility of cooperation and the peasants had given no support to the *foco*.

It is beyond the scope of this paper to explain why the *foco* failed to obtain peasant support. I may suggest some sociopolitical and economic characteristics which made the Bolivian and Cuban peasants different. In Cuba the *foco* and the peasants shared the same racial, cultural, and linguistic background. In Bolivia they did not. Second, despite progress made after 1952, political integration of the peasant in Bolivia was lower than in Cuba. Third, although the Cuban peasant was better off than the Bolivian, he was hungry for land, and the promise of agrarian reform made him side with the rebels [3: 29]. Fourth, many Cuban peasants participated in occupations that made them resemble more an alienated urban proletariat than a traditional peasantry as in Bolivia.[18] Although Guevara knew that such differences existed, he was convinced that it would be only a matter of time before the Bolivian peasant would be won over to the rebel cause [7: 18, 189].

It was also Guevara's luck that his troops were mistakenly divided and did not succeed in getting together again, thus offering the armed forces the chance of exterminating them piecemeal. The fact that a U.S. military mission helped train the Bolivian rangers to avoid the success of the *foco* is another element that should perhaps be taken into consideration in explaining the failure of the Bolivian *foco*. Communication lines essential for the life of the *foco*, especially during its formative period, were cut.

All these elements, however, are external circumstances which do not affect directly the doctrine of the *foco*. The facts, however, that the peasants did not respond and the *llano* leaders did not cooperate pose indeed more serious questions to the doctrine of the *foco*, which evidently the proponents of the doctrine have to answer if the theoretical value and applicability of such doctrine is to be maintained.

References

1. *Ché Guevara on Guerrilla Warfare*, New York: Frederick A. Praeger, Publishers, 1966.
2. 'Guerra de Guerrillas: Un Método', *Cuba Socialista*, Septiembre de 1963 (Ano III, No. 25), 1-17.
3. 'Cuba: Exceptional Case or Vanguard in the Struggle against Colonialism', *Verde Olivo*, April 9, 1961.
4. 'Notes for the Ideology of the Cuban Revolution', *Verde Olivo*, October 8, 1960.
5. *Relatos de la Guerra Revolutionaria*, Buenos Aires: Editora, Nueve 64, 1965.
6. *Reminiscences of the Cuban Revolutionary War*, New York: Monthly Review Press, 1968.
7. *El Diario del Che en Bolivia*, Mexico City: Siglo XXI Editores, S. A., 1968.
8. *The Complete Bolivian Diaries of Che Guevara and Other Captured Documents*, edited by Daniel James, New York: Stern and Day Pub., 1968.
9. *Che Guevara Speaks: Selected Speeches and Writings*, edited by George Lavan, New York: Grove Press, Inc., 1968.

Notes

The present paper is only part of a larger research project concerning social revolution in Latin America in which the author is presently engaged. The author wants to acknowledge the comments, criticisms and suggestions made for the first draft of this paper by Professors Gilbert Shapiro, William Delany, Carmelo Mesa-Lago, Gerhard Drekonja and by Miss Donna Maria Barbaro.

1. By revolutionary situation is understood here a state of affairs in which a *status quo* government is presented with a challenge from within which poses a credible threat to its ability to implement coercion and maintain stability. This situation is commonly preceded by the existence of some objective and subjective conditions and is often, but not necessarily, followed by an actual revolution or seizure of power by a group that seeks to implement basic structural changes.
2. The major works of Guevara used in this study are listed at the end of this paper. For the sake of brevity all references to these works will be put in brackets, the first digit indicating the number of such work in the list and the second pages.
3. The word *foco* is taken from the Spanish for focus, and is

described both in English and Spanish as 'the centre of activity, or area of greatest energy, of a storm, eruption, etc.' (*The Oxford Universal Dictionary*, 3rd ed.).

4. An excellent account of such conditions in modern theoretical approaches to the study of revolution is found in Harry Eckstein, 'On the Etiology of Internal War', in *History and Theory*, Vol. 4, 133-63, and in Lawrence Stone, 'Theories of Revolution', in *World Politics*, Vol. 18 (January 1966), 159-76. See also Chalmers Johnson, *Revolutionary Change* (Boston: Little Brown, 1966).

5. In 1961 Guevara described these conditions in the following manner: 'The objective conditions for struggle are provided by the people's hunger, their reaction to their hunger, the terror unleashed to crush the people's reaction, and the wave of hatred that the repression creates. [Latin] America lacked the subjective conditions, the most important of which is awareness of the possibility of victory through violent struggle . . .' (*Verde Olivo*, April 9, 1961).

6. No other specification is made by Guevara either in his manual or in his diaries as to what kind of ideology the guerrilla fighters should have. In the Bolivian Diary mention is made a few times of lectures on political matters given by him to the troops, without any further mention of political indoctrination. In an article published in *Verde Olivo*, October 8, 1960, Guevara pointed out that the Cuban Revolution followed the course of the Marxist theory, without its leaders 'professing or knowing such laws from a theoretical point of view'. In the same article he pointed out how the ideology of the Cuban fighters had been forged during the course of the struggle.

7. It might take months before new recruits from the city start arriving in any significant numbers. Although some city dwellers joined Castro in the first few months after the landing of the *Gramma*, it was not until March 16, 1957, that the first contingent of some 50 men organized by Frank País arrived in the mountains. By this time the attack on the presidential palace had taken place and Pelayo Cuervo, José A. Echevarria and others had been brutally assassinated by Batista's forces [5: 62].

8. It is clear from Guevara's writings that fear, mistrust and even betrayal may be expected from some peasants at first. Even at a later stage, when the peasants begin to cooperate with the guerrillas, some might be motivated by a petty-bourgeois ideology [3].

9. Régis Debray, *Revolution in the Revolution?* (New York: Grove Press, 1967), pp. 95-9.

10. Régis Debray, *op. cit.*, pp. 95-116, has spelled out in detail the strategic and tactical reasons for advocating a unified leadership in the Latin American context. The ideological differences between the so-called orthodox Marxists of the old school and

the new breed of self-made Marxists is emphasized. It is quite clear that such differences in the Marxist camp materialized only after the success of the Cuban Revolution encouraged the emergence of guerrilla *focos* in several Latin American countries, often without the blessing and sometimes with the opposition of the official Communist parties in such countries. See James Petras, 'Revolutions and Guerrilla Movements in Latin America', in James Petras and M. Zeitlin, *Latin America: Reform or Revolution?* (New York: Fawcett World Library Publishers, 1968), pp. 329-69.
11. José A. Moreno, *Barrios in Arms: Revolution in Santo Domingo* (Pittsburgh: University of Pittsburgh Press, 1970).
12. It is true that Guevara talks about 'developing a theory', but he also calls it a 'manual' [1: xxxi, 6] and a 'method' [2].
13. Perhaps the most significant indicator of group solidarity was shown in an incident in which Marcos was severely reprimanded by Guevara for his lack of discipline and warned that he could be expelled from the guerrillas. Marcos replied that he would rather be put to death by a firing squad than be expelled from the group [7: 10].
14. See the *New York Times*, June 4, 1965.
15. Altogether six Bolivians deserted the *foco*. None of the deserters, according to notes prior to their desertions, was a good fighter. Consequently, their loss was more psychological and political than military. However, most of the deserters when captured, furnished the army with accurate information that helped destroy the rebel band [7: 209, 212, 216, 232, 234].
16. See among others, Boris Goldenberg, *The Cuban Revolution* (New York: Frederick A. Praeger, Pub., 1965), and Theodore Draper, *Castroism: Theory and Practice* (New York: Frederick Praeger, Pub., 1966), and Theodore Draper, *Castro's Revolution: Myths and Realities* (New York: Frederick Praeger, Pub., 1962).
17. In a document found with Guevara's notes and evidently sent to him from La Paz on May 13, mention is made of a declaration published by the Communist Party and signed by Monje and Kolle in which they showed solidarity with the *foco* [7: 54, 256].
18. M. Zeitlin, *Cuban Workers and the Revolution*, Princeton University Press, 1967.

Neither Mao, Nor Che: The Practical Evolution of Revolutionary Theory
by Edward Friedman

An official, albeit implicit, Chinese critique of Che Guevara's methods in Bolivia in 1966–67 can be found in Peking's praise in 1969 for a Bolivian Communist Draft Peasant-Agrarian Program.[1] Peking stresses Bolivia's feudal character and its domination by American imperialism. That is, the revolution should ally with bourgeois nationalism. Consequently it must be based on a minimalist program to attract a maximum of support. The national Liberation Front in Vietnam follows that course. The Cuban revolution of Fidel Castro took that road. Nonetheless, Régis Debray insists that 'Cuba remembered from the beginning that the socialist revolution is the result of an armed struggle against the armed power of the bourgeois state'.[2] While *Peking Review* agreed on the need for an 'armed struggle . . . rely[ing] firmly on the peasants', it did not propose making the middle classes the main target of that fight. By stressing socialism instead of alliance with nationalistic capitalists, Che isolates himself from a potential source of support and forces those rich entrepreneurs to help reactionary militarists and landed oligarchs whom they may abhor. Nationalist anti-imperialism is a major issue in Latin America as most everywhere else in the world. Yet there is no necessary reason why narrowly based Che-style guerrillas must win the nationalist mantle. It has been the bourgeois parties and interests in Latin America which usually have expropriated Yankee businesses. It is the native national bourgeoisie who are most obviously hurt by Yankee competition. But if the guerrillas cannot appear as the nationalists, they will not win the thousands upon thousands of patriotic youth needed to educate, organize, lead and die. For Debray, however, Maoism, this 'theory of the alliance of four classes, including the national bourgeoisie' is 'old . . . outworn, discredited . . .'[3]

Alliance with patriotic middle strata wears better from Mao's point of view. Not even all rich peasants are enemies. One should 'guarantee the properties of those rich peasants who have abolished feudal exploitation'. Such a policy will reassure middle peasants. If patriotic and productive rich peasants are protected, then middle peasants will see a policy of 'protect and guarantee medium-sized property' not as a mere momentary, temporizing measure. They will not feel that they are next in line for expropriation. Consequently, when the Yankee is attacked and 'debts to imperialist agencies' are abrogated, then, as 'victims of

U.S. imperialism's economic competition', 'middle peasants . . . fight against the U.S. imperialists and their allies in Bolivia'. It is this group which is most consciously insecure against the palpable products, usurers and traders coming from foreign urban areas. They will be among the first to pledge allegiance to a revolutionary cause of local self-control and protection of wealth of local people making it on their own. Once they help organize the rural community on behalf of the revolution, it becomes more likely that their sons and, more important, the local landless will join the rebel army. Land reform thus is 'made an instrument of the struggle for national liberation'.

When Peking gently criticizes 'the guerrilla movement in the southeastern part of Bolivia', it does so on the basis of its errors towards the peasantry. Should one blame the Bolivian peasants for not welcoming Che?[4] Or should one blame Che, who hurried to fight battles against the ruling oligarchy's army instead of patiently educating, wooing and organizing the peasantry? Mao's experience was that reductions of interest and taxes and redistribution of land and goods led peasants to take the guerrillas seriously. *Peking Review* finds that the 'principal error' of the Bolivian guerrilla force 'was that it neglected the role of the peasants, failed to win their support and did not carry out intensive political work among them'. People are to be educated to the new possibilities. Mao believes, 'To overthrow a political power, it is always necessary first of all to create public opinion'. What the Chinese find fundamental, peasant support, Moreno finds 'beyond the scope of' his paper.[5]

Naturally the indigenous peasants mistrusted the outsiders. Of the 50 in Guevara's band, 42 percent were foreigners. It is not an 'external circumstance' of 'luck' that Che's troops were divided and could not re-join.[6] They lacked members who were local people who knew the territory and would be trusted by local dwellers as brothers. Can a foreigner, favoring the foreigners in his group, stressing the international character of the struggle, successfully lead a war of *national* liberation? Is it 'time' that is sufficient to win Bolivian peasants to such a cause? The Chinese would stress actual reforms, teaching people to read and write, to raise their estimate of themselves and on the basis of the new found dignity, asking these people to teach others.

Here then lies the most striking difference between Che and Mao. Che insisted on running the show. He distrusted the abilities of people around him except for his closest friends. He insisted on controlling everything.[7] This contrasts sharply with Mao's method of relying on decentralization, local initiative and immediate response. Mao is a populist with faith in the people. Che was a Leninist who believed in elite, tightly controlled organizations. Leadership for Mao is 'the example we set by our own work to convince and educate people . . . so that they willingly accept our proposals.' Mao could accept Chiang Kai-shek as national leader but work to displace him. Che insisted that others subordinate themselves to him. It was Che's distance from the Bolivian peasant that made institutional form more important than actual work.

Consequently it is a little difficult to understand Moreno's charge that Mao advocated 'guerrilla war . . . fought by a group of highly dedicated men, who receive precise orders from the party, whose leadership remains aloof from the battlefront'.[8] After all, virtually the entire Chinese leadership was attached in one form or another to armies. The army was the heart of the Chinese movement. Apparently Moreno, Debray, et al., have read foreign language editions of Mao's works and been impressed by the stress therein on the leadership of the Communist Party. But many of these claims to Communist orthodoxy were added after the heterodox victory of the Red Army.[9] Surely it should be obvious after the Cultural Revolution's attack on China's Communist Party that when Mao talks of politics in command he refers to the actions and beliefs of men, to ideological and not to institutional control. He has for years attacked and opposed Communist parties that do not act on the basis of revolutionary consciousness. The Party to Mao is an ideal in the process of creation through struggle. To demand that it lead is mainly to ask that its values be practiced.

Guevara, Debray and Moreno misunderstand the Chinese revolution. They see its essence as fixed guerrilla bases. But bases are not even mentioned in the *Peking Review* commentary, 'Rallying Peasant Masses for an Anti-Imperialist, Anti-Feudal Revolution'. That does not mean safe areas are not important. Revolutionary warriors need places to rest; the wounded need places to recuperate. But guerrillas who are at one with their people can, as in Vietnam, find those secure spots in the heart of enemy territory. They might also find refuge in remote areas or neutral neighbors, as Debray mentions.[10]

Debray's list of five unique Chinese characteristics do not actually distinguish China as a place especially suited to bases. (1) Contrary to Debray's notion, the guerrilla base areas did not have 'a high density of rural population'.[11] The Chinese Red Army rather tended to establish its headquarters in remote, poor, mountainous border regions which were sparsely populated. (2) Debray's view that China's revolution was aided by 'common borders with a friendly country'[12] is belied by the hundreds of thousands of Japanese, puppet and Nationalist Party troops which blockaded, surrounded and attacked the liberated areas. Stalin helped Chiang more than he helped Mao. Moreover, the continued stress on self-reliance as a theme in Lin Piao's writings is an indication of what the Chinese learned from their lack of aid from Moscow. In the final analysis, the revolutionaries must rely on themselves. (3) Rather than facing a 'numerical insufficiency of enemy troops',[13] as Debray insists, the red bases in the late 1920s and early 1930s were surrounded and attacked by a million or so soldiers gathered from all over China. One by one the bases were decimated and destroyed. Armed groups escaped each of these encirclements, found succour in more distant hinterlands, joined, mobilized and fought again. (4) Since his survival depended on 'an extensive territory'[14] in which he could maneuver, Mao concluded that China's vastness which reduced the effectiveness of the many pursuing troops so that they were in seeming short supply was 'a primary condition' for 'waging guerrilla warfare'. Mao found China's revolutionary potentialities great and special while 'small countries like Belgium which lacked this condition [room for maneuver] have few or no such possibilities'.[15]

In fact, even before the Red Army was forced into its fantastic long march for survival, Mao already believed that 'unusual phenomenon only to be found in China' and 'unusual phenomena . . . absent outside China' were the essence of the Chinese situation.[16] Mao was referring to the weakness of China's ruling classes who sought support from imperialist nations contending for the China market. An intensifying tangled warfare of the cliques among the ruling class, each backed by different elements of imperialism, resulted. This hyper-imperialism, as Sun Yat-sen called it, was essential. Waves of red political power in the sea of white reaction 'cannot occur in any imperialist country or in any colony under direct imperialist rule, but can only occur in China which is economically backward, and which is semi-colonial and under indirect imperialist rule. For this unusual phenomenon can occur only in conjunction with another unusual phenomenon, namely, war within the White regime'.[17]

Mao was wrong on this point. Revolutionary power in China grew after the Japanese invasion ended China's hyper-colonial status. Mao, of course, has recognized this. His 'view' has changed as a result of the 'changes in the situation'.[18] Japan's invasion of Southeast Asia and the popular forces formed to oppose the invader, the weakening of western European imperialism as a result of World War II, the relative strengthening of the U.S.S.R. and the emergence of revolutionary China now make it possible for colonial peoples 'in the East' to fight protracted revolutionary wars originating in the countryside which will culminate in nation-wide victory.[19]

Mao changes his views as the world changes. He is a revolutionary empiricist. He offers no ready-made revolutionary prescription for all times and places. However, Mao's realism is tempered by his patriotism. Invariably he finds a special role for China which does not clash with the on-going definition of Peking's international needs.

The retreat of France from Algeria and the victory of Castro in Cuba, both self-proclaimed guerrilla wars, showed the changed nature of the world 'since the end of World War II'. The 'sustained armed struggles' prove there are 'new historical circumstances'. The revolutionary forces are stronger and the imperialist forces weaker than previously suspected. The United States would have invaded Cuba if not for Russia. It would have invaded northern Vietnam if not for China. Consequently

> the conditions under which the people of various countries conduct guerrilla warfare today need not be quite the same as those which were necessary in the days of the guerrilla warfare waged by the Chinese people against Japan. In other words, guerrilla war can be victoriously waged in a country which is not large in territory, as for instance, in Cuba, Algeria, Laos and southern Vietnam.[20]

Lin Piao and Che Guevara seemed to agree that (1) true revolutionaries are known by their willingness to wage such popular wars, that (2) these wars must originate from the countryside and that (3) these wars can only succeed against a powerful, active, counter-revolutionary America when the wars are launched and sustained in many parts of Asia, Africa and Latin America so that U.S. forces must be divided, weakened, rendered defeatable.

But why is it so clear that urban slums cannot provide a basis for revolt? Won't the lack of a racially and culturally homogenous population in Latin America raise fundamentally new problems for a war of *national* liberation? If American counter-insurgency as a matter of policy prevents Mao-style guerrillas from a relatively tranquil organizing far from market oriented enclaves, then may it not be best to confront the imperial beast directly in his lair to call the nationalist credentials of the ruling oligarchy into question? Why is it obvious that the army can not be suborned? What is so wrong about choosing to try to win on the cheap through a coup? Will the thousands upon thousands of young people whose blood is needed join the rural movement until other courses fail? The revolution succeeded on the cheap in Cuba. Protracted popular war as fought by the Chinese—and as Che wanted to lead in Bolivia—brings death to millions, literally millions, of human beings. Mao could begin with a force of thousands of soldiers only because for prior decades others had attempted other kinds of coups and insurrections and the government had responded by recruiting and disbanding large armies. Revolution is usually a long, tortuous, bloody struggle. Those who were brought up on the myth of American troops walking befuddled into an empty Vietnamese village and who therefore believe that clever revolutionaries can run away to fight another day miss the point. Revolutionaries win because they run forward and in hand to hand combat kill the invaders. Revolutionary ambushes succeed because some people sacrifice their lives to entrap the enemy. The revolution continues and grows because more people run forward to replace the dead. Che never sufficiently answers the question of what will make people willing to fight and die. Revolution will succeed in Bolivia if others, many others, succeed Che as in China Hung Hsiu-ch'uan, the White Wolf, P'eng P'ai and their hundreds of thousands of supporters preceded Mao.

As the Chinese endlessly state, revolution cannot be exported. If a popular liberation struggle must take a national form, then only Bolivians can lead and make Bolivia's revolutionary war. Of course they should take whatever outside help they can get. It is natural that Bolivian revolutionaries try to apply the lessons of China, Algeria, Vietnam, Cuba. Yet it is necessary to innovate to succeed. Revolution is a creative act. As Mao put it in 1936,

> since each country or nation, especially a large country or nation, has its own characteristics, the laws of war for each country or nation also have their own characteristics, and here, too, those applying to one cannot be mechanically transferred to the other . . . [We] must oppose a mechanical approach to the problem of war.
> . . . All the laws for directing war develop as history develops and as war develops; nothing is changeless. [21]

Why defend Che's way for Bolivia, or for that matter Mao's? Fidel found a solution for Cuba. Ho Chi Minh led the Vietnamese. Che's running to the Congo or Bolivia for a national liberation struggle to lead is not the answer. Bolivians in terrible struggle and at monstrous costs must devise their own means to end the death-like conditions that make the risks of revolution preferable to the inhumanity of the status quo.

Notes

1. *Peking Review*, June 6, 1969, 20-3.
2. *Revolution in the Revolution?*, New York: Monthly Review Press, 1967, p. 19.
3. *Ibid.*, p. 89.
4. José Moreno, 'Che Guevara on Guerrilla Warfare: Doctrine, Practice and Evaluation', *CSSH*, 12: 2 (April, 1970), 124, 127.
5. *Ibid.*, 132.
6. *Ibid.*, 133.
7. *Ibid.*, 130.
8. *Ibid.*, 119.
9. Stuart Schram, *The Political Thought of Mao Tse-tung*, New York: Frederick A. Praeger, 1963.
10. Debray, p. 6l.
11. *Idem.*
12. *Idem.*
13. *Idem,*
14. *Idem.*

15. *Selected Military Writings of Mao Tse-tung,* Peking: Foreign Languages Press, 1966, p. 173. Debray's other distinguishing characteristic for China is the 'absence of airborne enemy troops'. (*idem*) The experience of Vietnam makes one wonder if this is a vital, unique circumstance.
16. *Ibid.,* pp. 65-6.
17. *Ibid.,* p. 13.
18. *Ibid.,* pp. 19-20.
19. *Ibid.,* p. 19. Nonetheless, at the founding of the People's Republic of China in 1949 the leadership was pessimistic about popular revolutionary prospects in Asia. With the exception of Indochina, potential rebels seemed weak and isolated.
20. *Ibid.,* p. 186. Nonetheless by 1963 a reassessment of the growing optimism of 1960-62 was called for. The Sino-Soviet split, India's break with China, the economic miracles in Europe and Japan with the consequent reincorporation of former colonies into neo-colonial dependencies, renewed American intervention in Southeast Asia, Cuba's dependence on the U.S.S.R., all made for a diplomacy of national maneuver rather than revolutionary change.
21. *Ibid.,* p. 80.

Political Mobilisation in Guinea-Bissau
by Lars Rudebeck

> We became so tired of the Portuguese . . . tired of being used and exploited . . . it is no good to work every day without getting anything in return, so we decided there had to be a change. Now we are working for ourselves and for our own interests. Therefore we don't get tired any more, in spite of difficulties and hard conditions.
>
> Pungana Nabila, *vice-president of a village committee in south western Guinea-Bissau*, November 1970.

In a recent theoretical essay I tried to demonstrate, among other things, the crucial importance of politics in any effort to overcome the underdevelopment of our own historical period; and I hope I have also shown how different kinds of political systems may be assumed to interact with various kinds of development strategies.[1] The most hopeful combination is where a dynamic political interaction is established between the developmental needs and asiprations of the masses, and a consciously applied strategy of challenge to the social, political, and economic *status quo* of underdevelopment. It is surely an important task of political science to attempt to define the conditions under which such a mobilising interaction may be established and sustained. They are probably minimal conditions of development for any Third World nation in the predominantly capitalist international system.

This article attempts to present some concrete facts about an important experiment in political mobilisation for development in one particular country. It is my hope that this study of political, administrative, and judicial innovations in the liberated areas of Guinea-Bissau will also improve our understanding of some of the more general political principles and mechanisms involved in development.

An Historical Perspective

In a very general way it is probably correct to say that skilful political leaders, in colonised and underdeveloped countries, have usually been able to release the historical process of decolonisation, whenever contradictions have grown to become intolerable between (i) the structure, institutions, and organisation of society, and (ii) the people's claim for dignity, self-respect, daily bread, a house to live in, a future for their

432 children, and a life of peace and security. But, precisely for this reason, it is also easy to understand that the process is unlikely to come to an end at the moment the formal powers of the colonial government are transferred to the dominating social stratum in the former colony. On the contrary, the pressure for change will continue to be strong, as long as holders of national and international power, and structures and institutions of a neo-colonial character, stand in the way of the resolution of the kinds of contradictions just mentioned. Viewed in this way, the process of decolonisation is just one important phase of a greater, anti-imperialist, historical process whereby the peoples and countries of the Third World strive to achieve development—in the sense of putting their own human and material resources to use for their own purposes.

If this view is accepted, it follows that any colonial power which obstinately resists formal political decolonisation, once the claim has been seriously raised, in a sense only accelerates the process of political development by provoking a higher degree of political mobilisation than would otherwise have been needed. The less there is of political organisation, and of mass consciousness and commitment, in order to achieve independence, the more there remains of the political struggle against underdevelopment afterwards, and *vice versa*. This is one perspective in which events in Guinea—Bissau since the end of the 1950s take on a significance that far surpasses the boundaries of this small country colonised by Portugal more than 500 years ago.

Guinea-Bissau is a West African peasant country covering about 14,000 square miles and inhabited by about 800,000 people. Historically and politically it is linked with the Cape Verde islands in the Atlantic. Approximately two-thirds of the mainland, and about half of the population, are today under the effective political and military control of the national liberation movement, *Partido Africano da Independencia da Guiné e Cabo Verde*.

This is not the place to tell the story of how the P.A.I.G.C. was founded in 1956 by a small group of comrades under the leadership of the present Secretary-General of the party, Amílcar Cabral, and how armed insurrection started in 1963 following several futile attempts to reach an agreement with the Portuguese by peaceful means.[2] Suffice it to say that the children in the P.A.I.G.C.-schools, hidden away in the woods, give a correct

picture of the situation in the liberated areas, when they sing the
following song about the distribution of power between the
modern colonial army and the people in Guinea:

> Up in the clouds the Portuguese hold power
> but on the ground the guerrilla decide...

This article is about the foundations and exercise of political
power 'on the ground' in the liberated areas of Guinea-Bissau.

Development and Politics in Guinea-Bissau

If underdevelopment in the Third World is to be conquered,
it is reasonable to assume that the underprivileged majority of the
people of each country must be mobilised into productive activity
through development policies which systematically take, as their
point of departure, the people's self-experienced and concrete
interests of a better life. For this to be possible, political confidence has to be established between the masses and their
leaders. Political power, in other words, has to be made
dependent upon the interests of the masses in overcoming underdevelopment, rather than upon various vested interests in
slow change or no change at all.

The beginnings of such a dymamic political process of
development are clearly discernible in Guinea-Bissau. The
following extract from a message to the party cadres by Amílcar
Cabral in 1965 demonstrates how clearly the mechanism of
political mobilisation is realised by the P.A.I.G.C. leaders:

> Remember always that the people do not fight for ideas, for
> things that only exist in the heads of individuals. The people
> fight and they accept the necessary sacrifices. But they do it
> in order to gain material advantages, to live in peace and to
> improve their lives, to experience progress, and to be able to
> guarantee a future to their children. National liberation, the
> struggle against colonialism, working for peace and progress,
> independence—all these will be empty words without
> significance for the people, unless they are translated into
> real improvements of the conditions of life. It is useless to
> liberate a region, if the people of the region are then left
> without the elementary necessities of life.[3]

An interesting, practical demonstration of this is the system of distribution introduced by the P.A.I.G.C. in the liberated areas, where the monetary system has been temporarily abolished, in order to break the dependence upon the Portuguese. Sixteen *armazéns do povo*, or 'people's stores', have been established, where rice and other local produce may be exchanged for such necessities as cloth, needles, sandals, and soap. The exchange value of, for example, cloth in terms of rice is established by the central party leaders upon the recommendation of the local cadres. This applied 'value theory' might be said to be derived from a kind of 'philosophy of natural rights'. Its basic principle is political, and its purpose is to satisfy elementary material needs in communities with extremely scarce resources, and in such a way that the people are encouraged to produce more.

The principle stated by Cabral is equally relevant, both to the direct war against colonial domination and to the more general struggle against underdevelopment, of which the former is only a part. In both cases, a key problem is the establishment of a political and organisational link between the developmental needs and aspirations of the people, and the systematic action of their leaders. In Guinea-Bissau this link is provided by the P.A.I.G.C.—it is supported by a judicial system based upon People's Courts in the villages, and is fortified through political education.

The Party as a State

In a recent book collectively written by leading members of the P.A.I.G.C., we find the following statement:

> By providing the people of Guinea with political, social, and economic institutions in accordance with the spirit of their traditions, and by facilitating progressive measures for their emancipation, and also by maintaining the control and administration of the liberated areas, the *party* P.A.I.G.C. has today established itself as a *state*.[4]

Amílcar Cabral made the same point in his October 1970 report, when he emphasised that the role of the party in Guinea-Bissau was becoming more and more that of a government ('direction d'un état') whose territory is partially occupied by foreign troops.[5] According to Cabral, the days when the P.A.I.G.C. was just a rebel movement had thus passed long ago. It is easy to confirm this opinion after having spent some time in the liberated areas of the country. There can be no doubt that the P.A.I.G.C. today is a revolutionary movement building a new society with broad popular support, and a small but well-organised people's army.

How does it look, then, in practice, when the 'party' functions as a 'state'? If it is true that one party legitimately represents the people and manages the affairs of the whole society, then it is obviously difficult to maintain a clear distinction between citizen and party member. Unless we think in terms of a superimposed elite party, the natural tendency in such a situation will be in the direction of some kind of fairly open mass party. This is also, very characteristically, the case in Guinea-Bissau at present. Membership is not a clearly defined concept' was the typical answer received when I discussed this problem with different persons within the P.A.I.G.C. Membership is rather something that comes 'naturally', through participation in the struggle for national liberation and development. The basic precondition for any work carried on by the party is the broad popular support which it enjoys; but this has to be maintained and reinforced continually through concrete action. From the sympathetic villager upwards there is a gliding scale of gradually-more-active membership to the totally-committed cadres and leaders who devote all their time and energy to the party. But it is impossible to state exactly where on this imaginary scale that 'membership' begins. On the other hand, people active in the P.A.I.G.C. sometimes talk about 'the party within the party'. This expression refers to the backbone of the emerging state, i.e. to all persons with special responsibilities and tasks, from the local village committees to the full-time cadres of the militia and the army.

What has been said so far refers explicitly to the *present* situation in the liberated areas of Guinea-Bissau. It is quite possible—and there is an awareness of this within the P.A.I.G.C.—that other solutions may have to be sought in the future. One possibility in a totally decolonised and independent Guinea-Bissau could be a growing division of functions between (i) the people in general, (ii) the party, with some minimal ideological criteria for membership, and (iii) the state administration. A tendency of this kind clearly could entail some new organisational problems of democracy which only exist in an embryonic form so far. But for the time being there does not seem to be any strong pressure in such a direction. On the contrary, the emerging state is closely identified with the mass party.

'Flexibility' and 'adjustment to the shifting conditions of the struggle' are key expressions in the organisational vocabulary of the P.A.I.G.C. In agreement with these principles, the general framework is also characterised by a combination of continuity and frequent minor adjustments. Certainly the structures seem to be more permanent than the terms employed to label them. As a result, although details change from time to time and even from place to place, the following description of the situation in 1970 provides a valid picture of the basic structural and political characteristics of the P.A.I.G.C. today.

Important changes in the political and administrative organisation of the party and the country were decided upon early in 1970. The formal decision was taken by a body of 30–50 leading members (the 'enlarged Political Bureau') meeting on 12–15 April. Their main purpose was to clarify the division of labour and responsibility at the higher organizational levels, which appear to be more difficult to handle than the basic party structures.

1. The Central Organs

The central organs of the national liberation movement are now identical for the army, the party, and the civil administration. At the highest formal level is the Party Congress, which so far has only met once—in 1964, in the southern part of the country. Nothing has been decided yet about the date of the next Congress; and the offical reason for this is understakable, that the risks involved in gathering simultaneously all active leaders of the movement at the same place, under conditions of war, would be too great.

Next in the P.A.I.G.C. hierachy is the *Conselho Superior da Luta*, or 'The Highest Council of the Struggle', with a fairly broad membership of about 70. This is roughly the equivalent of the Central Committee, which existed before the 1970 reorganisation. About one-third of its members comprise the *Comité Executivo da Luta*, in which are also found the members

Political Mobilisation in Guinea-Bissau

The Political and Administrative Organization of the P.A.I.G.C. in the Liberated Areas of Guinea-Bissau, 1970-71

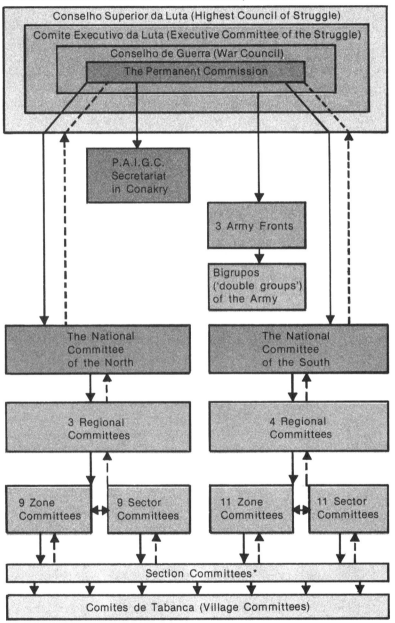

* Administrative four-member liaison committees functioning in many parts of the country, but probably not everywhere in early 1972.

of the seven-man war council, the *Conselho de Guerra*. The 'Executive Committee of the Struggle' is the nearest equivalent of what was previously called the 'enlarged Political Bureau'. The original 15-man Political Bureau has no place in the new organisational structure; but, in spite of this, the old terminology is still frequently used.

The day-by-day leadership of the P.A.I.G.C. is the responsibility of a powerful three-man executive, called the *Comissão Permanente*: Amílcar Cabral, the Secretary-General of the party, is responsible for the political and military leadership of the national liberation movement; Aristides Pereira is responsible for economy and security questions; and Luis Cabral is responsible for 'national reconstruction', which includes education, health, and other civilian tasks in building a new society. All three are also members of the *Conselho de Guerra*.

The secretariat of the party is located in Conakry, capital of the Republic of Guinea, and is under the direct supervision of the three-man *Comissão Permanente*. But these key members themselves carry on a great deal of their activities within the liberated areas of Guinea-Bissau.

It seems clear that the central organs of the party do not meet very regularly, and that even then not all members always manage to be present. This can be understood against the background of the difficult conditions imposed by the war upon all P.A.I.G.C. activities; but in the long run it also implies an undeniable risk of organizational top-heaviness. At present, however, possible tendencies in that direction appear to be counterbalanced by the systematic emphasis given to the concrete tasks of political mobilisation within the country. These are thus the key local levels which we have to observe, in order to understand how the 'party as a state' functions.

2. North and South: Regions and Sectors

The liberated areas of Guinea-Bissau have been administratively divided into the North and the South. These two main divisions are each headed by a twelve man 'National Committee'. The division of labour within each is the same as in the Permanent Commission, but with the important addition that the general political leadership is also responsible for production at this level. This is a 1970 innovation, and demonstrates the *political* importance attached to economic output and material improvement by the P.A.I.G.C. leaders.

The North and the South, in their turn both have four Regions, each further divided into Sectors or Zones (the terminology is not fixed), within which are found a large number of *tabancas*, or villages. Altogether in 1970 there were 20 Sectors, two to three in each Region. In the North, in addition to the nine Sectors, there are also areas which are not entirely liberated and therefore not yet included in the civilian organisation, but administered by the armed forces.

The political and administrative work in each Region is directed by a three-man Regional Committee with the same functional division of labour as in the two National Committees. If the need to do so is felt, any Regional Committee may also call in representatives of their Sectors for consultation and information.

Since no Party Congress has been held since 1964, the membership of the three levels dealt with thus far—at the centre, in the North/South National Committees, and in the seven Regions—are renewed by appointment and co-optation rather than by election. It is significant, for instance, that the 1970 reorganisation was decided by a body more or less equivalent to the present *Comité Executivo da Luta*. This does not mean that these central organs are not representative in a more general sense—they probably are—but it does mean that at present there is no institutionalised, formal procedure by which the people directly select their national leaders.[6] On the other hand, concrete formulation of policies that concern the people directly is carried on in close contact with the basic levels of the party.

The structure of the organisation in the Sector differs from that encountered at the higher levels. Within each of the 20 Sectors we find two distinct committees working together with each other. Both of these usually have five members, but the first is made up of 'ordinary people', each supposed to represent about one-fifth of the villages within the Sector, whereas the second committee consists of party-appointed functionaries. Since the introduction of the 1970 official terminology, the former should properly be called a Sector Committee and the latter a Zone Committee; but in daily speech it is still usually the other way around. The five appointed functionaries have clearly defined tasks: the Political Commissar is also in charge of supervising and directing the sector's economic production, and the other four have specific responsibility for security, the militia, education, and health. Their crucial task is to provide the focus for, or—to put it another way—the direct link between, the general policy aims of the party and the concrete needs of the people. This is naturally why the attempt has been made to introduce an element of direct representation at this level, by the creation of a parallel Sector Committee made up of village representatives.

3. The Village Committee

The basic unit, finally, in this political and administrative structure, is the *Comité de Tabanca*, or Village Committee. This is a popularly elected five-man committee that the P.A.I.G.C. attempts to establish in all liberated villages. The number of villages in an agricultural peasant country such as Guinea-Bissau is, of course, very great—in one of the Sectors I visited there were 44, for instance, and in another 23, with populations varying from about 100 to 300, or perhaps even more. It is impossible to know if these Committees really function actively everywhere. It is quite probable that they do not, and the P.A.I.G.C. leaders themselves readily concede that their political work is much easier in some areas than in others, depending upon the different cultural and social characteristics of the various ethnic groups.[7] Above all, it is a time-consuming job to mobilise previously quite isolated people into understanding the connection between colonialism and underdevelopment, and, consequently, the connection between national liberation and chances for development of one's own country. All I can verify, in the very strictest sense, is that in the villages I had a chance to visit (during several marches back and forth in the south and south-west), I was able to observe the existence of these Committees, organised and conscious of their tasks. But the success of the party in the political and military struggle against the Portuguese colonial Government, as well as the testimony of other visitors, gives us good reason to believe that the number of effective Village Committees in Guinea-Bissau is considerable.

The five members of each Committee have clearly defined functions which reflect, in a most revealing manner, the essential tasks of the national liberation movement. The *president* directs

and leads the activities of the village in general, including
agricultural production—naturally the most basic activity. At
his side stands the *vice-president*, with specific responsibility for
both security and the local militia—necessary to defend the
village in case of Portuguese attacks by, for example, helicopter-
borne troops. The *third member* is in charge of education, social
questions, health, and child care. The *fourth member* is
responsible for the organization of the village's deliveries to the
party and the army, and for the care of visiting soldiers or party
members who stop for rest during their marches. The *fifth
member* is in charge of the civil register, which includes data in
connection with the census planned and already partly carried
out by the P.A.I.G.C.[8] It should be noted that, according to the
party rules, at least two of the five members of these Village
Committees have to be women. This appears to be followed quite
scrupulously, and is consistent with other systematic efforts to
improve the traditionally subdued position of women.

It appears that once elected—by more-or-less unanimous
acclamation at a general meeting of the inhabitants—the
Village Committee remains in office until some concrete reason
for dissatisfaction arises. Thus the local leaders do not have to
stand for re-election at regular intervals. These villages have a
different kind of social structure from that which historically has
given rise to the elaborate procedures of liberal representative
democracy in the western world. The main point is that the
action of the Committee members becomes impossible *unless* they
are held in trust and confidence by their fellow villagers. This is
the mechanism through which local representativeness is
maintained.

4. Democratic Centralism

The P.A.I.G.C. leaders have tried to create a party structure
based on the principles of so-called 'democratic centralism'. At
each level of organisation the members and officials are held
responsible for specific tasks, which they are expected to carry
out on their own responsibility, and without much supervision.
At the same time, the population as a whole is expected—by
consultation and discussion—to help formulate the policy
decisions later made by the top-level leaders. [9]

It is, of course, easy to express doubts as to the effectiveness of the influence exercised by the people in this kind of political sturcture. 'Democratic centralism' is a term which has often been abused. The fact is, however, that policy execution and application in a rural country with difficult communications—such as Guinea-Bissau at present—necessarily will have to be quite decentralised. For this reason, among others, the leading policy- and decision-makers depend very much upon the free consent of the people and especially the local cadres for the *application* of their policies. It can thus be argued that democratic centralism is real to the extent that each of the two main components of this concept depends, for its own material existence, upon the existence of the other.

5. The Military Organization

The struggle for national liberation has also made a military organisation necessary. Those active within the P.A.I.G.C. like to say that they are not 'military people' but 'armed militants'.[10] It is quite clear that this is not a mere pun, but a good way of characterising a movement where (i) conscious application of the principle of rotation, (ii) decentralised responsibility for concrete action, and (iii) an egalitarian spirit—e.g. small material differences, and no military grades—all demonstrate a willingness to counteract tendencies in the direction of hierarchy and militarism. If this willingness will suffice in the future is of course a different question; but the fact is that it exists, both in theory and practice, under present conditions.

Nevertheless, for obvious practical reasons, it has proved advantageous to introduce a distinction between the military and civilian organisations at the level below that of the unified centre. Whereas the *regular army* needs to be mobile and able to operate effectively all over the country, the *representative and administrative organs* naturally have to be locally based. Similarly, the *militia* belongs on the civilian side of the organisational structure, under the local direction of the Sector Committees and of the vice-presidents of the villages, as already noted. But the distinction is always difficult to maintain in Guinea-Bissau, where the military and the socio-economic-political efforts are consistently regarded as merely different aspects of the same struggle.

The regular army is simply organised into three fronts, each under the leadership of a Commander, in his turn directly responsible to the *Conselho de Guerra*. The basic unit of the army is the *bigrupo*, or 'double group' of 32 men, operating with a great deal of autonomy under the dual leadership of one military and one political leader—*chefe militar* and *comissário político*. A realistic but very approximate estimate of the size of this mobile and moderately well-equipped army would be about 5,000 men. The Portuguese colonial army, which the P.A.I.G.C. seem to be driving out of their country, probably totals about 40,000.[11] The substantial disadvantage in numbers is more than made up for by the massive differences in political strength.

A New Judicial System

The legal framework of Portuguese colonial domination in Guinea and the Cape Verde islands has been thoroughly analysed by Amílcar Cabral himself. Some of his conclusions are worth quoting, since they follow logically from the facts:

> What has now been said demonstrates that the constitutional, political, legal, administrative, and judicial situation of 'Portuguese' Guinea is very far from being the situation of a 'province of Portugal'. It is, on the contrary, that of a non-autonomous country, conquered by force of arms, and dominated and administered by a foreign power. The economic, political and social life of the people of 'Portuguese' Guinea is subjected to laws and norms which differ from those applied to the people of Portugal. The people of Guinea have no political rights, do not take part in the functioning of their own country's institutions or in the elaboration of the legal texts they are obliged to respect and follow; they do not elect and can not appoint or dismiss the political and administrative leaders of their country; they do not enjoy either the most elementary of the rights of man or the fundamental liberties. The people of 'Portuguese' Guinea are therefore far from having a juridical personality of their own. They are, on the contrary, a colonised people, dependent and profoundly affected in their human dignity. They do not have the power to determine, either directly or indirectly, their own present or future destiny. It is thus an incontestable fact that the people

of 'Portuguese' Guinea have been deprived of their right to self-determination, a right proclaimed and established, for all peoples, in the Charter of the United Nations.[12]

Consequently, when the construction of a new society was begun in the liberated areas, there was not much the P.A.I.G.C. was prepared to take over from the Portuguese system with its special legislation and courts for the 'natives'—over 99 per cent of the population. It was necessary to start from scratch. After a period of pragmatic improvisation, a radical and systematic experiment was initiated in 1968.[13]

Originally the guerrilla leaders had the authority to settle both military and civilian disputes in their liberated areas. It soon became clear, however, that such a concentration of power was not desirable, and during 1965–6 the authority to settle civilian disputes was transferred to the Political Commissars of the Sector Committees. In 1966 the important step was also taken of promulgating the *Lei da Justiça Militar*—the first and, so far, only legal text adopted and published by the P.A.I.G.C. in its role of 'party as a state'.[14] This deals with the constitution of the courts and the formal legal processes. Various types of crimes and penalties are specified; the sentence of death is generally reserved for those guilty of, for example, espionage and the rape of minors, but it may also be applied to murderers. This law defines only the principles by which the most serious crimes are to be judged. It functions also, however, as a frame of reference and a guide for those charged with the task of judging minor offences according to traditional customary law.

In 1968 it was agreed to construct a new, revolutionary judicial system, the foundation of which is a series of popularly elected courts in the villages. The decisions of these People's Courts are registered by village teachers acting as clerks; special forms have been printed for this purpose, and are eventually filed in the party headquarters in Conakry. The hope is that gradually a codified synthesis will emerge of traditional customary law and modern principles of justice. But so far—it is said with some exaggeration, perhaps—there are almost as many codes of justice as there are People's Courts, although they all have to fit within the general framework established by the *Lei da Justiça Militar* and by party policy.

The People's Court has three members, elected in the same

way, and remaining in office according to the same principle, as the five members of the Village Committee. Their jurisdiction is limited to offences that cannot be punished by jail (corporal punishment is strictly prohibited), which means that they deal with thefts of, for example, cattle and rice, disputes between families, minor violence, and troublemaking. In cases of dispute, efforts are made to arrive at a voluntary reconciliation. The punishments that can be meted out by the People's Courts are limited to fines and repayments in kind. The party also makes a systematic effort to imbue the judges with a 'revolutionary spirit', which means that they are encouraged to be conscious of their political and educational responsibilities when admonishing any culprits.

Naturally it is not always easy to achieve the desired synthesis between traditional and modern ideas. Conflicts arise, for instance, over the customs and rules regulating marriage and family life. To give a typical example, a new rule being enforced by the P.A.I.G.C. states that no girl or woman can marry until she has notified the party that she does so voluntarily. The formal act of marriage is carried out by a party official, and both the husband and the wife can obtain a divorce by presenting reasons for such a step to the People's Court. Things like these are radical innovations in the traditional peasant society of Guinea-Bissau.

At the next level of the new judicial system we find a Zone Court with five members—a kind of combined regional and sector body, partially changing its membership depending upon where it is called to sit. This is designed to function, both as a court of appeal for people dissatisfied with village decisions (it appears, however, that such appeals are rare), and as a court with autonomous jurisdiction with regard to certain kinds of more serious cases.

The Zone Court is presided over by either the official in charge of politics and production at the *regional* level, or the official in charge of *regional* security. In the former case, the second member is the official in charge of security in the Sector where the court meets; but the Political Commissar of the Sector joins the court, if it is presided over by the regional security chief. The third member is a personal representative of the party's Secretary of Justice, and the remaining two members represent the population of the Sector. If the defendant is a soldier, one of the two people's representatives should also be a soldier.

The jurisdiction of the Zone Courts covers crime too severe to be judged in the villages, but punishable by less than four years in jail. This means mostly crimes of property and violence of medium gravity. My own impression—and it is only an impression—is that the Zone Courts have very few cases to judge, but that the People's Courts fulfil important social and political functions by resolving conflicts and introducing modern democratic authority into the villages.

The Supreme Court of the liberated areas of Guinea-Bissau is the *Tribunal de Guerra*—the War Tribunal, instituted by the *Lei da Justiça Militar* of 1966. This Court has five members: the military Commander of one of the three fronts, the official in charge of either security or political leadership in one of the two National Committees, the party's Secretary of Justice, and two representatives of the population. If the defendant is a soldier, one of the two people's representatives will also be a soldier, according to the same principal as in the Zone Courts. The composition of the Supreme Court also varies with the place where the crime has been committed. If this is in the North, the Commander of the Northern Front and a representative of the National Committee of the North act as judges; whereas the Commander of the Southern Front and a representative of the National Committee of the South join the Court, if the crime was committed in the South. The *Lei da Justiça Militar* also states that the Court, if need be, may be presided over by the Secretary General of the party.

As far as the jurisdiction of the *Tribunal de Guerra* is concerned, it functions (i) as a general court for offences committed by members of the armed forces and by party cadres, (ii) as a court of appeal from the Zone Courts, and (iii) as a court with autonomous jurisdiction for civilian crimes punishable by four years imprisonment or more.

This new judicial system has been gradually introduced in the liberated areas since 1968 by the P.A.I.G.C. in a systematic effort to create order in the emerging nation state, without violating either the principle of popular consent or the chances for dynamic social change. It should not be necessary to emphasise the difficulties of such political and developmental tasks, but the revolutionary experiment attempted in Guinea-Bissau indicates that they need not perhaps be altogether impossible.

Political Education

Education is a fundamental aspect of the entire P.A.I.G.C. programme, and great efforts have been made with extremely limited means. Village schools with attentive children, hidden in the woods to reduce the risk of attack by air, are an important part of the social infrastructure of the liberated areas. [15] It is obvious that the political implications of these educational developments are considerable, and it would be feasible to write a worth-while article about this subject alone. [16] My intention here, however, is only to present some facts about an unusual experiment in *direct* political education carried on at a party school located near the south-eastern border between the Republic of Guinea and Guinea-Bissau. This is called the *Centro de Instrução Politico Militar* (Centre for Political and Military Education), and its purpose is to provide the P.A.I.G.C. with competent cadres at the crucial lower levels of their organisation. The school has thus direct relevance for the functioning of the political and administrative structures analysed in this article.

At the time of my visit in November 1970, the primary purpose of the C.I.P.M., as defined by the director, [17] was twofold: (i) to give political and military training to the soldiers of the army and to the Political Commissars working in the liberated areas, particularly in the *bigrupos* of the army, and (ii) to give courses of introduction and readaptation to the realities of the present situation in Guinea to persons returning from abroad to join the struggle. A more general formulation of purpose is found in the internal statutes of the school, which contains the following:

> *The C.I.P.M. is above all a political school* (my italics). Its fundamental purpose is to train conscious party militants and fighting members, determined to struggle until the victory of our just cause of independence and progress for our people has been reached. The activities of our Centre may be summarised in three main points: 1, Political preparation. 2, Military preparation. 3, Alphabetisation. [18]

After a recent reorganisation, completed during the first half of 1970, the C.I.P.M. is now able to receive up to 300 students of varying age, background and experience—from boys to adults, and from illiterates to persons with university training. The students are organised in groups of 25, with autonomous responsibility for carrying out various practical tasks under the direction of a leader selected among themselves. The maximum number of teachers is said to be about 30. The length of a period spent at the school varies with the background of the individual concerned, but the normal period of training appears to be a few months, and in any case always less than one year.

The general programme of political studies covers a wide field of knowledge, ranging from the historical and present facts of colonial domination in Guinea and Cape Verde, to the moral code of behaviour that should guide the party militant. Important points in this systematically elaborated syllabus include: the party organisation and programme; the problems of regionalism, tribalism, general ignorance, and technical backwardness; the aims, strategy, and tactics of the enemy; the relations between the party and the people; the role in Africa of the national liberation movement in Guinea and Cape Verde; the historical process of decolonisation; the present international situation, and the role of the imperialists, the socialist countries, and the Third World.

The full implications of this kind of training have not yet made themselves felt. But the mere existence of such a highly organised institution, far away in the bush, is a remarkable fact in itself. It must be emphasised that it is of strategic importance for a mobilisation-oriented movement, such as the P.A.I.G.C., to be able to fill exactly the kind of lacuna which the graduates of the C.I.P.M. are expected to fill. A *sine qua non* of enduring political mobilisation is the existence of capable and competent cadres, able to identify with the people and thus to evoke their confidence, and at the same time aware of the broader perspectives defining the conditions of their own political work.

Conclusion

This article has dealt with a particular case of political mobilisation for development, and it is of course possible to point to some special reasons for the relative success so far registered here. The most important explanations are probably the small size of Guinea-Bissau; the lack of any ethnic group heavily dominating the others; remarkably skilful political leadership;

and the smaller economic importance to Portugal—as well as to the international capitalist system in general—of Guinea-Bissau as compared with Angola and Moçambique. In a future independent Guinea-Bissau, it is also possible to forsee real difficulties in maintaining the present degree of political mobilisation, social egalitarianism, and absence of bureaucracy. The concrete tasks of the direct struggle for national liberation will eventually give place to the less harsh, but in some respects perhaps more subtle and complicated, tasks of peaceful economic and social development.

But there are always particular factors involved in particular cases, and none of them reduce the significance of social facts and processes that are of general importance in themselves. The emphasis in this study has been on the description and analysis of some such particular facts and processes. Their general importance lies in the fact that they demonstrate in practice the possibility of an adequate political response to underdevelopment in at least one type of concrete situation.

Notes

1. Lars Rudebeck, 'Political Development: towards a coherent and relevant theoretical formulation of the concept', in *Scandinavian Political Studies* (Oslo), V, 1970, pp. 21–63; and 'Developmental Pressure and Political Limits: a Tunisian example', in *The Journal of Modern African Studies* (Cambridge), VII, 2, July 1970, pp. 173–98.
2. For the most comprehensive and penetrating account, see Basil Davidson, *The Liberation of Guiné: aspects of an African revolution* (London, 1969). Another useful study is by Gérard Chaliand, *Lutte armée en Afrique* (Paris, 1967).
3. Translated from *Palavras de ordem gerais do comarada Amílcar Cabral aos responsáveis do partido* (Conakry, 1969), p. 23.
4. Translated from *Guinée et Cap-Vert. Libération des colonies portugaises* (Alger, 1970), p. 24.
5. Translated from Amílcar Cabral, *Sur la situation de notre lutte armée de libération nationale, janvier--septembre 1970* (Conakry, 1970, mimeo), p. 7.
6. It should be noted, though, that the *Consalho Superior da Luta* decided in August 1971 to create 'the first National Assembly of the People of Guinea.' See Amílcar Cabral, *Rapport bref sur la situation de la lutte: janvier--août 1971* (Conakry, 1971, mimeo.), p. 17.
7. In a message from Amílcar Cabral to the people on the occasion

of the 14th anniversary of the foundation of the P.A.I.G.C., the need is emphasised to 'improve our political work and develop the participation of the people' (Conakry, 1970, mimeo.), p. 14. Cf. also Cabral's 'Brief Analysis of the Social Structure in Guinea', first presented at a seminar in 1964, and later printed in English in a collection of his speeches and articles: *Revolution in Guinea, an African People's Struggle* (London, 1969), pp. 46—61.

8. Documents I saw in Guinea-Bissau included a simple form for the registration of village marriages. This is used by women and men to notify the party of their free intention to marry, and contains the following words in Portuguese: 'Marriage in the form and with the effects established by our Party, during its excercise of the sovereignty of the People of Guiné and Cabo Verde, of whom it is the legitimate representative.' The forms also note that persons who do not know how to read and write may use their finger-prints as a signature. These documents are remarkable evidence of the 'state-building' going on in the liberated areas.

9. The abstract principle is explained to the cadres by Cabral in *Palavras de ordem gerais*, pp. 32–3.

10. In *Guinée et Cap-Vert*, p. 139, it is stressed that the fighting men of the party are 'pas de militaires mais des militants armés'.

11. Cf. also Paul M. Whitaker, 'The Revolutions of "Portuguese Africa"', in *The Journal of Modern African Studies*, VIII, 1, April 1970, especially pp. 24–7.

12. This extract is from a document presented to the U.N. Spiecal Committee for the Territories Administered by Portugal, in June 1962. The present quotation has been translated from a version prepared for the International Conference for Solidarity with the Peoples of the Portuguese Colonies in Rome, 1970, by Cabral; *Sur les lois portugaises de domination coloniale* (Conakry, 1970, mimeo.), p. 28.

13. The following account is based upon oral information, primarily from the man in charge of working out the new judicial system, Fidelis C. Almada, himself a jurist and acting as the party's Secretary of Justice. Complementary information was received from members of several courts, as well as from a number of people not specifically concerned with the judicial system.

14. *Lei da Justiça Militar de 19 de Setembro de 1966. Com as modificações introduzidas pelo Bureau Politico do Partido, na reunião de 20 a 23 de Dezembro de 1966* (Conakry, 1966, mimeo.). This is published in Portuguese, the official language of the P.A.I.G.C. and of the future independent state of Guinea-Bissau.

15. According to *Guinée et Cap-Vert*, p. 81, the number of village schools was 153 in 1967, and is said to have increased by 1970 to 162.

16. The same is true of the ambitious health services, which deserves study both from a socio-political and medical point of view. With few educated doctors and nurses, and very limited supplies of medicines, the programme reaches far out into the villages.
17. A former officer of the Portuguese army by the name of Silvino Manuel da Luz.
18. Translated from *Regulamento de diciplina interna, Quembra, 21 de Maio de 1970* (Quembra, 1970, typescript).

Selections from
The Struggle for Mozambique
by Eduardo Mondlane

In many ways the period of preparation puts more strains on a movement than does the time of action. Once fighting begins, solidarity is generated in the face of immediate danger from the enemy. Also, the movement is able to prove itself: it can show concrete results for its work and a practical justification of its policy. As it proves itself, the enthusiasm and confidence of its own members grows, while at the same time outside interest and support increase. During the time of secret underground work, however, there is little to be seen of the party except a name, an office and a group of exiles who claim to be national leaders but whose integrity is always open to question. It is then that a movement is especially vulnerable to internal dissension and outside provocation.

For FRELIMO in its first two years, the potential danger was aggravated by the inexperience of its leaders in working together. Many of its members also lacked an understanding of modern politics. On the other hand, the problem of maintaining unity was eased by the fact that there were no other parties in existence. After the union of 1962, our problem was not one of bringing together major rival groups but of preventing factions from developing within.

The heterogeneous nature of the membership carried certain dangers as well as advantages. We came from all over Mozambique and from all walks of life: different language and ethnic groups were represented, different races, different religions, different social and political backgrounds. The occasions for possible conflict were unlimited, and we found that we had to make a conscious effort to preserve unity. The main form this took was education. From the very beginning we carried on political education to combat tribalism, racism and religious intolerance. Portuguese was retained as the official language mainly for convenience, as no African language has the wide currency in Mozambique that Swahili, for instance, has in Tanzania. Work, however, is also often carried on in other languages, and the fact that people from different areas are constantly working together has encouraged the learning of these. From the outset, military units were always very mixed in composition, and the experience of working together with people from other tribes did a great deal to lessen tribal friction. FRELIMO is an entirely secular body; within it all religions are tolerated, and a great variety practised.

There was, all the same, shortly after the formation of FRELIMO, a tendency for individuals to claim to represent Mozambique and form bogus, splinter organizations. This seemed mainly due to the conjunction of certain personal ambitions with the maneuvers of the Portuguese and other interests threatened by the liberation movement. Early on, COSERU (*Comité Secreto de Restauraçao da* UDENAMO) appeared, and then gave place to a New UDENAMO, which then split into New UDENAMO-Accra, and New UDENAMO-Cairo; both of these have since disappeared. In addition, there emerged a new UNAMI (now vanished), a new MANU, and yet more variations on the theme. The individuals forming these different organizations were often the same. Then, in 1964, a group called MORECO (Mozambican Revolutionary Council) was formed, which later changed into COREMO and almost immediately underwent further upheaval, as the Chairman and National President, the General Secretary and the Plenipotentiary Secretary, each expelled the other. There is now a branch of COREMO in Lusaka and a branch in Cairo which seem to be separated by ideological differences. COREMO-Lusaka has recently undergone another split, which resulted in the formation of yet another group called *União Nacional Africana da Rombézia*. UNAR's programme aims at weakening the work of FRELIMO in the area between the two main rivers of north Mozambique, the Zambezi and the Rovuma. At the most charitable estimate, the leaders of the group must be naïve to take seriously the rumours, fanned by the Portuguese, that they would be ready to hand over the northern third of the country to Malawi if, by that maneuver, they would be assured of perpetual control over the two-thirds of Mozambique from the Zambezi southwards. It is significant that the headquarters of UNAR is in Blantyre, and that the leaders enjoy the protection and cooperation of some influential figures in the Malawi Congress Party.

COREMO-Lusaka is the only one of these groups which has attempted any action inside Mozambique: in 1965, COREMO supporters started military action in Tete, but they were crushed immediately. It seemed that no groundwork had been done on which to base such an action; as a result of the repression which followed, some 6,000 people fled to Zambia, and the government of Zambia at first assumed that, as the action had been instigated

Selections from
The Struggle for Mozambique

by COREMO, these were COREMO supporters. After questioning them, however, they found that they had not heard of COREMO, and that those who were attached to any party were members of FRELIMO.

None of these splinter movements was, fortunately, serious enough to interfere with work inside Mozambique, as most of them consisted of just an office and a small group of exile supporters. At that time, however, when all FRELIMO had to show outsiders was a number of officers, there was a risk that they might do some damage beyond. The proliferation of small opposition groups was an embarrassment to the countries which offered support to liberation movements, since it was not easy to tell at first which groups had real backing within Mozambique.

Another difficulty, particularly acute in the early stages of development, when many of the movement's members know little about one another, is the danger of infiltration by Portuguese agents. And this is connected with the problem of splinter groups, since these may use a member of the main organization to try to spread dissent, so as to bring over a section of the membership. The complexity of motives behind divisive conduct makes it the more difficult to guard against: individual neuroses, personal ambitions, real ideological differences are muddled up with the tactics of the enemy secret service. A movement cannot afford to become too paranoiac, or it will alienate potential support and fail to reconcile those real differences that somehow must be reconciled if its broad basis is to survive and develop. On the other hand, it must guard against the more dangerous type of infiltration organized by its enemies, inevitably expending time and energy in the process.

The best answer to splinter groups, agents, spies, inflammatory propagandists, is a strong movement. If the leadership is united and is based on mass support in the country; if the programme is realistic and popular, then the damage which such outside efforts can achieve will be marginal. In FRELIMO, although in some instances specific action may have to be taken, our general policy is simply to press on with the main work in hand, ignoring petty provocations. . . .

A number of factors have contributed to the success of FRELIMO forces against the much larger and better equipped Portuguese army.

On the military front, the Portuguese face all the problems

of a regular army combating a guerrilla force and a foreign army of occupation fighting in hostile territory. First of all, only a small fraction of the armed forces can be used in action. The colonial government must employ large numbers to protect towns, economic interests, lines of communication, and to guard the population confined in 'protected villages'. Thus, out of the 65,000 Portuguese soldiers in Mozambique, only about 30,000 are used against our forces in Niassa and Cabo Delgado; and not even all these are free to engage in action against us, since many are pinned down defending strategic points and population centres in the area. Secondly, the Portuguese are fighting on unfamiliar terrain against an enemy which belongs to that terrain and knows it well. Much of the land in these northern provinces is heavily wooded, providing good cover for the guerrillas and their bases. Often the only means of penetrating the bush is along narrow footpaths, where a body of men must walk in single file, a sitting target for ambush. In such conditions, heavy equipment like aircraft and armoured vehicles is of little use.

The political aspect is of even greater importance, for the struggle is essentially a political struggle in which the military is only one aspect. To justify their presence, the Portuguese must affirm that their army is defending Mozambique against outside aggression. Yet such a posture is impossible to maintain persuasively, for the FRELIMO forces are, without exception, composed of Mozambicans, whereas the Portuguese army is almost entirely composed of Portuguese troops and numbers little more than one thousand puppet African soldiers among its ranks. When it does use African troops, these are surrounded by Portuguese soldiers to guard against desertion.

Then the people themselves are in the main overwhelming hostile to the Portuguese. To prevent them cooperating with FRELIMO, the Portuguese army organizes them into 'protected villages', villages surrounded by barbed wire and guarded by Portuguese troops, the counterpart of the resettlement centres set up by the French during the Algerian war, and the strategic hamlets of the Americans in Vietnam. This may temporarily cut off the villagers from FRELIMO; but it does nothing to reduce hostility to the Portuguese, and when an opportunity arises, the population of such 'protected villages' revolts.

The war is also creating internal problems for the Portuguese

Selections from
The Struggle for Mozambique

government. It is faced not only with the war in Mozambique but is fighting on two other fronts, Angola and Guinea Bissau. At the same time it has to keep forces of repression in São Tomé, Cabo Verde, Macau, Timor and Portugal as well, where opposition to fascism, although severely weakened by forty years of repression, has never been completely crushed. The government's resources of manpower and money are being stretched to breaking point by wars thousands of miles from home, wars for which the population is paying but from which, for the most part, it cannot hope to gain anything. This adds fuel to internal opposition and at the same time weakens the government's defences against it. . . .

Because the Portuguese troops are tied down in defending various settled strategic positions, the guerrilla forces always have the initiative in choosing the time and place to mount an attack. FRELIMO forces are fighting on their own ground, in a terrain they know, among a people who know and support them. A defeat for the Portuguese means that the struggle is pushed into a new area, and that as a result they have to bring up more combat troops there, weakening still further their overall position. A defeat for FRELIMO is more easily retrievable, as it involves only a temporary reduction of strength in one area.

Any progress in the war means very much more to FRELIMO than a mere gain in territory. The war has altered the whole internal structure of the areas deeply affected by it: in the liberated zones, the various systems of exploitation have been abolished, the heavy taxes have gone, the repressive administration has been destroyed; the people are free to cultivate their land as they need to, campaigns against illiteracy have been started, schools and health services have been established, and the people are involved in political debate, in making their own decisions. However embryonic these developments may be, the change has been felt in some way or another by almost all those living in the zones, and given them that much more reason to fight. Each zone freed in this way is a reservoir of new recruits for the fighting forces. In the villages popular militias are formed which at once confirm the power of the people and relieve the ordinary FRELIMO forces of many defence tasks; by cooperation with the army, they also extend FRELIMO's overall offensive capacity.

The FRELIMO army and the population are closely linked;

the people are a constant source of information and supply for FRELIMO, while they are a further source of danger to the Portuguese. FRELIMO's forces live for the most part off what they produce in the fighting areas, and what has to be transported is taken on foot through the bush between the small centres which have been established. As a result, FRELIMO has no vulnerable supply lines, no military or economic strategic positions to defend. The loss of a single small base or area of crops is not very serious; it has no significance beyond the immediate loss of resources.

The longer that the struggle lasts, the more evident becomes its popular basis, the more support flows to FRELIMO, and the more confidence there is in FRELIMO's ability to succeed, while the less confidence the allies of Portugal have in her own prospects. As the struggle progresses, material aid for FRELIMO increases, while FRELIMO becomes itself more formidable. Thus every victory adds to our chances of winning yet further victories and reduces the ability of the Portuguese to counter our activities. . . .

To understand the real nature of the war, it is not enough to take into account these general factors, common as they are to nearly all popular guerrilla wars. More detailed points about the composition, organization and leadership of the army are important.

The army is representative of the population at large, in that the vast majority of the guerrillas are peasants initially uneducated, illiterate and often unable to speak any Portuguese; but there is also a scattering of those who have had some education within the Portuguese system. The majority naturally come from the areas at present affected by the fighting, because it is there that widespread campaigns of political education and training programmes are possible. There is, however, a continual stream of people from further south, from all over Mozambique, who escape in order to join the struggle; and at the beginning, many people from refugee camps, who had fled from every district of Mozambique to escape repression, joined as soon as a structure to contain them had been created. In the army, people from different areas accordingly mingle, so that each unit contains representatives from different tribes and different areas fighting together. In this way, tribalism is being effectively combatted within the forces, and an example is being set to the rest of the population.

This is not the only way in which the army leads the way to social change. By accepting women into its ranks, it has revolutionized their social position. Women now play a very active part in running popular militias, and there are also many guerrilla units composed of women. Through the army, women have started to take responsibility in many areas; they have learned to stand up and speak at public meetings, to take an active part in politics. In fact they do a great deal of important work in mobilizing the population. When a women's unit first visits a village which is not yet sufficiently involved with FRELIMO, the sight of armed women who get up and talk in front of a large audience causes great amazement, even incredulity; when the villagers are convinced that the soldiers in front of them really are women, the effect on the astonished men is often so forceful that the rush of recruits is very much greater than the army can cope with or than the area can afford to lose.

The army is helping to raise the standard of education as well as of general political consciousness. Recruits are taught wherever possible to read and write, and to speak Portuguese, and even where an organized teaching programme cannot be arranged, they are encouraged to help each other to learn these basic skills. Indeed, the Portuguese authorities are increasingly suspicious of ordinary peasants who speak Portuguese, because they know that these are more likely to have learned it in the FRELIMO army than in a Portuguese school. The army also organizes various specific training programmes such as radio work, accounting, typing, as well as in subjects more narrowly oriented to the war. Finally, the army cultivates and produces its own food wherever possible, thereby relieving the population of the burden of supporting it and at the same time spreading the lessons of its example.

In these respects the army leads the people; but more important yet is the fact that the army is the people, and it is the people who form the army. There are civilian members of FRELIMO engaged in all types of work among the population; but cooperation extends beyond this, to the large body of peasants who are not members of FRELIMO, but who support the struggle, looking to the army for protection and the party for assistance of various kinds. They in turn help the militants whenever they can.

These features are best substantiated from the words of the militants themselves:

> Joaquim Maquival, from Zambezia: I come from Zambezia, a Chuabo, and I have fought in Niassa where the people are Nyanjas, and they received me like a son. I have worked among Ajuas, Macuas, who received me as if I were their own son. (F.I.)
>
> Miguel Ambrósio a company commander from Cabo Delgado: I have fought in Zambezia and Niassa, far from my own region and my own tribe. I have fought in the country of the Chuabos and the Lomes. . . . The Chuabos, the Nyanjas and the Lomes received me even more warmly than if I had been from their own region. In Western Niassa, for example, I came across Comrade Panguene, and although he is from the south, you couldn't distinguish him from the people of the region: he is like a son of the region. The people understand that we are all Mozambicans. . . . The people are united and help us. Otherwise, for instance, we couldn't go into enemy areas; it is the people who give us all our information about the movements of the enemy, their strength and their position. Also, when we start working in an area where we have no food, because we have not yet had the opportunity to grow any, the people supply us and feed us. We also help the people. Until militias have been formed in a region, we protect the people in their fields against the action and reprisals of the colonialists; we organize new villages when we have to evacuate the people from a zone because of the war; we protect them against the enemy. (F.I.)
>
> Rita Mulumbua, woman militant from Niassa: In our units there are people from every region; I am with Ajuas, Nyanjas, Makondes, and people from Zambezia. I believe this is good; before we did not think of ourselves as a single nation; FRELIMO has shown us that we are one people. We have united to destroy Portuguese colonialism and imperialism.
>
> The struggle has transformed us. FRELIMO gave me the chance to study. The colonialists didn't want us to study, while now I am in this detachment we train in the morning and in the afternoon I go to school to learn reading and

writing. The Portuguese didn't want us to study, because if we did we would understand, we would know things. For this reason FRELIMO wants us to study so that we should know, and in knowing we understand better, we fight better and will serve our country better. (F.I.)

Natacha Deolinda, woman militant from Manica and Sofala: When I went into the army, FRELIMO put me though a course on youth organization and also gave me my military training. Then I went to work in Cabo Delgado province. Our detachment held meetings everywhere explaining the politics of our party, the reasons for the struggle and also the role of the Mozambican woman in the revolution.

The Mozambican woman participates in all revolutionary activities; she helps the combatants, she has an important role in production, she grows crops, she also has military training and fights, she joins the militias which protect the people and the fields. (F.I.)

It is clear from these comments that the role of the army goes far beyond simply fighting the Portuguese. Like the party, it is a nation-making force. It prepares not just soldiers but future citizens, who pass on what they learn to the people among whom they work.

Leadership is not based on rank but on the concept of responsibility; the leader of a certain body is referred to as the man 'responsible' for it. Many of those now 'responsible' had never been to school before they entered the army; they were illiterate, with no formal education, when they joined near the beginning of the war. They have acquired the ability to lead through their practical experience of fighting and political work, and through the education programmes of the army. There were some who had had a little schooling; but very few of these, even among those in the most important positions today, had gone beyond primary school.

Our experience, that of the militants and the leaders, has developed with the struggle. In 1964 the army comprised small groups of men, frequently ill-armed and ill-supplied, able to mount only ambushes and small-scale raids. The army was fighting against tremendous odds. The following account, by a man who is now national political commissar and a member of the Central Committee, gives some indication of what the war was like at the beginning, of the people who engaged in it, and of how they were able to expand their activities. Some of this man's earlier struggle with the Portuguese educational and economic structure have been related in previous chapters. The present account begins just after he was forced to flee from Mozambique.

Raul Casal Ribeiro: Some comrades from FRELIMO found me and educated me. . . . Three months later I asked to join FRELIMO. From then on I was a member of FRELIMO and began to work. I went to one of our party's bases to do my training, and since then I have been fighting. We have had to face many difficulties. There were times at the beginning when we didn't even have food. There were moments of hesitation, but the work of political education had taught me how to accept sacrifices and to struggle on.

The party had confidence in me and gave me responsibility. I studied hard. I was entrusted with the education of other comrades in the units. Then I was put in charge of sabotage on the Tete-Maturara railway line and other operations. We had a tiny detachment and very little equipment; the enemy sent a whole battalion to destroy us, but they failed. They attacked us, but they always suffered major losses. On one occasion they encircled us when we had only five bullets between us. They fired on us, but we had taken cover. Thinking that they had killed us, since we had not answered their fire, they advanced. When they came to within three or four metres of us, the comrades who had the bullets opened fire and killed one of them. The Portuguese were frightened and withdrew, giving us the opportunity to escape without them knowing. From a distance they went on firing for an hour, even shooting at each other. Afterwards we found the body of a South African Boer who had been with the Portuguese and had been killed by them.

This is how the enemy sows the wind and reaps the whirlwind. In this battle we captured one MG3, six loaded magazines, one offensive and two defensive grenades, and a knife. (F.I.)

It was through such small operations with courage and initiative in the face of difficult conditions, that the present size and strength of the army became possible. As an indication of the rapid growth of guerrilla action, here is a communiqué relating to an action which took place on 2 August 1967 and which was subsequently confirmed by a report on the Portuguese radio:

> Three aircraft and a store of ammunition completely destroyed; the fuel deposit burnt; nearly all houses near the airfield ruined; dozens of Portuguese soldiers killed or wounded: this happened at Mueda in a mortar attack launched by FRELIMO forces on 2 August. The fire raged for two days. (FRELIMO communiqué)

In a situation like this, where a country is in a state of war and the army inevitably has very extensive powers, there is a potential danger of conflict between the military and civilian organizations. In our system, though, this is minimized by the fact that they are both answerable to the political body of FRELIMO, which is itself composed both of military and civilian personnel. The relationship of the political, military and civilian bodies is not one which can be described in terms of a neat hierarchy, where one is subordinated to another. Policy decisions are made by the political body, the supreme organ of which is the Central Committee. The army, like the various departments, works in conformity with the decisions made by the Central Committee; but the army leaders themselves, as members of the Central Committee, also help to make these policy decisions. The meetings of the military command, which take place once a fortnight, are normally presided over by the President or Vice-President of FRELIMO, which ensures that between Central Committee meetings, close coordination of political and military decisions is maintained.

On the local level, in the field, the people's militia plays an important part in linking the civilian population and the army. These militias are formed from militant members of the civilian population, who carry on with their normal occupations and, at the same time, though not incorporated in the guerrilla army, undertake certain military duties. Their main task is the defense of their home region. If there is a danger of attack from the Portuguese forces, they can be mobilized as an additional armed force. While there is fighting in an area, they coordinate their activities with the guerrillas, reinforce them when necessary, and supply them with information about the particular locality. When the guerrilla forces have liberated an area, the militia can then take over the organization of defense, of production and supply, leaving the main forces free to move on to a new fighting area. In regions where there is not yet an active armed struggle, militias are formed in secret whose task is to prepare the ground for guerrilla fighting; to mobilize the people; to observe the Portuguese forces; to arrange supplies and assistance for the guerrillas as they move into the region.

In a sense, these people's militias are the backbone of the armed struggle. The guerrillas carry the main offensives and do most of the direct fighting, but it is the work of the militias which makes it possible for them to operate.

After the initial phase of our offensive was over and our forces had withdrawn to the two northern provinces, there followed a period of apparent stalemate which lasted through 1965 and most of 1966. During this period, FRELIMO controlled most of the country and villages in the northern area; the Portuguese controlled the towns and retained a number of fortified bases where they were relatively secure. The main roads were disputed, as the Portuguese were still trying to use them for the transport of soldiers and supplies, while FRELIMO was constantly mining them and mounting ambushes. The Portuguese were unable to mount an effective offensive because, when they left their bases and went into the bush to look for our forces, they were ambushed. On the other hand, FRELIMO was not yet strong enough to mount major attacks against the Portuguese positions. Yet all the while FRELIMO was gaining strength, consolidating the military and political position, training new recruits, and gradually eroding the Portuguese strength in many small actions. By the second half of 1966, the increased strength of FRELIMO was becoming apparent, and our forces were able to begin assailing the Portuguese bases themselves. Between September 1966 and August 1967, more than thirty Portuguese military bases were attacked; and at least ten more in the last three months of 1967. Many of these bases were badly damaged, and some were evacuated after the attacks. For example, the post of Maniamba (Western Niassa) was attacked on 15 August and evacuated; it was reoccupied, but abandoned again after a second attack 31 August; ten days after this a strong body of marines was sent to reoccupy it. On 13 September, the post of

Nambude (Cabo Delgado) was attacked, and buildings, three vehicles and the radio equipment destroyed. The airforce base at Mueda, an extremely important target and heavily guarded by the Portuguese, was bombarded twice and five planes were completely destroyed on the ground there.

During 1967 the area of fighting was extended in all regions. In Cabo Delgado our forces advanced to the river Lúrio and surrounded Porto Amelia, the capital, at the same time consolidating their position in the rest of the province, which is now almost entirely in our hands. In Niassa, our forces have advanced to the Marrupa-Maua line, and are approaching the frontiers of the provinces of Mozambique and Zambezia. To the south, they have gained control of Catur zone, between the provinces of Zambezia and Tete; while, to the west, they created the conditions necessary for reopening the struggle in Tete and Zambezia, a region of great importance for its agricultural and mineral resources.

The Portuguese have been working to improve their anti-guerrilla tactics and in particular have been trying to profit from the experience of their NATO allies: Britain in Malaya, the United States in Vietnam, and France in Algeria. Afonso Henriques Sacramento do Rio reported:

> This instruction is given to the Portuguese soldiers in the first part of their six months' training. The soldiers learned the theoretical base of anti-guerrilla warfare in courses tested by examinations. These courses are usually given by officers who have undergone special training, theoretical and practical. During the Algerian war, several Portuguese officers were given military training in Algeria by French specialists in 'subversive warfare'. Many other officers have been sent to the USA, where they took commando and marine courses and studied all the techniques used by the American army against the Vietnamese people.

One effect of this is that the Portuguese army now rarely operates in units below company strength, so that when these units are attacked, even if they suffer heavy losses, they retain a sufficient numerical strength to prevent the guerrillas from achieving one of their chief objectives: the capture of arms and ammunition.

Nonetheless, the Portuguese are still suffering heavy losses when they attempt to leave their bases, and are making little headway against the guerrilla forces, who simply retreat until such time as they can attack with advantage. As a result, the Portuguese have been turning more and more to aerial warfare, knowing that it is not easy for us to acquire and transport the type of heavy equipment needed against such action. They have carried out raids against bases, villages, schools, clinics; bombed areas of crops, and made some attempts in places to destroy the bush which gives cover to our guerrillas. The casualties caused by these raids have been almost entirely among the civilian population, and high priority has been given by us to organizing protection for the villagers. We are increasing our anti-aircraft strength; in October 1967 one of three planes which were bombing Marrupa zone was shot down, and the others were forced to withdraw.

Faced with a series of military reverses, the Portuguese authorities have been experimenting with various extra-military anti-guerrilla tactics, a mixture of terrorism and psychological warfare, with the main aim of forcing or persuading the population to withdraw support from FRELIMO. On the psychological side, in 1966 and 1967 they mounted propaganda campaigns on the radio and through a wide-scale distribution of leaflets. These leaflets were in general attractively printed on brightly coloured paper, with a parallel text in Portuguese and an African language, describing conditions of starvation and misery in FRELIMO areas, and of a prosperous, comfortable life under the Portuguese. They carried large pictures illustrating this contract or caricaturing FRELIMO as 'living it up' in exile at the expense of the rest of the population. In their propaganda they have also tried to exploit natural divisions in the population, accusing FRELIMO of fostering one tribe's ambitions against its neighbour.

The distance between the Portuguese and the African population, however, greatly diminishes the effect of such campaigns, since with the high level of illiteracy and the low material standards prevailing, pamphlets and radio cannot by themselves reach a wide audience. Besides this, the falsity of the content is not hard to see; the people remember well enough that there was no prosperity under the Portuguese, and where FRELIMO has been active they will have observed themselves

Selections from
The Struggle for Mozambique

that its members and leaders are drawn from different tribes and different religious groups. FRELIMO has the vast advantage that its political work is carried out mostly through personal contact, by word of mouth, by meetings, by example, and is convincingly undertaken by members of the ordinary population. Moreover, there is no attempt to distort the truth by promising impossibilities: we admit that the war may be long; that it will certainly be difficult; that it will not bring prosperity and happiness by magic; but that it is bringing some improvements and that it is the only way of eventually achieving a better life. The message issued to the Mozambican people from the Central Committee on 25 September 1967, for example, stated:

> ... There are many difficulties. The guerrillas sometimes have to spend whole days without eating, have to sleep out in the cold, and sometimes have to march days or even weeks in order to carry out an attack or an ambush. ... The people also suffer in this stage of the liberation struggle, for the enemy intensifies its repression to try to terrorize the population and prevent them from supporting the guerrillas. There are many difficulties. The battle for liberty is not easy. But that liberty we want to gain is worth all these sacrifices. ...

The work of mobilization is done essentially through direct contact, but it is backed up by literature and by the radio. Statements and messages such as the one quoted above are duplicated and distributed in the camps and at meetings. Simple duplicated leaflets are also circulated, depicting, for example, an exploiting 'boss' being driven out by FRELIMO. There are also regular broadcasts, organized through Radio Tanzania, which, since 1967, have been powerful enough to reach beyond the southern border of Mozambique. In the liberated areas, we have distributed radio sets to help the people hear these broadcasts. Programmes include: news in Portuguese and African languages; reports on the struggle; messages and political talks; educational programmes on hygiene and public health; revolutionary songs, traditional and popular music.

Having obtained very little result from direct propaganda, the Portuguese have been trying to develop more elaborate techniques. In 1967, for example, they set up a puppet African in

Tete Province as leader of a 'nationalist' party and arranged public meetings where he appeared with Portuguese officials and asserted that the Portuguese were prepared to grant independence peacefully to his party but not to the 'FRELIMO bandits'. This campaign had some initial success; but as explanations from FRELIMO militants were backed up by a lack of any signs of good faith on the part of the Portuguese, the people became sceptical and for the most part stopped going to the meetings.

Faced with the failure of both military action and 'persuasion', the Portuguese have been relying increasingly on terror, in an attempt to frighten away support from FRELIMO. Seeing the liberation forces living among the people like fish in the water they want to heat the water until it cooks the fish.

Since the war, all over Mozambique, and not just within the fighting areas, there have been drives to round up nationalist sympathizers, and thousands of 'suspects' have been arrested. The majority of these are peasants and manual workers, 'natives' by Portuguese terminology. They are not brought to trial, nor sentenced; they are imprisoned, interrogated, tortured and not seldom executed in complete secrecy. Even their families are not told anything definite: all they know is that the person disappears.

Among these 'suspects' there have also been a few intellectuals, people too well known beyond Mozambique to be allowed simply to vanish without causing an international outcry. Such have included the poets José Craveirinha and Rui Nogar; Malangatana Valente, the painter; Luis Bernardo Honwana, the short story writer. The Portuguese authorities brought just these few prominent men to trial, making their case public and trying to give the impression that all their proceedings against nationalists, saboteurs, etc., were conducted in the same open legal manner. But even such a show trial did not conform to the standards of legality laid down in non-fascist countries. When it first took place in March 1966, as many as nine of the thirteen accused were acquitted for lack of evidence; but the government refused to accept this verdict and ordered a new trial by military tribunal. This tribunal, acting under precise government instructions, convicted those who had been acquitted before and lengthened the sentences imposed on the other four. The sentences themselves are in any case meaningless, because they all

Selections from
The Struggle for Mozambique

included 'security measures', which mean that the term of imprisonment can be extended indefinitely. A delegation of international jurists and all foreign journalists were forbidden to attend this second trial.

Yet the Portuguese succeeded to some extent in their intentions, for the international outcry directed specifically against this trial and the fate of those thirteen intellectuals helped to draw attention away from the real issue: the much worse fate of the many undistinguished Mozambicans who have not been been given even a semblance of a trial, but have been either killed or else imprisoned under very much more severe conditions.

In the fighting areas, the campaign of terror is more widespread, and more indiscriminate, with reprisals carried out against the population as a whole. Where it cannot reach the villages, the Portuguese army resorts to air-raids, but where the soldiers are able to get to the people, more personal forms of terror and torture are used. The type of incident will be only too familiar to anyone who has studied the methods of fascist armies elsewhere.

Extreme brutality, however, does not as a rule have the desired result but confirms the people in their hostility to the Portuguese, and in fact often drives them to desperate acts of defiance.

Such a policy is thus not only cruel; it is tactically foolish. The FRELIMO army, in contrast, is firmly and constantly instructed to attack military and economic targets. The statements of militants indicate how they understand this policy.

> Joaquim Maquival: ... In our units and on our missions we have often come across unarmed Portuguese civilians. We didn't harm them. We asked where they were coming from; we explained our struggle to them, our sufferings; we received them kindly. We do this because our struggle, our war, is not against the Portuguese people; we are struggling against the Portuguese government, against those who turn their weapons on the Mozambican people; we are at war with those who injure the people. ... We know that it is not all the people of Portugal who exploit us, but only a minority, which is also exploiting the Portuguese people themselves. Among the Portuguese people there are also people who are exploited. FRELIMO cannot fight the

people, cannot fight against the exploited.

Miguel Ambrósio Cunumshuvi (company commander): We have never thought of murdering Portuguese civilians; we don't terrorize the Portuguese civilian population because we know who we are fighting and why. For that reason, we have never planned an attack against Portuguese civilians. If we wanted to, we could, the civilians live near us, we have every opportunity; but our objective, our target is the army, the police, the administration.

Our programme, our standing orders state clearly that we must not attack any civilian, only those people who are with the army, that is, accompanying and assisting it. The only terrorists in Mozambique are the colonialists.

This policy has importance for the future, when we shall be trying to establish a society that can absorb the different people living in Mozambique without racial bitterness; but it also has immediate practical advantages. At the beginning of the war, for instance, the Portuguese authorities distributed arms to the white *colonatos* and traders in certain areas for use against FRELIMO. These people then saw that unarmed civilians would not be harmed, but that those who carried arms against us would be treated as an appendage of the army; and the result was that most of the civilians refused to accept arms. The failure of the Portuguese forces to understand this policy has even sometimes roused the Portuguese civilians against them: in one instance the Portuguese forces entered a village where they knew that FRELIMO had been, and when they found that the Portuguese civilians had not been touched, they accused them of collaborating with FRELIMO and arrested and punished them, their own settlers.

Editor's note: Mondlane was assassinated in February, 1969, presumably at the instigation of the Portuguese secret police.

Chapter VI
Urban Guerrilla Warfare

Urban guerrilla warfare has been adopted as a revolutionary strategy more from the failure of other strategies than as a result of any achievements of its own. Revolutionaries in largely peasant-derived although urbanizing societies in Latin America, as well as in other areas, have transferred their activities to the cities, primarily because of failures in the countryside. By so doing they have reasserted a degree of initiative. Urban guerrillas may achieve greater publicity more readily. Urban guerrilla war, more than that in the countryside, relies on terrorism: kidnapping, assassination, robbery, bombings, etc. If the government does not react effectively, mass fear and apparent revolutionary success may neutralize or convert the populace, while a harsh over-reaction may alienate the populace from the regime. Advocates of this strategy would also argue, contrary to the Maoist model, that the revolutionaries should work in the cities because the people and the power structures are increasingly concentrated there.

Robert Moss analyzes the factors underlying urban guerrilla warfare. He argues that the failure of Maoist and Guevara models necessitated a shift in strategy. Urban guerrilla war subjects a society to stresses which expose its vulnerability, particularly a society professing to be democratic. Counter-strategy must avoid driving significant numbers and groups into the camp of the antagonists; but there is danger, too, of a negative reaction against the guerrillas. The author offers many examples illustrating the causes and impacts of urban guerrilla warfare.

The "Minimanual" of Carlos Marighella, published by Moss as an appendix, has been accepted by most scholars as the primary theoretical source. It is a detailed guide to principles, strategy and tactics. It may be noted that Marighella died in Brazil while practising what he preached.

Urban Guerrilla Warfare*
by Robert Moss

The kidnapping of ambassadors, the hijacking of aircraft and the bombing of company offices are likely to continue to be familiar hazards of life in the 1970s. Such incidents attract headlines, but they are only part of the repertoire of urban guerrilla warfare, and not the most important part. On the face of it, the phrase 'urban guerrilla' is a nonsense. From the time of Clausewitz, it has been generally agreed that guerrilla warfare can only be carried on where insurgents can range widely over the countryside and dispose of irregular, difficult terrain as a base-area. Most theorists of guerrilla warfare agree with Fidel Castro that 'the city is a graveyard of revolutionaries and resources'.[1]

But there has been a recent upsurge of revolutionary violence in western industrial cities; in Latin America, the heirs of Che Guevara have made the city their target; and Maoist groups in India have launched a terrorist campaign in New Delhi and Calcutta. In cities like Montevideo or Guatemala City, urban terrorism is in some ways the precise counterpart of rural guerrilla warfare—just as riots can be seen as the urban equivalent of spontaneous peasant uprisings. In military terms, the terrorists and the guerrillas are waging a campaign of harassment and attrition against superior, conventional forces. Their basic target is not control of territory, but control of men's minds. They are essentially political partisans, for whom success or failure will hinge less on what happens on the battleground than on their capacity to get their message across, to erode the morale of the forces of order, and to induce a general 'climate of collapse'.

Terrorism as a Political Weapon

Terrorism could be defined as the systematic use of intimidation for political ends: Lenin put it tersely when he said that the purpose of terror is to terrorize. In the colonial situations, the goal of political terrorists was to persuade the occupying power that it had become too costly to hold on. This was the technique successfully applied by the Irgun and the Stern Gang in Palestine, and by the EOKA in Cyprus. The desire to win world publicity, in the hope of enlisting outside support and of provoking a political debate inside Britain, was the key element in this strategy. The 'Preparatory General Plan' drawn up by General Grivas in Athens before the start of the EOKA campaign, defined the political objective in the following terms: 'The British must be continuously harried and beset until they are obliged by international diplomacy exercised through the United Nations to examine the Cyprus problem and settle it in accordance with the desires of the Cypriot people and the whole Greek nation.'[2]

The strategy of the Irish Republican Army (the IRA) in Northern Ireland today has some similar features. The IRA, who see themselves as victims of a colonial situation, are hoping that their campaign of selective terrorism against the British troops stationed in Ulster will lead, first, to a breakdown in army discipline and morale and, second, to a failure of the will in Westminster: the political decision to hand the province over to Dublin.

Within an independent state, the use of terror is more complex. It can be employed as a defensive or an offensive weapon, to preserve the *status quo* (the original *raison d'être* of the Ku Klux Klan, the Organization de l'Armee Secrète (OAS) in Algeria or the *esquadrão da morte* in Brazil) or to overturn the existing system. It can be used to erode democratic institutions and clear the way for the seizure of power by an authoritarian movement (like the Nazis) as well as to resist an absolutist government or a foreign invader. The *Narodniks* in Tsarist Russia regarded assassination as a means of 'warning off' members of the official hierarchy who sought to abuse their powers. Unlike most present-day terrorists, the *Narodniks* were acutely conscious of the moral dilemmas involved in the systematic use of political violence. They argued that they had been forced to use terror because the Tsarist regime had closed all possibilities of peaceful reform. The leaders of the *Narodnaya Volya* organization even promised that, if they ever saw signs of even 'the possibility of an honest government' they would then 'oppose terrorism, as we are now opposed to it in free nations'.[3] Nothing could be farther removed from the spirit and tactics of contemporary terrorists, like the Weathermen or Uruguay's Tupamaros. Both those groups, with differing success, have set out to undermine their countries' democratic institutions and to create the conditions for civil war by bringing about a polarization of political forces.

Most terrorists in modern history have alienated public sympathy by adopting gangster-style techniques. A good example of that was the reaction of the crowd in an Istanbul street in June 1971 after the Turkish police had managed to rescue a fourteen-year-old girl from her kidnappers in a brilliantly executed operation. The guerrillas, members of the Turkish People's Liberation Army, shouted to the crowd 'We are doing this for you' as they exchanged fire with the police from an upstairs window. But the crowd in the street broke through police barricades in an attempt to lynch the single terrorist who finally emerged alive. That is a fairly extreme example of the circumstances under which terrorist actions are purely counter-productive. . . . But the normal response to terrorism is revulsion. That is why the more successful urban guerrillas, in Latin America in particular, have gone to considerable pains to try to rationalize their crimes and have been very selective in choosing their targets. Terrorists can never win popular support unless they can explain their actions as something more than random criminal assaults or lunatic gestures.

The need to make converts also explains the exceptional importance of marksmanship for the urban guerrilla. There has to be some discrimination in the choice of targets. At least at the outset, the urban guerrilla is less concerned with intimidating the civil population than with proving that the government and the forces at its disposal are vulnerable to attack. It is only when a rebel movement has *already* established a secure grip on a significant part of the civil population that it can afford to use terror as a means of extorting aid and supplies, of conscripting new recruits and deterring potential defectors. The Viet Cong have perfected the system of 'repressive' terrorism since the late 1950s. Terrorism against neutral or anti-Communist elements in government-controlled areas has usually taken the form of preliminary warnings, followed by kidnapping or assassination. Over the four-year period between 1966 and 1969, American estimates place the total number of assassinations at 18,031. There were an estimated 25,907 kidnappings for indoctrination and other purposes. Terrorism in Communist-controlled areas has ranged from verbal intimidation through 'home surveillance' and 'thought reform' to execution.[4]

In Latin America, the terrorists have singled out individuals and installations that they can publicly identify with what they regard as an oppressive system. The Guatemalan terrorists, for example, have kidnapped an archbishop and a foreign minister as well as local businessmen and foreign envoys. The bombings in Montreal by the Front de Libération du Québec (FLQ) were primarily aimed at foreign enterprise and military installations. The IRA snipers in Ulster have made the British Army their prime target. Selective assaults on soldiers and policemen can bruise the morale of men who have to wear uniform (an essential precondition for any successful uprising) as well as eliminate enemies.

But indiscriminate terror also has a place in urban guerrilla warfare. The IRA bombing campaign in England that began in January 1939, was one of the most notorious examples. In the fifteen months that followed, bombs were exploded in station buildings, electricity plants, letter-boxes, cinemas, post offices, public lavatories, shops and telephone boxes. The campaign may not originally have been intended to cause injuries, but the targets chosen and the extraordinary incompetence of those who manufactured and deposited the bombs made that inevitable. . . . But the campaign was entirely counter-productive, coming at a time when the British Government was wholly preoccupied with Nazi expansion in Europe, and it only resulted in effective police action to curb the IRA.[5] In Algiers in 1955, Front de Libération Nationale (FLN) terrorism progressed from actions against men in uniform, to selective assaults on individual Europeans, to the depositing of bombs in public places where French Algerians were known to gather (cafés, restaurants and so on). The FLN used indiscriminate terror to dig an unbridgeable gap between the Arab and European populations and to provoke the kind of communal backlash that helped them to destroy their rivals for the Arab leadership—those they described as 'the party of the lukewarm'.[6] There was a period when the FLQ terrorists in Quebec went about the English–Canadian suburb of Westmount in Montreal dropping bombs into pillar-boxes. That was partly an attempt to intensify the feelings of distrust and mutual dislike between the English and French communities.

In the United States, the Weathermen and the radical 'bombers' set out to attack the entire capitalist system by a wave of assaults on property. There were 4,330 incendiary bombings in the fifteen months up to April 1970, and the targets included banks, company offices, high school buildings, and military installations. Attempts to rationalize the bombings were hardly designed to convert public opinion. A letter to *The San Francisco Chronicle* from a group calling itself 'The Volunteers of

America', after the bombing of the Bank of America's Santa Barbara branch in June last year, likened the role of the bank to that of 'the German financiers during the rise of Hitler'. A letter from another group, 'Revolutionary Force 9' declared that companies like IBM and Mobil Oil are 'the enemies of all life', responsible not only for the prolongation of the Vietnam war but for such diffuse crimes as 'encouraging sexism and the degradation of employees'.

The American bombings are the work of a lunatic fringe, not a case of terrorism used rationally as a political weapon. But it is a central goal of all urban guerrillas to break down the existing social framework and encourage a general feeling of insecurity and disorientation. It has been argued that the first task of the rebel is to 'disrupt the inertial relationship between incumbents and the mass'.[7] This terminology cannot be applied to western pluralistic societies, where relations between government and people are complex and multi-faceted. But conditions of general insecurity favor extremists in any society. The government is discredited because it cannot protect the civil population, and the people will finally be forced to side with whatever group is in a position to apply coercion or guarantee protection.

The Varieties of Urban Militancy

Terrorism is only one form of urban militancy. Unlike riots, political strikes, student demonstrations and ghetto revolts, terrorism is a minority technique, and the need to ensure security under urban conditions dictates a fairly standard form of organization: members of the terrorist group are divided into cells or 'firing groups' of from three to five men, with a link man in each. This clearly limits the possibility of betrayal or of police infiltration, but it also limits the possibility of political agitation.

The terrorist has a political tool; the urban guerrilla has a strategy for revolution (however utopian it may seem). The Brazilian guerrilla leader, Carlos Marighella—who was shot dead in a police ambush in São Paulo at the end of 1969—outlined part of that strategy in his *Minimanual of the Urban Guerrilla*. Marighella wrote that 'It is necessary to turn political crisis into armed conflict by performing violent actions that will force those in power to transform the political situation of the country into a military situation. That will alienate the masses who, from then on, will revolt against the army and the police and thus blame them for the state of things.'[8] That is one scenario for civil war. It might be called the strategy of militarization. Marighella's thesis, that by inviting repression the urban guerrillas will pave the way for popular revolt, seems to be working out in one part of Latin America—Uruguay. But Uruguay is an isolated case. In the rest of the continent, the urban guerrillas are learning to their cost that, if the government is sufficiently ruthless and can present a united front, effective repression is more likely than a popular uprising. The same is almost certainly true of western societies. Modern techniques of police control rule out the possibility of a successful urban uprising unless a political crisis cripples the government or the loyalty of the security forces is in doubt.

It is dangerous to generalize about the causes of urban revolts. Clearly, the reasons why men revolt in Guatemala City or Belfast are different. It is possible to define three main contemporary forms of urban guerrilla warfare: (i) 'Technological terrorism' in the industrial cities; (ii) Ghetto revolts and separatist uprisings; and (iii) Urban violence in the pre-industrial cities (notably Latin America). It is striking that, in the industrial countries, the groups that have resorted to urban terrorism draw their support from marginal social elements: middle-class student radicals or cultural and ethnic minorities. The increased frequency of this kind of political violence in western societies is bound up both with a romantic or nihilistic disenchantment with existing systems and with a curious resurgence of sectional loyalties. The Basque extremists who kidnapped the West German Consul at San Sebastián in 1970, the Flemings who take to the streets to protest about the dominance of the French language in Brussels and the Quebec terrorists who claim to be combating Anglo-Saxon imperialism are similar in their origins and the roots of their complaints. In the age of what Marshall McLuhan calls 'the global village', there is a new insistence on the *patria chica*.

A hypothetical revolution in a western country would have to be city-based; but it has come as a surprise to some observers that, since about the time of Che Guevara's death in the Bolivian hills in October 1967, his successors in Latin America have made the city their target. The reasons are fairly simple. The first was the patent failure of peasant uprisings and rural guerrilla movements in Latin America in the decade of the 1960s. Between 1965 and 1968, the Peruvian Army mopped up the remains of Hugo Blanco's peasant revolt; the rebel movements in Colombia and Venezuela melted away into insignificant frontier bands; and the Guatemalan Army waged a ruthlessly efficient campaign (including extensive resettlement and also, according to some reports, the use of napalm and defoliants) against guerrilla forces in the eastern hills. Equipped and guided by the Americans, Latin American security forces displayed a vastly increased capacity to handle rural uprisings. At the same time, it became apparent that a provincial revolt was rarely a direct threat to the government in countries where wealth and power was gravitating towards a few enormous metropolitan centres. Secondly, the guerrillas realized that it is much easier to win headlines by kidnapping a foreign envoy than by gunning down country gendarmes. Urban operations have an obvious attraction for an isolated extremist group bent on winning publicity, and, by the end of the 1960s, most guerrilla organizations in Latin America were cut off not only from the Moscow-line Communist parties, but from Castro as well. Finally, the shift to the cities was an attempt to take advantage of the continent's phenomenal urban growth. Latin America's cities are growing faster than any in the world, but industrialization has lagged behind, creating vast and unpoliceable slums.[9]

It remains to be seen whether the special factors that have conditioned the rise of urban guerrilla warfare in Latin America will influence the future course of insurgency in Asia or Africa. It is surprising that there has so far been little urban terrorism in Asian cities, given the political instability of many of the countries in the region. The Tet offensive in South Vietnam in 1968 and the Gestapu affair in Indonesia in 1965 (when the Indonesian Communist Party, with the collusion of President Sukarno, tried to stage a *putsch* by assassinating army leaders) might be classed among the rare attempts at urban insurrections. The Maoist groups in India recently broadened their tactics to include urban guerrilla techniques. In May 1970, one of the main organs of the Communist Party of India (Marxist-Leninist) (CPI(M-L)) announced that 'While the main task of armed struggle would be in the villages, the party would not allow towns and cities to become strongholds of bourgeois terror'.[10] In the same month, the Naxalites (a terrorist movement that takes its name from the Naxalbari uprising in 1967) made their first appearance in New Delhi, handing out leaflets and painting slogans on walls. One of their spokesmen promised that 'The red terror activities in cities and towns have come to stay.'[11] By the end of the year, the Naxalites were reported to have made 50-60 active recruits at Delhi University, and to have built up a support group on the campus of about 200.[12]

These figures, insignificant in themselves, were a sign of an attempt to transfer terrorist operations to the towns from their original base in the rural areas of West Bengal and Andhra Pradesh. This tactic was at odds with the Maoist ideology of the groups responsible. For example, Charu Mazumdar, the leader of the CPI(M-L), has remained insistent that 'the path of India's liberation is the path of people's war' and that the first step along that path is to create 'small bases of armed struggle' all over the country.[13]

He has also insisted that rural guerrilla warfare and agitation among the peasant poor is an essential apprenticeship for the young students and urban unemployed who have formed the backbone of the Maoist fighting squads.[14] A similar ideological rigidity has impeded the emergence of urban guerrilla warfare in other Asian countries.

The main thing that turned the Indian Maoists towards an urban campaign was a series of reverses in the countryside. The Naxalites, for example, were active for many years among the Girijan tribesmen, who occupy an area of some 500 square miles of hilly uplands in Srikakulum. It is good country for guerrillas, and the Naxalites also found a popular grievance to exploit. From the early 1960s, there was considerable unrest among the Girijans, stemming from the fact that merchants and money-lenders in the towns were acquiring tribal lands through usury. But government legislation to control the transfer of land placated the Girijans, the Naxalites were divided by personal jealousies and dissension, and, by mid-July 1970, the police were confident that they had eliminated all of the six original leaders in Srikakulum. As in Latin America, urban terrorism was employed by the Indian Maoists both in the attempt to offset their rural setbacks and as a means of tying up the security forces and adding to the political crisis. The Indian Minister of Home Affairs announced on 18 November 1970 that there had been 341 murders in West Bengal since March, of which 172 were political. He added that 25 policemen had been killed during 526 attacks on individual members of the force.

Events in East Pakistan, the influx of East Bengali refugees, and the presence of the Mukti Fauj (the East Bengali resistance movement) have opened new possibilities for India's Maoists. But there is also considerable scope for urban terrorist activity in other parts of Asia. In Thailand, the failure of the Thai Communist Party to develop urban guerrilla activity in Bangkok reflects the fact that the Thai insurgency is still largely bound up in the sectional grievances of ethnic minorities (encouraged, especially in the north east, by the Chinese and North Vietnamese). But the Thai police have reported increasing activity by the Communists in Bangkok, and this could eventually lead to a campaign of selective terrorism. In Malaysia, the remnants of Chin Peng's Communist guerrillas are based in the wild country up around the Thai border, although early in 1971 there were signs of a more aggressive forward movement towards the tin-mining region around Ipoh. But the fact that a future insurgency in Malaysia would almost certainly be bound up with deteriorating race relations, and that the west coast cities are overwhelmingly Chinese, means that the possibility of a future urban guerrilla campaign cannot be lightly passed over. In Singapore, some of the extremist elements associated with the *Barisan Socialis* (the main opposition to the ruling People's Action Party) have been pressing for a campaign of street violence. The city-state is exceptionally well-policed, but its total dependence on trade and foreign investment might encourage extremists to try to precipitate an economic crisis through political violence.

It is possible to make three general observations about the patterns of political violence in both the industrial and the pre-industrial cities:

(i) *The disruptive effects of population movements.* Internal migration has had an unsettling effect in both North America and the third world, for rather different reasons. The cities of the third world are like sponges, sucking in the surplus rural populations faster than they can absorb them. The visible effect of this process has been the mushroom-like spread of slums and shanty-towns. Each city is encircled by its 'misery-belt' of huts patched together out of odd bits of cardboard, tin and timber. . . .

The process has gone farthest in Latin America. More than two-thirds of the populations of Argentina, Uruguay, Venezuela and Chile now live in towns. For Mexico, Brazil and Colombia, the figure is over 50 per cent. In southern Asia, by comparison, some 14 per cent of the population of the region live in towns. The figure for black Africa is slightly lower—about 13 per cent. But the static population spread matters less than the startling rate at which urbanization has been taking place. Third world cities are growing at a rate of between 3 and 8 per cent a year. That means that most of them are doubling in size every 10 or 15 years.[15] The reason for this is internal migration rather than natural population growth, since there is evidence to suggest that the birth-rate in most cities is below the national average.

There are many reasons why peasants are leaving the land. Overpopulation or land-hunger is probably the most important, and it has been accentuated in some areas by mechanization and the application of modern techniques that have caused extra redundancies. Better roads and communications between city and countryside have facilitated population movements, and the fact that more people are going to school or listening to radios has influenced job expectations and helped to give the city a glamorous image in the minds of young villagers. It must be added that political disturbances and natural disasters (like the droughts that send hordes of starving peasants, or *retirantes*, out of northeastern Brazil toward the coastal cities every few years) have triggered off the most dramatic population shifts. South Vietnam's cities doubled in size between 1963 and 1968;[16] Phnom Pehn was swamped by rural refugees in the months after the fall of Prince Sihanouk in March 1970; and Calcutta's crowded streets have been swollen to bursting by the influx of Bengalis who have fled the war across the border. In Guatemala, peasant unrest and a sustained campaign of political terrorism drove the *hacenderos* (wealthy landowners) and the more affluent peasants, as well as the poor, to take the roads to the major towns in search of security in the course of the 1960s.[17]

Urbanization in the third world is often compared with the process of urban growth in Europe and North America in the nineteenth century, but it differs from that earlier model in two vital ways. Firstly, third world cities are growing faster. The average rate of urban growth in Europe between 1850 and 1900 was only about 2.1 per cent. Secondly, the rise of the European cities was related to industrialization. In most third world countries, the Urban Revolution was not preceded by an Industrial Revolution. To take one comparison, 12 per cent of India's population were living in towns in 1951, while 11 percent of the total work force were employed in industry. Austria had reached the same stage of urbanization by 1870, but in that year 30 per cent of the Austrian work force were employed in industry.[18] Failing to find jobs in industry, most of the rural migrants in third world cities have had to scrape a living in the service sectors—a polite phrase that usually means nothing more than a daily round of boot-blacking, begging for odd jobs, or cleaning public buildings. Some 60 percent of Chile's labour force are employed in jobs that do not produce goods. Whereas in England after the Industrial Revolution, factory owners were crying out for manpower, it seems likely that in most third world countries the gap between the number of rural migrants looking for work and the number of new jobs being created by local industry will become bigger, not smaller.

This makes it impossible to apply the comfortable view of the link between urbanization and political violence derived on European experience to third world conditions. According to the popular view, the life cycle of civil violence in western societies passed through three phases in the course of urban growth and industrial development: an early phase, in which political violence was a response to the social disruption and disorientation resulting from the rise of urban industry; a transitional phase, in which a militant union movement emerged and briefly posed a serious threat to bourgeois society; and a 'mature' phase, in which the organized working class was peacefully integrated into the new social system. As recent historians have demonstrated, that view of the past is inaccurate and simplistic even for western societies.[19] It simply does not fit the very different circumstances of the third world, where the flight of peasants to the towns has created a whole new social class that the Brazilians call *marginais* (or 'marginal people') and that Marx, who had a notoriously low view of their revolutionary potential, called the *lumpenproletariat*. Friedrich Engels claimed that the members of this 'underclass' were 'absolutely venal and absolutely brazen', wholly concerned with the routine of eking a living by petty theft or by performing minor services.

In contrast, Frantz Fanon, the psychologist from Martinique

who joined the Algerian revolution, saw them as the armies of future revolutions. He believed that this 'horde of starving men, uprooted from their tribe and from their clan, constitutes one of the most spontaneous and most radically revolutionary forces'.[20] Was he right? Are the people of the slums a potentially revolutionary force? It has been argued that recent migrants are too preoccupied with surviving from day to day to lend their support to a political movement; and that men who are always moving from one shack to the next without steady jobs are not easily organized by a party or a revolutionary group. It is certainly true that the 'bazaar system' in third world cities provides some kind of safety valve; unemployment is worse than underemployment, and the profusion of uneconomic service industries and petty retailing at least makes it possible for people to eat. Assuming that rural migrants have only modest expectations, the service industries are capable of absorbing new arrivals into what has been described as a 'system of shared poverty': a bufferzone between the traditional and the modern economies.[21] But it seems doubtful whether this constitutes any long-term solution, both because of rising expectations and because eventually the parasitic service sector will be saturated. This means that a rising proportion of the populations of third world cities will remain outside the modern economy and the present forms of social organization while becoming increasingly conscious of their plight . . . .

One fairly sophisticated analyst has argued that the slumdwellers are 'basically conservative so long as live is barely livable' but 'catapult to revolution the moment that life is no longer seen as livable for whatever reason'.[22] The slum fringes of the third world cities contain a volatile mass that may explode during periods of rapid social transition or economic recession. And the urbanization process has had other political side-effects. In Southeast Asia in particular, it has heightened racial tensions, usually at the expense of the Chinese who dominate local commerce. There have been anti-Chinese riots in several Asian cities, and it is worth noting that it was Malays who had migrated from the east coast who were responsible for the violent race riots in Kuala Lumpur in May 1969.[23] At the same time, the concentration of wealth and power in a few enormous cities means that, in Latin America in particular, a rebel movement cannot confine itself to the countryside if it hopes for success. Moises Moleiro a former Venezuelan guerrilla leader, pointed out in a recent article:'in Venezuela, it is just not possible to start a rural uprising that will end with the countryside encircling the town. The rural areas are marginal to the life of the country. . . . A peasant revolt is impossible, in the last analysis, because we are not a peasant people.'[24] In the major Latin American countries, it seems that a peasant revolution is no more possible than in the industrialized west.

In the United States, internal population movements have added to social tension in subtler ways. Recent statistics from the American Census Bureau show that half of the country's negro population is now concentrated in 50 cities. Fifteen of those cities account for a third of the total. While black Americans have been moving into the city centres, middle-class whites have escaped to the suburbs, taking new industry with them. But this is not a one-way process. A recent study of the Cleveland riots of 1967 showed that negroes with steady incomes were also moving out of the ghettos, leaving behind those at the very bottom of the social scale.[25] The black ghetto in Detroit or Chicago is not simply a racial enclave, but also an island of deepening poverty. And the fact that the ghettos are often located close to the traditional centres of commerce or command key services like railway lines or power stations makes racial unrest a threat to the normal functioning of the economy.

(ii) *The sense of relative deprivation.* Population movements in the Americas have sometimes added to the sense of relative deprivation. . . . Men do not rebel because they are deprived, but because they are conscious that they are deprived. De Tocqueville's celebrated argument that the French Revolution came about because things were getting better (so that people who had formerly accepted their lot became conscious of the possibility of changing it) works equally well in reverse. Uruguay is the one country in Latin America where armed revolution seems possible in the foreseeable future. Yet it has also been one of the most enlightened societies in Latin America, with a tradition of constitutional rule and much of the apparatus of a welfare state. Uruguay is a democracy that has come upon hard times. The economic crisis that began in the late 1950s has crippled public service salaries and led to cutbacks in social spending, while the activity of an exceptionally efficient guerrilla movement has forced the Government to resort to repressive measures. The reason why the Tupamaros have been able to count upon a substantial amount of middle-class support has been that people's expectations have been disappointed.

(iii) *The character of the terrorist.* But urban guerrilla warfare is essentially the work of a tiny self-styled revolutionary elite. That makes it important to consider whether there is not something in the apparent truism that 'it takes a rebel to rebel'.[26] The FLQ in Quebec, the National Liberation Action (ALN) in Rio, and the Weathermen in the United States draw their recruits from similar social sectors and share not only a certain range of guerrilla techniques, but a common faith in political violence and the theory of a global revolution. Frantz Fanon provided the most comprehensive version of the now-fashionable theory of violence as a liberating force. 'At the level of individuals,' according to Fanon, 'violence is a cleansing force. It frees the native from his inferiority complex and from his despair and inaction; it makes him fearless and restores his self-respect.'[27] The radical 'New Left' in western countries as well as the Guevarists in Latin America have tended to talk in similar terms.

What Fanon completely failed to analyse was the corrupting effect of the systematic use of political violence, and its reinforcement of the totalitarian impulse. He also ignored the attraction of a terrorist organization for some criminal elements. It is no accident that the IRA Provisionals in Belfast have drawn support from petty criminals in the Catholic slums and have set up their own protection rackets for extorting 'party funds'. . . . The FLQ in Montreal has recruited drifters and corner pickpockets from the *hangars* (or gang-territories); and the Moslem FLN in Algiers enlisted the services of professional thugs like Ali-la-Pointe. The frequent confusion between criminal and political motives is bound to be accentuated when guerrilla groups rely upon 'fund-raising' devices like bank-robberies for their finnance—and that is why Cuban leaders, as well as Russian, have criticized this kind of operation.

The irony is that the founding impetus of many urban guerrilla groups has come from young idealists: middle-class students and intellectuals who share a belief in a global revolution aimed primarily at the United States. The Tupamaros first signed their name to a manifesto protesting about the American involvement in Vietnam. The FLQ in Quebec wrote the slogan 'Long Live the Cuban Revolution!' on the bottom of a kidnap note; and Pierre Vallières, their spiritual leader, has a wild-eyed vision of a world-wide revolution that would take account of the 'cultural and ethnic origins' of workers as well as their proletarian 'character'.[28] With the American Weathermen, the idea of a global holocaust approaches sheer nihilism, an itch to tear down the class from which these middle-class rebels sprang and everything it stands for. The Weathermen are essentially derivative: they see themselves as the white auxiliaries of a revolution that would be made by coloured men through a great upheaval in the third world and an uprising by Black Power groups in the United States.

The theory of international solidarity, it must must be added, has not been accompanied by much interchange of cadres or resources. The Cubans, the North Koreans and the Palestinians have all provided a certain amount of training for urban guerrilla groups. Some of the 70 Brazilian political prisoners who were freed in exchange for the life of the Swiss Ambassador in January 1971 had received training in Uruguay,[29] and there are signs that the Tupamaros have close links with guerrilla movements in Argentina and Bolivia as well. But the urban guerrillas are almost entirely self-reliant in terms of arms and supplies, and the form of co-operation that counts for most is the borrowing of ideas. The most dramatic example of that was the wildfire spread of diplomatic kidnapping as a political technique after the Brazilians used it to secure the release of 15 prisoners in 1969. But terrorist groups have also copied methods of 'armed propaganda'. For example, in October 1969, on the second anniversary of Che Guevara's death, the Tupamaros occupied the town of Pando and held it for about 15 minutes while commando groups raided the police barracks and the banks. It was a brilliant publicity technique, and probably a decisive turning-point in their campaign of political terror— although a rearguard party of Tupamaros was intercepted by the Army on the way back to Pando and shot to pieces.[30] At any rate, the Argentine Revolutionary Armed Forces (FAR) were sufficiently impressed by the occupation of Pando to copy it in the following year, when they took over the town of Garín. . . .

490 The Arsenal of the Urban Guerrilla

There are four main urban guerrilla techniques that have been explored over the past few years, and largely explain the success of a group like the Tupamaros. They are (i) Armed propaganda; (ii) Political kidnapping; (iii) 'Stiffening' riots and strikes; and (iv) Subversion of the security forces. These will be briefly discussed in turn.

(i) *Armed propaganda* can be defined as the attempt to prove to the people, through successful military actions, that the government is weak and the guerrillas are strong. One of the central problems for all guerrilla movements is how to get their message across to the man in the street. As a Tupamaro mouthpiece put it, the problem is that 'for the urban guerrilla, discretion must take the place of the rural guerrilla's hideout in the jungle'.[31] Since the possibilities for normal political agitation are restricted (and since the urban guerrillas have normally held themselves aloof from the traditional political parties, including the Communists) 'armed propaganda' must take the place of polemics. . . . In the face of official censorship, the Tupamaros have tried to construct 'counter-media', including a private radio transmitter and the use of eletronics experts to break into normal broadcasts with special messages. They have also taken over public meeting-places like cinemas and workers' canteens to deliver impromptu harangues.

(ii) *Political kidnapping* has been used to capture publicity, to free political prisoners and to extract other concessions, and to provoke controversies within governments. The government of President Pacheco Areco in Uruguay was so deeply divided after the wave of kidnappings in August 1970, for example, that he was on the point of handing in his resignation before the police captured several Tupamaro leaders in a lucky strike.[32] The Brazilian experience shows just how dangerous it is for governments to give in to kidnappers. It cost the Brazilian Government 15 political prisoners to free an American Ambassador, but later the price was 70 for a Swiss Ambassador—rampant inflation by anyone's standards, and a sign that for kidnappers, as for other mortals, the appetite grows with the eating. The game of bluff that is being played out between governments and guerrillas is not over. The Tupamaros held Geoffrey Jackson, the British Ambassador to Uruguay, for eight months after his capture in January 1971 in what they grandiosely called a 'people's prison'. They also held a number of prominent Uruguayans, including a close friend of the President and a former Minister of Agriculture. The Tupamaros, secure in the knowledge that they can outfox the police, did not let themselves be panicked into murdering another hostage as they murdered the American Dan Mitrione last year. They learned that they could humiliate the Government and the police more effectively by playing a waiting game.

(iii) *'Stiffening' riots and strikes* is one way of establishing closer links between the terrorist organization and popular grievances. Recent experience of urban riots in Northern Ireland and the United States supports the idea that crowd disturbances can pass through several phases and can finally pass under the control of extremist organizers. Since mid-1970 for example, the rioting in Ulster has ceased to be a fairly spontaneous cycle of communal conflict and has assumed a more sinister character. The British troops, rather than members of the other religious community, became the prime targets for hostile mobs egged on and infiltrated by the IRA. Street violence was prolonged in Belfast and Londonderry for five or six nights on end. Members of the crowds were armed with fire-bombs and gelignite nail-bombs, snipers fired on the British troops from neighbouring buildings, and there was systematic arson and destruction of property. On the night of 27 June 1970, more than 100 fires were started in Belfast, and troops were fired on in Ballymacarett and the Crumlin Road by IRA snipers armed with machine-guns. The pattern of those riots was repeated in 1971. Similarly, in the United States, from early in 1968 there was evidence of much greater organization and increased sniping in negro riots, although many Black Power leaders were distrustful of the riot as a political weapon and the incidence of rioting fell away after 1968.

(iv) *Subversion of the security forces* was seen by Lenin as one of the essential preconditions for a successful urban uprising. 'Unless the revolution assumes a mass character and affects the troops, there can be no question of a serious struggle'[33] All serious rebel movements have attempted to demoralize and subvert the army and the police and, historically, revolution has only been possible when (for internal or external reasons) this has already succeeded. Urban guerrillas are bound to be outgunned unless they can at least manage to neutralize a majority of the security forces, and one of the reasons for the Tupamaros' remarkable capacity for survival has been that they have shaken the confidence of the men in uniform by alternately circulating propaganda and practising selective assassination, while infiltrating their own agents at all levels. (It was the presence of a Tupamaro agent on the nightwatch at the naval training barracks that enabled the guerrillas to occupy it last year.) Terrorists have two apparently contradictory means of subverting the armed forces: one is to appeal to individual soldiers or policemen as 'fellow-workers'; the other is to issue threats and carry out selective terrorism or harassment.

The process has gone further than is sometimes realized in some western countries. Subversion in the American Army is obviously bound up with opposition to the Vietnam war and resistance to conscription; underground GI news sheets are primarily anti-*Vietnam* publications. Eldridge Cleaver, the man who is now contesting the leadership of the Black Panthers from his exile in Algiers, has said in his quasi-apocalyptic way that 'the stockades in Babylon are full of soldiers who refuse to fight. These men are going to become some of the most valuable guerrilla fighters.' Perhaps this need not be taken too seriously; but it does seem that disaffected conscripts have supplied the American underground with arms, instruction and trained recruits. The Deputy Attorney-General of California announced in April 1971, for example, that his office had recovered 55 grenades, 94 bricks of plastic explosives, 10 bazookas, 52 rifles and 65 revolvers stolen from local army bases. The racial factor has added to the dissension in the ranks. Fighting between black and white GIs has become commonplace in Vietnam, but there have been similar incidents among the American forces stationed in Germany, and rioting by black soldiers at bases in the United States, including Fort Hood and the riot control training centre in Kansas. One of the prime techniques used by radical activists in the American Army has been to try to pit conscripts against professional soldiers. One of the group responsible for the bombing of a military police station in San Francisco last year, for example, declared: 'We consider the GI to be a civilian, whereas we consider the lifers and the military structure to be a structure which is evolving to a more Gestapo-type experience.' [34] In other contexts, terrorists have tried to isolate the 'elite' units and those directly responsible for counter-insurgency operations from the armed forces as a whole.

The Limits of Urban Violence

Are the urban guerrillas likely to achieve their goals? In western industrial societies, to ask this question is really to ask whether revolution is possible. In the third world, urban guerrilla warfare fits into pre-existing patterns of insurgency and political instability. The modern city is vulnerable to terrorist attack; but in the last analysis, success or failure hinges on the public reaction.

(i) *The vulnerability of the industrial city.* The complexity of the modern city makes it vulnerable to the forms of sabotage that might be called 'technological terrorism'. No extremist group has succeeded in causing serious disruption in transport and communications in a Western society, although in the United States there has been a wave of assaults on property (and in Puerto Rico this has been part of a concerted drive by the Armed Liberation Commandos to scare off outside investors).[35] But the possibility of programmatic sabotage of essential services cannot be discounted, and plans for such a campaign in the United States have been elaborated by the Revolutionary Action Movement (RAM)—an organization of black extremists founded by Robert Williams. One of Williams' supporters has argued: 'What we must understand is that Charlie's system runs like an IBM machine. But an IBM machine has a weakness, and that weakness is complexity. Put something in the wrong place in an IBM machine and it's finished for a long time.'[36] Williams pointed out that it is possible to use primitive techniques to disrupt sophisticated institutions. He advocated a black revolutionary organization divided into three sections: armed self-defense groups operating legally; underground guerrilla squads to be employed against the police during riots; and a system of autonomous 'fire teams' who would be responsible for programmatic sabotage. They would pose as 'moderates' or 'patriots' in order to infiltrate high-security zones. Their first targets would be transport and communications in the major cities, followed by random attacks on corporation buildings and military installations. The saboteurs would try to create general panic and urban chaos by diverse means. For example, they might scatter tacks or boards with protruding nails on turnpikes and at major intersections during rush-hour traffic. And Williams took an unhinged arsonist's pleasure in the prospect of 'strategic fires' started across the countryside by teams of roving guerrillas. The fires would be used as a diversion and 'to elicit panic and a feeling of impending doom'.[37] Williams, unlike most other Black Power leaders, believes in the possibility of a minority revolution in the United States. His lunatic schemes for 'liberation zones' in the deep South or his idea that American middle-class society is so soft that it would fall apart as soon as economic production fell need not be taken seriously. But he pointed out that a marginal extremist group does have the *technical* power to cause enormous damage. The political consequences are a different matter.

(ii) *Terrorism and public opinion*. Herbert Marcuse was right to insist that the most violent political groups in western societies are composed of marginal social elements: ethnic and cultural minorities, and middle-class radicals. That is the source of their weakness. If they push the confrontation of political forces too far through the use of violence, they will eventually be swamped by the majority groups.

The advocates of 'student power' feel differently. They argue that the events of May 1968 in France demonstrated that radical students can provide the trigger for a broader movement of social unrest in an advanced industrial society. They point to the occupation of factories by French workers between 14 and 17 May and to the overnight formation of strike committees as examples of popular 'spontaneity'. They argue that orthodox Communists are wrong to insist that a 'vanguard party' is a prerequisite for revolution. 'What we need,' according to Daniel Cohn-Bendit, 'is not an organization with a capital O, but a host of insurrectional cells, be they ideological groups, study groups— we can even use street gangs.'[38]

But in fact there has not been a follow-up to May 1968. The temporary alliance between students and workers that was achieved in France crumbled away once the Government made up its mind to grant some limited economic concessions (the highest wage-rise granted was about 14 per cent). While student theorists were talking of revolution, most of the workers who joined the rallies and participated in strikes were merely posing bread-and-butter demands that the system was able to satisfy. The decisive factor that helped to turn the tide in Paris was the hostility and distrust shown by the leaders of the French Communist Party and the trade unions towards the student movement. What made a real insurrection in France in 1968 impossible was the factor that the students had decided to neglect: the absence of a mass organization with a coherent strategy for the seizure of power. It was abundantly clear in May that the French Communists were not prepared to adopt this role. And the leaders of the Italian Communist Party (PCI) have swung towards an equally reformist position more recently.

Until the twelfth Party Congress in 1968, the Italian Communists had probably played a more militant part in political strikes and student protest than any other Moscow-line Communist party in Europe. In June 1969, the 'Manifesto' group (the left-wing extremists associated with the paper *Il Manifesto*) were excluded from the party on the grounds that they had acted as a divisive force and sapped the party's strength. The present tactical goal of the party leadership, according to Luis Magri, one of the 'Manifesto' rebels, 'is a convergence between the working class and the "advanced" wing of big capital, on a common economic programme for the elimination of parasitism and the development of social services, that will harmoniously reconcile the exigencies of productivity and the needs of the workers within the system'.[39] It remains to be seen whether the new orientation of the party executive (which is being encouraged by Aldo Moro and others within the Christian Democratic Party who have floated the idea of achieving an 'understanding' with the PCI) will turn out to be more than a tactical ruse. The 'Manifesto' group still hopes to inspire a revolt within the party ranks and argues that continuing labour unrest, student radicalism and the resurgence of extreme right-wing groups like the Italian Social Movement (MSI) are all leading towards the polarization of political forces, not to any lasting form of 'convergence' between capitalist and Communist.

The important thing to note is that, in the two European countries with the highest incidence of political violence, the Moscow-line Communist parties seem to have placed themselves *hors de combat* as far as student rebellion and insurrectionary tactics are concerned. Something similar has happened in Japan, where student militants have remained fairly isolated from unionized workers—apart from the few thousand who have joined the Youth Committees against the War, founded in 1965, and the radical railway workers who have joined in political strikes against the Japanese–American alliance. The nationalism and devotion to duty that are built into the Japanese social system have discouraged widespread protest movements, and the fragmented character of Japanese trade unions (each corporation has its own trade union) has deterred attempts at nation-wide political strikes.

In the United States, student rebels are almost completely divorced from the union movement, although there was a strong faction inside the Students for a Democratic Society (SDS) group that argued that effective political action would only be possible through off-campus agitation. But that faction was outvoted by those who formed the Weathermen in 1969, and the SDS strategy was narrowed down to support for minority groups and third world revolutionaries through a campaign of terrorist violence. 'Winning state power,' according to the first important Weatherman manifesto, 'will occur as a result of the military forces of the US extending themselves around the world and being defeated piecemeal.'[40]

The student radicals have declined in political importance because events showed that they were incapable of cementing a broad front with workers or the traditional left-wing parties. The nature of New Left protest limits its popular appeal. It is partly a *qualitative* protest against the life-styles of bourgeois society and the problems of living in an advanced industrial country (centralization, urbanization, pollution and so on); and partly a *vicarious* protest in sympathy with deprived minority groups of guerrillas and peasant rebels in the third world. It has rarely touched upon the everyday problems of the ordinary man. It is interesting to note that as student radicals have rediscovered their basic isolation as a political force, they have become more violent and more 'professional'. The way that Japanese student militants have organized themselves for street-fighting is a dramatic example of that.[41] . . . Similar rituals of violence have become familiar in Paris. . . . The case of the Weathermen was the supreme example of how one radical student movement, failing to strike a responsive chord among the nation as a whole, took to underground violence.

It is tempting to judge these *groupuscules* and would-be guerrillas in the light of one of Lenin's more acute observations: 'Serious politics begins where millions of men and women are.' Popular attitudes about politics and violence in most western societies mean that most people will tend to view a terrorist group like the Weathermen with incomprehension or anger—although it is important to note that the attitude of minority groups are somewhat different. The 'Angry Brigade' terrorists who bombed the flat of Mr. John Davies, the British Secretary for Trade and Industry, on 31 July 1971 (supposedly in sympathy with Upper Clyde Shipyards workers threatened with redundancy) were not likely to win much of a hearing in a society where the legitimacy of the private use of political violence is not generally accepted. In the United States, a country with a record of much greater civil violence, the report of a recent national commission of violence was undoubtedly right when it said: "The historical and contemporary evidence of the United States suggests that popular support tends to sanction violence in support of the *status quo*; the use of public violence to maintain public order; the use of private violence to maintain popular conceptions of social order when government cannot or will not.'[42] Put in cruder terms, this means that an increase in left-wing or revolutionary political violence is likely to mobilize the 'law and order' majority and drive the government to take progressively tougher measures.

In short, the failure to mobilize popular support is the weakness of most of the contemporary urban guerrilla movements. Where they can find this support they have a chance of success: where they can't, they fail. And failure is frequent. This stems partly from the fact that the movements are estranged from the major left-wing parties (Asia is an exception), but also from something more fundamental. The terrorist is a man who refuses to compromise, to explore the possibilities of peaceful change. It is part of his task to convince his potential supporters that there are no prospects for constitutional change or non-violent reform. Hence the dilemma of a group like the Chilean Movement of the Revolutionary Left (MIR), which advocates violent revolution, after a Marxist President, Dr. Salvador Allende, was voted into power in September 1970. It is also the dilemma of armed extremists in the western democracies. Their common tactic is to try to erode public confidence in the constitutional system by creating disorder in the streets, economic chaos and polarization of political forces around the 'law and order' issue. As Che Guevara observed, 'Where a government has come to power through some popular vote, fraudulent or not, the guerrilla outbreak cannot be promoted since the possibilities of peaceful struggle have not yet been exhausted'. . . .[43]

498   (iii) *The problems of response.* Experience has shown that most modern governments can contain urban terrorism, so long as they can count on the loyalty of the security forces. The question is at what cost. Experience also leads to the sorry conclusion that police-states are the most efficient of all in suppressing terrorist groups. No one is anticipating a wave of urban guerrilla activity in the Soviet Union or in South Africa.

Venezuela and Brazil are good examples of strong handling. In Venezuela, this was combined with the use of an election to swing public opinion away from the guerrillas.[44] President Betancourt was very astute in handling the armed forces and in dealing out repression. By studiously cultivating his senior officers, Betancourt won back the support of the security forces (which had been notoriously faction-ridden) while crushing the 1962 mutinies ruthlessly. He took great pains to show his respect for the due legal process, and exceptional measures were applied only when moderate opinion was already convinced of the need for them. For example, a vicious attack by the National Liberation Front (FLN) on an excursion train in September 1963, provided the perfect justification for the tough measures and emergency laws that were applied immediately afterwards. Finally, by holding a free election (where 90 per cent of the voters turned out despite the FLN instruction to boycott the polls) Betancourt imposed a shattering political defeat on the rebels. After December 1963, the insurgents were hopelessly divided and the Communists returned to the theory of 'peaceful coexistence' that was formally reinstated as party policy in 1967. The Venezuelan insurgency is now confined to a few roaming guerrilla bands in the hills of Falcón province who have so far managed to survive but have no impact on the politics of the country as a whole.

The example of Venezuela is important, because the guerrillas came closer to realizing the conditions for a successful urban insurrection than any later movement has managed to come. Looking back on 1962, Teodoro Petkoff, a leader of the Venezuelan Communist Party, still insists that 'we could have won.'[45] Betancourt's formula for urban counter-insurgency may not be relevant to all contemporary situations. The Brazilians, operating within a very different political framework, have tried something cruder. The military response of the Brazilian Government to the urban offensive was to eliminate the terrorist bases in the cities and to force them to do battle in situations where they were bound to be outgunned. This tactic depended (as did the French operation in Algiers in the late 1950s) on the use of mass interrogation—including the frequent and often irresponsible use of torture—to track down the guerrillas. . . . [46] In the last quarter of 1970, there were six street battles in Rio and São Paulo, from which the security forces emerged the clear victors. By the end of 1970, several urban guerrilla groups had been decisively crushed, including the Revolutionary Armed Vanguard-Palmares and the Revolutionary Communist Party of Brazil, and successive leaders of the National Liberation Action (ALN) had been captured or killed. Several attempts by the People's Revolutionary Vanguard (VPR) to found a rural base had been defeated by classic methods of encirclement, and urban terrorist operations in the first half of 1971 were confined to insignificant robberies and acts of random terrorism. In the course of the campaign, it became clear that the terrorists partly succeeded in their tactic of driving President Garrastradú Medici's Government towards more repressive measures, and there is no doubt that these served to alienate important sections of the Brazilian middle class as well as liberal opinion abroad. On the other hand, the cohesion of the Brazilian armed forces, the size and complexity of the country, and a period of sustained economic growth all helped the Government to overpower its local opposition.

But urban guerrillas can succeed in producing a polarization of political forces in such a way that the situation cannot easily be untangled, and this is precisely what extremists of both sides are counting on. Tom Hayden, one of the founders of the SDS, gave an American version of Marighella's strategy of militarization when he wrote: 'The coming of repression will speed up time, making a revolutionary situation more likely—We are creating an America where it is necessary for the government to rule behind barbed wire, for the President to speak only at military bases and, finally, where it will be necessary for the people to fight back.' [47]

It is in this sense that the urban terrorist in industrial societies should be seen as a political catalyst. It is arguable that, in most cases, urban guerrillas are dangerous less for what they do than for what they inspire: the erosion of the consensus, a hardening of the political battle lines, and a backlash that strikes back too hard and too indiscriminately. Terrorism and street violence were used by the Nazis to help break down the fabric of Weimar Germany; the assassination of leading moderate politicians was used by ultra-nationalist groups in inter-war Japan to swing government policy towards a programme of military expansion. What worked for the far right between the wars is likely to have rather different consequences for the far left today, although the tactic is similar. In the advanced industrial societies, political terrorists are unlikely to win support except in conditions of extreme social and economic crisis. On the other hand, as Mr. Pierre Trudeau, the Prime Minister of Canada, observed after the Cross–Laporte kidnappings in October 1970: 'It only takes a few fanatics to show us just how vulnerable a democratic society can be when democracy is not ready to defend itself.'[48] The full logic of that statement seems to be working itself out in Uruguay.

The Uruguayan Government of President Pacheco Areco has faced a sustained offensive from the National Liberation Movement (Tupamaros) since 1968. (The organization was founded in 1963.) The security problem and the emergency measures that Pacheco applied to deal with it have deeply divided the Government and in June 1971 there was an unsuccessful attempt to impeach the President by Congress. The Tupamaros have shown signs of having infiltrated the armed forces, the police and the civil service. Although the relatively powerful Uruguayan Communist Party and the trade union leadership have refused to declare public support for the guerrillas, they may have to revise their attitude if the left wing *Frente Amplio* is defeated in the November 1971 presidential elections. The success of the Tupamaros in winning popular support has owed something to the country's continuing economic crisis (stemming from the drop in the world price of pastoral products) and to their very selective methods. Uruguay is a small and vulnerable democratic society that has come upon hard times—a welfare state that failed. But the most important factor was that the Government of President Pacheco Areco was weak and divided, while the security forces at his disposal were tiny and without experience of counter-guerrilla operations. There are only about 12,000 men in the Uruguayan armed forces and 22,000 in the police. At the same time, President Pacheco found his hands tied by public opinion when it came to dealing with the guerrillas. Congress resisted the requested reintroduction of emergency measures. The weightiest obstacle to a revolution is the possibility of military intervention by one of the giant neighbours, Brazil or Argentina.

The situation in Ulster is a special case, which might be defined as *quasi*-colonial in the sense that the IRA bases its hopes on the calculation that the British Government will eventually respond to the human and economic costs of maintaining order in the Province by pulling out altogether. The IRA has set out to undermine successive moderate Unionist governments in Stormont in order to provoke a right-wing Protestant backlash, or direct rule from Westminster (which has sometimes been regarded as a step towards the unification of Ireland). The IRA's strength in the current campaign has stemmed from the measure of support it can command from the Catholic part of the population rather than from any degree of military competence or ingenuity. The IRA's campaign of terror in 1970-71 had some success, helping to precipitate the fall of the Northern Ireland Prime Minister in March 1971, and then, in August 1971, leading to the introduction of internment. From its original role of keeping the peace between the Catholic and Protestant communities, the British Army moved over to an offensive intended to root out the IRA as a fighting force. Although the new tactics produced military results, they helped to polarize opinion in Ulster and enabled Catholic critics to represent the army as a repressive force. In this sense, IRA terrorism succeeded. It led to a situation where the British Army, which began as the referee between the two communities in Ulster, appeared as a party to the quarrel. The chaos it engendered helped to postpone the application of social reforms designed to get to the root of the problem and thus eroded Catholic faith in solutions within the existing framework.

Whatever their hopes of success, the tactics employed by the urban guerrillas pose a direct threat to the international order. The theory of global revolution that has been used in the attempt to rationalize crimes against diplomats and other foreign citizens is a flat rejection of the principles that have traditionally guided relations between sovereign states. The claim to a right to rebel under intolerable social and political conditions cannot be used to sanction this type of international crime. It is also clear that there is an indirect link between civil violence and the strategic balance.

In the United States, the immediate effect of mounting civil dissent (for which the common platform is opposition to the Vietnam war) has been to impose constraints on foreign policy. Dissent within the American armed forces has made it increasingly probable that the Army will have to dispense with the draft and 'to professionalize'—which will clearly limit the country's capacity to intervene in outside conflicts and will also make it difficult to maintain current troop levels in Europe. The likelihood that dissent will continue to take violent forms in the United States adds to the possibility that the Americans will enter a new isolationist phase in their attitude to the world. Although race relations and Black Power violence remain the most obvious threats to civil peace in America, these are unlikely to boil over into full-scale ghetto revolts. It is hard to imagine that any rational Black Power leader would expose his followers to the risks involved in an uprising in a limited area that could be easily encircled—or that any American government would fail to take vigorous action against it.

Terrorism may prove to have the most dangerous effects in western industrial societies. A revolution in Uruguay, after all, would hardly alter the strategic balance in any significant way. On the other hand, ghetto revolts in the United States could disrupt the most powerful economy in the world and impose severe constraint on America's capacity to act as a great power. A sustained campaign of urban terrorism in Europe might undermine popular faith in the democratic system and raise the prospect of a more repressive form of government. Terrorists, of course, rarely make revolutions. In Latin America, for example, the most radical social changes in recent years have been brought about by a reformist military junta, in Peru, and a freely elected Marxist government in Chile. These are likely to be the patterns for future bloodless revolutions. And the Chilean formula could apply to Italy, where the prospect of a governing coalition including Communists is much more immediate, and more serious, than a revolution of the streets. This leads to the conclusion that the urban guerrilla is a political catalyst whose actions can radicalize a society and bring about the kind of social and economic confusion that will lead to a decline in popular belief in peaceful solutions. The end results may be indirect and will often take forms that neither the guerrillas nor the government anticipated.

Notes

*Reprinted with minor omissions.

1. See Carl von Clausewitz, *On War*, trans. J.J. Graham (London: Routledge and Kegal Paul, 1956), Book VI, chapter 26; Peter Paret and John W. Shy, *Guerrillas in the 1960s* (New York: Praeger, 1962), pp. 11-15; Regis Debray, *Revolution in the Revolution?* (Harmondsworth: Penguin, 1968), p. 67.
2. Reprinted as Appendix I in *The Memoirs of General Grivas*, Ed. Charles Foley (London: Longmans, 1964), p. 204.
3. See Feliks Gross, *The Seizure of Political Power* (New York: Philosophical Library, 1957), pp. 109-10; Gross, 'Political Violence and Terror in Nineteenth and Twentieth Century Russia and Eastern Europe' in *Assassination and Political Violence:* A Staff Report to the National Commission on the Causes and Prevention of Violence (New York, 1970), pp. 516-44; Stepniak, *Underground Russia* (New York: Scribners, 1892).
4. See Stephen T. Hosmer, *Viet Cong Repression and Its Implications for the Future* (Lexington, Mass.: RAND Corporation, 1970), pp. 63-111.
5. See Tim Pat Coogan, *The IRA* (London: Pall Mall, 1970), pp. 150-73.
6. See, *inter alia*, Roland Gaucher, *Les terroristes* (Paris: Albin Michel, 1965), pp. 255-77, and Edgar O'Ballance, *The Algerian Insurrection* (London: Faber, 1967), pp. 53-54.
7. See Thomas Perry Thornton, 'Terror as a Weapon of Political Agitation' in Harry Eckstein (Ed), *Internal War* (New York: Free Press of Glencoe, 1964), p. 74.
8. See 'On Principles and Strategic Questions' reprinted in *Les Temps Modernes* (Paris) November 1969. See also Appendix, pp. 40-41.
9. See the author's *Urban Guerrillas in Latin America* (London: Institute for the Study of Conflict: Conflict Studies No. 8, 1970).
10. *Times of India*, 19 May 1970.
11. *Indian Express*, 13 May 1970.
12. *The Hindustan Times Weekly*, 17 January 1971.
13. *Forum* (Dacca), 7 November 1970.
14. *Times of India*, 19 May 1970.
15. See D. J. Dwyer, *Urbanization as a Factor in the Political Development of South-East Asia* (Discussion paper at Pacific Conference, Vina del Mar, Chile, 27 September-3 October 1970).
16. On Vietnam, see Samuel P. Huntington, 'The Bases of Accommodation' in *Foreign Affairs* (New York), 46, 1968, p. 648.
17. See Bryan Roberts, 'Migration and Population Growth in Guatemala City' in Roberts and Lowder, *Urban Population*

Growth and Migration in Latin America (Liverpool: University of Liverpool Press, 1970).
18. B. F. Hoselitz, 'The Role of Urbanization in Economic Development' in Roy Turner (Ed), *India's Urban Future* (Berkeley: University of California Press, 1962), pp. 164-67.
19. See, for example, Charles Tilly, 'Collective Violence in European Perspective' in *Violence in America: Historical and Comparative Perspectives*, A Report to the National Commission on the Causes and Prevention of Violence (New York, 1969), esp. pp. 33-37.
20. Frantz Fanon, *The Wretched of the Earth*, trans. C. Farrington (Harmondsworth: Penguin, 1970), p. 103.
21. One of the more convincing arguments along these lines is T.G. McGee and W.R. Armstrong, 'Revolutionary Change and the Third World City' in *Civilisations* (Paris), Vol. XVIII, No. 3, 1968.
22. Martin Oppenheimer, *Urban Guerrilla* (Harmondsworth: Penguin, 1970), p. 42.
23. See, for example, John Slimming, *Malaysia: Death of a Democracy* (London: Murray, 1969), pp. 25-60, and T.G. McGee, *The Urbanization Process in the Third World* (London: Bell, 1971), pp. 64-89; 149-72.
24. Moises Moleiro, 'Las Ensenanzas de la Guerra Revolucionaria en Venezuela' in V. Bambirra *et al.*, *Diez Anos de Insurreccion en America Latina* (Santiago: Prensa Latino-Americana, 1971), Vol. 1, p. 173.
25. See Walter Williams, 'Cleveland's Crisis Ghetto' in Peter H. Rossia (Ed), *Ghetto Revolts* (New York: Trans-Action Books, 1970), pp. 13-30.
26. Brian Crozier, *The Rebels* (London: Chatto and Windus, 1960), p. 9.
27. Fanon, op. cit., p. 74.
28. Pierre Vallieres, *Negres blancs d'Amerique* (Montreal: Editions Parti Pris, 1969), pp. 66-67.
29. See *O Jornal do Brasil*, 12 January 1971.
30. See Maria Esther Giglio, *La guerrilla tupamara* (Havana: Casa de las Americas, 1970), and *Revolucion y Cultura* (Havana) No. 21, December 1970.
31. *Granma* (Havana), 8 October 1970.
32. *Latin America* (London), 26 February 1971; *Prensa Latina* (Havana), 7 April 1971.
33. Lenin, 'Lessons of the Moscow Uprising' in *Collected Works* (Moscow, 1967), Vol. II, p. 174.
34. Warren Hinckle, *Guerrilla War in the USA* (unpublished manuscript, New York, 1970), p. 151.
35. See 'Porto-Rico: Le Reveil en Armes' in *Africasia* (Paris), 26 April 1971; and interview with Alfonso Beal reprinted in John Gerassi (Ed), *Towards Revolution* (London: Weidenfeld and Nicolson, 1971), pp. 641-44.
36. Max Stanford, 'Black Guerrilla Warfare: Strategy and Tactics' in

The Black Scholar (San Francisco), November 1970, p. 37.
37. Robert F. Williams, 'The Potential of a Minority Revolution: Part 2' in *The Crusader* (Havana), August 1965.
38. Daniel Cohn-Bendit, *Obsolete Communism: The Left-Wing Alternative*, trans. A. Pomerans (Harmondsworth: Penguin, 1968), p. 256. See Daniel Singer, *Prelude to Revolution: France in May, 1968* (London: Cape, 1970) for an interesting attempt to use *les evenements* as a basis for prediction.
39. See Lucio Magri, 'Italian Communism Today' in *New Left Review* (London), No. 66 March-April, 1971, p. 49.
40. See 'You Don't Need a Weatherman to Know Which Way the Wind Blows' in Harold Jacobs (Ed), *Weatherman* (New York: Ramparts, 1970), p. 53.
41. For a useful discussion of changing techniques, see Bernard Beraud, *La gauche revolutionnaire au Japon* (Paris: Editions du Seuil, 1970), pp. 131-37.
42. *Violence in America*, op. cit., pp. 813-14.
43. Che Guevara, *Guerrilla Warfare* (Harmondsworth: Penguin, 1969), p. 14.
44. This account largely follows Moises Moleiro, *El MIR de Venezuela* (Havana: Guaivas Instituto del Libon, 1967), and Richard Gott, *Guerrilla Movements in Latin America* (London: Nelson, 1970), pp. 93-165.
45. See interview with Petkoff in *World Marxist Review* (Moscow), April 1968.
46. See, *inter alia*, Joao Quartim, 'La guerilla urbaine au Bresil' in *Les Temps Modernes* (Paris), November 1970, pp. 838-74.
47. Tom Hayden, *Rebellion and Repression* (Cleveland, Ohio: Meridian, 1969), pp. 14 and 16.
48. *Le Monde* (Paris), 20 October 1970.

Minimanual of the Urban Guerrilla
by Carlos Marighella

A Definition of the Urban Guerrilla

The chronic structural crisis characteristic of Brazil today, and its resultant political instability, are what have brought about the upsurge of revolutionary war in the country. The revolutionary war manifests itself in the form of urban guerrilla warfare, psychological warfare, or rural guerrilla warfare. Urban guerrilla warfare or psychological warfare in the city depends on the urban guerrilla.

The urban guerrilla is a man who fights the military dictatorship with arms, using unconventional methods. A political revolutionary and an ardent patriot, he is a fighter for his country's liberation, a friend of the people and of freedom. The area in which the urban guerrilla acts is in the large Brazilian cities. There are also bandits, commonly known as outlaws, who work in the big cities. Many times assaults by outlaws are taken as actions by urban guerrillas.

The urban guerrilla, however, differs radically from the outlaw. The outlaw benefits personally from the action, and attacks indiscriminately without distinguishing between the exploited and the exploiters, which is why there are so many ordinary men and women among his victims. The urban guerrilla follows a political goal and only attacks the government, the big capitalists, and the foreign imperialists, particularly North Americans.

Another element just as prejudicial as the outlaw and also operating in the urban area is the right-wing counter-revolutionary who creates confusion, assaults banks, hurls bombs, kidnaps, assassinates, and commits the worst imaginable crimes against urban guerrillas, revolutionary priests, students, and citizens who oppose fascism and seek liberty.

The urban guerrilla is an implacable enemy of the government and systematically inflicts damage on the authorities and on the men who dominate the country and exercise power. The principal task of the urban guerrilla is to distract, to wear out, to demoralize the militarists, the military dictatorship and its repressive forces, and also to attack and destroy the wealth and property of the North Americans, the foreign managers, and the Brazilian upper class.

The urban guerrilla is not afraid of dismantling and destroying the present Brazilian economic, political, and social system, for his aim is to help the rural guerrilla and to collaborate in the creation of a totally new and revolutionary social and political structure, with the armed people in power.

The urban guerrilla must have a certain minimal political understanding. To gain that he must read certain printed or mimeographed works such as:

Guerrilla Warfare by Che Guevara
Memories of a Terrorist
Some Questions about the Brazilian Guerrillas
Guerrilla Operations and Tactics
On Strategic Problems and Principles
Certain Tactical Principles for Comrades Undertaking Guerrilla Operations
Organizational Questions
O Guerrilheiro, newspaper of the Brazilian revolutionary groups.

Personal Qualities of the Urban Guerrilla and How He Subsists

The urban guerrilla is characterized by his bravery and decisive nature. He must be a good tactician and a good shot. The urban guerrilla must be a person of great astuteness to compensate for the fact that he is not sufficiently strong in arms, ammunition, and equipment.

The career militarists or the government police have modern arms and transport, and can go about anywhere freely, using the force of their power. The urban guerrilla does not have such resources at his disposal and leads a clandestine existence. Sometimes he is a convicted person or is out on parole, and is obliged to use false documents.

Nevertheless, the urban guerrilla has a certain advantage over the conventional military or the police. It is that, while the military and the police act on behalf of the enemy, whom the people hate, the urban guerrilla defends a just cause, which is the people's cause.

The urban guerrilla's arms are inferior to the enemy's, but from a moral point of view, the urban guerrilla has an undeniable superiority.

This moral superiority is what sustains the urban guerrilla. Thanks to it, the urban guerrilla can accomplish his principal duty, which is to attack and to survive.

The urban guerrilla has to capture or divert arms from the enemy to be able to fight. Because his arms are not uniform, since what he has are expropriated or have fallen into his hands in different ways, the urban guerrilla faces the problem of a variety of arms and a shortage of ammunition. Moreover, he has no place to practice shooting and marksmanship.

These difficulties have to be surmounted, forcing the urban guerrilla to be imaginative and creative, qualities without which it would be impossible for him to carry out his role as a revolutionary.

The urban guerrilla must possess initiative, mobility, and flexibility, as well as versatility and a command of any situation. Initiative especially is an indispensable quality. It is not always possible to foresee everything, and the urban guerrilla cannot let himself become confused, or wait for orders. His duty is to act, to find adequate solutions for each problem he faces, and not to retreat. It is better to err acting than to do nothing for fear of erring. Without initiative there is no urban guerrilla warfare.

Other important qualities in the urban guerrilla are the following: to be a good walker, to be able to stand up against fatigue, hunger, rain, heat. To know how to hide and to be vigilant. To conquer the art of dissembling. Never to fear danger. To behave the same by day as by night. Not to act impetuously. To have unlimited patience. To remain calm and cool in the worst conditions and situations. Never to leave a track or trail. Not to get discouraged.

In the face of the almost insurmountable difficulties of urban warfare, sometimes comrades weaken, leave, give up the work.

The urban guerrilla is not a businessman in a commercial firm nor is he a character in a play. Urban guerrilla warfare, like rural guerrilla warfare, is a pledge the guerrilla makes to himself. When he cannot face the difficulties, or knows that he lacks the patience to wait, then it is better to relinquish his role before he betrays his pledge, for he clearly lacks the basic qualities necessary to be a guerrilla.

The urban guerrilla must know how to live among the people and must be careful not to appear strange and separated from ordinary city life.

He should not wear clothes that are different from those that other people wear. Elaborate and high fashion clothing for men or women may often be a handicap if the urban guerrilla's mission takes him into working-class neighborhoods or sections where such dress is uncommon. The same care has to be taken if the urban guerrilla moves from the South to the North or *vice versa*.

The urban guerrilla must live by his work or professional activity. If he is known and sought by the police, if he is convicted or is on parole, he must go underground and sometimes must live hidden. Under such circumstances, the urban guerrilla cannot reveal his activity to anyone, since that is always and only the responsibility of the revolutionary organization in which he is participating.

The urban guerrilla must have a great capacity for observation, must be well informed about everything, principally about the enemy's movements, and must be very searching and knowledgeable about the area in which he lives, operates, or through which he moves.

But the fundamental and decisive characteristic of the urban guerrilla is that he is a man who fights with arms; given this condition, there is very little likelihood that he will be able to follow his normal profession for long without being identified. The role of expropriation thus looms as clear as high noon. It is impossible for the urban guerrilla to exist and survive without fighting to expropriate.

Thus, within the framework of the class struggle, as it inevitably and necessarily sharpens, the armed struggle of the urban guerrilla points toward two essential objectives:

(a) the physical liquidation of the chiefs and assistants of the armed forces and of the police;

(b) the expropriation of government resources and those belonging to the big capitalists, latifundists, and imperialists, with small expropriations used for the maintenance of individual urban guerrillas and large ones for the sustenance of the revolution itself.

It is clear that the armed struggle of the urban guerrilla also has other objectives. But here we are referring to the two basic objectives, above all expropriation. It is necessary for every urban guerrilla to keep in mind always that he can only maintain his existence if he is disposed to kill the police and those dedicated to repression, and if he is determined—truly determined—to expropriate the wealth of the big capitalists, the latifundists, and the imperialists.

One of the fundamental characteristics of the Brazilian revolution is that from the beginning it developed around the expropriation of the wealth of the major bourgeois, imperialist, and latifundist interests, without excluding the richest and most powerful commercial elements engaged in the import-export business.

And by expropriating the wealth of the principal enemies of the people, the Brazilian revolution was able to hit them at their vital centre, with preferential and systematic attacks on the banking network—that is to say, the most telling blows were levelled against capitalism's nerve system.

The bank robberies carried out by the Brazilian urban guerrillas hurt such big capitalists as Moreira Salles and others, the foreign firms which insure and reinsure the banking capital, the imperialist companies, the federal and state governments—all of them systematically expropriated as of now.

The fruit of these expropriations has been devoted to the work of learning and perfecting urban guerrilla techniques, the purchase, the production, and the transportation of arms and ammunition for the rural areas, the security apparatus of the revolutionaries, the daily maintenance of the fighters, of those who have been liberated from prison by armed force and those who are wounded or persecuted by the police, or to any kind of problem concerning comrades liberated from jail, or assassinated by the police and the military dictatorship.

The tremendous costs of the revolutionary war must fall on the big capitalists, on imperialism, and the latifundists and on the government too, both federal and state, since they are all exploiters and oppressors of the people.

Men of the government, agents of the dictatorship and of North American imperialism principally, must pay with their lives for the crimes committed against the Brazilian people.

In Brazil, the number of violent actions carried out by urban guerrillas, including deaths, explosions, seizures of arms, ammunition, and explosives, assaults on banks and prisons, etc., is significant enough to leave no room for doubt as the the actual aims of the revolutionaries. The execution of the CIA spy Charles Chandler, a member of the US Army who came from the war in

Vietnam to infiltrate the Brazilian student movement, the military henchmen killed in bloody encounters with urban guerrillas, all are witness to the fact that we are in full revolutionary war and that the war can be waged only by violent means.

This is the reason why the urban guerrilla uses armed struggle and why he continues to concentrate his activity on the physical extermination of the agents of repression, and to dedicate twenty-four hours a day to expropriation from the people's exploiters.

Technical Preparation of the Urban Guerrilla

No one can become an urban guerrilla without paying special attention to technical preparation.

The technical preparation of the urban guerrilla runs from the concern for his physical preparedness, to knowledge of and apprenticeship in professions and skills of all kinds, particularly manual skills.

The urban guerrilla can have strong physical resistance only if he trains systematically. He cannot be a good fighter if he has not learned the art of fighting. For that reason the urban guerrilla must learn and practise various kinds of fighting, of attack, of personal defence.

Other useful forms of physical preparation are hiking, camping, and practice in survival in the woods, mountain climbing, rowing, swimming, skin diving, training as a frogman, fishing, harpooning, and the hunting of birds, small and big game.

It is very important to learn how to drive, pilot a plane, handle a motor boat and a sail boat, understand mechanics, radio, telephone, electricity, and have some knowledge of electronic techniques.

It is also important to have a knowledge or topographical information, to be able to locate one's position by instruments or other available resources, to calculate distances, make maps and plans, draw to scale, make timings, work with an angle protractor, a compass, etc.

A knowledge of chemistry and of colour combination, of stamp-making, the domination of the technique of caligraphy and the copying of letters and other skills are part of the technical preparation of the urban guerrilla, who is obliged to falsify documents in order to live within a society that he seeks to destroy.

In the area of auxiliary medicine he has the special role of being a doctor or understanding medicine, nursing, pharmacology, drugs, elementary surgery, and emergency first aid.

The basic question in the technical preparation of the urban guerrilla is nevertheless to know how to handle arms such as the machine gun, revolver, automatic, FAL, various types of shotguns, carbines, mortars, bazookas, etc.

A knowledge of various types of ammunition and explosives is another aspect to consider. Among the explosives, dynamite must be well understood. The use of incendiary bombs, of smoke bombs, and other types are indispensable prior knowledge.

To know how to make and repair arms, prepare Molotov cocktails, grenades, mines, mome-made destructive devices, how to blow up bridges, tear up and put out of service rails and sleepers, these are requisites in the technical preparation of the urban guerrilla that can never be considered unimportant.

The highest level of preparation for the urban guerrilla is the centre for technical training. But only the guerrilla who has already passed the preliminary examination can go on to this school—that is to say, one who has passed the proof of fire in revolutionary action, in actual combat against the enemy.

The Urban Guerrilla's Arms

The urban guerrilla's arms are light arms, easily exchanged, usually captured from the enemy, purchased, or made on the spot.

Light arms have the advantage of fast handling and easy transport. In general, light arms are characterized as short barrelled. This includes many automatic arms.

Automatic and semi-automatic arms considerably increase the fighting power of the urban guerrilla. The disadvantage of this type of arm for us is the difficulty in controlling it, resulting in wasted rounds or in a prodigious use of ammunition, compensated for only by optimal aim and firing precision. Men who are poorly trained convert automatic weapons into an ammunition drain.

Experience has shown that the basic arm of the urban

guerrilla is the light machine gun. This arm, in addition to being efficient and easy to shoot in an urban area, has the advantage of being greatly respected by the enemy. The guerrilla must know thoroughly how to handle the machine gun, now so popular and indispensable to the Brazilian urban guerrilla.

The ideal machine gun for the urban guerrilla is the Ina 45 calibre. Other types of machine guns of different calibres can be used—understanding, of course, the problem of ammunition. Thus it is preferable that the industrial potential of the urban guerrilla permits the production of a single machine gun so that the ammunition used can be standardized.

Each firing group of urban guerrillas must have a machine gun managed by a good marksman. The other components of the group must be armed with .38 revolvers, our standard arm. The .32 is also useful for those who want to participate. But the .38 is preferable since its impact usually puts the enemy out of action.

Hand grenades and conventional smoke bombs can be considered light arms, with defensive power for cover and withdrawal.

Long barrel arms are more difficult for the urban guerrilla to transport and attract much attention because of their size. Among the long barrel arms are the FAL, the Mauser guns or rifles, hunting guns such as the Winchester, and others.

Shotguns can be useful if used at close range and point blank. They are useful even for a poor shot, especially at night when precision isn't much help. A pressure airgun can be useful for training in marksmanship. Bazookas and mortars can also be used in action but the conditions for using them have to be prepared and the people who use them must be trained.

The urban guerrilla should not try to base his actions on the use of heavy arms, which have major drawbacks in a type of fighting that demands lightweight weapons to insure mobility and speed.

Home-made weapons are often as efficient as the best arms produced in conventional factories, and even a cut-off shotgun is a good arm for the urban guerrilla.

The urban guerrilla's role as gunsmith has a fundamental importance. As gunsmith he takes care of the arms, knows how to repair them, and in many cases can set up a small shop for improvising and producing efficient small arms.

Work in metallurgy and on the mechanical lathe are basic

skills the urban guerrilla should incorporate into his industrial planning, which is the construction of home-made weapons.

This construction and courses in explosives and sabotage must be organized. The primary materials for practice in these courses must be obtained ahead of time to prevent an incomplete apprenticeship—that is to say, so as to leave no room for experimentation.

Molotov cocktails, gasoline, home-made contrivances such as catapults and mortars for firing explosives, grenades made of tubes and cans, smoke bombs, mines, conventional explosives such as dynamite and potassium chloride, plastic explosives, gelatine capsules, ammunition of every kind are indispensable to the success of the urban guerrilla's mission.

The method of obtaining the necessary materials and munitions will be to buy them or to take them by force in expropriation actions especially planned and carried out.

The urban guerrilla will be careful not to keep explosives and materials that can cause accidents around for very long, but will try always to use them immediately on their destined targets.

The urban guerrilla's arms and his ability to maintain them constitute his fire power. By taking advantage of modern arms and introducing innovations in his fire power and in the use of certain arms, the urban guerrilla can change many of the tactics of city warfare. An example of this was the innovation made by the urban guerrillas in Brazil when they introduced the machine gun in their attacks on banks.

When the massive use of uniform machine guns becomes possible, there will be new changes in urban guerrilla warfare tactics. The firing group that utilizes uniform weapons and corresponding ammunition, with reasonable support for their maintenance, will reach a considerable level of efficiency. The urban guerrilla increases his efficiency as he improves his firing potential.

The Shot: The Urban Guerrilla's Reason for Existence

The urban guerrilla's reason for existence, the basic condition in which he acts and survives, is to shoot. The urban guerrilla must know how to shoot well because it is required by his type of combat.

In conventional warfare, combat is generally at a distance

with long-range arms. In unconventional warfare, in which urban guerrilla warfare is included, the combat is at close range, often very close. To prevent his own extinction, the urban guerrilla has to shoot first and he cannot err in his shot. He cannot waste his ammunition because he doesn't have large amounts, so he must save it. Nor can he replace his ammunition quickly, since he is part of a small group in which each guerrilla has to take care of himself. The urban guerrilla can lose no time and must be able to shoot at once.

One fundamental fact, which we want to emphasize fully and whose particular importance cannot be overestimated, is that the urban guerrilla must not fire continuously, using up his ammunition. It may be that the enemy is not responding to the fire precisely because he is waiting until the guerrilla's ammunition is used up. At such a moment, without having time to replace his ammunition, the urban guerrilla faces a rain of enemy fire and can be taken prisoner or be killed.

In spite of the value of the surprise factor which many times makes it unnecessary for the urban guerrilla to use his arms, he cannot be allowed the luxury of entering combat without knowing how to shoot. And face to face with the enemy, he must always be moving from one position to another, because to stay in one position makes him a fixed target and, as such, very vulnerable.

The urban guerrilla's life depends on shooting, on his ability to handle his arms well and to avoid being hit. When we speak of shooting, we speak of marksmanship as well. Shooting must be learned until it becomes a reflex action on the part of the urban guerrilla.

To learn how to shoot and to have good aim, the urban guerrilla must train himself systematically, utilizing every apprenticeship method, shooting at targets, even in amusement parks and at home.

Shooting and marksmanship are the urban guerrilla's water and air. His perfection of the art of shooting makes him a special type of urban guerrilla—that is, a sniper, a category of solitary combatant indispensable in isolated actions. The sniper knows how to shoot, at close range and at long range, and his arms are appropriate for either type of shooting.

The Firing Group

In order to function, the urban guerrillas must be organized in small groups. A group of no more than four or five is called *the firing group*.

A minimum of two firing groups, separated and sealed off from other firing groups, directed and coordinated by one or two persons, this is what makes a *firing team*.

Within the firing group there must be complete confidence among the comrades. The best shot and the one who best knows how to manage the machine gun is in charge of operations.

The firing group plans and executes urban guerrilla actions, obtains and guards arms, studies and corrects its own tactics.

When there are tasks planned by the strategic command, these tasks take preference. But there is no such thing as a firing group without its own initiative. For this reason it is essential to avoid any rigidity in the organization in order to permit the greatest possible initiative on the part of the firing group. The old-type hierarchy, the style of the traditional left doesn't exist in our organization.

This means that, except for the for the priority of objectives set by the strategic command, any firing group can decide to assault a bank, to kidnap or to execute an agent of the dictatorship, a figure identified with the reaction, or a North American spy, and can carry out any kind of propaganda or war of nerves against the enemy without the need to consult the general-command.

No firing group can remain inactive waiting for orders from above. Its obligation is to act. Any single urban guerrilla who wants to establish a firing group and begin action can do so and thus become a part of the organization.

This method of action eliminates the need for knowing who is carrying out which actions, since there is free initiative and the only important point is to increase substantially the volume of urban guerrilla activity in order to wear out the government and force it onto the defensive.

The firing group is the instrument of organized action. Within it, guerrilla operations and tactics are planned, launched, and carried through to success.

The general command counts on the firing groups to carry out objectives of a strategic nature, and to do so in any part of the country. For its part, it helps the firing groups with their difficulties and their needs.

The organization is an indestructible network of firing groups, and of coordinations among them, that functions simply and practically with a general command that also participates in the attacks; an organization which exists for no purpose other than pure and simple revolutionary action.

The Logistics of the Urban Guerrilla

Conventional logistics can be expressed by the formula CCEM:

C —food (*comida*)
C —fuel (*combustivel*)
E —equipment
M —ammunition (*munições*)

Conventional logistics refer to the maintenance problems for an army or a regular armed force, transported in vehicles with fixed bases and supply lines.

Urban guerrillas, on the contrary, are not an army but small armed groups, intentionally fragmented. They have no vehicles nor fixed bases. Their supply lines are precarious and insufficient, and have no established base except in the rudimentary sense of an arms factory within a house.

While the goal of conventional logistics is to supply the war needs of the gorillas to be used to repress urban and rural rebellion, urban guerrilla logistics aim at sustaining operations and tactics which have nothing in common with a conventional war and are directed against the military dictatorship and North American domination of the country.

For the urban guerrilla, who starts from nothing and has no support at the beginning, logistics are expressed by the formula MDAME, which is:

M —mechanization
D —money (*dinheiro*)
A —arms
M —ammunition (*munições*)
E —explosives

Revolutionary logistics takes mechanization as one of its bases. Nevertheless, mechanization is inseparable from the driver. The urban guerrilla driver is as important as the urban guerrilla machine gunner. Without either, the machines do not work, and as such the automobile like the machine gun becomes a dead thing. An experienced driver is not made in one day and the apprenticeship must begin early. Every good urban guerrilla must be a good driver. As to the vehicle, the urban guerrilla must expropriate what he needs.

When he already has resources, the urban guerrilla can combine the expropriation of vehicles with other methods of acquisition.

Money, arms, ammunition and explosives, and automobiles as well, must be expropriated. And the urban guerrilla must rob banks and armouries and seize explosives and ammunition wherever he finds them.

None of these operations is undertaken for just one purpose. Even when the assault is for money, the arms that the guards bear must also be taken.

Expropriation is the first step in the organization of our logistics, which itself assumes an armed and permanently mobile character.

The second step is to reinforce and extend logistics, resorting to ambushes and traps in which the enemy will be surprised and his arms, ammunition, vehicles, and other resources can be captured.

Once he has the arms, ammunition, and explosives, one of the most serious logistics problems the urban guerrilla faces at any time and in any situation, is a hiding place in which to leave the material and appropriate means for transporting it and assembling it where it is needed. This has to be accomplished even when the enemy is on the lookout and has the roads blocked.

The knowledge that the urban guerrilla has of the terrain, and the devices he uses or is capable of using, such as guides expecially prepared and recruited for this mission, are the basic elements in the solution of the external logistics problem the revolutionary faces.

The Technique of the Urban Guerrilla

In its most general sense, technique is the combination of methods man uses to carry out any activity. The activity of the urban guerrilla consists in waging guerrilla warfare and psychological warfare.

The urban guerrilla technique has five basic components:

(a) one part is related to the specific characteristics of the situation;

(b) one part is related to the requisites that match these characteristics, requisites represented by a series of initial advantages without which the urban guerrilla cannot achieve his objectives;

(c) one part concerns certain and definite objectives in the actions initiated by the urban guerrilla;

(d) one part is related to the types and characteristic modes of action for the urban guerrilla;

(e) one part is concerned with the urban guerrilla's method of carrying out his specific actions.

The technique of the urban guerrilla has the following characteristics:

(a) it is an aggressive technique, or in other words, it has an offensive character. As is well known, defensive action means death for us. Since we are inferior to the enemy in fire power and have neither his resources nor his power force, we cannot defend ourselves against an offensive or a concentrated attack by the gorillas. And that is the reason why our urban technique can never be permanent, can never defend a fixed base nor remain in any one spot waiting to repel the circle of reaction;

(b) it is a technique of attack and retreat by which we preserve our forces;

(c) it is a technique that aims at the development of urban guerrilla warfare, whose function will be to wear out, demoralize, and distract the enemy forces, permitting the emergence and survival of rural guerrilla warfare which is destined to play the decisive role in the revolutionary war.

The Initial Advantages of the Urban Guerrilla

The dynamics of urban guerrilla warfare lie in the urban guerrilla's violent clash with the military and police forces of the dictatorship. In this clash, the police have the superiority. The urban guerrilla has inferior forces. The paradox is that the urban guerrilla, although weaker, is nevertheless the attacker.

The military and police forces, for their part, respond to the attack by mobilizing and concentrating infinitely superior forces in the persecution and destruction of the urban guerrilla. He can only avoid defeat if he counts on the initial advantages he has and knows how to exploit them to the end to compensate for his weaknesses and lack of *matériel*.

The initial advantages are:
(a) he must take the enemy by surprise;
(b) he must know the terrain of the encounter better than the enemy;
(c) he must have greater mobility and speed than the police and the other repressive forces;
(d) his information service must be better than the enemy's;
(e) he must be in command of the situation and demonstrate a decisiveness so great that everyone on our side is inspired and never thinks of hesitating, while on the other side the enemy is stunned and incapable of responding.

Surprise

To compensate for his general weakness and shortage of arms compared to the enemy, the urban guerrilla uses surprise. The enemy has no way to fight surprise and becomes confused or is destroyed.

When urban guerrilla warfare broke out in Brazil, experience proved that surprise was essential to the success of any urban guerrilla operation.

The technique of surprise is based on four essential requisites:

(a) we know the situation of the enemy we are going to attack, usually by means of precise information and meticulous observation, while the enemy does not know he is going to be attacked and knows nothing about the attacker;
(b) we know the force of the enemy that is going to be attacked and the enemy knows nothing about our force;

(c) attacking by surprise, we save and conserve our forces, while the enemy is unable to do the same and is left at the mercy of events;

(d) we determine the hour and the place of the attack, fix its duration, and establish its objective. The enemy remains ignorant of all this.

Knowledge of the Terrain

The urban guerrilla's best ally is the terrain and because this is so he must know it like the palm of his hand.

To have the terrain as an ally means to know how to use with intelligence its unevenness, its high and its low points, its turns, its irregularities, its regular and its secret passages, abandoned areas, its thickets, etc., taking maximum advantage of all this for the success of armed actions, escapes, retreats, cover, and hiding places.

Its impasses and narrow spots, its gorges, its streets under repair, police control points, military zones and closed-off streets, the entrances and exits of tunnels and those that the enemy can close off, viaducts to be crossed, corners controlled by the police or watched, its lights and signals, all this must be thoroughly known and studied in order to avoid fatal errors.

Our problem is to get through and to know where and how to hide, leaving the enemy bewildered in areas he doesn't know.

Familiar with the avenues, streets, alleys, ins and outs, and corners of the urban centres, its paths and shortcuts, its empty lots, its underground passages, its pipes and sewer system, the urban guerrilla safely crosses through the irregular and difficult terrain unfamiliar to the police, where they can be surprised in a fatal ambush or trapped at any moment.

Because he knows the terrain the guerrilla can go through it on foot, on bicycle, in automobile, jeep, or truck and never be trapped. Acting in small groups with only a few people, the guerrillas can reunite at an hour and place determined beforehand, following up the attack with new guerrilla operations, or evading the police circle and disorienting the enemy with their unprecedented audacity.

It is an insoluble problem for the police in the labyrinthian terrain of the urban guerrilla, to get someone they can't see, to repress someone they can't catch, to close in on someone they can't find.

Our experience is that the ideal urban guerrilla is one who operates in his own city and knows thoroughly its streets, its neighborhoods, its transit problems, and other peculiarities.

The guerrilla outsider, who comes to a city whose corners are unfamiliar to him, is a weak spot and, if he is assigned certain operations, can endanger them. To avoid grave errors, it is necessary for him to get to know well the layout of the streets.

Mobility and Speed

To insure a mobility and speed that the police cannot match, the urban guerrilla needs the following prerequisites:

(a) mechanization;
(b) knowledge of the terrain;
(c) a rupture or suspension of enemy communications and transport;
(d) light arms.

By carefully carrying through operations that last only a few moments, and leaving the site in mechanized vehicles, the urban guerrilla beats a rapid retreat, escaping persecution.

The urban guerrilla must know the way in detail and, in this sense, must go through the schedule ahead of time as a training to avoid entering alleyways that have no exit, or running into traffic jams, or becoming paralysed by the Transit Department's traffic signals.

The police pursue the urban guerrilla blindly without knowing which road he is using for his escape.

While the urban guerrilla quickly flees because he knows the terrain, the police lose the trail and give up the chase.

The urban guerrilla must launch his operations far from the logistics base of the police. An initial advantage of this method of operation is that it places us at a reasonable distance from the possibility of persecution, which facilitates the evasion.

In addition to this necessary precaution, the urban guerrilla must be concerned with the enemy's communications system. The telephone is the primary target in preventing the enemy from access to information by knocking out his communications system.

524   Even if he knows about the guerrilla operation, the enemy depends on modern transport for his logistics support, and his vehicles necessarily lose time carrying him through the heavy traffic of the large cities.

It is clear that the tangled and treacherous traffic is a disadvantage for the enemy, as it would be for us if we were not ahead of him.

If we want to have a safe margin of security and be certain to leave no tracks for the future, we can adopt the following methods:

(a) purposely intercept the police with other vehicles or by apparently casual inconveniences and damages; but in this case the vehicles in question should not be legal nor should they have real license numbers;
(b) obstruct the road with fallen trees, rocks, ditches, false traffic signs, dead ends or detours, and other ingenious methods;
(c) place home-made mines in the way of the police, use gasoline, or throw Molotov cocktails to set their vehicles on fire;
(d) set off a burst of machine-gun fire or arms such as the FAL aimed at the motor and the tyres of the cars engaged in pursuit.

With the arrogance typical of the police and the military fascist authorities, the enemy will come to fight us with heavy guns and equipment and with elaborate maneuvers by men armed to the teeth. The urban guerrilla must respond to this with light weapons easily transported, so he can always escape with maximum speed, without ever accepting open fighting. The urban guerrilla has no mission other than to attack and retreat.

We would leave ourselves open to the most stunning defeats if we burdened ourselves with heavy arms and with the tremendous weight of the ammunition necessary to fire them, at the same time losing our precious gift of mobility.

When the enemy fights against us with cavalry we are at no disadvantage as long as we are mechanized. The automobile goes faster than the horse. From within the car we also have the target of the mounted police, knocking him down with machine gun and revolver fire or with Molotov cocktails and grenades.

On the other hand, it is not so difficult for an urban guerrilla on foot to make a target of a policeman on horseback. Moreover, ropes across the streets, marbles, cork stoppers are very efficient methods of making them both fall. The great disadvantage of the mounted policeman is that he presents the urban guerrilla with two excellent targets: the horse and its rider.

Apart from being faster than the horseman, the helicopter has no better chance in persecution. If the horse is too slow compared to the urban guerrilla's automobile, the helicopter is too fast. Moving at 200 kilometres an hour it will never succeed in hitting from above a target lost among the crowds and the street vehicles, nor can it land in public streets in order to catch someone. At the same time, whenever it tries to fly low, it will be excessively vulnerable to the fire of the urban guerrilla.

Information

The possibilities that the government has for discovering and destroying the urban guerrillas lessen as the potential of the dictatorship's enemies becomes greater and more concentrated among the popular masses.

This concentration of opponents of the dictatorship plays a very important role in providing information as to moves on the part of the police and men in government, as well as in hiding our activities. The enemy can also be thrown off by false information, which is worse for him because it is a tremendous waste.

By whatever means, the sources of information at the disposal of the urban guerrilla are potentially better than those of the police. The enemy is observed by the people, but he does not know who among the people transmits information to the urban guerrilla. The military and the police are hated for the injustices and violence they commit against the people, and this facilitates obtaining information prejudicial to the activities of government agents.

The information, which is only a small area of popular support, represents an extraordinary potential in the hands of the urban guerilla. The creation of an intelligence service with an organized structure is a basic need for us. The urban guerrilla has to have essential information about the plans and movements of the enemy, where they are, and how they move, the resources of the banking network, the means of communication, and the secret moves the enemy makes.

The trustworthy information passed along to the urban guerrilla represents a well-aimed blow at the dictatorship. It has no way to defend itself in the face of an important leak that jeopardizes its interests and facilitates our destructive attack.

The enemy also wants to know what steps we are taking so he can destroy us or prevent us from acting. In this sense the danger of betrayal is present and the enemy encourages betrayal or infiltrates spies into the organization. The urban guerrilla's technique against this enemy tactic is to denounce publicly the traitors, spies, informers, and *provocateurs*.

Since our struggle takes place among the masses and depends on their sympathy—while the government has a bad reputation because of its brutality, corruption, and incompetence—the informers, spies, traitors, and the police came to be enemies of the people without supporters, denounced to the urban guerrillas, and, in many cases, properly punished.

For his part the urban guerrilla must not evade the duty—once he knows who the spy or informer is,—of wiping him out physically. This is the correct method, approved by the people, and it minimizes considerably the incidence of infiltration or enemy spying.

For the complete success of the battle against spies and informers, it is essential to organize a counterespionage or counter-intelligence service. Nevertheless, as far as information is concerned, it cannot all be reduced to a question of knowing the enemy's moves and avoiding the infiltration of spies. Information must be broad, it must embrace everything, including the most insignificant matters. There is a technique of obtaining information and the urban guerrilla must master it. Following this technique, information is obtained naturally, as a part of the life of the people.

The urban guerrilla, living in the midst of the people and moving about among them, must be attentive to all types of conversations and human relations, learning how to disguise his interest with great skill and judgment.

In places where people work, study, live, it is easy to collect all kinds of information on payments, business, plans of all types, points of view, opinions, people's state of mind, trips, interiors of buildings, offices and rooms, operation centres, etc.

Observation, investigation, reconnaissance, and exploration of the terrain are also excellent sources of information. The urban guerrilla never goes anywhere absentmindedly and without revolutionary precaution, always on the lookout lest something occur. Eyes and ears open, senses alert, his memory engraved with everything necessary, now or in the future, to the uninterrupted activity of the fighter.

Careful reading of the press with particular attention to the organs of mass communication, the investigation of accumulated data, the transmission of news and everything of note, a persistence in being informed and in informing others, all this makes up the intricate and immensely complicated question of information which gives the urban guerrilla a decisive advantage.

Decision

It is not enough for the urban guerrilla to have in his favor surprise, speed, knowledge of the terrain, and information. He must also demonstrate his command of any situation and a capacity for decision without which all other advantages will prove useless.

It is impossible to carry out any action, however well planned, if the urban guerrilla turns out to be indecisive, uncertain, irresolute.

Even an action successfully begun can end in defeat if the command of the situation and the capacity for decision falter in the middle of the actual execution of the plan. When this command of the situation and a capacity for decision are absent, the void is filled with vacillation and terror. The enemy takes advantage of this failure and is able to liquidate us.

The secret for the success of any operation, simple or complicated, easy or difficult, is to rely on determined men. Strictly speaking, there are no easy operations. All must be carried out with the same care exercised in the case of the most difficult, beginning with the choice of the human element, which means relying on leadership and capacity for decision in every test.

One can see ahead of time whether an action will be successful or not by the way its participants act during the preparatory period. Those who are behind, who fail to make designated contacts, are easily confused, forget things, fail to complete the basic elements of the work, possibly are indecisive men and can be a danger. It is better not to include them.

Decision means to put into practice the plan that has been devised with determination, with audacity, and with an absolute firmness. It takes only one person who vacillates to lose all.

Objectives of the Urban Guerrilla's Actions

With his technique developed and established, the urban guerrilla bases himself on models of action leading to attack and, in Brazil, with the following objectives:

(a) to threaten the triangle in which the Brazilian state system and North American domination are maintained in Brazil, a triangle whose points are Rio, Sao Paulo and Belo Horizonte and whose base is the axle Rio-Sao Paulo, where the giant industrial-financial-economic-political-cultural-military-police complex that holds the entire decisive power of the country is located;

(b) to weaken the local guards or the security system of the dictatorship, given the fact that we are attacking and the gorillas defending, which means catching the government in a defensive position with its troops immobilized in defense of the entire complex of national maintenance, with its ever-present fears of an attack on its strategic nerve centres, and without ever knowing where, how, and when that attack will come;

(c) to attack on every side with many different armed groups, few in number, each self-contained and operating separately, to disperse the government forces in their pursuit of a thoroughly fragmented organization instead of offering the dictatorship the opportunity to concentrate its forces of repression on the destruction of one tightly organized system operating throughout the country;

(d) to give proof of its combativeness, decision, firmness, determination, and persistence in the attack on the military dictatorship in order to permit all malcontents to follow our example and fight with urban guerrilla tactics. Meanwhile, the government, with all its problems, incapable of halting guerrilla operations in the city, will lose time and suffer endless attrition and will

finally be forced to pull back its repressive troops in order to mount guard over the banks, industries, armouries, military barracks, prisons, public offices, radio and television stations, North American firms, gas storage tanks, oil refineries, ships, aircraft, ports, airports, hospitals, health centres, blood banks, stores, garages, embassies, residences of outstanding members of the regime, such as ministers and generals, police stations, and official organizations, etc.;
(e) to increase urban guerrilla disturbances gradually in an endless ascendancy of unforeseen actions such that the government troops cannot leave the urban area to pursue the guerrillas in the interior without running the risk of abandoning the cities and permitting rebellion to increase on the coast as well as in the interior of the country;
(f) to oblige the army and the police, with the commanders and their assistants, to change the relative comfort and tranquility of their barracks and their usual rest, for a state of alarm and growing tension in the expectation of attack or in search for tracks that vanish without a trace;
(g) to avoid open battle and decisive combat with the government, limiting the struggle to brief and rapid attacks with lightning results;
(h) to assure for the urban guerrilla a maximum of freedom of maneuver and of action without ever relinquishing the use of armed violence, remaining firmly oriented towards helping the beginning of rural guerrilla warfare and supporting the construction of the revolutionary army for national liberation.

On the Types and Nature of Action Models for the Urban Guerrilla

In order to achieve the objectives previously enumerated, the urban guerrilla is obliged, in his technique, to follow an action whose nature is as different and as diversified as possible. The urban guerrilla does not arbitrarily choose this or that action model. Some actions are simple, others are complicated. The urban guerrilla without experience must be incorporated gradually into actions and operations that run from the simple to the complex. He begins with small missions and tasks until he becomes a completely experienced urban guerrilla.

Before any action, the urban guerrilla must think of the methods and the personnel at his disposal to carry out the action. Operations and actions that demand the urban guerrilla's technical preparation cannot be carried out by someone who lacks that technical skill. With these cautions, the action models which the urban guerrilla can carry out are the following:

(a) assaults;
(b) raids and penetrations;
(c) occupations;
(d) ambush;
(e) street tactics;
(f) strikes and work interruptions;
(g) desertions, diversions, seizures, expropriations of arms, ammunition, explosives;
(h) liberation of prisoners;
(i) executions;
(j) kidnappings;
(k) sabotage;
(l) terrorism;
(m) armed propaganda;
(n) war of nerves. . . .

The urban guerrilla is engaged in revolutionary action in favor of the people and with it seeks the participation of the masses in the struggle against the military dictatorship and for the liberation of the country from the yoke of the United States. Beginning with the city and with the support of the people, the rural guerrilla war develops rapidly, establishing its infrastructure carefully while the urban area continues the rebellion. . . .

Chapter VII
The Problems of the Defenders

Defense against revolution requires an efficient and understanding government; but the very existence of a revolutionary environment is a manifestation of political, social and psysychological weakness, and effective counter-revolutionary action must derive from the same vulnerable environment. Foreign powers attempting to assist threatened regimes face a difficult task, for external power usually has but limited influence over the sources of vulnerability. This chapter explores some of the ramifications of these dilemmas.

In an example of the application of modern social science techniques, in this case orginating in economics, Leites and Wolf look at revolution and counterrevolution as competing systems in a kind of marketplace of force. The authors develop counter-insurgent strategies based on the goal of diminishing the revolution's capacity to meet the demands upon it. In addition to this supply-demand analysis, they also suggest a cost-benefit calculus designed to affect the choices of individuals and groups confronted with rival appeals. The reader may wish to ponder carefully the implications of this approach.

Tanham and Duncanson believe the defenders have a choice only of lesser evils. They examine the contradictions between policy alternatives, pointing out that the government's efforts may divert resources from modernization and thus reduce its appeal; on the other hand, "whatever is done to help peasants . . . will also help insurgents." Such dilemmas are inherent in vulnerable societies.

Edward Gude focuses on popular perceptions of the legitimacy of the actions of the rivals. Comparing Batista's Cuba and Venezuela under Betancourt, he concludes that Batista acted prematurely and over-harshly and thus sacrificed his legitimacy. In contrast, Betancourt waited patiently while the revolutionaries alienated initially undecided segments of the public.

Amrom Katz, critical of the U.S. performance in Vietnam, concludes that the military repeated its mistakes over and over instead of learning from mistakes. He urges the need for open and innovative minds. The reader may wonder about the deeper causes of these shortcomings, given the traditional American reputation for inventiveness.

The final selection, by former State Department official Charles Burton Marshall, raises the problem of dissidence in a liberal counterrevolutionary power. Marshall regards any war that cannot be readily justified to be a difficult proposition in a pluralistic democracy. The diversion of resources from domestic needs added to the complications of the situation in Vietnam. To attacks upon the legitimacy of the regime in South Vietnam, the author replies, "what this formula amounted to was that we should never assist any country except the ones that would not need any assistance."

Selections from *Rebellion and Authority: An Analytic Essay on Insurgent Conflicts* *
by Nathan Leites and Charles Wolf, Jr.

The hearts-and-minds view of rebellion is that of the outsider looking in. In its stress on popular sympathies and economic conditions, it concentrates on the environment that evokes R and causes it to emerge and grow, more or less spontaneously. Its emphasis is on the demand side of the problem. To transpose an analogy from economics, the hearts-and-minds view is a *demand-pull* version of the process, whereas the view we shall be presenting is more in the nature of a *cost-push* version. Our view will, to a greater extent, emphasize factors within the insurgent organization which influence its capabilities and growth. It will thus place somewhat greater emphasis on the supply (production) side of R's growth, and the bearing of supply considerations on the prevention or defeat of R.[1]

Of course, behavior depends on interactions *between* supply and demand. Both need to be considered in understanding population behavior in the insurgency context, no less than consumer behavior in the marketplace. We offer two reasons for placing somewhat more emphasis on supply. One is that, while both demand and supply are important, we feel that in most discussions supply factors have either been neglected or misconstrued. In the theory of consumer behavior, to revert to the economic analogy, it is customary to distinguish between the effect of consumer preferences (demand conditions) and the possibilities for buying different commodities as reflected by their relative costs (supply conditions). The interaction between them determines market behavior. By contrast, the hearts-and-minds analysis focuses principal attention on the preferences, attitudes, and sympathies of the populace (demand), to the neglect of the opportunities and costs required to indulge these preferences. Similarly, in discussions of campus rebellions, principal attention is often focussed on student demands and grievances, rather than on the actions (or inaction) of administrators and faculty that lower the costs and facilitate the organization and radicalization of student rebellion.

The second reason is that supply conditions are probably more clastic (responsive)—at least in the short run—to programs and policies than are demand conditions, especially from A's point of view. Dealing with the demand conditions in the less developed countries involves the massive problems of modernization, and in the more developed countries the problems of reform that are only less massive in a relative sense. It is important and necessary to grapple with these problems (among other reasons, so that A can sustain its own sense of rectitude and purpose). Nevertheless, the problems are apt be unyielding in the short run. The progress that can realistically be aimed for will probably leave the demand for R fairly strong, especially if—as seems likely—progress lags behind promises. This prospect presents an asymmetrical advantage to R. It may be much easier for R to activate and enhance a potential demand for itself than for A to reduce this demand. Thus, demand may be harder to shift downward than upward. Hence, while *both* A and R must attend sharply to the supply or production side of the problem, A may have less leverage on the demand side than R. Hence, it may be efficient for R to allocate relatively more resources to influencing the demand side, and for A to allocate more to the supply side.

Fundamental to our analysis is the assumption that the population, as individuals or groups, behaves "rationally": that it calculates costs and benefits to the extent that they can be related to different courses of action, and makes choices accordingly. Apparent irrationalities can be explained by mistakes; uncertainties; misinformation; a shortage of information on the part of the population; or a misunderstanding on the observer's part of how the population weighs different things in its calculations. Consequently, influencing popular behavior requires neither sympathy nor mysticism, but rather a better understanding of what costs and benefits the individual or the group is concerned with, and how they are calculated. The rationality assumption is admittedly an oversimplification. Its justification hopefully lies in helping to analyze a subject that has often been treated in an obscure, if not obscurantist, way.

The following discussion will describe our alternative approach in terms of three elements: (1) the environment of the less developed countries, (2) the insurgency—R—as a system, and (3) the individual or group in relation to R. Finally, contrasts are drawn between the alternative approach and the hearts-and-minds view discussed in Chapter 2.

The Environment

Traditional societies that have begun to change provide, by the process of change itself, opportunities for insurgent movements.[2] (And societies in which the structure of traditional authority remains intact potentially provide the same apportunities, to the extent that change lies ahead of them.) Endemic, if latent, cleavages and antagonisms tend to be inflamed once the transition to modernization has begun—antagonisms between landlords and tenants; between urban and rural areas; among ethnic, racial, religious, and linguistic groups. Inequities in the distribution of wealth, income, education, and opportunity are chronic and widespread, and the pain that accompanies them is often felt more acutely as modernization begins to open up the possibility of remedies and evoke promises and aspirations that move ahead of the remedies. Resentment against the privilege and status enjoyed by foreigners as a colonial legacy, or by domestic elites as a legacy of traditional society, is often acute or easily aroused. Such patterns of bitterness and resentment are as much a part of the realities of transitional societies as are low income levels, and they are very likely to intensify as income levels rise—other things being equal—at least up to some threshold. As one experienced observer, Robert Thompson, summarizes the point:

> Every insurgency . . . requires a cause. [But] there is always some issue which has an appeal to each section of the community, and, even if dormant, an inspired incident may easily revive it in an acute form. . . . All governments are vulnerable to criticism, and every grievance, shortcoming or abuse will be exploited.

Although the preceding point applies with particular force in the less developed countries, it is relevant in the more developed countries, too. Thus Sidney Hook comments on disruption in university campuses in the United States:

> On every campus there are always some grievances. Instead of seeking peacefully to resolve them through existing channels of consultation and deliberation, the SDS [Students for a Democratic Society] seeks to inflame them. Where grievances don't exist, they can be created. In one piece of advice to chapter members, they were urged to sign up for certain courses in large numbers, and then denounce the University for its large classes!

540   Another characteristic of the less developed countries that enhances their vulnerability to insurgency is the mutual isolation of their component parts. Less developed countries are "plural" economic and social entities in the sense that they contain units that are physically, as well as functionally and technologically, remote from one another. Villages, districts, towns, provinces, and cities are in imperfect and intermittent contact. They are often in isolation from one another and particularly from the capital city and the institutions of the central government concentrated there. Thus, flows of commodities, information, and people from place to place are extremely limited. Because the links and contacts among these enclaves, and between them and the center, are meager, the ability of an A to maintain surveillance and establish control over an inchoate insurgency is accordingly limited. The difficulty (that is, the high cost) of obtaining reliable and timely information—which A needs more than does R—is highly correlated with many other structural characteristics of the less developed countries—for example, per capita income, urbanization, literacy, longevity, industrialization, and political participation. But from the standpoint of the circumstances that facilitate R's emergence, the high cost of information and communication may be considerably more significant than other typical attributes of the environment in less developed countries.[3]

Given these characteristics, it is a truism to say that transitional societies are vulnerable to insurgency. Changing the characteristics is complex and time-consuming. Moreover, the *process* of modernization itself by no means reduces the vulnerabilities in question, although that is more likely to be the *outcome* of modernization in the longer run. For these reasons, it is wise to separate the analysis of R from that of development and modernization in general. To analyze and understand R in the less developed countries, we need to factor it out of the wider set of modernization problems to which it is related. Focusing on R leads to viewing it as a system.

Rebellion As A System

What does it mean to view an insurgent movement as a system? The alternative approach to be explored here starts with the observation that insurgent movements, as operating systems,

Selections from *Rebellion and Authority:*
An Analytic Essay on Insurgent Conflicts

require that certain inputs—obtained from either internal or external sources—be converted into certain outputs, or activities. These activities characterize the stage to which R has progressed.

In general, insurgency requires inputs of recruits, information, shelter, and food—almost always obtained from the internal environment (endogeny)—and cadres, publicity, material, and initial financing—often provided from external sources (exogeny). The "mix" between endogeny and exogeny is variable: it differs between different Rs, and in the same R at different times. To obtain inputs from the local environment, R relies on various persuasive as well as coercive (damage-threatening or damage-inflicting) techniques. In practice, both persuasion and coercion are important as well as intimately linked. Severe coercion is often combined with a considerable and effective persuasive effort by Rs.[4]

Persuasion may take many forms: ideological preparation, education, discrediting of established authority and practices,[5] and payment (rewards). Coercion may also take many forms: the threat and carrying out of kidnapping, assassination, torture, forcible tax collection, and destruction or confiscation of property, including crop and land seizure. Often coercion and persuasion are mixed, as, for example, in compulsory assemblies for group criticism and selfcriticism. Again, the actual and the efficient combination between persuasion and coercion are important to study, in order to understand both the organization and operation of R and the problem of countering R. Certain hypotheses can be examined concerning mixes of coercion and persuasion that may be effective in influencing different types of individuals or groups. For example, coercion may be more effective in obtaining compliance from the "haves," who initially are relatively favored and hence have something appreciable to lose; while persuasion and inducements may be more effective in obtaining compliance from the disavantaged, who have little to lose and may therefore tend to cherish, and perhaps magnify, any gains by comparison. Of course, a mixture of the two may be more effective than either alone, but the proportions in the mix will vary with the circumstances of the intended target.

The inputs acquired by combining persuasion and coercion are converted into outputs by the insurgent organization. As with many organizations, R tends to organize personnel, financial, logistics, intelligence, communications, and operations branches to manage the conversion of inputs into activities; and it uses a wide range of incentives (recognition, reward, promotion) and penalties (criticism, isolation, demotion, and physical punishment) to spur the operations of these branches.

The outputs or activities of R include acts of sabotage, violence against individuals, public demonstrations, small-scale attacks, and eventually larger attacks and mobile warfare, on the military side. But R's outputs also include the exercise of administrative and governmental jurisdiction (village aid projects, education and training, formation of youth and other organizations concerned with group action programs). The aim of R's activities is to demonstrate that A is immoral, incompetent, and impotent—that A is, in other words, undeserving and a loser.

The view of insurgency described here can be summarized in Figure 1.

The systems view of insurgency enables one to distinguish four methods of counterinsurgency which will be summarized here and elaborated later. This first is to raise the cost to R of obtaining inputs, or reduce the inputs obtained for given costs: the aim is input-denial. The second is to impede the process by which R converts these inputs into activities—that is, to reduce the efficiency of R's production process. The third is to destroy R's outputs. And the fourth is to blunt the effects of R's outputs on the population and on A—that is, to increase A's and the population's capacity to absorb R's activities.

The first two methods may be termed "counterproduction," which hinders R's production of activities by either denying inputs or changing the production coefficients so that smaller outputs are generated from given inputs.

Examples of the first method, input-denial, include interdiction by air, ground, or naval action; construction of barriers that impede the movement of people or supplies from a source to a destination; and preemptive buying programs that try to engage the available suppliers of particular inputs (such as rice) so that these goods are less readily available to R.

Selections from *Rebellion and Authority:
An Analytic Essay on Insurgent Conflicts*

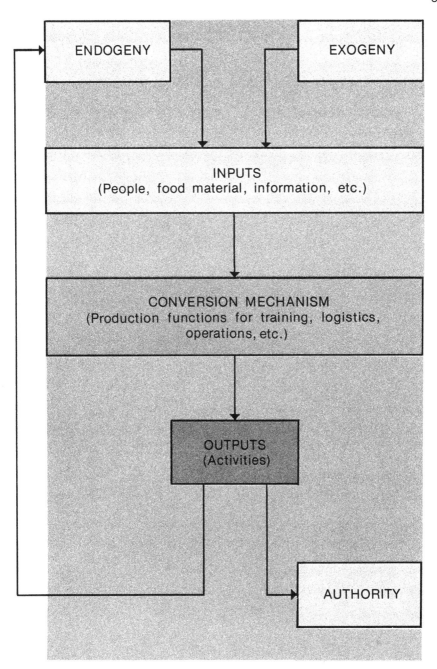

Efforts by A to reduce R's productive efficiency (the second method) include creating distrust and frictions within R's organization by planting rumors; attracting defectors (particularly those from the higher ranks in R's civil and military organization); disseminating credible misinformation about the behavior of R's leadership; and generally raising the noise level in R's information system.

The third method is the traditional counterforce role of military action. Besides the application of firepower from ground and air, it depends especially on accurate intelligence, so that targeting error in the use of such firepower is reduced. Otherwise, such error is likely to be high because targets are closely collocated with the people. (The importance of intelligence to reduce targeting error in counterforce operations can hardly be overemphasized, and we shall return to it later.)

The fourth method, increasing A's and the population's capacity to obsorb the outputs of R, is analogous to passive and active defense in strategic analysis.[6] Its passive-defensive aspects include such measures as building village fortifications ("hardening"), and relocating villagers so that they are less accessible to R (evacuation). Its active-defensive aspects involve creating or strengthening local paramilitary and police units with increased capacity to provide local defense against small unit actions by R. In the realm of political action, such capacity requires (1) A's adherence to law and order in contrast to R, and (2) its demonstrated ability to complete announced programs, thereby certifying that it *should* govern because it *is* governing.

How different is this approach from the one associated with the hearts-and-minds doctrine? Admittedly the differences are of degree rather than kind. But the differences of degree involve an important degree of difference. One contrast is to lay greater stress, in dealing with problems of counterinsurgency, on the supply side of insurgency (for example, on how the R system obtains its inputs, from what sources, in what quantities, in return for what persuasive, coercive, or inducement measures, how it manages these inputs and converts them into the system's outputs) rather than on the demand side (how receptive the feelings of the population are to an insurgency).

The supply side of the problem relates to the difficulty or cost of producing R_s activities; the higher these costs, the lower the scale or the probability of R. The demand side of the problem relates to what people are willing to pay (or contribute) for R's activities. The more they want an insurgency, the higher the price they will pay for these activities; hence, the greater the scale or the probability of R.[7] But for given preferences or desires, the price people will be willing to pay depends also on the resources they have available and the terms under which contributions toward an insurgency might be made (that is, the risks of damage or hopes of gain that enter into their calculations).

When counterrebellion operates on the supply side, the aim is to make the cost of R exceed the price that its internal or external supporters are willing to pay to support it, especially at high levels of activity. When counterrebellion operates on the demand side, it tries to reduce what people are willing to pay for R activities. Stressing the supply side means trying to raise the costs of producing R's activities, hence raising the costs of reaching a given scale or probability of rebellion. The analysis presented here places relatively greater emphasis on the supply than on the demand side, while the reverse applies to the hearts-and-minds orientation.[8]

While the demand-supply distinction helps clarify the contrasts between orientations, there are important interactions between demand and supply that should not be overlooked. The difficulty or cost of operating R and increasing its strength depends, as discussed earlier, on its access to various inputs provided by the population. The population's demand function influences that access and the terms on which it is obtained. For example, if demand rises, the costs of information and recruits may get lower (the supply function may fall). Conversely, if costs are increased, the demand function may fall. In effect, demand excercises an influence on supply, and vice versa.

This problem is also familiar in economics, although there too the usual demand-supply dichotomy often ignores interactions between the two.[9] But several particular points should be noted about the demand-supply interaction in the insurgency context. The demand that is operative in the sense of influencing R's supply function may, as discussed earlier, be confined to a small segment of the population. And the adroitness of R itself, as well as A's maladroitness, can activate and stimulate popular demand. What is at work is a network of positive feedbacks: the population's effective demands influence the costs and effectiveness of R's activities, and R's activities influence (by the manipulation of both persuasion and coercion) the population's demands. Conversely, clumsy reactions and overreactions by A to provocation from R can intensify popular demand. The discussion of provocation by R and "hot" violence by A, in Chapter 6, provides examples of this type of interaction.

Another contrast lies in the different view of endogeny and exogeny which emerges. Hearts-and-minds stresses nearly pure endogeny,[10] whereas the systems approach views the problem in terms of tradeoffs between the two. The inputs that the R system requires can be provided from internal or external sources, in combinations that may vary at different times in the same insurgency, and in different insurgencies. Internal sources can be primary, in the sense that they provide a larger (or more valuable) share of the total input than does exogeny, or they can be secondary.[11] Moreover, the value of external (or internal) inputs cannot be inferred from their bulk, or their market prices. For example, external provision of leadership, money, intelligence, training, sanctuary, propaganda, and diplomatic pressure may have an importance in the emergence and growth of R which is not adequately measured by the flow of tons of supplies, or numbers of people, across a contiguous border.[12] Thus, while the problem of internal versus external sources is more likely to arise in terms of the mix between two sources of inputs, successful counterrebellion has always required either the absence of significant external support (for example, the Philippines and Malaya) or the shutting off of such support (Greece and Algeria). This is consistent with the fact that there have been cases of successful insurgency *without* such external support (Cuba), where the authority was weak and ineffectual. Curtailing exogeny is necessary but not sufficient for successful counterrebellion.

Even if one assumes the primacy of endogeny, the systems approach leads to different implications from those associated with the hearts-and-minds approach. The central questions include not only popular attitudes, but also R's operations: how R obtains its supplies; what forms of coercion and persuasion are used to influence the population; how R makes payments and raises revenues. Whether one wants to control R, or to strengthen or replicate it, the *inside* of R is what needs to be studied. While one wants to know something about the market within which R operates, under the systems approach one is especially concerned with how R operates within that environment, and with the difference between a successful and a less successful R in such operations (that is, an "interfirm" contrast).

Consider the analogy between two firms, F_1 and F_2, producing the same product in two noncompeting markets, M_1 and M_2. If, at the end of a period, F_1 shows high output, low cost, and high profit, should we say that the explanation for its success relative to F_2 is due to differences in conditions *within the market* M_1, compared with M_2?

Sometimes this may be so, and if it is we would look principally to differences in demand conditions in the two markets—hence, in consumer preferences and income—for the explanation. But our analysis would be incomplete if we did not look as well to possible differences *within the firms*, F_1 and F_2, to account for their different degrees of success. For example, we might find that management in F_1 is superior to that in F_2, or the labor productivity in F_1 surpasses that in F_2, or that wage rates and labor incentives differ in the two firms, or that the speed of delivery or the quality of product differs. Market conditions may not differ at all, or not by enough to explain differences in performance.

In other words, even within the framework of a purely endogenous explanation (in the sense of *conditions within the country* rather than assistance from outside the country), we should make a distinction between factors accounting for R's success which are to be found *within* R itself and factors prevailing within the country but *outside* R.

Thus, endogeny needs to be further subdivided: endogenous with respect to the country, and endogenous with respect to the R movement itself. On this basis, one can accept pure endogeny without accepting the hearts-and-minds view that it is conditions prevailing in the country that explain successful R. It is possible to assert, on the one hand, that the success of R may be determined by factors inside its area of operation, and to deny, on the other, that its fortune depends decisively on the amount of sympathy for R and the extent of deprivation to which the bulk of the population is subjected. In this light, the subject of rebellion and counterrebellion should be considered as much a problem in organization and management as in political-economic development.

A comment very much in this spirit is made by George Kennan. Discussing the Bolshevik revolution's conquest of the Tsarist regime, he observes that the revolution's success depended on

> ... the extraordinary discipline, compactness and conspiratorial tightness of the Communist Party; the magnificent political leadership . . . [of] Lenin; and the driving, unrelenting military leadership which the Party gave to the Red Army units in the civil war. . . . The cutting edge of these qualities was of far greater effectiveness than any of the shifting, undependable winds of popular sympathy.[13]

The Population Between R and A

The basic importance of the population to R is as a principal—though not exclusive—source of inputs on which the insurgent system depends. This role is not necessarily less important than that ascribed to it in the hearts-and-minds view, but the role is different. What are some of the differences?

One difference is that the required size of the population that provides the needed inputs can be, as noted earlier, quite small. Depending upon the size of the R system and the stage of its activities, the inputs of food, personnel, weapons, and information that it needs can be more or less limited, and consequently the subset of the population that is involved can be extremely limited. In other words, a small popular minority can be operationally a quite satisfactory underpinning for R, with a generalized impact that may be relatively large.

As a source of inputs, the important characteristic for scrutiny in this minority of the population is behavior or conduct not sympathies or preferences. Conduct is, of course, affected by both preferences (goals) and opportunities (options). But there are at least two reasons that suggest the analysis of opportunities may be more rewarding than that of preferences. The first is that opportunities are more readily and reliably observable than preferences. Economy of effort would generally warrant seeking explanations that are readily available before looking for those that are elusive. The second reason is that the particular set of preferences to which the behavior of the population is relevant

The Problems of the Defenders

may have relatively little to do with sympathy for, or identification with, either contesting side—the insurgents or the authority. A pervasive, and probably frequent, passivity of feeling toward both sides is quite consonant with popular behavior that is highly beneficial to one side. As we have argued earlier, limiting damage or enhancing gain may be a sufficient explanation for the behavior of the population, without recourse to more elusive explanations concerning putative preferences or sympathies.

According to the alternative approach we are describing, it is appropriate to view an individual or group within the population as a rational decisionmaker who assesses opportunities and consequences of alternative actions.[14] The assessment involves a set of preference functions in which feeling for A or R may be relatively unimportant, or may even take a different direction from that obtained by attributing the burden of explanation for popular behavior to sympathetic feelings alone.

Moreover, the time horizon over which the calculations of this hypothetical and rational *decisionmaking* unit extends may be extremely short. The need to avoid today's damage may overwhelm considerations of long-run preference, or cumulative long-run gain, associated with a different course of action. (The time discount for the population, searching for a path to survival between pressures of R and A, may be extremely high.)

As an example of behavior from pure profit-maximization, note the following description by a Viet Cong defector of the reasons for his action:

> Question: What made you decide to rally [that is, defect]?
> Answer: . . . I thought that in fighting on the GVN side, a soldier may be happy because he has a good salary and even though he dies on the battlefield, he dies with a full stomach. On the contrary, a VC soldier usually eats at 3 p.m. a rice bowl as small as that [the subject described it with his fingers] and he walks all night long to fight and to die with an empty stomach.[15]

Or again the following statement by a Viet Cong prisoner (or defector?):

> I do not know which side is winning . . . I did not think

about which side was winning. I take the side which can do the most for me.[16]

Frequently, of course, pure profit-maximizing or damage-limiting influences may be less operative than a mixture of the two. For example, both influences may merge when the population is astute enough to comply, or seem to comply, with *both* A and R. Thus, in the Philippines during the Aguinaldo rebellion, a picture of jointly compliant behavior emerges in the following account by General Adna Chaffee:

> Throughout these islands, wherever a *presidente* of a *pueblo* or *cabeza* or a *barrio* was appointed or elected under American authority, he, with few exceptions . . . acted in the same capacity for the insurgents. . . . This dual form of government existed everywhere, in strongly garrisoned towns like Manila and in the smallest *barrio*. . . . [They] now commenced the difficult task of serving two masters. In all lawful matters, they served with due appearance of loyalty to the American government, while at the same time . . . they secretly levied and collected taxes . . . from the people. . . . They held communications with the enemy, and in all ways open to them gave the guerrilla bands aid and comfort.[17]

Notwithstanding the earlier point about the high time discount for the population and the probably overriding necessity of choosing today's safety at the cost of tomorrow's welfare, there is presumably a negative correlation between a population's belief in the eventual victory of a particular side (whether A or R) and the level of immediate threat required from that side to obtain a given degree of compliance. If I expect a particular side to lose—that is, myself to be ultimately at the mercy of its enemy—I will need a higher instant threat to offset the forecast of future damage at the hands of the other side.

Note that in the preceding discussion of the importance of profit-maximizing—in both the pure- and mixed-motivation examples—there would appear to be an inconsistency with the earlier discussion of the limited effectiveness of raising income and alleviating deprivation in securing compliant behavior. Resolution of the apparent inconsistency can be put in the following terms: considerations of gain have a more certain effect on income than on preferences; to the extent that a given side can manage the rate of exchange between gains and compliant behavior—that is, the substitution effect—its access to compliance is likely to be enhanced. But if the terms of exchange are not manipulated at the same time as income is raised, the benefactor may very well be himself adversely affected by the benefits he is providing, which may redound instead to the advantage of the other side.

To recapitulate the main points of contrast between the role of the population in the approach we have been describing and its role in the hearts-and-minds view, let us set down four principal points:

1. As a source of critical inputs needed by R in its growth and progress, the proportion of the population that is important can be a small minority, rather than a plurality or majority.
2. In discussing the population, emphasis should be placed on behavior, rather than on attitudes and sympathies. Attitudes, in the sense of preferences, affect behavior but are not identical with it; nor in most cases are they the primary influence on it.
3. In addition to attitudes and feelings, what influences behavior are the opportunities available to the population for choosing. In the population's calculations of the options available, *predictions* of the consequences of alternative actions may be crucial. Such predictions determine the estimates of profit (gain) or damage (loss) which influence behavior.
4. Moreover, the predictions within which profit-maximizing or damage-limiting calculations are made are very likely to give heavy weight to short-term as against long-term prospects—that is, to be accompanied by a high implicit time discount.

On each of these four points, the message usually conveyed by the hearts-and-minds view is distinctly different from, if not opposite to, that which we have been advancing. To be sure, our approach does not deny that there are those within R and A (and in the population, generally) who are disposed to disregard personal considerations on behalf of loyalty to a cause. Often R has an edge over A in this respect. But frequently feelings about a cause begin to merge with calculations of gain and loss. And where dissonance between them arises and endures, the result is often a change of feelings, rather than acceptance of repeated loss.

Appendix: Cost-Benefit Calculations and Behavior

The demand and supply formulation can also be described in terms of the costs and benefits of rebellion, as the population views them. Consider the following diagram in which costs (as calculated by an individual or group) are measured vertically, benefits horizontally.[18]

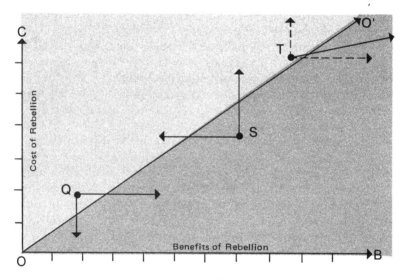

For all points lying along OO', costs and benefits are equal. For A, the desirable region is above OO', for R below OO'. At any given time, an individual's calculations may locate him at a particular point in the field. For example, an individual at Q is a supporter of A; or, more accurately, a nonsupporter of R. Toward such an individual, R's objective should be to shift him east or south; conversely, from A's point of view, it is desirable to shift an individual located at S west or north. When both R and A engage in such efforts, it is the resultant that matters. Resultant vectors that are flatter than OO' will tend to strengthen R; those that are steeper will tend to strengthen A. The diagonal vector at T is an example of the former.

In demand-supply terms, eastward (westward) movements represent an upward (downward) shift in demand for R; northward (southward) movements represent increases (decreases) in costs, hence a fall (rise) in the supply of R.[19] Our prior discussion (and some of what follows in Chapter 5, below) suggests that A's efforts are perhaps more likely to be efficiently expended on raising costs than in lowering demand, while R's efforts, with nearly equal likelihood, may be efficiently expended on either. Yet, if concentration on raising costs causes A to overlook R's efforts to raise demand, the resultant may be flatter than OO', to A's disadvantage. Indeed, if R is astute and A clumsy, R may turn A's efforts to raise costs into increased benefits instead. Various examples of this "judo effect" (provoking A to overreact, decoyed reprisals, coercion based on poor [or no] intelligence) are presented in Chapter 6.[20]

Notes

Reprinted with minor omissions.
1. R refers to the revolution, rebellion, or insurgency. A designates the existing authority.
2. That change exposes and intensifies vulnerabilities to insurgency is, of course, not confined to traditional societies. Thus, in the United States during the past decade, the most rapid improvements in civil rights since the Civil War have been followed or accompanied by the most violent resistance to the residual, if declining, discrimination. Eric Hoffer has eloquently and exhaustively examined the phenomenon in his various works. See, for example, *The True Believer* (New York: Harper & Bros., 1951); *The Passionate State of Mind* (New York: Harper & Row, 1955); and *The Ordeal of Change* (New York: Harper & Row, 1963). For another penetrating exposition, see Robert Waelder, *Progress and Revolution* (New York: International Universities Press, 1967).
3. Although the characteristics we have been describing typify the less developed, transitional countries, they are not entirely excluded from the more developed countries. Watts and Appalachia are LDC pockets within an MDC garment. While opportunities for insurgency are more limited in the MDCs, they are not absent.
4. R's effective use of persuasion is closely related to the asymmetrical quality of the demand for rebellion, alluded to

Selections from *Rebellion and Authority:*
An Analytic Essay on Insurgent Conflicts

earlier: such demand may be easier to shift upward than downward. See p. 29.

5. For any set of implementers of authority (officials, policemen, military personnel) there will always be a lower-performing segment whose discredit is easier and more appropriate for R to target.
6. Cf. Robert Levine, *The Arms Debate* (Cambridge: Harvard University Press, 1963), pp. 229-233, 240-243, 309; Herman Kah, *On Thermonuclear War* (Princeton: Princeton University Press, 1960), pp. 126-144, 303-304, 518-521.
7. See the earlier discussion of income and substitution effects in Chapter 2, and the Appendix to chapter 3.
8. In formal terms, the distinction between demand and supply relates to two functions:
$$D = D(p, x)$$
$$S = S(c, y)$$

    D is the quantity of R activities that will be bought; p is the price per unit. S is the R activities produced: c is the cost per unit; x_i and y_j are other influences on demand and supply, respectively. (Both D and S can be disaggregated into endogenous and exogenous components.) The intersection between D and S determines the scale of R, or, from another standpoint, the probability of R.
9. The demand curve is likely to be "kinked" at both high and low price levels (because of a shortage of wholly committed, ardent supporters at high levels, and the "bandwagon" effect at low levels), and hard to shift (inelastic with respect to policies and programs). The supply curve may have an inflection point and later a negative second derivative because of economies of scale and efficiency from "learning-by-doing."

    The interactions can operate both through the effects of shifts in demand on the supply function, and through movements along a given demand function. For example, shifting demand functions may stimulate (or discourage) research and development, and the emergence of technological change that influences supply. And movements along a given demand function can cause suppliers to accelerate (or deccelerate) cost-reducing innovations. Where sellers and buyers are numerous and atomistic, the interactions are weakened.
10. Cf. *passim* various works by Senator J. W. Fulbright, Roger Hilsman, and David Halberstam.
11. It is another question whether the level of, and changes in, the exogeny/endogeny ratio may not be highly important for U.S. policy. It may be the case that those Rs in which U.S. political interests are most involved are likely to be cases in which the ratio is large, or is rising. However, one must be careful about imputing too much significance to this ratio, inasmuch as it can

change over time. Furthermore, there are likely to be considerable lags between the achievement of a particular level and the flow and processing of information relating to it. Hence, what was at one time a high exogeny/endogeny ratio may have, by the time the relevant information reaches a decision point in A's bureaucracy, already become substantially lower.

12. See *New York Times*, January 10, 1967, p. 3; John Randolph article in *Los Angeles Times*, April 2, 1967; Richard L. Clutterbuck, *The Long, Long War* (New York: Praeger, 1966). p. 74.
13. *Foreign Affairs*, Vol. 46, No. 1, October, 1967, 7.
14. The Appendix to this chapter extends this idealized view of the individual as a rational calculator in the insurgency context.
15. From a series of RAND interviews with former Viet Cong members.
16. *Ibid.*
17. Quoted by Leon Wolff in *Little Brown Brother* (Garden City: Doubleday, 1961), p. 334.
18. We assume (conveniently) that nonmaterial and probabilistic elements in benefits and costs can be handled through a Von Neumann-Morgenstern decision-theoretic process that individuals in the population engage in, or simulate in an approximate way. Cf. Howard R. Raiffa, *Decision Analysis* (Reading, Mass.: Addison-Wesley, 1968).
19. The cost-beneift formulation can also be related to the discussion of preference effects, substitution effects, and income effects, mentioned in Chapter 2 (see Appendix to Chapter 2). The preference effect represents movements along the horizontal axis (from A's viewpoint, westward movements; from R's, eastward). The substitution effect implies vertical movements (north, from A's viewpoint; south from R's). The income effect may move individuals southward, to R's advantage, because the costs of rebellion relative to income now seem lower than before. Or it may move them northward, to A's advantage, because they fear the loss of their increased income as a result of rebellion.
20. We are indebted to Daniel Ellsberg for this analogy. See also George K. Tanham and Dennis J. Duncanson, "Some Dilemmas of Countersurgency," *Foreign Affairs*,Vol. 48, No. 1, October, 1969, pp. 119-121.

Some Dilemmas of Counterinsurgency
by George K. Tanham and Dennis J. Duncanson

A reviewer of Senator Fulbright's indictment of President Johnson's policy in Viet Nam has pointed out that "it is possible to argue that the false starts of American policy in Asia and elsewhere have been at least as much due to the illusions of liberalism as to the 'arrogance of power'."* Obviously, particular policies and actions may be judged as making a bad situation worse, but they may not be the cause of its being bad in the first place. Much of the hawk-versus-dove dispute stems from shared misconceptions about communist insurgencies in Southeast Asia and therefore also about the resulting counterinsurgency actions—misconceptions which form part of an ethos largely inapplicable to that troubled region today.

In this article, it is proposed to put aside this debate over the rights and wrongs of Viet Nam and the counterinsurgency effort made by successive American administrations and to dwell instead on certain counterinsurgency dilemmas. They are intractable enough to tax the most statesmanlike minds. It is doubtful that these problems involve ethical issues; therefore, choosing the right solution cannot be said to confer any special righteousness upon the solver or wickedness upon one who makes an unsuccessful choice. Furthermore, the article does not try to offer solutions but rather attempts to delineate a few of the more perplexing and practical dilemmas faced by those governments dealing with communist insurgents.

All measures of defense taken by a government can be represented as a choice between evils, in so far as it stands to gain nothing by them, only to avoid defeat, and this invariably at considerable cost. In the broadest sense, counterinsurgency is a branch of defense and therefore inherently also a choice between evils. In Southeast Asia the defense is against communist revolutionary warfare, and the choices present themselves to the governments of the region as a series of acute dilemmas, precisely because war and revolution are intertwined, yet may demand different solutions. Naturally, the communist strategists try to make things as difficult as they can, deliberately contriving to add to the dilemmas of the government they wish to overthrow. This makes it harder for the defense to choose a solution and to pursue it consistently; the result is often vacillation and hesitation, muddle and even chaos. And in the longer term the strength of the defense will be dissipated and worn down as much by its own efforts and consequent internal strains as by the damage the insurgents can inflict on it directly.

A well-known and much discussed military dilemma illustrates the nature of the problem. Should the primary military effort be directed to search-and-destroy operations, the usual military strategy, or to the more passive secure-and-hold tactics? If the former, the elusive enemy may lead the pursuing force a long chase without itself being destroyed, while no permanent friendly base can be established because adequate forces will not be available to protect it. Yet if too many forces are allocated to secure-and-hold operations, the enemy may also secure and hold his area, and this may result in the de facto loss of part of the country—an outcome the counterinsurgent government is trying to prevent. The debate over this dilemma in Viet Nam need not detain us here, but the problem will be faced in other areas where insurgency breaks out.

The dilemma of the police chief faced with civil disturbance is a familiar one. Usually the margin of error in judging the precise degree of force to employ in restoring order is very narrow; and if, after it is all over, he seems to have used either too much or too little he stands to be roundly condemned.

This dilemma, which has arisen many times in Southeast Asia in the past, is a crude one when compared with the more subtle applications of the underlying principles developed by the Marxist-Leninists in recent times. These turn mainly on "front" tactics. The communists will try to penetrate and dominate other organized bodies, such as peasant cooperatives, women's organizations, schools and trade unions; and in Viet Nam use has been made of certain religious bodies. Because of the very nature of fronts and the lack of accurate intelligence, it is often difficult to tell which are true voluntary organizations of the people and which have been cleverly subverted or created by the communists for their own ends. The communists may, of course, allow organizations they dominate to continue their overtly useful functions, and yet mobilize them skillfully for certain unannounced objectives. Some of the activities may serve to promote communist goals but they will be portrayed as normal protest—strike demonstrations, mass protests and even sit-ins. These may not only sap a government's resources but also, and even more importantly, lead it into unfortunate actions which cause it to lose support at home and abroad. It seems to be almost a rule of public opinion that people are more prone to believe that a government is deliberately repressing worthy causes than to believe that these causes are being exploited for undisclosed political ends. If an organization is found to be a communist front, what is the best way of dealing with it, both in terms of effectiveness and with regard to worldwide and local psychological impact?

II

Dilemmas in other fields hamper the actions of governments, regardless of whether the states are newly emancipated or have never been colonized and whether or not they seek foreign aid and advice. At the head of the list are those which arise as a result of the communist strategy of trying to gain power first in the countryside. In most of the countries of Southeast Asia the issues involved here are very complicated and include long-term political factors as well as immediate administrative ones. Southeast Asian countries exhibit only to a limited extent the characteristics of the modern nation-states of the West. Their traditional notions of government rest on the dichotomy between central sovereignty and peripheral self-administration. Frontiers were not conceived as fixed lines until modern conditions produced maps requiring sharp definition or the state developed into a more identifiable entity. In China itself, the state consisted historically of a central throne whose authority in local affairs tended to diminish with distance; beyond the borderlands lesser states and petty principalities reflected China's image to a greater or less degree. Some broad areas between the various realms were in effect rulerless, inhabited by minority peoples lacking a central authority of their own but lending allegiance from time to time to one neighbor or another. Sometimes an energetic prince, or the Son of Heaven himself, would impose his authority for a time, but radical changes in the political balance were rare and short-lived.

Colonial administrators tended either to step into the shoes of traditional rulers, usurping their sovereignty outright, or to leave them undisturbed except to establish a paramount suzerainty which similarly found a place in local tradition. In either event, the relationship between subject and ruler remained essentially unchanged—especially over such matters as taxation (or tribute), labor and military service, and, in most of Southeast

Asia, the regulation of irrigation.

All this is fully understood by the communists of Southeast Asia, if not by their "revisionist" brethren in Europe. After the general failure in the 1930s to achieve "proletarian revolution," the Asian communists formulated the idea of capturing the periphery and working back toward the centers of power. In other words, the revolution would now seek its "contradictions" and exploit conditions and issues in the borderlands and the outlying villages where a political vacuum prevailed and where the Leninist dynamic of struggle could be cultivated. "Contradictions" were not to be found exclusively in class conflicts, but wherever any type of social tensions could be built up—in religious differences, minority grievances, tribal rivalries, local ambitions or personal feuds. The revolutionary apparatus would find its bases far from the capital and gradually work inward against the government in the hope that after a number of years it would be strong enough either to assault the capital itself or bring the government to its knees in a protracted struggle. The application of this doctrine of "using the countryside to surround the towns" was perfected in the lands of Indochina during the 1950s and 1960s.

The first of the dilemmas it poses is that if urgent steps are taken by the government to fill the political vacuum in the borderlands by providing officials who can promote the welfare of the people as well as enforce law and order, the communists can represent this response as a policy of oppression. Because of the tradition of local autonomy and suspicion of outsiders, problems will arise even where officials are able and dedicated, which of course they frequently are not. On the other hand, if no such steps are taken and the area is left unoccupied by government forces, the vacuum, though relatively harmless in the past, will provide a field in which the revolutionaries can operate unhindered and where in addition they can claim that the government is ignoring and neglecting the people.

The second dilemma arises when a government resolves the first and decides it must take action in rural areas. In addition to the general need for modernization and public welfare felt in underdeveloped countries, defense of the state and the people in the face of a Marxist-Leninist bid for power necessitates new administrative undertakings. Should these be organized by the central government or entrusted to village or local government?

If the administration is undertaken from the center there may be problems. Officials, possibly many from new agencies, with no experience and no precedents, little local knowledge and no established relationships, may descend on village leaders and the rural population. The resulting inefficiency may be compounded by resentment against the intruders; and there will be opportunities for corruption which may be difficult to detect in the novel situation. The whole attempt may become raw material for communist propaganda and blackmail.

Yet if the task is left to local authorities, in order to foster local government, inefficiencies of another sort are almost bound to occur: policies may not be fully understood; there may be no control to ensure that policy, if understood, is properly carried out; and almost certainly local officials will be inadequately trained to formulate and execute programs in accordance with the overall policy and strategy. Furthermore, there may be local corruption, which is no less reprehensible and no more acceptable publicly than under a centralized system and, in any case, is equally helpful to the communist cause.

While most governments usually resolve the first dilemma by deciding that some action must be taken in the rural areas, the second one is not so easily determined. There is no body of principle or experience to help a government decide to centralize or decentralize its effort, or to determine what the blend should be. The uncertainty often leads to switches of emphasis from one to the other with consequent loss of momentum and continuity—all to the advantage of the communists.

III

Another dilemma which receives considerable attention because of its ideological implications stems from a government's desire to obtain support while engaging in activities which may not be popular but may be necessary. It is assumed in the West that popular support can be won and held only by arousing mass enthusiasm for a cause. The communists, like most of the governments they are trying to overthrow in Southeast Asia, know that this assumption is largely untrue in their type of society. They aim to obtain total support in practical matters as they bring areas progressively under their control. They want to leave no time, energy or material resources for allocation to private priorities, no opportunity for individuals to opt out. More often than not, individual actions spring from fear or the hope of personal gain and less frequently from principle or belief. Even in the dedicated cadre, ambition and personal needs may be powerful motivations. The greater the anxiety not to be the odd man out, the more total popular support can be made. The involvement is often slow and subtle; contributions in money or kind, performance of errands and collection of intelligence grow ultimately into full commitment and enrollment in fighting units. In many cases, in Southeast Asia at least, conversion to the communist cause follows subversion to the movement.

Southeast Asian governments have a tendency, strongly encouraged by the Western world, to hold elections in order to prove that they are popular with the people and thus preferable to the communists, who also claim popular support. But parliamentary democracy is still relatively new to most people in Southeast Asia, and conditions of insurgency are not generally propitious for introducing it. If the elections are completely free, in the sense that the government does not get the candidates (as communist governments get theirs), then the insurgents can put up their own men and let it be known that, even though the government may guarantee secrecy of the individual ballot, there will be trouble for constituencies which do not elect them. What would actually happen in a Southeast Asian country that did try to introduce parliamentary democracy during counterinsurgency operations is a matter for conjecture, for it has never yet been achieved. It remains a dilemma how much democracy should and can be started or expanded by a government while it is engaged in a protracted conflict with a strong monolithic enemy.

IV

Dilemmas also exist in development, both economic and social; in practice they are difficult to separate from those that are political, and they are even more numerous. That the social and economic development processes summed up in the word "modernization" are good in themselves is not seriously disputed. But in practice, ordinary peasants often resent the details of its realization because of strong local beliefs, traditions and leadership which run counter to modern ideas. This dilemma is generally ignored or denied.

Some Dilemmas of Counterinsurgency

To many Western observers communist insurgents owe their hold over the masses to their promises to redress such grievances as social neglect, ethnic discrimination or economic exploitation. There is truth in this view, but it is only one among many factors. Many Southeast Asians, including ignorant peasants, understand better than Westerners that the communists who come into their midst are interested in their grievances not so much to right them as to exploit them as pretexts for seizing power. Moreover, not all alleged grievances are real; hence their redress is unlikely by itself to avert or halt a communist insurrection.

And yet, whatever the truth about "popular aspirations" and "popular support," a government which proposes to rule by consent of the ruled, signified even in the most general and vague fashion, must offer positive attractions of some kind. The political philosophies of ancient China and ancient India, which between them have shaped the outlook of Southeast Asia, are at one with modern Western thought on this point. The need for a government to appear responsive and popular would seem to be greater than ever when it inaugurates a new regime—one recently emancipated from colonialism, or seated in power by a coup d'état, or incorporating in itself unfamiliar institutions such as an elected parliament and a ministerial organization.

One path that may seem to avoid the dilemmas reviewed so far is to offer free social services or free bounty; that is to say, to make the fruits of development available to the outlying population without drawing on local resources and without introducing any measure that smacks of control. It is this contention and this policy which have provided, after the requirements of military defense, the most plausible argument in favor of massive American aid to the countries of Southeast Asia which are menaced by communist insurgency and unable to afford such a remedy for themselves. Yet it leads straight to yet another dilemma involving governmental civic action, perhaps the most confusing of all those considered here.

To see clearly its implications, we need to look again at the manner in which the communist fights. It is true that at times he fights like a boxer, but at other times he is more like a wrestler. The secret of wrestling is to fell one's adversary, not by means of his weakness and one's own strength, but by means of *his* strength, which by cunning and skill can be turned against him. In its early stages, a communist insurgency usually draws most of its money

564 and its material resources, including weapons and ammunition, from or through the population it is subverting or intimidating. In Southeast Asia, especially in the borderlands, the population has been accustomed by centuries of experience both to a host of petty exactions, material and otherwise, and to buying off bandits and secret societies. It is a tradition that insurgency prospers on. But substantial external aid across a friendly frontier can be expected only after the insurgency has become sufficiently well established to make good use of such aid and to be able to cover up, through enforcing silence on its own people, the source of supply. It is a widespread misconception that because the inspiration, and even the direction, of an insurgency may be external, the source of supplies must also be external; or conversely, that when supplies clearly are local, it follows that the direction must also be indigenous and not from outside.

In reality, the Marxist-Leninists of the East not only take pride in winning power with a minimum contribution from their own side, but have found from experience that this is a more reliable road to victory. Besides concealing the directing hand, it leaves the door open to repudiation of unsuccessful insurgencies—like the one in Malaya. Also, surprising as this may sound at first, it has the psychological merit of appearing more convincing in peasant eyes.

It is the probable, not the desirable, that commands peasant support. If the conflict appears to be between two forces both sustained logistically from afar, then the outcome is likely to turn on the size of the investment each backer is prepared to put into it. What each will amount to cannot be told until the day of decision, so the peasant in his rice field will be cautious about committing himself. If, however, it is clear that one side is dependent on supplies from afar, and that the insurrectionists can leech off a fairly constant proportion by purchase, theft or capture, then it becomes only a question of patience and endurance (the true meaning of "protracted" in the Chinese phrase translated as "protracted war") until the cleverer insurgents exhaust the morale and the resources of their opponents.

The dilemma facing the defenders, who frequently cannot tell friend from foe, is now plain: whatever is done in the hope of convincing peasants that they have a vested interest in the victory of the established government, whatever improvements are made in social services or however much outright bounty is distributed, it will be seen that the insurgents are also benefiting; and thus the operation will be stripped of some of its persuasiveness. If measures are taken to avoid this situation, in which the insurgents are actually being supplied by their opponents—and they are most difficult and costly—they may be so oppressive or so poorly executed as to spoil the desired effect on the population. The dilemma is clear but not easily resolved: how to gain active popular support and help the people without helping the insurgents also.

V

The dilemmas inherent in civic action programs are compounded when one considers foreign aid. In the first place, what we have just said about a Southeast Asian government's efforts to win the peasant's heart by providing benefits applies with even greater force to schemes visibly financed by a distant nation. The foreigners may leave at any time; they cannot play as enduring a role as that of the insurgents, who are local actors. Furthermore, foreign aid is just as subject to the leeching-off process as are the government's locally derived resources, and it may even be used as a means to avoid tax collection. This merely deprives the peasant of an excuse for not contributing to the insurgents' exchequer, for whatever he does not pay to the government he will be expected to pay, as a levied "voluntary" contribution, to the insurgents.

Another difficulty with foreign aid for an already harassed government is that it places on the administrative machinery an additional burden which in nearly every case is more than can immediately be borne. "Foreign aid corrupts," we are often told by critics. But corruption is unhappily a superficial symptom of the deeper problem resulting from accelerating the adjustment of the native culture, the native bureaucracy and traditional relations between administrators and administered to modern ways. Modernization is made both more difficult and more urgent by the communist challenge. A remedy may be sought in setting up parallel government agencies alongside the traditional ones, to operate in fields and at a tempo designed to meet that challenge and make use of foreign technical advice. But the psychological result will be that the impermanence of the operation will be emphasized in the public mind. Two additional

The Problems of the Defenders

dangers are that the ad hoc agencies will encounter envious opposition from the traditional bureaucrats, and that they themselves will acquire a vested interest in the perpetuation of the insurgency they are being employed to terminate.

There are no preordained outcomes to revolutionary warfare: recent history contradicts the faith of Mao Tse-tung in his own historical "science." Insurgency is a form of struggle which presents almost endless fresh chances to both sides. So long as the defense is alive to the true nature of the dilemmas described here, and is not diverted by falsely romantic views about how Marxism-Leninism works, the ability to match cunning with skill remains. If the difficulties are not understood, they are insuperable; once they are understood they can be resolved. All the dilemmas are practical and as neutral in an ethical sense as the laws of physics; only pragmatic statecraft and the instincts of the indigenous leadership can spot the means of solving them,˙ case by case, moment by moment. And yet, idealism has played its part. Without it, American aid would not have flowed to developing and harassed countries and we would have no policy in Southeast Asia to argue about; the Marxist-Leninists would probably long since have had their way by default. Perhaps, in the end, the hardest of the dilemmas of counterinsurgency turns out to be the most ancient political dilemma of all: how to marry idealism to statecraft.

Notes

* Coral Bell, "Power and Anguish," *International Journal* (Canadian Institute of International Affairs). Toronto, v. XXIII, n. 3, 1968, p. 475.

Batista and Betancourt: Alternative Responses to Violence
by Edward W. Gude

Governmental response to violence is of both theoretical and practical importance. On the most obvious level, it takes two parties to generate a cycle of violence—insurgents and government forces. Police-type action in response to terrorism is at one end of the coercive response scale, military counterinsurgency at the other. All such responses interact with those of the insurgents in the revolutionary processes. On the practical level, governmental response is the side of the equation over which governments obviously have some control.

Much analysis has focused upon the strategy and tactics of revolutionaries, as a means for assessing how governments ought to respond. This focus assumes that government forces are reactive, neglecting the important effect the government forces themselves have on the course of strategy and tactics of insurgents. In actuality, revolutions are made by both insurgents and governments. Mistakes or excesses committed by government forces are as responsible for the collapse of established regimes as are the tactics of the insurgents. This essay examines one aspect of the revolutionary process—governmental responses—in two cases: Cuba under Batista and Venezuela under Betancourt.

Theoretical Considerations

It can be argued that revolutionary politics are but an extension of violent politics of less severity. Revolutionary politics exaggerates the implications of governmental actions in such a way as to make clearer the implications of this part of the equation. Some argue that violence signals the end of politics and that force becomes the sole arbiter. This military view of politics sharply differentiates the normal nonviolent processes of governing societies from those involving force. This view usually does not differentiate between the means, in this case violence, and process, in this case politics. Certainly when violence is introduced, a new form of communications and bargaining emerges on both sides. Symbols are partially displaced by violence as the means and substance of politics, but the processes of maintaining legitimacy continue to be the task of governing.

In the military view of violence in politics, the most important criterion of success is usually violent confrontation of military and paramilitary forces. When this becomes the overwhelming objective, the political implications of the use of

570 force become lost. Kill ratios are substituted for support ratios. In the political view, conflict is basic to politics and is handled in more peaceful times by the process of allocation of resources in response to various demands in such a way that a government maintains support, a sense of legitimacy, and an adherence to these more peaceful means. This allocation of resources involves both substantive and symbolic rewards. While in these more peaceful times the allocation of symbolic and substantive rewards is accomplished by bargaining and communication whose primary content is symbols, in less peaceful times violence becomes part of this process. The goals of government, however, remain the same: to maintain support and a sense of legitimacy.

Before violence can become a factor, there must be a governmental failure to maintain support or legitimacy; otherwise there would be no basis for a serious threat to it. Even a few hundred dedicated revolutionaries would be captured in short order if no portion of the community was willing to provide them some protection. Any willingness to give such protection is an indication of decline of support and legitimacy of government. Any government in such a situation must conduct itself so as not to increase the loss of support and legitimacy. Organized violence in this sense is a signal of political trouble within regimes, not only a signal of the existence of a group dedicated to changing the means and content of the political process. It signifies that a government has failed to maintain adherence to more peaceful means.

In this perspective a government must do more than eliminate a few revolutionaries, for if there has been a decline in popular support and legitimacy, they can be easily replaced. A government so threatened by violence must continue to seek its political goals. It is potentially disastrous to lose sight of political objectives in the effort to stop violence.

Two important concepts have a bearing on this problem: legality and legitimacy. Officials naturally tend to see problems in legal perspective, in terms of the prescribed rules governing peaceful political bargaining as well as personal behavior. A legal system can function, however, only so long as there is popular consensus and legitimacy for that system of laws and rules. The introduction of violence as a means in the political process indicates that there has been at least a partial breakdown in support. To view events only in conventional legal terms at such a time is to exclude from consideration the attitudes of a significant portion of the polity. In a situation in which the legal system has partially broken down, it is important to shift the perspective from the formal legal level to the underlying sense of legitimacy. Only when legitimacy becomes the locus of concern is it possible to pursue the political objective of maintaining support.

If actions are categorized as they are in the following table, there is an inherent bias against recognizing the broader threat of violence. The legal perspective leads to consideration only of cells I and IV in the chart: Government actions are assumed to be legal, insurgent actions illegal. This is further complicated by the

	Acts of	
	Government	Insurgents
Legal	I	II
Illegal	III	IV

fact that governments are primarily responsible for the promulgation of law.

Violence becomes a factor in politics either when some individuals recognize the illegality of their acts yet consider them legitimate, or when they recognize the legality of goverment acts yet consider them illegitimate. By replacing the concept of legality with legitimacy, we can more easily see the problem facing a regime:

	Acts of	
	Government	Insurgents
Legitimate	I	II
Illegitimate	III	IV

In this formulation it is possible to define relatively peaceful political times as those in which most acts of violence are perceived to be on the I and IV axis. But the greater the number of people who attribute legitimacy to the acts of insurgents and illegitimacy to the acts of government, the more threatening the situation is for government. This formulation is independent of the actual level of violence; violence is significant when citizens' perceptions fall in the II-III axis. Thus a shift in popular attitudes from the I-IV axis to the II-III axis signals the introduction of violent politics. This can be summarized in a proposition:

572 Political stability varies directly with the proportion of citizens whose perceptions fall on the governmental axis (I-IV), and inversely with the proportion whose perceptions are on the insurgent axis (II-III). The greater the rate of change from the governmental to the insurgent axis, the greater the decline in stability. It is the nature and rate of change in perceptions of legitimacy that are crucial for stability, not the absolute level of violence. This relationship may be stated more precisely, as follows: Political stability is equal to the rate of change of perceptions from the I-IV axis to the II-III axis, or

$$PS = \frac{d}{dl}[\frac{(I+ IV)}{(II + III)}]$$

This formulation provides a basis for evaluating the impact of governmental and insurgent actions in the course of violent politics. It subsumes several other important factors such as outside support for either party, degree of organization of the revolutionary movement, particular ethnic or class differences within the polity, terrain, ethical tradition, and many others. For the purposes of comparative analysis, they might well be tested only as they affect the overall perceptions by significant groups of the legitimacy of the acts of governments and of insurgents. The one other factor that might be singled out for attention is the degree of revolutionary organization and skill, which is important for assessing whether a government was lucky or skillful in defeating an insurgent movement. After all, revolutionaries can make critical mistakes as well as governments.

One way to look at the impact of the organization and skill of the revolutionaries is to assess the impact of this factory on the probability of successful revolution. To do this consider the figure below. The probability of successful revolution is a function of the position of the curve, which in turn is dependent on the degree of revolutionary skill and organization. At a given level of political stability (A) we have differing probabilities of successful revolution (B, C), depending on the curve for a particular level of skill and organization. The upper curve denotes a higher level of skill and organization than the lower.

As we have developed these concepts, it is obvious that a government has more control over the level of political stability than revolutionary organization. Thus it is less able to affect the position of the curve than the position of point A. Effective police and military tactics, especially intelligence, can affect the skill and organization of the revolutionaries, but the fundamental fact is that these must be carried out in such a way as to keep point A stationary or move it to the right of the chart. Governmental actions frequently do reduce the effectiveness of revolutionaries, but if this is accompanied by a shift toward governmental illegitimacy, and a consequent shift in level of political stability in the direction of D, the probability of revolutionary success is not reduced.

The primary focus of this essay, then, is to examine the political and military actions of governments under threat of violence, with reference to their effects first on the degree of political stability and on revolutionary organization and skill, and finally on the overall probability of successful revolution. This is at least as much a political problem as a military one: the determining factor, according to this analysis, is the perception of the legitimacy of acts by significant groups within the polity.

Violent politics characteristically involve only a few active participants. The number of revolutionaries is never large in comparison with the population as a whole. Similarly, the number actively involved on the governmental side is small. The large majority of citizens are not mobilized to the struggle. One

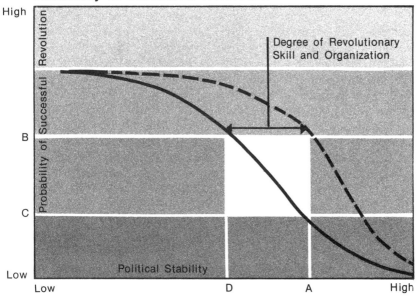

of the primary tasks of a regime is to avoid losing further support; the goal of the insurgents is the converse. The decline in political stability that characterizes the revolutionary situation inclines many people to take a hesitant and more critical attitude toward the egime than is usually the case. They frequently develop a skeptical "wait and see" attitude, which makes the task of government that much more difficult. A regime needs positive support if it is to get the necessary intelligence and cooperation needed to weather a revolutionary situation. The insurgents, on the other hand, need only for the majority to withdraw support from the government. In the early phases they do not need positive revolutionary support.

In pursuit of these objectives, insurgents attempt through violence to demonstrate that the government cannot provide public safety. This is a relatively simple task for terrorists. A more important objective involves the attempt to induce the regime to commit acts against the uncommitted so that they will withdraw their support from the government. This involves inducing a government to overrespond and commit acts that citizens will consider illegitimate (even though they may be technically legal). In the battle for differential political mobilization, the task of the government is enormously delicate and requires political leadership of tremendous sophistication to avoid falling into the traps set by revolutionary violence.

If the assumptions that underlie this analysis are correct, then the political tasks of a government faced with revolutionary violence must determine and take precedence over the military tasks. Excessive military action can lead to a reduction in the organization and skill of an insurgent movement, but if done at the cost of increasing the number of people who perceive the acts of government as illegitimate, it is a pyrrhic victory.

The cases of Cuba under Batista and Venezuela under Betancourt provide a useful comparative test of these ideas. The analysis in this essay is of necessity incomplete, but nevertheless sheds some light on the relationship of violence to politics. The relationship of political processes dominated by largely symbolic communications and bargaining to those in which violence becomes a factor is most clearly illuminated by discussing extreme cases of revolutionary potential, which these are. Similar analysis could be used to discuss the earlier phases of violent politics, in which the shift to violent means initially occurred. The same demands, pitfalls, and opportunities exist.

Cuba Under Batista

When Fulgencio Batista seized power in the coup of March 1952, shortly before a scheduled election, he set in motion the violent processes that led to his flight on New Year's Day in 1959. Batista was an obscure sergeant when he led his first revolt in 1933, against the regime of Gerardo Machado. He dominated the government for the next 10 years, until his voluntary retirement in 1944. In 1952, during his attempt at a legal comeback, the election polls found him a poor third. This apparently motivated his trip to Camp Columbia, Cuba's most important military base, to lead his second coup. Although the popular outcry was great, little opposition immediately developed. Fidel Castro, who had been a candidate for Congress when the coup interrupted the campaign, attempted to challenge the legality of the move but the courts refused to consider the issue. These events convinced many that the peaceful route to social or political change was rather a losing proposition. The attack on the Moncada barracks on July 26, 1953, was the first step in Castro's rise to power. Although the attack was an absolute failure militarily, it became the symbol of resistance and provided as well the name of Castro's organization, the July 26th Movement.

Much has been written of how Castro, after release from prison, went to the hills and organized a peasant revolution. This myth has been most recently enshrined in the writings of Regis Debray. While it is true that Castro landed with 82 men in 1956, after a period of training and organization in Mexico, only 12 of this group reached the mountains. The others were killed or captured and tortured. In the Sierra Maestra base he was able to build a group of some 180 men by the spring of 1958—certainly not a formidable band compared with the 30,000 American-trained troops at Batista's disposal. Even late in 1958 the total rural guerrilla force numbered about a thousand, with some 7,000 in the urban underground. It was this latter group that became the backbone of the revolutionary movement.

The Castro forces themselves did engage in some terrorism and raids on supply stores, but they did not constitute a serious military threat. In fact, even to the end they avoided a large-scale military confrontation with the Army. It is certainly true that Castro became an important symbol of the revolution via his frequent radio broadcasts over clandestine stations and other publicity he received both in and out of Cuba. In order to achieve even this base of support, it was necessary for the guerrilla band to develop friendly relations with local peasants for logistics and intelligence. This Castro certainly did.

How could a band of several hundred, finally a thousand, guerrillas bring down a government? Clearly they did not do so alone. Operating in almost all urban areas were significant underground organizations led by numerous leaders, mostly unheralded. These forces harassed and terrorized the government, provoking the most brutal of responses. It is clear that counterterror became the strategy of the Batista government.[1] Everyone suspected of the slightest disloyalty was subjected to the threat of arrest, torture, and even death. The urban underground was not comprised solely of middleclass ex-students, the mainstay of the Castro forces. Its members included much of the political and professional elite of the country, as well as many skilled workers.

Batista's forceful accession to power had deprived him of much of the legitimacy normally associated with a constitutional government.[2] With marginal support from the start, he apparently did not think it necessary to seek a broader base of political support. Through his police he succeeded in alienating much of the middle class, who otherwise were not totally unsympathetic to the government's efforts to restore law and order. When police efforts focused on the sons of the middle class, this potential support evaporated. Batista acted as though he accepted Mao's dictum that power comes out of the barrel of a gun.

The record of the Batista police and military is impressive if only for its thoroughness. Approximately half of the government forces were tied down with the urban problem, some 15,000 troops. It has been estimated by some that as many as 20,000 civilians were tortured or killed by the government from 1956 until the end of 1958. This is a victor's figure, but even if exaggerated by a factor of 2 or 3, it represents a tremendous number for such a small country. It is not difficult to see how so many families could be affected. Throughout this struggle the question of legality seemed to lose importance—there was little legality anywhere. The overriding consideration became the legitimacy of guerrilla terror and governmental counterterror. The rebel forces took great care to develop tactics that alienated as few people as possible. Thus when burning sugarfields brought great hardship on the peasants as well as the landowners, it was discontinued. The government lacked this sensitivity and continued to use tactics that offended many.

The acts of the insurgents were not so much considered legitimate, as the governmental acts illegitimate. This affected the shift from the stable axis (I-IV) to the unstable axis (II-III): the political consequences of the governmental use of force significantly increased the probability of successful revolution. At the same time the rural and urban resistance groups gained experience that increased their ability to operate in a police state and their sensitivity to popular reactions and cooperation.

This process must be understood in the context of full-scale revolutionary violence. In addition to the selective raids for military supplies, the insurgents carried out numerous assassinations of particularly hated figures of the Batista regime and other serious acts of terrorism. Students made an abortive attempt on the dictator's life in March 1957 that brought particularly severe reprisals. Other acts of revolutionary terrorism were carried out to demonstrate that the government could not provide physical security for the population.

Another revolutionary stratagem was a general strike called for April 8, 1958. The strike was a debacle, since it exposed so much of the revolutionary organization to police reprisals. The refusal of Communists to cooperate in the strike insured its failure. It was not until the pact of Caracas, signed in July 1958, that the various revolutionary groups were able to agree to a degree of cooperation necessary to avoid such failures. The agreement brought the several groups under the nominal leadership of Castro, by then the best known of the leaders. The weakness of this coalition is revealed in the rapid factionalization that occurred after January 1959, in which the rural (but decidedly not peasant) faction gained tentative ascendancy.

The actions of the police during 1957 and 1958 were significant in that they served to alienate significant factions from the government. Because the revolution was primarily middle class in composition, counterterrorism particularly affected that class: This was particularly counterproductive because the police and military themselves were soon affected. Military morale and discipline rapidly declined to the point that the spring offensive of 1958, which pitted 10,000 or more troops against the few hundred Castro forces in the mountains, failed completely. This failure seemed to signal to the populace that the government was no longer viable.³

Given its internal divisions, lack of meaningful coordination, poor communications, and extremely limited numbers, it is difficult to see how the revolutionary movement succeeded except as a consequence of the ineptitude of Batista. He repeatedly fell into the trap set by the insurgents by reacting viciously with his own forces. The insurgents demonstrated that the government could not prevent violence; the police and military then proceeded to engage in counterterror perceived by many as illegitimate. In the race for differential mobilization, the regime fared badly.

Having only limited popular support after the 1952 coup, Batista tried "free" elections, amnesty, and similar moves to bolster his regime. All of these failed to attract significant popular support, however, and as a result many Cubans were uncommitted before revolutionary violence began. The significant events of the revolution were the steps by which this uncommitted group shifted increasingly to the side of the insurgents.⁴ The Batista regime appeared unable to control the revolutionary violence, no matter how violent its own response, promising only civil war for the future. The popular relief at Batista's flight probably reflected as much a hope for an end to violence as it did positive support for the insurgents, but this just as surely represents the failure of government.

In summary, the Cuban revolution can be interpreted as a case in which a middle-class insurgency in both urgan and rural areas, without substantial peasant or working class participation, brought down a regime. The trigger of terrorism and violence of the insurgents opened the floodgates of police and military reprisals, which alienated the significant uncommitted segment of the population. This sealed the fate of the regime by tipping the scales of political mobilization in favor of the insurgents.

Venezuela Under Betancourt

Romulo Betancourt gained power under the most difficult of circumstances in February 1959. The dictatorship of Perez Jimenez had brought both economic and political havoc to the country. The tens and probably hundreds of millions of dollars lost in graft had left the nation in a state of bankruptcy.

Batista and Betancourt:
Alternative Responses to Violence

Oil resources had been squandered through new concessions that never reached the national treasury, public works projects had been established to appease labor with no meaningful program of payment, and there was lavish spending on the military to maintain their support. The country was in the most serious of crises in terms of maintaining its ability to govern.[4] When students and others finally brought down the Perez regime in early 1958, the intervening junta was hesitant to act decisively since its members had programed a return to civilian rule within the year. This meant an additional year in which conditions did not improve.

In addition to the difficult economic position, the factions that had momentarily united against Jimenez reappeared. A considerable faction within the military felt threatened by the end of the military rule. Throughout the immediate postcoup period there were recurrent rumors of plots and several actual attempts at overthrow. Castro Leon, almost pathologically opposed to the possibility of rule by Romulo Betancourt and the Acción Democrática (AD) Party, which won the election of December 1958, was particularly active in anti-government activity. There were serious attempts in July and September of 1958, October of 1959, and April 1960. All of these attempts involved the classic techniques of the military coup. They did not involve the mobilization of the population; the dynamics of political violence as we have defined them were not involved. Such seizures of government, if successful, present the public with a fait accompli with little chance to intervene.

When the Acción Democrática Party was first in power, in 1945, Betancourt was not sufficiently sensitive to support from the military and the party was thrown out in 1948. In 1959 the newly elected President was wiser, carefully courting the advice and support of the various leaders in the military. It was his success in holding the majority of the military to his side that led to the failures of some rightwing military leaders.

Leftwing political violence dates from a split in the AD party. A younger faction of the party had been active in the underground during the Jimenez regime, while leaders such as Betancourt were in exile. In this period members of the faction had worked actively with the radical faction of the Communist Party. Becoming disenchanted with the pace of reform under Betancourt, this group broke with the party in 1960 and formed

580 the Movement of the Revolutionary Left (MIR). Members of this group became convinced that violence was the only route to the type of reform they demanded.[6] Against the advice of more seasoned and older Communist leaders, the MIR almost immediately launched a campaign of urban violence, which reached its peak in November 1960. The failure of this attempt at fomenting large-scale riots and strikes resulted in a plan for developing a full revolutionary situation in 1962. The Communist leadership attacked the assumptions, timing, and general plan of the MIR insurgents, causing considerable conflict within the radical movement. Nevertheless the frequency of attacks increased during the ensuing months. The 46 reported serious attacks attributed to it in 1960 may not appear numerous, but the MIR had become a factor to be reckoned with in Venezuelan politics.[7] These early attempts were quite random, probably being used as training missions. Random terrorism takes less planning and skill than robbery or sabotage. Table 20-1 indicates the types of violence used by the MIR during the 1960-63 period. Random terrorism declined in proportion over this period and

Leftwing Violence in Venezuela, 1960-63 [a]

Type	Percentage distribution of reported leftwing political violence by means employed			
	1960	1961	1962	1963
Riot	17	3	13	5
Assassination	4	9	12	6
Robbery	0	6	9	12
Terrorism	63	60	50	42
Sabotage	5	7	8	2
Other	11	15	10	33
	100	100	100	100
Cases reported	46	33	120	181

(a) From *El Universal*

more purposive violence such as robbery and sabotage increased. The initial terrorism, in addition to possibly providing training, communicated most distinctly the concerns and commitment of this group.

Throughout 1961 the government was active in attempting to capture the terrorists, though the tactics terrorists used made detection particularly difficult. However, Betancourt was careful not to employ police tactics in an indiscriminate manner so as to alienate innocent victims. During this period there was also the development of significant leftwing support within the military itself. This was manifested in June 1962 with serious risings at Carupano and Puerto Cabello. These risings were used as a pretext by the government to suspend some constitutional guarantees and ban the Communist Party and the MIR. However, the government waited to carry out large-scale public measures against the insurgents until the public had been sufficiently frightened by the seriousness of the uprisings at Carupano and Puerto Cabello. That is, Betancourt waited until the illegitimacy of the insurgent actions was well accepted before making a major move. In this manner he was able to maximize support for the government action. In addition, he did not use the opportunity of suspended constitutional guarantees to attack other political opposition. On the 31st of July 1962, the guarantees were restored.

Continued insurgent activity led, however, to a second suspension on the 7th of October. This time the government went out of its way to enlist the support of major groups within the society. The strong measures against the insurgents were accompanied by successful appeals for support from the military and groups such as labor. Betancourt accelerated action against the insurgents only as the public demands for such action warranted, and was extremely careful to maintain popular and institutional support for his use of counterforce against the insurgents. In this manner he avoided the problem of alienating the innocent who might otherwise have been affected by repression.

The Communist Party became more and more disenchanted with the program of violence, which they regareded as counterproductive.[8] The threat from the left actually increased support for the government. Many factions that might otherwise have been more vocal in opposition united under the threat of violence. In fact, Betancourt was not unmindful of this phenomenon and used it repeatedly in appeals for unity, action, and support. One of the strategies advanced by the insurgents was an attempt to stimulate a military coup against the AD government. They judged that a return to military dictatorship would make recruitment into the revolutionary ranks easier, since it would unite the moderate left with the violent left.[9] Again, Betancourt used the existence of this plan to convince the military that they must at all costs remain loyal to avoid falling in with the plans of the insurgents themselves. Restraint again was the characteristic of the government action. Throughout the entire period of violent threat, the reported frequency of government acts of violent reactions was closely correlated with the reported acts of the insurgents. This apparently contributed to the impression of legitimacy of the police and military responses. The insurgents during 1962 did not significantly increase the base of their support and the government did not suffer significant defections.

As the elections of December 1963 approached, violence increased rapidly. The insurgents felt that it was necessary either to induce a military coup or to force the government into overreaction before there was a successful transfer of democratic power. Again Betancourt was careful not to overstep his support, and he appeared in public to be responding less forcefully than many would have wished. His was an unstable tightrope to walk, because he was exposed to failure from insufficient action as easily as from excessive action. Because the insurgents were unsuccessful in expanding their scope of operation, the threat against the government did not increase as rapidly as it did in Cuba. A vicious attack on an excursion train in September 1963 provided an opportunity for a very forceful response with large-scale popular support. Betancourt took advantage of this situation to arrest MIR and Communist Party deputies, to use regulary military troops in urban areas, and to pass emergency measures. These actions were carried out with the strong support of the military and the public at large. As previously, the Betancourt response to insurgent activity appears to have been commensurate with public judgments about what was appropriate.

As in the case of Cuba, violence remained primarily a middle-class phenomenon. The labor movement, though its members had voted against the AD Party and were a potential base of insurgent support, remained neutral or pro-government during the entire period in spite of the fact that the Communist Party had gained control of the union movement during the Jimenez era.[10] The significant difference in the experience of the two countries lies in the continued support for the government from the Venezuelan middle class, including the military and police.

While the insurgents in Venezuela were probably less highly skilled than the Cubans, the government did not make their job easier by overresponding and thus alienating important middle-class support. Had the government lost the support of this important sector, it is entirely possible that the skill and success of the insurgents would have increased sufficiently to overthrow the government or stimulate a military coup.

The cadre of the insurgents was about a thousand men, more than Castro had in the mountains until the fall of 1959 but less than the combined urban and rural force of insurgents in Cuba. The split between Castro and the Communists was similar to the split in Venezuela. The revolutionary testing ground of the Jimenez period was similar to that of Batista. Given all these similarities it is difficult to assess finally the causes of the failure of the insurgents in Venezuela. It is clear that maintaining support from the middle class as well as labor and the peasants was critical for Betancourt's success. With less political adroitness and more intensive police response, this support could easily have been lost. The AD government was successful in maintaining popular perceptions that the acts of violence fell on the I-IV axis of stability rather than the II-III axis of instability. In addition, the government responses did not lead to significant defections that could have increased the skill and organization of the revolutionary movement. There is one last qualification to this analysis. It is entirely possible that Betancourt was able to minimize the factionalism of his own coalition, the moderate left opposition, and the military by the very threat of the insurgents. Had that threat not existed, Betancourt might not have been able to govern successfully. If so, it is strange irony for the insurgents.[11]

Concluding Comments

In this essay we have examined two cases of revolutionary violence. In one instance—Cuba—the insurgents were successful, in the other—Venezuela—unsuccessful. We have looked at the tactics of the respective governments in terms of the underlying strategy, either political or military; the perception of the legitimacy of acts of violence; and the differential political mobilization that occurs as a consequence. If this analysis is correct, it suggests that a predominantly military strategy of eliminating insurgents at all costs is fraught with pitfalls that can benefit the insurgents. It suggests that primary issues for scholarly and policy analysis are popular perception of the legitimacy of the acts of violence committed by insurgents and responded to by government forces, the impact of governmental actions on the organization and skill of the insurgents, and the continuance or withdrawal of support of the largely uninvolved sectors of the population. There are, of course, other relevant factors that would be necessary for a complete comparison of these cases. Outside support, popularity of the existing regime, terrain, objective economic conditions, and others are all of importance. What is argued here is that they are not so critical to the dynamics of the political processes involving violence as the ones chosen for this analysis.

The cases selected were, in addition, ones in which the objectives of the insurgents were the complete overthrow of the regime and system in power. In such cases the dynamic relationship between the violence of both sides and the political process is clearly defined. In cases in which violence is used as a tool of reform rather than revolution, it is likely that the same dynamic relationship exists, although not so clearly evident. The argument is based on the assumption that the critical variable is the perception of the legitimacy of the act of violence, not its legality. Violence in this sense may represent a breakdown in the legal system but not necessarily of politics. No political process is as delicate as one involving significant levels of violence. To lose sight of the political implications of force and violence is to toy with dangerous consequences. Governments can no more hide behind the legality of their acts that insurgents can assume they will have large-scale support. In this dynamic relationship, the violence of the governments is more amenable to control than that of insurgents. At the least a government should be able to avoid counterproductive consequences of its own actions.

Notes

1. David D. Burks, *Cuba Under Castro* (New York. Association, 1964), p. 8.
2. George I. Blanksten, "Fidelismo and Its Origins: Fidel Castro in

Latin America," in Robert Tomasek, ed., *Latin American Politics: 24 Studies of the Contemporary Scene* (Garden City, N.Y.: Doubleday, 1966), p. 369.
3. See Federico Gil, "Antecedents of the Cuban Revolution," *The Centennial Review*, vol. VI (Summer 1962), p. 383.
4. Douglas P. Bwy, "Discerning Casual Patterns Among Conflict Models: A Comprehensive Study of Political Instability in Latin America," unpublished MS., p. 35.
5. Robert J. Alexander, *The Venezuelan Democratic Revolution* (New Brunsiwck: Rutgers University Press, 1964), pp. 42-43.
6. Atlantic Research Corp., *Castro-Communist Insurgency in Venezuela* (Alexandria, Va,: author, Dec. 31, 1964), pp. 39-53 (mimeo).
7. As reported in the Caracas daily, *El Universal*.
8. See Atlantic Research Corp., *op. cit.*, p. 38.
9. For an important statement of the thinking of the leftists, see Walter H. Slote, "Case Analysis of a Revolutionary," in Frank Bonilla and Jose A. Silva Michelena, *The Politics of Change in Venezuela*, vol. 1: *A Strategy for Research on Social Policy* (Cambridge: M.I.T. Press, 1967), pp. 241-311.
10. Alexander, *op. cit.*
11. See Edward W. Gude, "Political Violence in Venezuela: 1958-1964," in James C. Davies, *When Men Revolt—and Why* (New York: The Free Press, forthcoming), for a fuller treatment of this point.

An Approach to Future Wars of National Liberation
by Amrom H. Katz

. . . Vietnam is a large, undeclared war, a war without a front line, a war in which the prominence of the military activities should not have obscured the other, equally important, nonmilitary components of the war.

Among many other things, Vietnam became a huge cracking plant. It has ingested strategic theories, tactics, money, prestige, time, blood, and conventional wisdom. It has distilled off many anomalies—and some insights. It has left residues of unsolved and ill-understood problems. There are numerous asymmetries. One of these is that military success (for our side) is necessary, but not sufficient, to win. However, military failure is sufficient for us to lose. I continue to be astonished and surprised by otherwise distinguished academic and political leaders who still fail to distinguish between *necessity* and *sufficiency*.

I will consider, briefly, three of the many problems that arise from the military portion of the war, but which find their counterparts in the nonmilitary parts of the war. Elsewhere, I have suggested that Vietnam is an interdisciplinry war.[1]

Statistics and Consequences. Even a casual tourist of the Vietnam morass cannot but be impressed—and overwhelmed— by the mass of charts, briefings, statistics, tables, and data. It is usual form, and perhaps comforting, to blame Washington. "Washington needs statistics," or "We must feed the computers in the Pentagon." But one is easily persuaded that even if the demand did not originate in Washington, the statistics would be kept anyway.

By and large, the statistical presentations one gets in Vietnam, or about Vietnam elsewhere, are not data which suggest *how* we are doing in Vietnam. Rather, these are data about *what* we are doing. Confusion on and misunderstanding of this fundamental point has prevented and delayed the search for meaningful and usable insights. We will get nowhere on the understanding curve until we are able to distinguish between "inputs" and "outputs."

If one visits a large air base and inquires as to *how* the war is going, he gets a briefing which describes how may aircraft are assigned, the operation-readiness rate, the sortie production rate per operational-ready aircraft, the number of tons of bombs, ammunition, fuel expended, the number of missions in various parts of that theater, etc., etc. We now believe that asking how the war is going at such levels is inappropriate. *Why?* Because these people are engaged in production, not analysis. It is as if one were to go to the production manager of a Chevrolet assembly plant in Detroit and inquire about the philosophy of transportation, the meaning of his work, consumer satisfaction, or the impact of cars on the economy and its social structure. The man in charge of this plant may have fifty thousand or more employees working for him. He is concerned with the delivery of parts from thousands of subsuppliers and with the well-being of his production workers, the tight scheduling of parts, and so on. . . .

Questions about "how the war is going" are not only meaningless, but entirely irrelevant and hence inappropriate. Such questions reveal a deficiency in the questioner, one which I may now retrospectively admit. Hence, at any of the large air bases or other military centers in Vietnam, one should not ask how things are going, but rather, "What are you doing?" To this question, answers *are* available. But the answers are, in turn, only the necessary beginning of a higher order analysis.

A front line is sorely missed in Vietnam. A statistical substitute for a front line is unsatisfactory. In the paper cited earlier, it was said that:[2]

> "In the absence of a front line, we are left with statistics—incidents, destruction, defection, weapons lost and captured, kill ratios. And the statistical 'front line' constructed from and balanced on these statistics is a poor and unconvincing substitute for a real one. But it's the only one we have, and in the absence of either conspicuous and overwhelming defeat or victory, its equivocality counts for much of the travail and argument about the war."

Unfortunately, we failed to standardize on a common analytical and statistical base with the enemy. In the absence of a common data base (and, of course, a common front line is an outstanding example of a noncooperatively agreed-on common data base), it is possible and plausible that North Vietnam and the Viet Cong are using different graphs, different parameters of performance, and different indices of success. The availability of the data we gather is insufficient to guarantee its relevance.

Consider—as a thought exercise—what kind of briefings Ho Chi Minh was getting on the course of the war, and compare these with the briefings that were given Secretaries Rusk and McNamara, and President Johnson (or are now given to Secretaries Rogers, Laird, and President Nixon). It could have been simultaneously true that Ho's graphs, if he used any, were going up while ours also were going up. If we want to persuade the leaders of North Vietnam that they are "losing" the war, we had better (1) find out what measures, indicators, or graphs they are using; and (2) operate in such a way as to make these graphs, etc., go down.

Some of the parameters are not ours to affect. But certainly, as a first priority measure, we had better find out what the North Vietnamese high command is getting in their briefings. We must stop being mesmerized by our own extensive and largely irrelevant statistics, graphs, and data. We had better launch an all-out intelligence attack on this problem. It suggests also that we should compose a North Vietnamese and VC "high command" in Washington composed of people *whose sole assignment is to think full-time* like the leaders of North Vietnam and the VC. It is not sufficient for hard-working Americans to take ten minutes off during lunch to speculate about this subject. We need a competent full-time group, totally immersed in this problem.

Surely we can find a small number of people with long-term backgrounds in Indochina, with language skills and with diverse specialties, people who are empathetic (if not sympathetic) with the goals, ways of thought, actions, and reactions of the other side. Such a group, free of all distracting duties, could track the other side, and within a fairly short time predict what the other side will do.[3] Such people can be found and such a group can be formed. It should have been done long ago. Doing it is still a useful idea. That it is inexpensive is not enough to condemn it.

Learning and Remembering. General O.P. Weyland makes the following comment about how we fought the Korean War:[4]

> "An astounding facet of the Korean War was the number of old lessons that had to be relearned . . . it appears that these lessons either were forgotten or were never documented—or if documented, were never disseminated."

Because we did not always take Vietnam seriously, because Vietnam is, in all respects, a very novel problem for us, because we have not declared war, because of our one-year rotation

The Problems of the Defenders

policy for most Americans assigned to Vietnam—for these and for many related reasons, we do not seem to have been learning from Vietnam at a rate commensurate with our expenditures in prestige, blood, dollars, and time.

We have gone through a tremendous effort to interrogate captured VC prisoners. It is inconvenient to interrogate them, and the results are far from unequivocal. Yet the effort, correctly, is judged worthwhile. *How come we have not mounted a corresponding effort to interrogate Americans?* They don't have to be captured, they're willing to talk, there's no language problem, and there's plenty to be learned.

Why should we interrogate Americans? Every American who has gone to Vietnam has learned something; many of them have learned much. Many returnees assert that they have not been debriefed systematically; further, after say a year, their sharp, keen remembrances of what they learned, how they failed or succeeded in Vietnam, degenerate and erode into a collection of anecdotes or "war stories."

What do we mean by learning? The image shared by many observers of United States' participation in World War II is one of steady growth and improvement. We started that war slowly, inexperienced, out of focus. By the end of the war, some three and a half years after we started, the U.S. armed forces were a sharply tuned, finely honed instrument of extreme power, power that could be applied with precision and focus. It was just a few years from wooden rifles on maneuvers in Louisiana to VJ Day. The image one gets of progress in the Vietnam engagement is that of a saw-tooth wave. It builds up, steps down, builds up again, steps down, the stepdown being closely associated with the rotation cycle. Now when we say this, we know full well that units stay in Vietnam; nevertheless, people change. The mechanism for passing lessons from one group to another is unclear; in most cases, it is either nonexistent or inadequate.

This point can be illuminated by a couple of typical stories. Of course, such stories also verify the enormous adaptability and versatility of the Americans sent to Vietnam. One captain who had flown well over five hundred missions as a forward air controller (just before his year tour was up) was asked, "Where is the book you were supposed to have produced?" When he asked, "What book?" I said, "Well, you've been flying around Pleiku here for a year. You've learned a lot about the VC, their habits, their characteristics, various signs and indicators useful for intelligence, useful for your replacement. Now the next man is coming soon, isn't he?" And he said, "Yes, I haven't met him yet." "Where's the book that would let him get off to a better start than you did?" He said, "I haven't written it. Let him start where I did. It's only fair." We knew he hadn't written it, and were using this merely as an argumentative question. But what was true was that the man who had replaced him would have to start at the same absolute zero level where this captain had started. The knowledge he had accumulated would be of no use to him in his next assignment at Air Defense Command Headquarters.

A lieutenant colonel in charge of a Direct Air Support Center (DASC) was asked what training he had for that job before he got there. He replied, "Absolutely none." We asked how much overlap there was between him and his predecessor on the job. He gave a loud and uproarious laugh, and said, "The day I got there, that man was leaving. He had his hat and coat on, threw me the key and said, 'There's the shack. Good Luck. Every day is different around here.' That's all the training I had."

Similar stories can be repeated at length, but with the same point. People arrive with little or no training, which is understandable. But what is not understandable is that in many cases their replacements must perforce start at practically the same low level as they did. The same sort of stories apply to provincial advisors, sector advisors, and so on. Of course, there is no adequate preparation for such a job, because we have never had jobs like that in either civilian or military life. The question remains, Why must the replacements start so close to zero? One mechanical reason that there is, typically, no overlap between, say, the forward air controller and his replacement (who, given several weeks of flying in the back seat of the FAC airplane, would have read the "book"), is chargeable to the personnel system. Overlap might have resulted in going over the legal limit of troops in Vietnam.

The lack of a collective memory, the lack of an information base, the lack of ready libraries with relevant data, force people to repeat errors which have already been committed. If we need a single policy for Vietnam, it should be, "Don't make an old mistake." It is not enough, of course, to suggest that we must learn the lessons and pass these lessons on; they have to be the right lessons. Getting agreement on who learned what, and which are the right lessons is no downhill slalom. This is a tough job; but it must be tackled, with energy and speed. And soon.

Technology and the War in South Vietnam. Alternatively, this section could be titled, "The State of the Art vs. the Art of the State."

No one can—or should—be opposed to technology. Indeed, we are now using advanced technology in Vietnam in many ways: communications, airlift, aerial bombardment, and so forth. But there is an enormous predilection on the part of the scientific, technological, and military apparatus in this country, when confronted with a new problem, to look first (and, sometimes, only) toward technology for the answer.

The search is usually directed toward some weapon system, some effect, some gadget, or some form of technological magic which, if it works, will contribute immediately and obviously to winning the war. Such gadgetry does not exist—for Vietnam. Much more important, its pursuit is self-defeating, because it diverts the eye from the task at hand, of using the equipment already in hand, of inventing suitable tactics and techniques, and of coming to a full understanding of the problem in Vietnam. With the Holy Grail of technology as a target, we shovel all that off to one side; and instead, depend on the vast scientific apparatus of the United States to spit up an answer. Technology has become the opiate of America. To the lay person, or even to one who is fairly sophisticated, it does seem odd that as long ago as 1964 we were able to impact the moon and transmit pictures while at the same time we were unable to find Viet Cong by technical and classic reconnaissance methods in Vietnam. The much more complex task of landing men on the moon and returning them, in 1969, sharpens the point, demonstrating enormous progress since 1964—in astronautics.

It is, of course, important to distinguish between these tasks. The moon, no matter that it is distant, is visible, predictable, and essentially cooperative; at least, it does not offer any opposition. It does not take evasive action after we publish our plans, it is not camouflaged, it has not infiltrated the solar system. Furthermore, the task of sending a package to the moon, although inordinately complicated and expensive, is *definable*, and in that sense, *straightforward.* The solution did not require radical inventions—just devilishly hard, careful work and lots of money, time, and luck. Our American style fits *orderly* problems, no matter how big and how difficult. We need to develop a style that fits *ornery* problems.

The difficulty with the predisposition to turn to technology for answers is that it takes one's eye off the ball. An almost therapeutic sense of relief occurs if one can say he needs a technological solution, because we have a large, competent, and enthusiastic scientific and technical apparatus ready, able, eager to work on such problems.

But we already have such marked technological superiority over the Viet Cong in Vietnam that it is difficult to believe that any new output of our technological machine will make a significant difference. As only one example, consider plotting one aspect of the operating characteristics of the Viet Cong. Let us plot VC speed against VC altitude. (Other parameters could have been chosen, but these are suggestive). By and large, a Viet Cong (unless he has climbed a tree) cannot see anything from more than five feet off the ground. We have aircraft that can fly at almost any altitude. The Viet Cong cannot move faster than about four miles per hour in South Vietnam. *We started with, and kept, a tremendous and obvious speed and mobility advantage.* We cannot be short of speed and altitude. Even the oldest propeller aircraft in Vietnam are infinitely faster than any enemy aircraft—in South Vietnam. The "Solution" lies in tactics, techniques, style, and, above all, among a set of parameters and indices which have little or nothing to do with technology, but include such things as stability of the government, credibility of the government, and so forth. To these, technology has little, if anything, to offer.

I certainly am not saying that technology should be turned off. It should be left alone, on the side, in parallel with the "problem" and not in series with it. A clear recognition of its proper place will hasten attention to the meaningful and hard problems, and thus enhance the chances for progress toward a solution.

Major General Moshe Dayan, Israeli Chief of Staff during the 1956 Sinai Campaign, visited Vietnam in 1966, and published a perceptive series of articles based on his experiences. Reporting on an attack by a Viet Cong regiment on a Korean company,[5] Dayan described the enormous support furnished by artillery, aircraft, tanks, and so forth. He said, "What happened to the Viet Cong soon after their attack was that along the two hundred yard wide strip between jungle and fence, the American support units laid down no less than 21,000 shells!" He added, parenthetically, "This is more than the total volume of artillery fire expended by the Israeli Army during the Sinai Campaign and the War of Independence together!" Dayan was impressed, as anyone should be, on whatever side. However, neither the moral nor the lesson of such observations is self-evident.

Albert Wohlstetter, in a useful piece,[6] ably demolishes (what he asserts to be powerfully held) beliefs about the inverse relation between military strength and distance. Costs of seaborne transport, the facts about economy of scale (for example, supertransports), and other useful data are reviewed. I had myself dealt with this topic earlier:[7]

> "Modern communication and transportation—of people, ideas, hardware, and weapons—have killed the nineteenth century notion of contiguous, nonoverlapping "spheres of influence." However, this obsolete notion still haunts the columns and speeches of well-known pundits. It is as misleading as the Mercator projection, that distorted map which was probably a major factor contributing to American isolationism. The sphere of influence of the US is a ball eight thousand miles in diameter—the Earth."

Our superb logistics system, backed up by the world's greatest production facilities, our ability to build remote bases, depots, storage dumps, and fill them with tools, machines, and lubricants of war, have impressed many observers. The data in Wohlstetter's article will persuade anyone that we can win any logistics Olympics where the emphasis is on long distance and large volume, that is, what we could call "wholesale" delivery.

But in the process of explicating these matters, a major point gets lost among the statistics. *Our ability to deliver large amounts of materiel and men over large distances at low costs causes us to do just that.* No one can beat us at *our* game. So why should we expect others to compete in our game? Sure we can beat Hanoi in delivering war supplies to South Vietnam, via either the "wholesale" delivery described above, or via "retail" delivery to the final user in the field. So what? Secretary of Defense Robert S. McNamara, testifying in the 1967 Hearings on the Air War Against North Vietnam,[8] said that "Intelligence estimates suggest that the quantity of externally supplied materiel, other than food, required to support the VC/NVA forces in South Vietnam at about their current level of combat activity is very, very, small. The reported figure is about fifteen tons per day." Even if we multiply this estimate by six, we get less than a hundred tons per day. Now the right logistics question should be asked: Is it easier for the NVA supply system to supply its troops with a hundred tons per day than for us to supply our troops *at our rates?* *A clue to these rates is the fact that at the period of maximum US presence in Southeast Asia, we were shipping about ten times as much to the post exchange system as the upper estimate (a hundred tons per day) required by the VC/NVA forces to fight the war.!*

We are better at a game that the other guy isn't playing.

John P. Roche, commenting on the development and defects of the doctrine of limited war, said[9]

> "We assumed, naively as it turned out, that the knowledge that the US could transport a hundred thousand men twelve thousand miles in 47 hours and 32 minutes (or some such logistical triumph) would itself act as a deterrent. Discussions of military strategy began to sound more and more like seminars in game theory. There was a kind of antiseptic quality permeating the atmosphere; one often had the feeling he was attending a chess match.
>
> "This, in part, was the source of many later problems. . . ."

There has been much disquiet, much international rumbling, much moralistic preachment in the United States and elsewhere about the conduct of the military operations in Vietnam. Charges of unrestricted, indiscriminate, senseless, and immoral bombing and artillery fire have been made with high frequency. It is difficult to answer the questions and charges with enough hard fact. The nagging residue dilutes, mutes, and otherwise inhibits both support and understanding here and abroad. The black shadow of My Lai, anomaly or aberration though it be, detunes and blurs all discussions of the war.

Even among experienced and full-time local observers of the Vietnam scene, such as province advisors at both the MACV and AID levels, the polarity of opinion about the bombing and artillery strikes is amazing. Opinion at a recent conference found participants at one end or the other, with nobody occupying middle positions. Some were sure that we were making VCs, others denied this effect of village attacks. Further, all opinions were firmly held and strongly expressed. Part of these differences may be explained by differences (for example, from corps to corps) within Vietnam itself. Differences in value judgments also explain part of this variation. It seems clear that an unequivocal, generally accepted collection of data in an easily interpretable form does not exist.

The US Strategic Bombings Survey of World War II followed immediately upon the conclusion of the war. But precisely because there are no front lines in South Vietnam, much of that country is accessible *during* the war. It would be useful to find out exactly what we have done in Vietnam, and document it in a way which does not emerge from standard reports or from suspect statistics. Besides, most statistics are gathered for specific, and often narrow, purposes, which do not include the purpose of this proposed survey. An *independent,* high-level part-Department of Defense, part-civil investigation of the bombing should provide data to either dispel, or confirm, the charges which have played so important a part in both domestic and international politics. Until and unless some actions similar to those proposed here are taken, the charges will continue to sap the intellectual and moral energies of those responsible for prosecuting the war, and will continue to inhibit more support from friends and allies.

The impact of bombing and heavy artillery fire on the pacification program is not readily gauged from statistics confined to military actions alone. A high-level, competent group could bring insight and skills varied and deep enough to bridge the problem of the interaction of pacification, with all that is subsumed in this overworked word, and military actions. We need, and deserve, a clearer understanding of what happens to a population in a country like Vietnam under protracted, heavy military action. We need to extend our information sources—as well as the breadth of our inquiry—beyond that generated by, and limited by, prisoner interrogations. We need this information as an input to the future. . . .

South Vietnam has been emphasized in this discussion. It deserves priority attention. But we have been bombing elsewhere, too. These other aspects of our strike operations have been given extensive analysis; our prestrike and poststrike reconnaissance and accompanying photointerpretation, in the case of North Vietnam, at any rate, has been superb. However, these classic tokens and inputs to analysis may give unwarranted confidence in our model or synthesis of the North Vietnamese economy and social structure, and our understanding of the control mechanisms, resilience of the regime, and staying power of North Vietnam. Fundamental aspects of our encounter with a country and a system about which our understanding is imperfect, are still not in hand.

For example, I would expect the proposed survey to clarify the notions and relationship between effectiveness and political acceptability for various kinds of military operations. Let me explain. Even without a precise and accepted definition or measure of "military effectiveness," it is plausible that different kinds of military operations vary in effectiveness. Some are better *impedance matches* to the problems they address than are others. Similarly one can, in principle, score operations on their political acceptability, their counterproductive value, or even (and I assume the word has not yet gone out of the vocabulary) their cost. What we might attempt to do is to relate these two variables, whose use can be understood, even if they are not clear and quantifiable.

Some operations may be described as "counterproductive." Despite the ease and frequency of use of this word, it is hard to define—except in a strict academic sense—wherein the *cost* exceeds the *benefit*. Unfortunately, in most real problems, massive doses of judgment are needed. The costs may be in one kind of "currency," and the effects, benefits, or profits may be in another kind of "currency," with exchange rates or conversion factors not well known or even understood. Counterproductive—as we use it here—means too much dollar cost, too much adverse publicity or polemics, or too much political cost compared to the value of the operation in advancing our objectives.

Our presence and all our activities draw uniform invective and polemics from the world Communist community, whether there is a single Communist bloc, or several, or none. They, and others who share their views, are not open to discussion, nor do they see differences between various kinds of military operations. But both among Americans and among our nominal allies and

598 friends in the Free World, some kinds of military activities inhibit and mute support for our entire enterprise. This is the minimum cost; remaining silent and hesitant is "better" than active opposition to our policies and our presence. I have the visceral feeling that most of the potential support that has soured into active opposition is directly related to a noisy and visible minority of operations and activities.

I can imagine, but I cannot believe, that we will finish in Vietnam without having conducted broad inquiries such as the particular survey outlined above. We must make them, or suffer in the future. . . .

Will the United States find itself involved in future Vietnams? The question, put this way, admits of only one answer: No. But the question more sharply: How will the United States respond to future wars of national liberation? Now we have a discussable subject.

Some—perhaps a large number of people—argue that our poor performance in Southeast Asia is evident proof that we should not have gone in. Putting aside the question of fact about our performance, surely the "Rightness" of our behaviour must depend on more than retroactive judgement based on performance. I find it useful to separate these two, performance and "rightness" (no, I don't mean "correctness," which carries overtones of appearance and modalities), as a guide to discussion.

We can visualize a graphical plot of positions about Vietnam and, say, our intervention in the Domincan Republic. Some might agree (with me) that we were high on the "rightness" scale for going into Vietnam, but we were low on performance scale. One could argue—it has been done—that even though "things worked out" in the Dominican intervention, we were wrong to go in. I leave the construction, plotting, and interpretation of other combinations of these two (necessarily qualitative) parameters, and their indentification with other situations and viewpoints, as an exercise for the reader.

Writing in 1961, John P. Roche suggested:[10]

> "The choice between 'intervention' and 'nonintervention' is a moral and political fiction. No matter what we do, we are intervening in someone's behalf. To refuse to recognize this fact is itself a form of intervention on behalf of the *status quo*. We must, therefore, turn to the question

which exists in the real world: By what methods and to what ends should the United States' power be utilized?"

We can hope that the trauma of Vietnam does not force us to choose only between the spectral extremes, either isolationism or nuclear weapons. We can hope; but, in addition, we had better construct viable positions between the extremes.

No one can quarrel with Brigadier F. P. Serong's maxim: There is only one good counterinsurgency operation—the one that never had to start. This statement implies early participation, before an insurgency becomes a forest fire. I still argue that the word "counterinsurgency" means that we have conceded the first blow to the insurgents. It conveys too much of a reactive, defensive *status quo* approach. It should be exised from our vocabulary before it gets rooted in the dictionary. What is needed is a concept, attitude, and program which does not exclude, in its title, possible support of insurgents in some future situation. We need to support freedom and independence, not just "counter" someone else's initiatives. I do not have a list of insurgents or movements that I would like to support at present; but I do not want to foreclose such support in advance on semantic grounds alone.

"Revolutionary democratic internationalism" is not yet part of our policy, and we are not operating a *Demintern*, to use the felicitous term of a colleague. But we cannot cope with the new problems by traditional and orthodox techniques applicable to other kinds of situations. To many old hands and to many beginning students of these new problems, the organization of the present American effort to anticipate, detect, identify, and respond to combustible situations seems ineffective and insufficient. Clearly there are enough resources, but they need focusing and correlation.

The many activities now identified and lumped under the "counterinsurgency" umbrella are minor, and usually unwelcome, guests in both military and civilian agencies. We have discovered that Vietnam, as all derivative or related conflicts will be, is an *interdisciplinary war*, meaning that there are a large number of simultaneous activities (an admixture of military, police, political, economic, and other) going on, with an even larger number of operating agencies. The Country Team concept is alleged to reach its zenith there, where all agencies—AID, the military, the CIA, and so forth—are supposed to work for our ambassador. This may be a nominal truth; but for a long time, perhaps even now, each member of this team looked to his own agency, US-based, for promotion and for his next assignment....

This suggests that a new agency devoted to these new problems on a full-time basis needs to be established. The title of the organization should reflect the earlier comment about counterinsurgency; that word should be dropped. A suggested name for such an organization could be the National Independence Support Agency (NISA). It should probably be in the Executive Office of the President. The agency need not be large, but should be big enough to make effective use of the talented, dedicated men who now find no useful continuous career. It would be a place where the interdisciplinary nature of the problem is recognized by using all the various skills and techniques....

The United States continues to pay heavily in blood, treasure, prestige, and credibility for its participation in Vietnam. We have lost self-confidence and momentum. It would be cruel and wasteful not to learn how to do better or differently.

Notes

* Reprinted with omissions.
1. Amron H. Katz, *The Short Run and The Long Walk,* prepared for the Wingspread Symposium on Southeast Asia, September 17, 1965. Part I published in *Air Force/Space Digest* (June 1967). Part II in *WAR/PEACE Report* (December 1965).
2. Ibid.
3. This idea is part of a larger thesis—the need for devil's advocacy at all levels of government.
4. From General O. P. Weyland's official report at the end of the Korean fighting, as quoted in R. F. Futrell, *The U.S. Air Force in Korea, 1950-1953* (New York: Duell, Sloan and Pearce, 1961), p. xvii.
5. *Los Angeles Times,* October 16, 1966.
6. Albert Wohlstetter, "Illusions of Distance," *Foreign Affairs,* vol. 46, no. 2 (January 1968), pp. 242-55.
7. "The Short Run and The Long Walk."
8. Hearings Before the Preparedness Investigating Subcommittee of the Committee on Armed Services, U.S. Senate, August 25, 1967, Part 4, pp. 277, 298.
9. John P. Roche, "Can A Free Society Fight a Limited War?" *New Leader* (October 21, 1968), pp. 6-11.

10. John P. Roche, "Further Thoughts on Intervention," *New Leader*, June 19, 1961. *See* also his essays, "Confessions of an Interventionist," *ibid.*, May 15, 1961; "Uses of American Power," *ibid.*, March 2, 1964; and "Can a Free Society Fight a Limited War?" *ibid.*, October 21, 1968.

Morality and National Liberation Wars*

by Charles Burton Marshall

Vietnam will not be the final reckoning on the doctrine of so-called national liberation wars. More of them are to come. The United States government will be disposed to take opposing actions.[1] It will wish to justify what it may do—in the sense both of squaring with a plausible frame of values, and with giving an account of its actions adequately persuasive to rally support. What I have been asked to do is to treat the rationale "on a philosophical basis" emphasizing considerations of "morality," particularly in view of a tendency among many young Americans and perhaps others to doubt whether the United States has the moral right so to act in situations where Communist insurgency seeks to overthrow what, for lack of a better phrase, is a legitimate government.

I own to a measure of diffidence about the topic. It is an exercise in abstraction founded on sketchy hypotheses—akin to trying to figure what card to lead before picking up the hand. It is a hard undertaking for a self-regarded empiricist. Edmund Burke's words are pertinent: "I must see with my own eyes, I must, in a manner, touch with my own hands, not only the fixed, but the momentary circumstances, before I could venture to suggest any political project whatsoever." To put the matter in Burke's words again: "Circumstances . . . give in reality to every political principle its distinguishing color and discriminating effect. The circumstances are what render every civil and political scheme beneficial or noxious to mankind." . . .

The tone of certitude is sometimes ideologically derived—as in the doctrine of national liberation wars itself. Morality is implicitly cognate with Marxism-Leninism. The view of mankind is Faustian. The human spirit is premised as basically unflawed. What ails the human situation is taken to be a defect in environing conditions susceptible of technical correction over a period of time, whereupon mankind will achieve amalgamation on a utopian plateau of contentment, prosperity, and perpetual tranquility—heaven brought into the realm of time. That outcome is inevitable by a dictate of historic laws. These laws are the determinants of all morality. Marxism-Leninism alone is consonant with them. It also asserts exclusive insight into the technical corrections necessary to achieve the perfect future and, as a corollary, has a monopoly of justice. Whatever moves with history's tides toward the perfect future thus perceived is legitimate, and whatever would impede them is seen as inherently illegitimate. Force used on the side perceived to be favorable to Communist interests is thus purportedly legitimized. As a corollary, any use of force to oppose it is historically illicit.

There are other versions of ideological invocations of morality besides the Marxist-Leninist. It is pointless to labor them here because there is nothing much to be done about them directly in argument. Ideological arguments are based on circular logic, tend to be impenetrable by any counterargument, and so far as their users are concerned are irrefutable. They focus on ends. In light of some putative goal, they dispose of all questions regarding means. An opposite kind of argument focuses on means exclusively and thereby puts aside all consideration of ends. A proponent focuses on a single value, to him the touchstone of morality and justice. Anything that impinges on it is wrong, no matter what other values may be at stake. Often the import of the argument is that a policy undertaking is immoral and unjust if it entails any regrettable consequences whatsoever. . . .

I stress the point of morality not merely as a conceptual good or exalted purpose but as a mode of conduct. Conduct essentially involves questions of means as well as ends. In this vein, Samuel Taylor Coleridge pointed to the linkage between morality and prudence. "In general," he wrote, "morality may be compared to the consonant; prudence to the vowel. The former cannot be uttered (reduced to practice) but by means of the latter." I would carry the point further by asserting prudence—that is, the application of sagacity in the management of affairs—to be inseparable from morality in its pertinent dictionary meaning. Regarding morality—like justice—as a matter involving apportionment of restricted means among ends which tend to exceed them, I agree with Robert Briffault's stricture against absolutism in morality: "a guarantee of objectionable morals in the same way as absolutism in government is a guarantee of objectionable government."

This view entails a considerably restricted view of the scope of foreign policy. It entails a precept quoted from Leopold Tyrmand: "Among other things, civilization means abiding by a convention according to which we human beings agree not to burden each other with our excessive humanity." It enjoins a proper consideration for practicability.

A colleague of mine recently remarked that a proposition

needed only to be conceptually good to qualify as a goal of foreign policy. I took exception. In my contrasting view, any proposed action in foreign policy must pass three preliminary tests in order to be eligible for consideration. Is success a logical possibility? Would the putative results, if achieved, be desirable in the measure of a plausible scheme of value? Finally, and critically, is the undertaking feasible for me? . . . I would still have to weigh the putative gains against the probabilities and the costs. It is the same way in foreign policy. A national purpose might be logically possible and abstractly desirable, but still not achievable by us, or, even if achievable, still too chancy and costly to be worth undertaking.

This view of foreign policy values experience above abstraction. It grants a presumption in favor of established norms as distinguished from merely national alternatives. The outlook eschews a current wide tendency to avow change, in the abstract, as inherently good. It acknowledges that mere universal change, with no continuities whatever, would by definition amount to chaos. Therefore, it seeks to preserve as well as to alter, and attributes no inherent virtue to mutability or mere tradition.

This view acknowledges, among the things owed unto Caesar, an obligation to avoid ways of thinking, even abstractly benign ways, which conduce to Caesarism. The point has a bearing on the question of legitimacy—another of the abstractions in the frame of my topic. Legitimacy in relation to political authority refers specifically to hereditary or established right to govern. More broadly, the term expresses a relationship between a regime and a determining number of the populace subject to its jurisdiction. In a condition of legitimacy, those in magisterial positions feel sure of their title to authority and are confident regarding acknowledgement of that title by an effective preponderance of the populace; and the preponderance does so regard the matter, with a result that the rightfulness of the regime is not an issue or a point of anxiety. In any event, the quality is one inhering in relationships between those who govern and those within their sphere of jurisdiction. The acknowledgement of legitimacy may be articulated in institutional procedures or granted in a tacit way. However bestowed, it is an internal attribute. According to a corollary principle in international law, it does not fall to outsiders to determine the legitimacy of a regime. In Emerich de Vattel's classic and, I think, still sound expression of the principle, "No sovereign state may inquire into the manner in which a sovereign rules, nor set itself up as a judge of his conduct, nor force him to make any change in his administration. If he burdens his subjects . . . or treats them with severity, it is for the nation to take action; no foreign state is called on to mend his conduct and to force him to follow a wiser and juster course."[2]

This sober view of the possibilities of foreign policy—eschewing ideology, counting on no transformation of human nature as a deliverance from the vicissitudes of nations, nonutopian, willing to temporize with imperfection—seems to me to be in keeping with correct canons of morality in general affairs. I prefer it to pushy, strident, all-encompassing alternative designs to redo the whole human situation. But I am not at all sure that it will work. In this sense, I am pessimistic.

My pessimism was sharpened in an appearance a short time ago before a group of Latin American military officers at the Inter-American Defense College in Washington. In my main discourse, I tried to put into perspective some of the problems currently confronting the United States in world affairs. In the question period, my listeners pressed me to give the elements of an ideology to counter the Communist version. In answering, I dealt analytically with the inherent nature of ideology—its total claim on historic legitimacy, its profession of having the key to everyone's future, the rigidity of its purported system of thought, the unscientific character of its pretensions to scientific reasoning, its religiosity about secular concerns, its inherent tendency toward authoritarian and totalitarian government, and so on. No one in the audience disagreed. Still I was pressed to reveal the elements of a new ideology to counter the prevalent one. Without one, defeat was inevitable, according to the argument. I explained the impossibility of propounding a new ideology, especially by someone categorically and devoutly a disbeliever in schemes for temporal, universal human perfection. I drew an analogy from the case of a man who wrote Voltaire inquiring how to set about establishing a new religion. Voltaire wrote back that the first thing to do was to get oneself crucified and then to rise from the grave. Whipping up an ideology like a pudding at a moment's notice was too much. The issue, I insisted, was one of nonideology versus ideology. In a competition among ideologies, I added, communism had a headstart and the inside track. I left my questioners perplexed. They still professed to want an ideology.

A second point of misgiving relates to a line of thought in an article by Jay Forrester in *Technology Review* for January 1971. Taken at face value, the article has profound implications for United States foreign policy. It portrays the United States as "setting a pattern that other parts of the world are trying to follow," a "pattern that is not sustainable." Thus our "foreign policy and our overseas commercial activity seem . . . contrary to overwhelming forces . . . developing in the world system." Industrialization, with medicine and public health as components, "may be a more fundamentally disturbing force in world ecology than is population." For present underdeveloped countries to reach the standards of livelihood in present industrialized nations may prove simply impossible in view of the pollution load and natural resource drain entailed. The present may prove to be the Golden Age, with the quality of life destined to slip henceforth. A high level of industrialization as exemplified in the United States may well prove self-extinguishing. Efforts for administered population control as distinguished from population control by catastrophe may prove "inherently self-defeating." The next century may well present a choice among suppression of modern industrial society by a natural resource shortage, collapse of world population from pollution factors, and population limitation by hunger, war, disease, or social stresses due to overconcentration of people. Over a like span, the underdeveloped countries may prove "closer to the ultimate equilibrium with the environment than are the industrialized nations," and hence in better condition than economically advanced societies for coping with "the forthcoming worldwide environmental and economic pressures."

The article, based on computerized calculations beyond my technical competence to criticize, corresponds to my intuitive assumptions, but no one can prove futurity. I refer to the theme because it raises a question as to whether the image of the future portrayed by the United States to the rest of the world is plausible.

The question has a bearing on the competition between American purposes and Communist aspirations epitomized in the problem of so-called national liberation wars. For coping with stringency, a rigorous totalitarian system of rulership enjoys distinct advantages over an accountable system of government, in my estimate. A remarkable aspect of the Communist record over more than half a century is an ability not merely to maintain continuity of control but, indeed, to turn willfully caused shortages into a source of strength for Communist regimes.[3] If the Forrester prospectus is even only approximately correct, the model which the United States has portrayed for emerging nations may be fated for irrelevancy. That idea, however, is a worry for a longer future—though the adumbrations may have effects in a shorter span. . ¡. .

In reality, almost all projects in foreign policy are speculative. Mostly they manage at best to postpone trouble or transform problems without basically alleviating them. To say such a thing is not a reproach to policy. As Roderick Seidenberg has written in *Posthistoric Man,* "the doubts, problems, and antinomies of one age are not so much resolved as supplanted by those of another. Man moves not from solution to solution but from problem to problem." Most of us learn to cope with the parking problem without ever finally solving it, pamper our health while growing older day by day, and carry on a protracted but ever losing skirmish with the Internal Revenue Service. The trouble is that many expect foreign policy to transcend the human situation. Perhaps it is necessary for those who govern especially to share such expectation. In any event, the process of recurringly generating something near a one hundred percent commitment to a 51 percent proposition requires a considerable doping-up of the body politic. It is probably unimaginable that any sizable segment of public affirmation could be adduced, or consent and funds in quantity elicited at the Capitol, if proponents of a policy were to own that the underlying information was chancy, the likelihood of fulfilling objectives sixty percent or under, and the time span of the putative benefits only a few years at best.

Moreover, the public media reflect a professional and pecuniary interest in finding out more than anyone can possibly know. Those in public authority, reluctant to confess their own ignorance and maybe even unaware of it in a particular case, cater to the demand. Yet inevitably a reckoning follows, and the so-called credibility gap is the consequence. It is not so much due to prevarication as to an understandable desire to yield to pressure for clarity regarding matters inherently unclear and for prediction about the unpredictable.

Something else borne on the information flood pertains to the coloring of the news—I mean this literally, not figuratively—on TV. Jenkin Lloyd Jones has written of the "burning villages; the hysterical widows; the napalmed babies; the sightless corpses sprawled in hideous repose; the excited voice of the television reporter painting in bloody colors the havoc that the gunships and halftracks leave behind them." He observes: "Our enemies are spared this. They see only gory propaganda pictures of what our side allegedly did." He adds that such "coverage of the war in Southeast Asia has no doubt contributed to the peace-at-any-price syndrome that afflicts so many citizens." He compares this output to "what sheltered Americans got in past wars, the impression that only the enemy could produce evil, and that for our side the band played, the angels smiled, and death, if it came, was a neat little hole in the breast." He sees "the awful immediacy . . . here to stay," and speculates that "it is just possible that if we hope to survive as a free people we all may have to become a little tougher-minded than our ancestors." The stress may be alleviated somewhat in future conflicts where United States involvement will be vicarious rather than direct, but it probably will not be eliminated altogether.

There is, however, another problem—conceptual rather than sensory—of stomaching unpleasant realities in connection with such wars. For some people, it is hard to affirm commitment to a military cause except on a premise of unequivocal good arrayed against unmitigated evil.[4] In combined efforts, this notion requires postulating an immaculate ally. The trouble is that immaculate allies are fabled rather than real.

This observation applies particularly to countries beset by insurgency. Well-integrated societies, with a high level of civic rectitude, are relatively immune to such threats. Countries under domestic threat, with analytic near certainty, will be marked by civic imperfections enough to prompt a spate of denunciation among idealists. The point applies not only to matters of pecuniary rectitude but also to justice. Societies with a shaky basis of order are likely to be rife with injustice.[5] A curious aspect of the Vietnamese experience has been a recurrence of demands for civic and commercial purification within the beset country as a precondition for continued United States assistance. A few years ago, an idealistic Under Secretary of State, relating to me the criteria which the United States should require to be fulfilled before vouchsafing any military assistance, uttered a list of civic excellencies beyond the reach of almost all of the societies I could think of. What his formula amounted to was that we should never assist any country except the ones that would not need any assistance. . . .

As a pertinent point, the farther a matter is along the abstraction scale, the more difficult it is to arouse public interest so as to gain support for having a policy and allocating resources. Roughly speaking, for each step along the progression, an additional syllogism of logic and argument is required for making a case for public action. Maybe the practicable limit in generating wide support is in the order of three steps along the scale of abstraction. That is to say, one should not expect to arouse affirmation with a line of argument requiring more than three syllogisms in a series—a point akin to what Burke had in mind when he said, "Refined policy has ever been the parent of confusion."

Here the focus is on matters of concern for national security. On a pertinent scale Pearl Harbor would have been at point one. The missile deployment to Cuba would be at point two. That is to say, in neither instance was any labored explication required. They were instances where, in the cliché, the facts spoke for themselves. One of the difficulties about the Vietnam interposition has been the complexity of the reasoning. I, for one, have understood the case made on behalf of the policy, but it has been necessary to work at it. . . . As a truism, any war for which the rationale cannot be simplified is likely to go begging.

One factor of complicacy in this connection has been the circumstance of trying, at once, to prosecute a war and to foster détente. Take my own case. I never aligned with opposers of the Vietnam War. Neither have I been enlisted among the vehement supporters of it. My attitude toward it has been one of unaccustomed velleity. Candidly, it has been beyond me to generate much affirmation in face of being told at one and the same time that the era of confrontation is over and that of negotiation has begun, and that destiny and the hopes of free men ride on the outcome of a confrontation as to which the enemy is manifesting no disposition whatever to negotiate.[6]

The ambiguity will persist. The point is reinforced by an item in the *New York Times* for March 9, 1971, reporting an

Morality and National Liberation Wars

interview in which President Richard Nixon speculates that the current war in Vietnam will probably prove to be the very last one in the American experience. The brink of perpetual peace is not a novelty in presidential discourse. The nation has been transported there many times. In my own mind, the discourse aroused a recollection from Clausewitz's ponderings. One of his great insights was to the effect that the option of war invariably lies with the defender. The party aggressing would invariably prefer to prevail without having to make a fight of it. The point is logically sound. Hence the only way to be sure of an absolutely warless future would be on the basis of an unconditional resolve to yield to every challenge. The assurance about a warless future can rest, it seems to me, on one of two propositions: either that no adversary will ever again formulate objectives contrary to our preferences (in which case there will never again be an adversary), or that we shall hereafter always relinquish our preferences under challenge. It is difficult to find a cogent reason, consistent with either premise, for committing support to any country beset by a national liberation war.

My final thought concerns a matter of weighing among competing demands. Surely estimates of relative exigency have a bearing on morality considered as embracing the principles of right conduct in human affairs. An essential question in connection with any undertaking of involvement or risk or any commitment of national resources is not merely whether a desirable purpose is involved, but how also does the purpose rate in competition with other demands against resources.

A factor in the situation in the United States is a consensus that present rates of diversion of resources to the public sector is relatively fixed. That is to say, the present level of taxation is a ceiling. Both major parties concur. Another tendency of the times is that of expressing every aspiration in terms of rights, and of having them generally accepted as such, so that public authority is pressed to grant fulfillment or stand judged in default of its duty. The result is a vastly expanded concept of the general welfare, each aspect of which entails a huge charge upon the revenue. As a result of the concurrent circumstances described, the general welfare—and a hugely expansive version of it—has become, for perhaps the first time in our national experience, a claimant against the common defense. This has produced a widely prevalent fiction—and that it is demonstrably fictitious

612 under a moment's analysis does not detract from its apparent credibility—to the effect that skyrocketing defense expenditures are the great impediment to national achievements of utopian scope and quality. If, as already indicated may be the case, the general factors affecting quality of life are heading toward regression, the pressure to bank great social programs by drawing upon allocations for strategic security will be increasing in the years ahead. The prospect augurs increasing difficulty for the maintenance of an adequate national defense. As a necessary corollary, an effective case for allocating resources to secure remote positions against Communist penetration will be all the harder to make. Here I mean effective in a sense of prevailing rather than merely making good points in logic.

Notes

* Reprinted with omissions.
1. The term "intervention" is sometimes applied to such actions. That term in international law means "interference by one state in the affairs of another for the purpose of compelling it to do or to forbear from certain acts or to maintain or to alter an internal condition." There is a general and longstanding precept against intervention in the discourse of governments, and the practice is forsworn in no end of treaties. So the term is accusatory and pejorative. Accordingly, those who speak for weaker governments incline to invoke the word to characterize any suasion attempted by a stronger government. Moreover, those who intervene—whether in Kashmir, Hungary, Egypt, or Guatemala—are most likely to disavow the term as pertinent. Those at the Capitol who belabor the United States course in Vietnam as intervention never apply the term to their own proposals for sacking the Saigon regime or forcing accommodation with its enemies. I forego considering whether hypothetical actions by the United States in face of the contingencies here postulated would constitute intervention in the rigorous sense of the word. Undoubtedly they would be called such by critics.
2. That principle is consonant with self-determination conceived as self-determination of states. That version of self-determination is noninterventionist in tendency. It contrasts with the idea of self-determination of peoples, a notion interventionist in tendency. Most governments are muddled in their thought and discourse. They tend to like self-determination of states in application to themselves, and to dabble on occasion in self-determination of peoples when others are involved. The United States record is no less muddled. The prevailing fraction of Americans favored

self-determination of peoples in the Âmerican Revolution and self-determination of states in the Civil War. President Wilson lumped the contrasting ideas together. He drew on a long tradition in assuming that peoples were inherently friendly to each other, and that only regimes caused tension and fomented war. In the eighteenth century, Diderot and Condorcet believed likewise—on empirical evidence as slim as Wilson's. Wilson founded his concept of an organized world order on the principle of nonintervention. He combined this precept with a normative notion that democratically-governed states would be peace-loving and peace-supporting. Self-governed states would inherently tend to be democratic. Hence self-determination of peoples. Two opposed notions were intertwined in his thinking. Such intertwining comes easy. How muddled the thinking and discourse of governments may be on such matters may be illustrated by citing former President Lyndon B. Johnson's State of the Union Message of 1966. He described "support of national independence—the right of each people to govern themselves and to shape their own institutions" as "the most important principle of our foreign policy." He added: "For a peaceful world order will be possible only when each country walks the way that it has chosen to walk itself." He summed up: "We follow this principle abroad . . . by continued hostility to the rule of the many by the few."

3. I wrote at length on this theme in the *New Republic,* May 25, 1963. Vice President Johnson had just made a speech making a big thing of the idea that Cuba would prove to be a showcase of Communist economic shortcomings. "What Communist country has not been?" I asked. "The rub is that such a showcase is likely to display something else—namely, the feasibility of taking over a society, deliberately depressing consumption, and using stringency in combination with monopoly of supply as an instrument of control. Communism has subsisted not on a reputation for generating affluence but on its effectiveness in holding on to power once installed."

4. This phenomenon may be enhanced by the characteristics of a Germanic-derived language in contradistinction to a Latin-derived one. Germanic languages make no adequate distinction between public and private enemies or between public friends and private friends. Latin tongues do. Somehow English usage makes it difficult to discern that an ally in war may not necessarily be someone you would want for a son-in-law or a member of the club.

5. Another touchstone related to what might be called the factor of acceptability in connection with foreign military involvement and commitment in the minds of some portion of the population concerns collective support. A collectivity of governments in support of some undertaking is supposed to endow it with

respectability not obtained when a government acts on its sole responsibility. I am not aware of any conclusive empirical basis for this idea, which persists anyway. The United Nations and, to a lesser extent, regional organizations serve as institutions for registering such multiple support. As instruments for symbolizing moral reassurance, they can scarcely be counted on in an instance of the sort pertinent here. The Declaration on the Principles of Friendly Relations Among States, adopted by the United Nations General Assembly in connection with the institution's 25th anniversary in the fall of 1970 as the culmination of an eight-year effort at negotiation, is equivocal on its face. What it twice explicity forbids it twice explicitly condones in the matter of intervention by one state in the affairs of another. At appropriate points, it registers absolute forbiddance, in keeping with the principle of self-determination of states. Again, however, it upholds the principle of self-determination of peoples and seems to authorize intervention when that concept has been invoked, as it so easily can be as a rationalization for any so called national liberation war. The Organization of American States also seems to be becoming equivocal on such matters. In early 1971, it proved impossible to get it to resolve a consensus on such a matter as international cooperation in suppressing acts of political terrorism.

6. I have often heard my own students express a willingness to fight for the United States "if the chips were down," or "if it were Armageddon," or "if it were obviously an issue of life and death for the country," and so on, while insisting on scorn for any marginal war in which the strategic stakes were of a secondary sort. I have repeatedly pointed out to them a corollary to this line of argument. An accountable government would, on such premises, be incapacitated for conducting a prudently conceived war. Every issue then would have to get down to a choice between giving in or carrying the conflict to the level of the final holocaust, and so on. The Armageddon-or-nothing argument becomes implausible if subjected to analysis. Nevertheless, the attitude is widely held. The circumstance is a debilitating one for accountable government.

Chapter VIII
Sources for the Study of Revolutionary Guerrilla Warfare
by A. Thomas Ferguson, Jr.

The liturgy of scholarship assigns a preeminent place to the imperative of a comprehensive bibliography. Familiarity with the literature is an accepted hallmark of scholarship, and a tradition of pedagogy in tenuous descent from Hegel promotes the construction of comprehensive bibliographical surveys. So, it is mildly surprising to encounter one division in the army of scholarship that is notably unequipped for the retrieval of antecedent literature. This corps, whose nonchalance it is the object of this essay to partially remedy, comprises nearly all those researchers working in the area of revolutionary guerrilla warfare.

In the past, bibliography of revolutionary guerrilla warfare may have been neglected because of wide gaps in the literature. In this connection we may note Harry Eckstein's observation of a decade ago that the pulse of research in this area was decidedly feeble.[1] But this defense no longer holds. In the course of the war in Vietnam, American social scientists, no less than other defense contractors, materially upgraded their estate. Research related to guerrilla warfare, and its prophylactic complement, counterinsurgency, proliferated. For a few golden years the money tree was in continuous bloom and research reports ripened copiously. There were several specialized projects undertaken to collect and order them, but none approached success, and the ablest remained very little known. The more general guides to the literature were also of little help. Those devoted generically to military affairs and national security were too panoramic to be very useful for the more restricted subject of revolutionary guerrilla warfare.[2] The same is true of the conventional guides to research sources in political science and the sporadic collections initiated by isolated individuals or military commands.[3]

This essay attempts to correct this confused situation. It inventories the significant bibliographies and the most useful general reference tools for locating current research in periodicals. The coverage is focused as finely on revolutionary guerrilla warfare as that blurred category permits, excluding materials devoted to more general themes, such as violence and modernization, although they are often apropos. Counting heavily in the final selection were availability and accessibility. Although a conscious effort was made to transcend the monoglot provincialism of most American social science, the bulk of the listings are of English language items. This circumstance, it is believed, reflects the state of world scholarship rather than the parochialism of the compiler. Finally, because of the transience of most writings in the field, the survey does not reach back further than 1962.

By far the most ambitious and systematic inventory of the literature on guerrilla warfare is also the least publicized. For this the vagaries inherent in the dissemination of scholarship are only exiguously to blame; more significant were the sponsoring agency's narrowly partisan aims and its employment of imaginative aliases.

In 1961 an agency called the Special Operations Research Office (SORO), sponsored nominally by American University of Washington, D.C. but funded by the U.S. Army, published the first of a series of bibliographies on topics of urgent relevance to national defense. SORO's lavish budget ensured an ample supply of capable researchers, but in 1965, with Project Camelot, the agency overreached itself.[4] The fortuitous revelation of that avowedly counterrevolutionary project and the ensuing barrage of public criticism sent SORO scrambling for cover. SORO metamorphosed into the innocuous Center for Research in Social Systems (CRESS), and its nascent Counterinsurgency Information Analysis Center (CINFAC) was blandly relabeled the Cultural Information Analysis Center. This change, perhaps, gulled only librarians and researchers, who had to scrutinize the publication listings of SORO, CRESS, and two different CINFAC's. The mass production of bibliographies seems eventually to have ceased, and the whole truncated CRESS operation finally disappeared from American University, reportedly into Maryland.

While the agency lasted, however, its output was prodigious, including two separate series of bibliographies devoted exclusively to revolutionary guerrilla warfare. Comprising large initial bibliographies and a program of periodically updated supplements, these are collectively perhaps the best source existing for the retrieval of recent literature, popular and scholarly. SORO's legacy also includes several other special bibliographies pertinent to the study of revolutionary guerrilla warfare. All are listed below, and are inventoried conventionally in library catalogs under issuance of American University, with a note to whatever appellation SORO was camouflaging itself under at the time of publication.

A Counterinsurgency Bibliography, by D. M. Condit, Barbara Reason, Margaret Mughisuddin, Bum-Joon Lee Park, and Robert K. Geis (Washington, 1963), is the master compilation. It was followed by a series of supplements of which #6 apparently never appeared. All were compiled by Margaret Mughisuddin, #7 with Heidi Berry: *Counterinsurgency Bibliography, Supplement*, No. 1, December, 1963; No. 2, March, 1964; No. 3, June, 1964; No. 4, September, 1964; No. 5, December, 1964; No. 7, with Heidi Berry, June, 1965.

An earlier SORO series on unconventional warfare contains much material of interest. There were two master compilations: *Unconventional Warfare: An Interim Bibliography, 1961*, and *A Selected Bibliography on Unconventional Warfare*, Miller and Lybrand, eds., 1961. These were followed by *Unconventional Warfare Bibliography, Supplement:* No. 1, by N. A. Gardner, January, 1964; No. 2, by Gardner, April, 1964; No. 3, by Gardener, July, 1964; No. 4 by Gardner, October, 1964; No. 5, by Nancy Currier, January, 1965; No. 6 by Gardner, April 1965; No. 7, by Currier, July 1965.

Later supplements to both of these series were combined with the supplements to several other SORO publications, retitled, and issued collectively, with the title changing once more after No. 9. (The entries for each series were separately maintained, making it unnecessary to sort through unrelated entries. Some supplements were delayed; their publication dates consequently are not in numerical sequence.) They are: *Bibliography on Counterinsurgency, Unconventional Warfare, and Psychological Operations, Supplement:* No. 8, by Margaret Mughisuddin *et al.*, February, 1966; No. 9, by Margaret Bittick *et al.*, May, 1966; *CINFAC Bibliographic Review, Supplement*, by Margaret Bittick *et al.*: No. 10, October, 1966; No. 11, November, 1966; No. 12, June, 1967; No. 13, May, 1967; No. 14, August, 1967; No. 15, November, 1967; No. 16, February, 1968; No. 17, June, 1968.

Three other pertinent bibliographies were produced by SORO. *A Selected Bibliography of Crowd and Riot Behavior in Civil Disturbances* (second edition, 1968), compiled with annotations by Adrian Jones, Margaret Bittick and Nancy Currier, surveys scholarly and popular sources and theses and dissertations. The scholarly sources in Skaidrite Malius Fallah, *A Selected Bibliography on Urban Insurgency and Urban Unrest in*

Latin America and Other Areas (1966), are less alloyed with popular entries. This work is unique in its attempt to treat urban guerrilla warfare. Its coverage is weighted heavily toward Africa, an imbalance which, Fallah asserts, reflects trends in scholarly literature. A final work, focused on counterinsurgency, is Willis M. Smyser et al., *Internal Defense: An Annotated Bibliography* (1968).

In the literature on guerrilla warfare there is one outstanding review essay surveying recent scholarship: chapter two of Henry Bienen, *Violence and Social Change* (published for the Adlai Stevenson Institute of International Affairs, Chicago: University of Chicago Press, 1968). As is typical of American writers on guerrilla warfare, Bienen is preoccupied with the literature in English. There is much in the book's other chapters of tangential interest to students of guerrilla warfare, and the whole lucid treatment constitutes an exceptionally capable introduction to its general subject.

Notable only for its coverage of non-English sources is Roger Cosyns-Verhaegen, *Guerres Révolutionnaries et Subversives: Selection Bibliographique* (Brussels: Les Ours, 1967). It begins with a brief inventory of classical sources and proceeds through the literature on revolution and subversion, with excursions into military history, coups, and related subjects. The material covered is wildly uneven, ranging from *Das Kapital* to articles from *Reader's Digest*. There is also undoubtedly the most extensive index extant to the writings on guerrilla warfare of Roger Cosyn-Verhaegen.

Ronald H. Chilcote, *Revolution and Structural Change in Latin America: A Bibliography on Ideology, Development, and the Radical Left, 1930-1965* (2 vols., Hoover Institution Bibliographical Series, XL; Stanford: Hoover Inst., 1970) is an outstanding compilation of sources, chiefly in English and Spanish, on Latin America. Listings include eyewitness reports and popular magazines as well as scholarly sources.[5]

Unique in its preoccupation with Southern Africa is Francis A. Kornegay, Jr., "A Bibliographic Essay on Comparative Guerrilla Warfare and Social Change in Southern Africa," *A Current Bibliography on African Affairs*, III (New Series), No. 2 (1970), pp. 5-20. It covers Rhodesia, Southwest Africa, Angola and Mozambique, as well as the Union of South Africa, and is limited to scholarly works published since 1966. Also indexed are some important original source documents.

Even less well known than SORO's bibliographies but also very useful is Vol. IV, *Bibliography on Counterinsurgency Topics*, part of the multivolume *Anthology of Related Topics on Counterinsurgency* prepared at the Lackland, Texas Military Training Center in 1966. Carrying no Library of Congress catalog number, this work is exceptionally comprehensive, very well annotated, and minutely subdivided. It concentrates almost exclusively on literature in English.

Focusing on counterinsurgency is Marcel Vigneras, *Bibliography on Counterinsurgency and Allied Subjects* (McLean, Va.: Research Analysis Corporation, Operational Logistics Division, 1965). This is a revised edition of a bibliography first published in 1962, and is very difficult of access.

Also useful is Miles D. Wolpin, *United States Intervention in Latin America: A Selected and Annotated Bibliography* (AIMS Bibliographical Series, No. 8; New York: American Institute for Marxist Studies, 1971). The introduction presses for a broad definition of "intervention," arguing that an excessively narrow construal of the term squeezes significant economic and political phenomena out of scholarly view. Included are chiefly works from English language sources.

In the absence of future special bibliographies, students of the ongoing literature must consult the general reference sources in political science. Of these, the very comprehensive Vol. II of Lubomyr Wynar, *Guide to Reference Materials in Political Science* (Rochester: Libraries Unlimited, 1968), indexes virtually all the abstracting services, digests, yearbooks, directories, periodical indexes, and statistical sources that the most ardent bibliomaniac could desire.

Four other general reference sources deserve mention, three because they have been neglected, and one because it has been heavily publicized. This last is the regularly updated *Universal Reference System: Political Science, Government and Public Policy Series* (Princeton: Princeton Research Publishing Co., 1967). Its sophisticated classification system has been promoted by its inventor, Alfred de Grazia, as a unique advance in information retrieval.[6] This index is remarkable only for its method, however, being relatively prosaic and circumscribed in its coverage.

Much less well known are two excellent continuing indexes which contain much relevant material. The Radical Research Center of Toronto, Canada publishes an annual *Alternative Press Index*, which is the only available source for many useful left wing publications of limited circulation. At the other end of the political spectrum is the *Air University Index*, issued by Air University Library, Maxwell Air Force Base, Alabama, which covers diverse subjects and regions and is published regularly. Lastly, Allan Millett and B. Franklin Cooling, *Doctoral Dissertations in Military Affairs* (Bibliography Series, No. 10; Manhattan, Kansas: Kansas State University Library, 1972) supplements the conventional registers of dissertations.

Students in search of English language sources can avail themselves of the Library of Congress subject catalog. This is not so obvious as it may seem, since for most topics this inventory is woefully incomplete; but the guerrilla warfare entries for the last decade have been uncharacteristically extensive. Subject catalogs (of uneven quality) are also available at the national libraries of most large countries.

Finally, in a genre conventionally omitted from bibliographies, a classic work is *The Battle of Algiers* (1966), the film directed by Gillo Pontecorvo. A later film, based on recent Uruguayan experience, is *State of Siege* (1972), directed by C. Costa-Gavras. Both of these motion pictures were co-authored with the directors by Franco Solinas.

Notes

1. Harry Eckstein, "Introduction. Toward the Theoretical Study of Internal War," in *Internal War*, ed. Eckstein (New York: Free Press, 1964), pp. 1-33.
2. E.g., Kurt Lang, *Military Institutions and the Sociology of War: A Review of the Literature with Annotated Bibliography* (Beverly Hills: Sage, 1972), which misses nearly everything of magnitude on revolutionary guerrilla warfare, or "The New Politics of National Security: A Selected and Annotated Research Bibliography," by Major Raoul H. Alcala and Douglas H. Rosenberg, in Bruce M. Russett and Alfred Stepan, eds., *Military Force and American Society* (New York: Harper & Row, 1973), pp. 196-371, which also omits many works cited in this essay.
3. E.g., Frederick L. Holler, *The Information Sources of Political Science* (Santa Barbara: American Bibliographic Service, 1971), and Clifton Brock, *The Literature of Political Science*

(New York: Bowker, 1969). There are some articles of limited usefulness, notably Heinz Eulau, "Political Science," in *A Reader's Guide to the Social Sciences,* ed. Bert Hoselitz (revised edition; New York: Free Press, 1970), pp. 129-237. Of little aid is a brief listing by K. R. Sachdeva: "Guerrilla Warfare—A Bibliography," *Journal of the United Service Institution of India,* LXXXXII, No. 386 (1962).

4. On Project Camelot, cf. "The Rise and Fall of Project Camelot," introductory essay by Irving L. Horowitz in *The Rise and Fall of Project Camelot,* ed. Horowitz (Cambridge: M.I.T. Press, (1967). A notable review of this work is A. L. Madian and A. N. Oppenheim, "Knowledge for What? The Camelot Legacy," *British Journal of Sociology,* XX, No. 3 (1969), pp. 326-336.

5. In the case of Cuba, Chilcote's inventory can now be supplemented by Jaime Suchlicki, *The Cuban Revolution: A Documentary Bibliography* (Coral Gables, Florida: Center for Advanced International Studies, Research Institute for Cuba and the Caribbean, University of Miami, 1968). A second edition of this work is reportedly under way.

6. Cf. his "Continuity and Innovation in Reference Retrieval in the Social Sciences," *American Behavioral Scientist,* X, No. 6 (1967) pp. 1-4.

355.02184
R454

96433

$11.77